New Approaches to Spatial Planning and Design

Murat Özyavuz

New Approaches to Spatial Planning and Design

Planning, Design, Applications

PETER LANG

Bibliographic Information published by the Deutsche Nationalbibliothek
The Deutsche Nationalbibliothek lists this publication in the Deutsche Nationalbibliografie; detailed bibliographic data is available online at http://dnb.d-nb.de.

Library of Congress Cataloging-in-Publication Data
A CIP catalog record for this book has been applied for at the Library of Congress.

ISBN 978-3-631-78274-3 (Print)
E-ISBN 978-3-631-78397-9 (E-PDF)
E-ISBN 978-3-631-78398-6 (EPUB)
E-ISBN 978-3-631-78399-3 (MOBI)
DOI 10.3726/b15609

Peter Lang – Berlin · Bern · Bruxelles · New York · Oxford · Warszawa · Wien

This publication has been peer reviewed.

www.peterlang.com

Contents

List of Contributors

Assoc. Prof. Dr. Funda K. Açıcı
Karadeniz Technical University, Faculty of Architecture, Interior Architecture Department, Trabzon, Turkey

Dr. Kofi Agyekum
Kwame Nkrumah University of Science and Technology, Department of Construction Technology and Management, Kumasi, Ghana

Res. Ass. Duygu Akyol
Karadeniz Technical University, Faculty of Forestry, Landscape Architecture Department, Trabzon, Turkey

Assoc. Prof. Dr. Sennur Akansel
Trakya University, Faculty of Architecture, Department of Architecture, Department of Building Science

Murat Akten
Department of Landscape Architecture, Faculty of Architecture, Süleyman Demirel University 32260 Isparta, Turkey (e-mail: muratakten@sdu.edu.tr)

Ayhan Akyol
Department of Forest Engineering, Faculty of Forestry, Isparta University of Applied Sciences 32260 Isparta, Turkey (e-mail: ayhanakyol35@gmail.com)

Sibel Akten
Department of Landscape and Ornamental Plants, Isparta University of Applied Sciences 32500 Eğirdir-Isparta, Turkey (e-mail: sakten@gmail.com)

Assist. Prof. Dr. Nurgül Arısoy
Selçuk University, Agricultural Faculty, Department of Landscape Architecture, Konya, Turkey

Assoc. Prof. Dr. Aslı Altanlar
Amasya University, Faculty of Architecture, Urban and Regional Planning Department, Amasya, Turkey

Assoc. Prof. Dr. Egemen Aras
Bursa Technical University, Faculty of Engineering and Natural Sciences, Department of Civil Engineering, Bursa, Turkey

Assist. Prof. Dr. Aylin Aras
Bursa Technical University, Faculty of Architecture and Design, Department of Architecture, Bursa, Turkey

Dr. Oğuz Ateş
İnönü University, Faculty of Fine Arts and Design, Landscape Architecture Department, Malatya, Turkey

Prof. Dr. Fatih Aydın
Karabuk University, Faculty of Letters, Department of Geography, Karabük, Turkey

Assoc. Prof. Dr. Canan Cengiz and Res. Asst. Aybüke Özge Boz
Bartın University, Faculty of Forestry, Department of Landscape Architecture, Bartın, Turkey

Assoc. Prof. Dr. Seda H. Bostancı
Namık Kemal University, Faculty of Economics and Administrative Sciences, Political Science and Public Administration Department, Tekirdağ, Turkey

Ezgi Ceyda Büyüktaşkın
Urban Planner, İzmir, Turkey

Assist. Prof. Dr. Ayça Yeşim Çağlayan
Istanbul University—Cerrahpaşa, Faculty of Forestry, Department of Landscape Architecture, Bahçeköy, Istanbul

Assoc. Prof. Dr. Suat Çabuk
Tekirdağ Namık Kemal University, Faculty of Fine Arts, Design and Architecture, Department of Urban and Regional Planning, Tekirdağ, Turkey

Graduate Student Ayşe Gülan Çelebi
Inönü University, Faculty of Fine Arts and Design, Department of Landscape Architecture, Malatya, Türkiye

Asst. Prof. Dr. A. Esra Cengiz
Department of Landscape Architecture, Faculty of Architecture and Design, Çanakkale Onsekiz Mart University, Çanakkale, Turkey

Nadim Kamel Copty
Boğaziçi University, Institute of Environmental Sciences, 34342, Bebek, İstanbul, Turkey)

Assoc. Prof. Dr. Demet Demiroğlu
Department of Landscape Architecture, Faculty of Engineering and Architecture, Kilis 7 Aralık University, Kilis, Turkey

Assist. Prof. Dr. Fulya Üstün Demirkaya
Karadeniz Technical University, Faculty of Architecture, Department of Architecture, Trabzon, Turkey

Assist. Prof. Dr. Pınar Öztürk Demirtaş
İstanbul Okan University, Faculty of Art, Design and Architecture, Industrial Product Design Department, Istanbul, Turkey

Assoc. Prof. Dr. Ferhat Hacialibeyoğlu
Dokuz Eylül University Faculty of Architecture, Department of Architecture, İzmir, Turkey

Assoc. Prof. Dr. Deniz Dokgöz
Dokuz Eylül University Faculty of Architecture, Department of Architecture, İzmir, Turkey

Assoc. Prof. Dr. Kemal Demir
Erciyes University, Faculty of Architecture, Department of Architecture, Kayseri, Turkey

Assoc. Prof. Dr. Zühal Dilaver
Ankara University, Faculty of Agriculture, Landscape Architecture Department, Ankara, Turkey

Assist. Prof. Dr. Yasin Dönmez
Karabuk University, Faculty of Forestry, Department of Landscape Architecture, Karabük, Turkey

Assist. Prof. Dr. Tuba Rastgeldi Dogan
Harran University, Faculty of Engineering, Department of Environmental Engineering, Şanlıurfa,Turkey

Assoc. Prof. Dr. Barış Ergen
Erciyes University, Faculty of Architecture, City and Regional Planning Department, Kayseri, Turkey

Asst. Prof. Dr. Şehriban Eraslan
Süleyman Demirel University, Faculty of Architecture, Landscape Architecture Department, Isparta, Turkey

Assoc. Prof. Dr. Şebnem Ertaş
Karadeniz Technical University, Faculty of Architecture, Interior Architecture Department, Trabzon, Turkey

Assoc. Prof. Dr. Ercan Gökyer
Bartın University, Faculty of Forestry, Department of Landscape Architecture, 74100, Bartın, Turkey, egokyer@bartin.edu.tr

Dr. Meltem Güneş
Namık Kemal University, Faculty of Fine Arts, Design and Architecture, Department of Landscape Architecture, Tekirdağ, Turkey

Azat Gül
İnönü University, Faculty of Fine Arts and Design, Landscape Architecture Department, Malatya, Turkey

Assoc. Prof. Dr. Özge İslamoğlu
Karadeniz Technical University, Faculty of Architecture, Department of Interior Architecture, Turkey

Assoc. Prof. Dr. Timur Kaprol
Namık Kemal University, Faculty of Fine Arts, Design and Architecture, Architecture Department, tkaprol@nku.edu.tr

Assoc. Prof. Dr. Aybike Ayfer Karadağ
Düzce University, Faculty of Forestry, Department of Landscape Architecture, Düzce, Turkey

Assoc. Prof. Dr. Dr. Ayşe Durukan Kopuz
Namık Kemal University, Faculty of Corlu Engineering, Civil Engineering Department, Corlu/Tekirdağ, Turkey

Prof. Dr. Aslı Korkut
Namık Kemal University, Faculty of Fine Arts, Design and Architecture, Landscape Architecture, 59030, Tekirdag, Turkey

Prof. Dr. Tuğba Kiper
Namık Kemal University, Faculty of Fine Arts, Design and Architecture, Landscape Architecture, 59030, Tekirdag, Turkey.

Ömer Lütfü Çorbaci
Recep Tayyıp Erdoğan University, Faculty of Fine Arts, Design and Architecture Department of Landscape Architecture, Rize, Turkey

Türker Oğuztürk
Recep Tayyıp Erdoğan University, Faculty of Fine Arts, Design and Architecture Department of Landscape Architecture, Rize, Turkey

Burcu Özdemir, Ph.D.
Istanbul Technical University, Faculty of Architecture, Urban and Regional Planning Department, Istanbul Technical University, İstanbul, Turkey

Lecturer Dr. Betül Atakan Öznam
Istanbul University—Cerrahpaşa, Vocational School of Forestry, Cultivation of Ornamental Plants Department, Bahçeköy, Istanbul.

Assist. Prof. Dr. Doruk Görkem Özkan
Karadeniz Technical University, Faculty of Forestry, Landscape Architecture Department, Trabzon, Turkey

Assist. Prof. Dr. Mahire Özçalık
Kahramanmaras University Faculty of Forestry Department of Landscape Architecture/Turkey

Melih Öztürk
Bartın University, Faculty of Forestry, Department of Landscape Architecture, 74100, Bartın, Turkey, melihozturk@bartin.edu.tr

Assist. Prof. Dr. Hülya Öztürk Tel
Harran University, Sanliurfa Technical Sciences Vocational School Department of Architectural Restoration, Sanliurfa, Turkey

Assist. Prof. Dr.Aylin Salıcı
Hatay Mustafa Kemal University, Faculty of Architecture, Landscape Architecture Department, Antakya/Hatay, Turkey

Dr. Burcu Salgın
Erciyes University, Faculty of Architecture, Department of Architecture, Kayseri, Turkey

Res. Ass. Demet Ülkü Gülpınar Sekban
Karadeniz Technical University, Faculty of Forestry, Landscape Architecture Department, Trabzon, Turkey

Dr. Barbara Simons
Kwame Nkrumah University of Science and Technology, Department of Construction Technology and Management, Kumasi, Ghana

Assist. Prof. Dr. Elif Sönmez
Altınbaş University, Faculty of Fine Arts and Design, Interior Architecture and Environmental Design Department, İstanbul, Turkey

Sümeyra Betül Tayfur
Namık Kemal University, Faculty of Fine Arts, Design and Architecture, Industrial Product Design, 59030, Tekirdag, Turkey.

Assist. Prof. Dr. Aysun Tuna
Inönü University, Faculty of Fine Arts and Design, Department of Landscape Architecture, Malatya, Türkiye Department, Trabzon, Turkey

Prof. Dr. Bülent Cengiz and Oğuz Erdi Yakan
Bartın University, Faculty of Forestry, Department of Landscape Architecture, Bartın, Turkey

Assoc. Prof. Dr. Şengül Yalçınkaya
KaradenizTechnical University, Faculty of Architecture, Interior Architecture Department, Trabzon, Turkey

Prof. Dr. Rüya Yılmaz
Namık Kemal University, Faculty of Art Design and Architecture,Landscape Architecture Department, Tekirdağ, Turkey

Assoc. Prof. Dr. Okan Yılmaz
Canakkale 18 Mart University, Faculty of Art, Design, and Architecture, Landscape Architecture Department, Çanakkale, Turkey

Assoc. Prof. Dr. Ülkü Duman Yüksel
Gazi University, Faculty of Architecture, Department of Urban and Regional Planning, Ankara, Turkey

Assoc. Prof. Dr. Aslı Altanlar

1 Perception of Quality of Life on the Axis of Tourism-Focused Local Development, Example of Amasya

Abstract: In this chapter, we study the effects that tourism creates on quality of life, as being important for the sustainable improvement of tourism. For this reason, this chapter is focusing on defining the behaviors and perceptions of the local public in order to understand the effect tourism has on the quality of life and trying to understand the relation in between the perception of quality of life of the local public and support given for tourism-focused development policies. The empirical part of the chapter covers the historical neighborhoods in the central district of Amasya. The survey was conducted between November and December of 2017 and sampled 460 people living in Amasya. Principal components analysis is applied to statistically measure structural validity of the scale. Non-parametric test methods Spearman's Correlation and Kruskal Wallis were used in the study. It was found that factors that determine how neighborhood inhabitants perceive the effects for tourism are, respectively, "effects of tourism oriented decisions," "positive effects of tourism on economic environment," "negative effects of tourism on physical environment," and "negative effects of tourism on social environment." It has also been determined that the dimensions determining the district residents' general support of tourism are "motive of participation in tourism activities" and "motive of support of tourism activities." The research data collected indicated that opinions regarding tourism development reflected differences according to the perceived impact and effects of tourism on the socio-cultural, physical, and tourism-oriented decisions. In the study, it has been found that the economical positive effect of the improvement of tourism is increasing the quality of life at the same level. In a similar manner, it has been found that the positive effects of tourism-oriented decisions on every individual are increasing the quality of life at the same level.

Keywords: Tourism development, Tourism support, Tourism perception, Impacts of tourism, Quality of life, Amasya

1 Introduction

As from the second half of 20th century, tourism has become one of the most rapidly developing and expanding sectors in the world economy and has often started to be used as a means of regional and national development as many other sectors [1]. While annual average increase in the world's goods production remained at about 3% in the years 1991 and 2000, annual increase in tourism was realized as 6.6%. For this reason, in the solution of national and international economic problems that the economies of countries faced and in overcoming the bottlenecks by the end of 20th century, tourism had been deemed like a point of exit with its dynamic and economic characteristics [2]. For this reason, tourism is being deemed as one of the most significant sources of growth and improvement in our country too, as in many other countries [3].

The improvement of tourism in a region may have significant economical effects such as increasing the infrastructure and superstructure investments like construction, food and beverage, entertainment, transportation, etc., thus decreasing the problem of unemployment by creating employment opportunities, and most important of all eliminating the interregional economic imbalance by reviving the local economies [2, 4, 5]. But as tourism has contributions to country's economy, it also has effects which may be deemed negative [4]. Dependency on foreign countries, decreased economic profitability, increased cost of living and inflation, benefit of a small part of the society from tourism may be listed among the main harmful effects of tourism on economic structure [6].

Tourism is not just an economic fact, but also an international act playing a significant role in the society and social structure, and with the social and cultural aspects of the society [4]. For this reason, the main point on which both the opponents of tourism and the ones who want to improve tourism focus is the fact that tourism is causing significant changes in the socio-cultural structure of the countries being subject to tourism [6]. The ones opposing the improvement of tourism are asserting that the common cultural structure of the society they live in such as traditions, customs, moral structure and life style will be negatively affected as the result of acts of tourism. These negative effects that tourism forms on social structure may be able to occur through changes in social life, family relations, moral attitudes, personal behavior, other systems, manners and customs, traditional ceremonies, life style, language and art, and in organization of the society [6,7,8].

Tourism is also able to make the natural, cultural, and historical values of a society a part of the commercial life [9]. Along with transformation of cultural characteristics to commercial meta, the interaction of cultures may sometimes be destructive. In this case, it is ending up with disappearance of local identity and authenticity [10]. Along with the improvement of tourism, environmental problems such as insufficiency of infrastructure, environmental and noise pollution, insufficiency of services, crowding, destruction of fertile lands, increasing of unearned income groups, destruction of natural and cultural environment, transformation of rural areas to an urban style are also being encountered [4].

In order to refer to a sustainable improvement in tourism, it is required to preserve the environment that the people interact or don't interact with without deteriorating or changing it, to sustain the cultural integrity, ecological processes, biological diversity and the systems maintaining life; to plan all the resources as to meet the economic, social and aesthetical requirements of the people and tourists in the visited area; and to ensure the balance of protection and use in a manner as to enable the future generations to meet their same requirements [11]. Within this scope, maximizing the positive effects of tourism while minimizing its ecological, social, cultural and environmental negative effects should be the main objective [10]. And the main condition of the actualization of this objective will be ensuring effective contribution of the society, administrators of the society and all stakeholders in the tourism sector at the phase of development of tourism policies and strategies [12].

For this reason, there are many national and international studies done in the recent two decades for understanding the effects of tourism. It is being observed that these studies made at the intersection point of tourism and local community focus on how the economic, cultural, social and environmental effects of tourism are being perceived by the local public; in which direction the referred effects affect the support given by the local public for tourism; and how these effects sensed in the recent few years affect the lives of the public living in the area [13, 14, 15, 16]. Such researches are focusing on defining the behaviors and perceptions of the local community in order to understand the effect tourism has on quality of life, and trying to understand the relation in between the perception of quality of life of local public and support given for tourism-focused development policies [13].

According to the literature on the subject, the positive effects of tourism are increasing the quality of life of the local community, and causing the local community to support tourism and to develop a positive attitude. And the negative effects of tourism are causing the local community to be dissatisfied with the improvement of tourism in the region [17]. Moreover, the increasing socio-cultural and re-creative activities, and interaction with the visitors from different cultures, are contributing to the socio-cultural improvement of the individuals. There are also studies which allege that participation in the referred touristic activities has direct and indirect positive effects on satisfaction from life by ensuring happiness, healthier life, longer period of residence and high self-confidence, and thus on more satisfaction of life [15,17]. On the other hand, there are studies that reveal that the crowds, traffic volume, increase in crime rate, environmental pollution, high cost of living, corruption of traditional culture and changes in daily life styles that are created by tourism have negative effects on the quality of life of the local community [17].

Quality of life may be briefly expressed as a sense of content of an individual regarding her/his life, or satisfaction of an individual arising from her/his experiences [13]. Along with the presence of consensus on the subject that it is required for the definition of quality of life to cover many dimensions of life, which dimensions they should be is still a subject being discussed. Even if the definitions made regarding quality of life generally turn around "subjective perception of the individual regarding her/his life," the quality of life is being examined in two aspects as being objective and subjective. While the objective indicators are income, education, profession, health, state of the lodging lived, etc., the satisfaction and content that the individual feels from these is the subjective indicator of quality of life [18]. Though there are various definitions and models on quality of life, it can be said that many researchers are like-minded regarding that quality of life has a multidimensional, interactive structure covering the lives of people and the environments they live in [13]. When the difference among studies of quality of life and attitude and perceived effect is considered, while the studies of attitude and perceived effect are mainly focusing on how the people perceive the effects of tourism on society and environment, the studies of quality of life are mainly dealing with how these effects affect the satisfaction of life of the individual and family including the society, neighborhood and personal conditions [13].

Up to this point it has been referred of positive and negative changes occurring at regions where their development is being tried to be based on tourism, and it has been dwelled on the issues required to be considered regarding a planning comprehension which doesn't decrease the quality of environment and people. When the relevant literature is examined, it is being observed that support of local community and local enterprises is required for the improvement of sustainable tourism. Secondly, it is being clearly understood that it is required to reveal how the local community perceive economic, social, cultural and environmental effects as the tourism sector improves and as the number of tourists increase, and the positive and negative results that these effects create on quality of life. For this reason in this chapter it has been tried to determine (1) the effectiveness level of the referred effects of tourism in the formation of the support of local community for tourism, (2) the perception of the effects of tourism on the quality of life by local community, (3) whether there is a significant difference in between the perception of the effects of tourism by the participants as per their income states, periods of residence and ownership statuses and their support for tourism.

2 Material and Method

In this chapter, the province of Amasya, county of Merkez, which purports to be a focus point regarding culture tourism, and which faces a renewal and transformation process the interferences made, has been determined as the workspace. This study couldn't be addressed in a manner as to cover the whole of province of Amasya, county of Merkez, due to constraints of time and cost. The empiric part of the study is covering the historical districts at the province of Amasya, county of Merkez. For this reason, the most significant constraint of the study is that it is inconvenient to generalize the obtained results for the whole community (including the ones living at the rural area).

Many houses which are reflecting the examples of Turkish civil architecture and which had not lost their traditional pattern, and districts of Hatuniye, Sofular, Nergis, Dere, Fethiye, Gökmedrese, Savadiye, Şamlar, Şehirüstü, Üçler at the county of Merkez drawing attention with their fountains, small squares, Islamic-Ottoman social complex and mosques, authentic natural beauties and traditional district culture constitute the subject of the study.

Number of dwellings of year 2017 of the ten districts, constituting the sample structure of the study, constitutes the universe of the research (Turkish Statistical Institute, 2017). Total number of sample has been determined as 384 for 3336 dwellings with a confidence interval of 95%, and sampling error of ±0.05. After determining the total sample number, the size of questionnaire to be conducted at each district had been determined in the direction of representation rates in the total. The questionnaires had been conducted in between November–December 2017 on weekdays and weekends by using the "simple random sampling method." In this process, 460 dwellings in total had accepted to respond to the questionnaire.

The questionnaire form is consisting of four sections. In the first section, there are questions for determining the demographic characteristics of the participants. The second section is consisting of 34 propositions regarding the economic, social, cultural and environmental effects of tourism, and effects of tourism-oriented decisions. And in the third section of the questionnaire, a scale consisting of 17 propositions for measuring the satisfaction of participants from general living conditions (4 propositions), total effect of tourism (3 propositions) and support of tourism (10 propositions) had been used. The level of participation of participants in the propositions had been assessed on a 5-point Likert scale as (1) I definitely don't agree, (2) I don't agree, (3) I'm indecisive, (4) I do agree and (5) I definitely agree [19–22].

Principal Component Analysis (PCA)—from among exploratory factor analysis—had been applied in order to determine the construct validity of the scale, and Cronbach's alpha reliability test had been applied in order to determine the reliability. In order to understand whether the scale is suitable for factor analysis, Kaiser–Meyer–Olkin (KMO) and Bartlett's Test of Sphericity had been applied. Shapiro-Wilk test had been used in testing the normality of the scale's scores; Sparman correlation test had been used for determining the direction and severity of the linear relationship among the scales; and Kruskal Wallis test had been used in the comparison of characteristics of the participants and the scales.

3 Findings

3.1 Descriptive Statistics of Household

About 9.8% of the participants were of ages 18–20, 21.3% of the participants were of ages 21–34, 29.3% were of ages 35–48 and 39.5% of them were over the age of 49; 53.9% of the participants were men, 67.8% of them were married, 21.1% of them were single and 10.2% of them were engaged. About 77.0% of the participants didn't have an income generating job, 17% of them were retired and 6.1% of them were working on a salary basis. About 22.7% of the participants were primary school graduates, 18.8% of them were high school graduates, 3.6% of them were academy graduates and 2.5% of them were holding a license degree. About 3.2% of the participants were living at the district for less than one year, 17.2% of them were living at the district for 2–6 years, 16.8% of them were living at the district for 7–12 years, 9.7% of them were living at the district for 13–17 years and 62.8% of them were living at the district for 18 or more years. About 67.4% of the participants were homeowners and 96.5% of the participants were generating income from tourism.

About 42.4% of the participants were living as four or more people in their dwelling, 23.9% of them were living as two individuals, 26.3% of them were living as three individuals and 7.4 % of them were living alone. About 25.8% of the dwellings had an income level below the subsistence wage. It had been determined that 10% of the participants with a household size of four or more individuals had an income level of 1404 TRY and less, 10.40% of them had an income level in between 1405–1568 TRY, 18.2% of them had an income level in between 1569–3569 TRY and 3.1% of them had an income level of 3570 TRY and more. According to the obtained findings, it had been determined that 42.40% of the participants living as four or more individuals in the dwelling had income level below 5152.71 TRY—which was declared as the limit of poverty for November 2017— and that 20.40% of the participants had an income level below 1567.45 TRY—which was declared as the limit of hunger [23]. It had been determined that 73 % of the participants were not satisfied with the total income of the dwelling ($\bar{x}$:3.08, std:1.04575), 74 % of them were not satisfied with their lives in general sense ($\bar{x}$: 3.99, std: 0.80777), 71.8 % of them were not satisfied with the way they sustain their lives ($\bar{x}$:4.05, std:0.79461).

3.2 Determination of the Relation in between the Level of Perception of the Effects of Tourism by the Local Community and the Support of Tourism

In order to be able to determine how the residents of a district perceive the effects of the improvement of tourism, first KMO and Bartlett Sphericity test had been applied on the scale of effects of tourism. According to the data obtained, the value of KMO had been found to be 0.873. As the result of Bartlett Sphericity test, it had been concluded that there is not significantly high relations among the variables, and that the data was suitable for applying factor analysis (X^2:6718.784, sd:561, p<0,00).

According to the Scree plot graph obtained as the result of the principal components analysis made, it had been decided to constitute the scale by four dimensions. Following the determination of the number of dimensions, varimax vertical rotation had been performed in order to determine the distribution of the propositions to factors. Total variance percentage of four factors observed in the cumulative variance section is 51.232. In social sciences, variance rates varying at the range of 40%–60% are being deemed ideal [24].

Factor loads of the first dimension are consisting of 11 articles varying in between 0.869 and 0.501. The rate of defining of total variance by the dimension is 20.760%, and reliability coefficient is 0.934. According to this, the factor's reliability level is very high. Factor loads of the second factor are consisting of 12 articles varying in between 0.669 and 0.463. The rate of defining of total variance by the dimension is 13.382%, and reliability coefficient is 0.871. Factor loads of the third dimension are consisting of 7 articles varying in between 0.788 and 0.481. The rate of defining of total variance by the dimension is 10.241%, and reliability coefficient is 0.813. Factor loads of the fourth dimension are consisting of 4 articles varying in between 0.757 and 0.606. The rate of defining of total variance by the dimension is

Tab. 1: Factors Affecting the Behaviors and Attitudes of the Residents of District against Tourism, and Results of Reliability Analysis

Dimensions	Component	Eigenvalues	Variance	Item	Factor Load	Cronbach's Alpha
1	Effects of tourism-oriented decisions	7.058	% 20.760	11	0,869–0,501	0.934**
2	Positive effects of tourism on economic environment	4.550	% 13.382	12	0,696–0,463	0.871**
3	Negative effects of tourism on physical environment	3.482	% 10.241	7	0,788–0,481	0.813**
4	Negative effects of tourism on social environment	2.329	% 6.849	4	0,757–0,606	0.814**

Cronbach's alpha (α) ≥0 .9 is perfect;** 0.9 > α ≥ 0.8 is good, *0.8 > α ≥ 0.7 is acceptable

10.241%, and reliability coefficient is 0.814 (Tab. 1). According to the data obtained, the reliability level of the dimensions is very high.

The first dimension is "effects of tourism oriented decisions" as it is covering the articles expressing the measures taken and interferences made by the central and local administrations for the improvement of tourism; the second dimension is "positive effects of tourism on economic environment" as it is covering the positive expressions such as increasing of income level and living standards, increasing of job opportunities and increasing of the number of local and independent enterprises; the third dimension is "negative effects of tourism on physical environment" as it is consisting of negative expressions such as the historical environment becomes destroyed by the improvement of tourism, the facilities constructed for the tourists dispels the authentic identity of the district, tourists cause problems such as noise, pollution and traffic jam; the fourth dimension is "negative effects of tourism on social environment" as it is covering negative expressions such as neighborhood relations weaken by the improvement of tourism, crime rates increase in the district, the peace of the district decreases and the authentic identity of the district spoils (Tab. 1).

Tab. 2: Criteria Determining the General Tourism Support of the Residents of District

Dimen-sions	Component	Eigenvalues	Variance	Item	Factor load	Cronbach's Alpha
1	Motive of participation in tourism activities	3.366	42.078	4	0.924–0.888	0.933
2	Motive of support of tourism activities	2.509	31.361	4	0.835–0.713	0.790

*Cronbach's alpha (α) ≥0 .9 is perfect; **0.9 > α ≥ 0.8 is good, *0.8 > α ≥ 0.7 is acceptable

According to the results of KMO and Bartlett Sphericity test made in order to be able to reveal the general tourism support of the residents of district, it had been found that the KMO value of the scale is 0.769. As the result of Bartlett Sphericity test, it had been concluded that the data is proper for the application of factor analysis (X^2: 2294,488, sd:28, p<0.00). Two dimensions had been determined as the result of the PCA operation performed on the scale (Tab. 2). Total variance percentage of these two dimensions is 73.439. It can be said that the obtained total variance rate of 73.439% is pretty satisfactory. Factor loads of the first dimension are consisting of four articles varying in between 0.924 and 0.888. The rate of defining of total variance by the dimension is 42.078%, and reliability coefficient is 0.933. According to this, the factor's reliability level is very high. Factor loads of the second factor are consisting of four articles varying in between 0.835 and 0.713. The rate of defining of total variance by the dimension is 31.361%, and reliability coefficient is 0.790 (Tab. 2).

The first dimension has been conceptualized as "motive of participation in tourism activities" as it is covering the expressions of "I would rent my house for enterprises such as hotel, guest house, restaurant" ($\bar{x}$:2.96, std:2.9609), "I would support my neighbors for them to rent their houses for enterprises such as hotel, guest house, restaurant" ($\bar{x}$:3.1522, std: 1.47698), "I would support my neighbors for them to rent a part of their houses to the tourists" ($\bar{x}$: 3.05, std:1.42025) and "I would rent a part of my house if the tourism is improved" ($\bar{x}$: 2.88, std:1.51112). It has been named as "motive of support of tourism activities" as it is covering the expressions of "I'm supporting the entertainment programs performed for the tourist groups" ($\bar{x}$: 4.17, std:0.75275), "I'm supporting the opening of enterprises such as hotel, guest house, restaurant at the district for the tourists" ($\bar{x}$:4.07, std: 0.99398), "I'm supporting the historical and cultural activities performed for the tourists" ($\bar{x}$:4.32, std:0.69033), "I'm supporting the physical changes on historical structure made with the purpose of increasing the service quality of enterprise such as hotel, guests house, restaurant" ($\bar{x}$: 4.12, std:0.93052).

Tab. 3: Results of ShapiroWilk Test Made on the Factors Determining the Effects of Tourism and Support

Tests of Normality						
	Kolmogorov-Smirnov[a]			**Shapiro-Wilk**[b]		
	Statistic	**df**	**Sig.**	**Statistic**	**df**	**Sig. (p)**
Effects of tourism-oriented decisions	0.236	401	0.000	0,770	401	0.000
Positive effects of tourism on economic environment	0.081	401	0.000	0,976	401	0.000
Negative effects of tourism on physical environment	0.091	401	0.000	0,953	401	0.000
Negative effects of tourism on social environment	0.155	401	0.000	0,825	401	0.000
Motive of participation in tourism activities	0.141	401	0.000	0,907	401	0.000
Motive of support of tourism activities	0.113	401	0.000	0,904	401	0.000

[a]Lilliefors Significance Correction
[b]$p<0.05$ no normal distribution, $p>0.05$ normal distribution

Shapiro-Wilk test, performed in order to determine whether the scores of scale indicate a normal distribution or not, had been applied. As all the scores don't indicate a normal distribution ($p<0.05$) according to the results of Shapiro-Wilk, Spearman correlation test—from among the non-parametric tests—had been applied on the variables in order to determine the direction and severity of the relation in between the effects of tourism perceived by the local community and quality of life (Tab. 3).

Spearman correlation test had been made in order to determine the direction and severity of the relation in between the perceptions of the residents of district regarding the improvement of tourism and general support of tourism (Tab. 4).

According to the data obtained, there is a weak relation in positive direction in between the score of factor of effects of tourism-oriented strategies and factor of positive effects of tourism on economic environment and motive of participation in tourism activities, and there is a very weak relation in positive direction in between effects of tourism-oriented strategies and motive of support of tourism activities. And there is a medium severity relation in positive direction in between score of positive effects of tourism on economic environment and motive of support of tourism activities. It is being determined that there is a very weak relation in positive direction in between the score of negative effects of tourism on physical

Tab. 4: Relation of Scores of Scale Determining Effects of Tourism and Support of Tourism

n=460		Positive effects of tourism on economic environment	Negative effects of tourism on physical environment	Negative effects of tourism on social environment	Motive of participation in tourism activities	Motive of support of tourism activities
Effects of tourism oriented decisions	r	0.200**	0.041	-0.071	0.219**	0.121*
	p	0.000	0.411	0.157	0.000	0.015
Positive effects of tourism on economic environment	r	1.000	-0.032	0.001	0.072	0.592**
	p	.	0.517	0.980	0.153	0.000
Negative effects of tourism on physical environment	r		1.000	0.135**	-0.491**	0.126*
	p		.	0.007	0.000	0.011
Negative effects of tourism on social environment	r			1.000	-0.162**	0.187**
	p			.	0.001	0.000
Motive of participation in tourism activities	r				1.000	-0.062
	p				.	0.186

**$p<0.01$ *$p<0.05$ meaningful relationship, $p>0.05$ no meaningful relationship
$0.00<r<0.20$ too low, $0.20<r<0.40$ low, $0.40<r<0.60$ medium; $0.60<r<0.80$ high; $0.80<r<1.00$ too high.

environment and score of negative effects of tourism on social environment and motive of support of tourism activities, and that there is a medium severity relation in negative direction in between the score of negative effects of tourism on physical environment and score of motive of participation in tourism activities. It is being determined that there is a very weak relation in positive direction in between the score of negative effects of tourism on social environment and score of motive of participation in tourism activities and motive of support of tourism (Tab. 4).

3.3 Determination of the Relation among the Level of Perception of the Effects of Tourism by the Local Community, Support of Tourism and Quality of Life

Results of Spearman correlation test made in order to be able to determine the direction and severity of the relation in between the variables determining the quality of life perceptions of residents of district and scales determining their manners of perception of effects of tourism's improvement are being provided in Tab. 5.

Tab. 5: Relation of Effects of Tourism, Scales of Support of Tourism and Perception Level of Residents of District Regarding Quality of Life

		Effects of tourism oriented decisions	**Positive effects of tourism on economic environment**	**Negative effects of tourism on physical environment**	**Negative effects of tourism on social environment**	**Motive of participation in tourism activities**	**Motive of support of tourism activities**
I'm satisfied with the total income of our dwelling	r	0.354**	0.151**	-0.089	-0.014	0.201**	0.128*
	p	0.000	0.002	0.076	0.779	0.000	0.010
I'm satisfied with my life in general sense	r	0.257**	0.215**	-0.028	0.121*	0.140**	0.192**
	p	0.000	0.000	0.572	0.015	0.005	0.000
I'm satisfied with way I sustain my life in general sense	r	0.228**	0,318**	-0.005	0.102*	0.103*	0.230**
	p	0.000	0.000	0.924	0.040	0.039	0.000

**$p<0.01$ *$p<0.05$ meaningful relationship, $p>0.05$ no meaningful relationship
$0.00<r<0.20$ too low, $0.20<r<0.40$ low, $0.40<r<0.60$ medium; $0.60<r<0.80$ high; $0.80<r<1.00$ too high

A weak relation in positive direction is being determined in between the score of the expression of "I'm satisfied with the total income of our dwelling" and factor of effects of tourism-oriented strategies, and very weak relation in positive direction is being determined among the factors of positive effects of tourism on economic environment, motive of participation in tourism activities and motive of support of tourism activities. There is a weak relation in positive direction in between the score of the expression of "I'm satisfied with my life in general sense" and factor of effects of tourism-oriented strategies, and there is a very weak relation in positive direction among the factors of positive effects of tourism on economic environment, negative effects of tourism on social environment, motive of participation in tourism activities and motive of support of tourism activities. It is being determined that there is a weak relation in positive direction in between the score of the expression of "I'm satisfied with way I sustain my life in general sense" and factor of positive effects of tourism on economic environment, and that there is a very weak relation in positive direction among the factors of effects of tourism-oriented strategies, negative effects of tourism on social environment, motive of participation in tourism activities and motive of support of tourism activities (Tab. 5).

According to the results of Kruskal Wallis test made for testing whether there is a significant difference among the monthly average income of the participants and averages of scale scores, significant relation has been found in between income level of dwellings and positive effects of tourism on economic environment

(X^2:10.605, p=0.014; p<0.05), negative effects of tourism on physical environment (X^2:10.466, p=0.015; p<0.05), negative effects of tourism on social environment and motive of participation in tourism activities (X^2:11.219, p=0.011; p<0.05) and motive of support (X^2:24.051, p=0.000; p<0.001). It has been determined that there is no significant difference in between income groups and only the effects of tourism-oriented strategies (X^2:3.465, p=0.0325; p>0.05). When the sequence averages of income groups in Tab. 6 are considered, it is being observed that the participants with an income level in between 3570–5106 TRY perceive the positive effects of tourism on economic environment more, that they perceive the negative effects of tourism on physical and social environment less compared to the other groups and that their motives of participation in tourism activities and motives of support of tourism activities is more compared to the other income groups (Tab. 6).

It has been determined that there is no significant difference among the ownership status of the dwellings that the participants reside, and effects of tourism-oriented strategies (X^2:5.314, p=0.070; p>0.05), negative effects of tourism on physical environment (X^2:2.901, p=0.235; p>0.05), negative effects of tourism on social environment (X^2:1.908, p=0.385; p>0.05), and motive of support of tourism activities (X^2:5.642, p=0.60; p>0.05). And it has been determined that there is a significant difference among the ownership status of the dwellings that the participants reside, and positive effects of tourism on economic environment (X^2:6.448, p=0.04; p<0.05), and motive of participation in tourism activities (X^2:8.425, p=0.015; p<0.05). It is being observed that the motive of participation in tourism activities of the participants being homeowners is relatively more compared to the others.

According to the results of Kruskal Wallis test made in order to test whether there is a significant difference in between the period of residence of the participants at the district and averages of the scores of scale, it has been determined that there is no significant difference among the period of residence of the participants, and effects of tourism-oriented strategies (X^2:1.381, p=0.926; p>0.05), negative effects of tourism on physical environment (X^2:1.877, p=0.866; p>0.05), negative effects of tourism on social environment (X^2:4.664, p=0.385; p>0.05), and motive of support of tourism activities. It has been determined that there is a significant difference only in between the period of residence of the participants at the district, and motive of support of economic tourism activities (X^2:11.765, p=0.038; p<0.05). It is being observed that the motive of participation in tourism activities of the participants being homeowners is relatively more compared to the others. Consequently, it is being understood that the positive effects of tourism on economic environment is being perceived less, and that the motive of support of tourism activities is decreasing as the period of residence increases.

4 Results and Recommendations

Support of tourism by local community is being shaped as per the economical, social, cultural and environmental consequences that tourism is creating at the touristic area. In case the perception of community is positive regarding the effects

Tab. 6: Comparison of Scale Scores of Income Groups by Kruskal Wallis Test

Factors	Income level of dwellings	N	Mean rank	Chi-square (X^2)	Asymp. sig. (p)
Effects of tourism-oriented decisions	Lowest thru 1404 TRY	193	202.67	3.465	0.325
	1568-2568 TRY	137	190.86		
	2569-3569 TRY	46	193.63		
	3570-5106 TRY	22	237.70		
Positive effects of tourism on economic environment	Lowest thru 1404 TRY	193	181.93	10.605	0.014*
	1568-2568 TRY	137	208.98		
	2569-3569 TRY	46	224.87		
	3570-5106 TRY	22	241.52		
Negative effects of tourism on physical environment	Lowest thru 1404 TRY	193	188.87	10.466	0.015*
	1568-2568 TRY	137	215.91		
	2569-3569 TRY	46	220.22		
	3570-5106 TRY	22	147.25		
Negative effects of tourism on social environment	Lowest thru 1404 TRY	193	192.92	11.219	0.011*
	1568-2568 TRY	137	220.59		
	2569-3569 TRY	46	192.61		
	3570-5106 TRY	22	140.34		
Motive of participation in tourism activities	Lowest thru 1404 TRY	228	227.84	10.000	0.019*
	1568-2568 TRY	148	220.19		
	2569-3569 TRY	54	208.27		
	3570-5106 TRY	23	306,46		
Motive of support of tourism activities	Lowest thru 1404 TRY	228	199,02	24.051	0.000***
	1568-2568 TRY	148	245,20		
	2569-3569 TRY	54	268,40		
	3570-5106 TRY	23	290,11		

*p<0.05,**p<0.005, ***p<0.001

of tourism, their support of tourism is increasing, and in case it is negative, their support of tourism is decreasing. The fact that the positive effects of tourism that it creates on local community is more than negative effects is closely related to contribution of community to tourism, and their inclusion in the improvement process of tourism. In this context, the effects of tourism on the quality of life are very important for the sustainable improvement of tourism. For this reason, this study is focusing on defining the behaviors and perceptions of the local public in order to understand the effect of tourism on quality of life, and trying to understand the relation in between the perception of quality of life of local public and support given for tourism-focused development policies.

Though there are numerous studies for determining the perceptions of local community in Turkey regarding the effects of tourism, the studies in which its affects the quality of life are being examined are limited. In this context, it is being thought that this research will contribute to the relevant literature. In a city such as Amasya, in which the traditional pattern and life style is being preserved, determining the effects of tourism on local community and the contributions created by tourism are important. In this context, it is important to understand how the effects created by tourism are being perceived by the community of Amasya, and to understand their attitudes for the sustainable improvement of tourism. In this regard, by this study it is being understood that factors being effective in the formation of total tourism perception of local community are related to tourism-oriented decisions, positive effects of tourism on economic environment, tourism'snegative effects on physical and social environment. The obtained findings are coinciding with the result of Cengiz and Kırkbir regarding that tourism's economical, social, cultural and environmental dimensions are determining the total effect of tourism at different degrees, and that the total effect of tourism is shaping the support of local community for tourism [25]. Altanlar and Enlil had determined that the most significant factor determining the support of tourism is the perception of political environment [16]. The findings obtained from this study are also supporting the study of Altanlar and Enlil [16]. It has also been determined that the dimensions determining the district residents' general support of tourism are motive of participation in tourism activities and motive of support of tourism activities.

In the study, it has been found that the economical positive effect of the improvement of tourism is increasing the quality of life at the same level. It has been determined that the local community perceives more the positive effects of tourism on economic environment as their income level increases, and that they perceive less the negative effects of tourism on physical and social environment compared to other income groups. The findings obtained from the research are consistent with the studies of Andereck and Nyaupane made in Arizona, Sarı and Özdemir made in Eskişehir Odunpazarı, Küçük made in Konya Beyşehir [13,15,26]. But it is not consistent with the findings of Kim and Khizindar regarding that the economic effects of tourism are not affecting, or negatively affecting, the quality of life of community [13,15,21,26,27]. In a similar manner, it has been found that the positive effects of tourism-oriented decisions on an individual are increasing the quality of life at the same level. The findings obtained are revealing that policies for the improvement of tourism products cover many factors that the local community may like such as festivals; social, cultural and artistic events; and policies such as providing job opportunities, providing infrastructures that will be able to meet the current and future requirements, organizing trainings and seminars for raising awareness regarding abstract and concrete cultural heritage, providing tourism-oriented incentives and loans may affect the local community's perception of quality of life in positive direction. Moreover, the study is revealing that the motive of participation in tourism activities of participants owning a property is relatively more compared to the others. In addition to these, it is being understood that the positive effects of tourism on economic environment is being perceived less, and that the motive of support of tourism activities is decreasing as the period of residence at the district increases. The findings obtained are revealing the importance of supporting the attempts for the inclusion of community in tourism activities in economic sense, and of informing the public on these subjects. By this purpose, encouraging policies should be adopted for increasing the employment of the local community in tourism sector, and for the local community to invest in tourism sector. Thus, it may be assisted in increasing the quality of life of local community by ensuring the domination of domestic investors in the sector.

In the study, it has also been determined that there is a significant relation among income level of local community and motive of participation in tourism activities and support of tourism. This fact is once more revealing the importance of encouraging policies that are developed for the local community to invest in tourism sector. Moreover, this study is also revealing a relation in positive direction among the perception of quality of life, and tourism-oriented decisions, perceived effects of tourism on social and physical environment, and motives support of tourism activities and participation in tourism activities. In this context, it is also revealing that tourism may affect the quality of life in a negative direction through its negative effect on social environment such as crowdedness in physical environment, problems of traffic and parking, weakening of neighborhood relations, decreasing of peace at district, increasing crime rates, deterioration of identity and authenticity of district and changes in the life style of the local community. For this reason, it is a significant requirement to address the required measures in the planning phase in order to eliminate such effects while determining the strategies relevant to the type, size and improvement of tourism.

Consequently, it required to include the local community in planning operations for a sustainable improvement of tourism, and to ensure the planning to be shaped in the direction of their recommendations. In order to increase the positive effect of the improvement of tourism on quality of life, it is required to address it within a context which is not exciting mass tourism, which will eliminate the social and economic inequalities of the area, and which is corporate and regulatory.

References

1 Hepaktan, E., & Çınar, S. (2010). The Effects of Tourism Sector on the Economy of Turkey. *Journal of Social Sciences*, 8 (2): 135–154.

2 Kozak, N., Kozak, M., & Kozak, M. (2013). *General Principles Of Tourism- Concepts.* Ankara: Detay Broadcast.

3 Çeken, H. (2008). A Theoretical Study into Effect of Tourism on Regional Development. *Afyon Kocatepe University, İ.İ.B.F. Dergisi*, 10 (2): 293–306.

4 Avcıkurt, C. (2009). *General and structural analysis of tourism sociology.* Ankara: Detay Broadcast.

5 Butler, G. (2013). *An Assessment of the Social and Economic Impacts of Tourism Development in Dullstroom, Mpumalanga.* fosaf.co.za: http://www.fosaf.co.za/documents/Dullstroom-Tourist-Study-Socio-economic-Impacts.pdf received from address.

6 Doğan, H. Z. (2004). *Socio-Cultural Foundations of Tourism.* Ankara: Detay Broadcast.

7 Duran, E., & Özkul, E. (2012). Residents' Attitudes toward Tourism Development: A Structural Model via Akcakoca Sample. *International Journal of Human Sciences*, 2 (9): 500–520.

8 Oter, Z., & Ozdoğan, O. N. (2005). Destination Image of Cultural Tourists: The Case of Selçuk-Ephesus. *Anatolia: Journal of Tourism Research*, 16 (2) 127–138.

9 Öztürk, Y., & Yazıcıoğlu, İ. (2002). A Theoretical Study on Alternative Tourism Activities for Developing Countries. *Gazi University Journal of Commerce and Tourism Education Faculty*, 2: 183–195.

10 Uslu, A., & Kiper, T. (2006). Effects of Tourism on Cultural Heritage Awareness of Local People in Beypazarı, Ankara. *Journal of Tekirdag Agricultural Faculty*, 3 (3): 305–314.

11 Çakılcıoğlu, M. (2013). A Proposal for a Method for Sustainable Tourism-Oriented Development. *Design+Theory*, 9 (16): 27–47.

12 Bolzoni, M. 2. (2013). What Tourists Ignore Ambivalences, conflicts and compromises in a Changing Neighbourhood. *RC21 Conference 2013, Session 17: Resistance and Protest in the Tourist City*, (s. 1–25). Berlin: http://www.rc21.org/conferences/berlin2013/prog-17.php. received from address.

13 Andereck, K. L., & Nyaupane, G. P. (2011). Exploring the Nature of Tourism and Quality of Life Perceptions among Residents. *Journal of Travel Research*, 50 (3): 248–260.

14 Kim, K., Uysal, M., & Sirgy, M. J. (2013). How Does Tourism in a Community Impact the Quality of Community Residents? *Tourism Management*, 36: 527–540.

15 Sarı, Y., & Özdemir, C. (2014). The Effect of Tourism Development on the Quality of Life of Eskisehir Odunpazarı Residents. *15. National Tourism Congress without Barriers Tourism 13–16 November 2014* (s. 241–253). Ankara: Gazi University, Faculty of Tourism, pp. 241–253

16 Altanlar, A., & Enlil, Z. (2016). Research to Measure Bosphorus Region Residents' Support for Tourism. *Megaron*, 11 (3): 352–397.

17 Türker, N., Selçuk, Ş., & Özyıldırım, A. (2016). The Effect of Tourism in Residents' Quality of Life: The Case of Safranbolu. *Journal of Social Sciences*, 6 (1):1–13.

18 Aydıner Boylu, A., & Paçacıoğlu, B. (2016). Quality of Life and Indıcator. *Journal of Academic Researches and Studies*, 8 (15): 137–150.

19 Yoon, Y. (2002, 09 19). Development of a Structural Model for Tourism Destination Competitiveness from Stakeholders' Perspectives. Virginia, USA: (Unpublished doctoral dissertation), Virginia Polytechnic Institute and State University.

20 Yoon, Y., Gursoy, D., & Chen, J. S. (2001). Validating a Tourism Development Theory with Structural Equation Modeling. *Tourism Management*, 22: 363–372.

21 Kim, K. (2002). The Effects of Tourism Impacts upon Quality of Life of Residents in the Community. Virginia, Blacksburg, USA: (Unpublished doctoral dissertation), Virginia Polytechnic Institute and State University.

22 Altanlar, A. (2015, 06 07). A Conceptual Model and a Method for the Measurement of Cultural Tourısm Development Capacıty: The Case of Bosphorus, İstanbul, Turkey: (Unpublished doctoral dissertation),Yıldız Technical University Institute of Science and Technology, Urban Planning Doctoral Program.

23 Syndicate of Energy, Water and Gas Workers of Turk. (2017, 12 12). Kasım 2017. *November 2017 poverty and hunger Limit.* http://www.tes-is.org.tr/.

24 Çokluk, Ö., Şekercioğlu, G., & Büyüköztürk, Ş. (2012). *Multivariate statistics for Social Sciences: SPSS and Lisster Applications.* Pegem Academy: Ankara.

25 Cengiz, E., & Kırkbir, F. (2007). How Effect's the Local Residents Perceptıon of Tourısm Impacts Level to the Tourism Supports: Dalyan Sample. *Journal of Social Sciences*, 1:19–37.

26 Küçük, M. (2014). Tourism Sector and Its Effects in the Eyes of Local People: A Survey on the Beyşehir District of Konya province. *15. National Tourism Congress without Barriers Tourism 13–14 November 2014* (s. 254–271). Ankara: Gazi University Faculty of Tourism, pp. 254–271.

27 Khizindar, T. M. (2012). (Journal of Hospitality Marketing & Management). Effects of Tourism on Residents' *Quality of Life in Saudi Arabia: An Empirical Study*, 21 (6): 617–637.

Assoc. Prof. Dr. Aybike Ayfer Karadağ and Asst. Prof. A. Esra Cengiz

2 Investigation of Water Process Analysis Method

Abstract: Water is one of the most important natural resources on earth. All living things need water for a healthy and quality life. In this context, the conservation of usable water is very important. In this chapter, the water process analysis method which is developed to contribute to the protection of water resources has been examined. The method aims to identify vulnerable areas to water conservation. In the scope of the method, the infiltration of the precipitation water in the vegetation cover, soil and rock layers are evaluated and are mapped. The findings obtained in the context of the method contribute to the spatial plans developed on the basis of protection of water resources. In addition, these findings contribute to the determination of ecological land use decisions and the determination of vulnerable landscape. In this section, nature-based solutions, ecological processes, hydrologic cycle which are the basis of the method are examined.

Keywords: Water process method, Nature-based solution, Ecological process, Hydrologic cycle

1 Introduction

The world runs on water. Water is the most abundant molecule on Earth. It is the only one that can be found naturally in solid, liquid, and gas and is essential to all life on Earth [1,2]. Therefore, clean and safe water supplies are vital for industry, agriculture, and energy production. Decisions about water affect many areas of life [3,4,5]. In addition, water-related issues are the cornerstones of life. These topics include sufficient water of adequate quality for drinking and sanitation services, biodiversity, food production, energy generation, inland water transport, and water-based recreational activities, as well as sustaining healthy water-dependent ecosystems and protecting the aesthetic and spiritual values of lakes, rivers, and estuaries [3,6,7]. They also involve overseeing water-related dangers such as floods, drought, and water pollution. Thus, it is essential for every biological organism in the world to have water for sanitation, cleanliness, and survival [5].

The reason why people have started consuming more water can be argued to be increasing population, economic pressures, and a change in people's dietary habits (consuming more meat, for example) [8,9,10]. With this rapid population growth in mind, approximately 1.8 billion people will face absolute water scarcity by 2025. Also it is estimated that the world will confront a 40% water supply deficit by 2030. Moreover, these water-related problems such as lasting water shortage, hydrological vulnerability, and extreme weather conditions are viewed as the chief dangers for global welfare. The role of water in global scale escalation of threat zones is accepted by many nowadays. Sustaining a population of 9 billion by 2050 will necessitate a 60% increase in agricultural cultivation, which will further consume more resources and lead to more water consumption. According to statistics, 40% of the population is already living under severe water shortages, which is equal to quarter of world's GDP. Thus, water security presents itself as a major problem that the countries have to face today. More than that, floods, unpredictable rains, droughts, and similar problems due to changing climate conditions will become other aspects that add up to the problem at hand. As an extra problem, water basins act as natural borders between countries, thus leading to another water security issue. Currently 276 water basins flow between 148 countries and they make up 60% of the world's fresh water supply. Furthermore, 300 aquifers are located in-between boundaries and thus 2 billion people are at the mercy of ground water [11]. This picture clearly shows the present situation of water problem in the world. Thus, to have a secure future, this issue should be addressed by a global cooperation which needs to solve water infrastructure, information systems, and smooth water management at a local and global scale. Since water is a natural resource, the solution should include nature-based assets.

The aim of this chapter is to reveal the water process analysis which is a nature-based evaluation. In addition, the concepts of nature-based solution, ecological process, and hydrologic cycle (water cycle), which constitute the basic idea of the analysis, were examined.

2 Nature-Based Solutions and Ecological Processes

Nature is becoming increasingly aware of the solutions produced by using natural ecosystems. These solutions can offer sustainable, cost-effective, multi-purpose, and flexible alternatives. It also leads to a more efficient, competitive, and greener economy. It can also help create new jobs and economic growth, rather than consuming natural capital through the production and delivery of new products and services [12]. Emerging in the last decade, Nature-based Solutions (NbS) focus on reducing the impact of climate change, sustainable resource utilization, natural ecosystems, and biodiversity, unlike traditional engineering solutions [13].

NbS are defined by The International Union for the Conservation of Nature (IUCN) as "actions to protect, sustainably manage, and restore natural or modified ecosystems that address societal challenges effectively and adaptively, simultaneously providing human well-being and biodiversity benefits" [14]. NbS are solutions inspired by nature or completely copied from it. NbS uses complex system processes of nature, such as its ability to store carbon and regulate water flow. These solutions provide simultaneous environmental, social, and economic profits [12,15]. NbS aims to manage the benefits of nature and society in a balanced way [16]. In this regard, the concepts of NbS, ecosystem services, and green infrastructure jointly reinforce each other and emphasize the dependencies between society and nature [17]. The NbS aims to achieve society's development goals and protect human well-being. It also strengthens the resilience, renewal capacity, and service delivery of ecosystems.NbS covers a variety of approaches that use natural processes and species to solve social challenges, and it is a rapidly evolving concept [18]. NbS is designed to address major societal challenges such as food security, climate change, water security, human health, disaster risk, and social and economic development. In fact, NbS is an umbrella term for ecosystem-related approaches. NbS has eight principles which are as follows [13].

- It embraces the principles of nature protection.
- It can be applied alone or integrated with other solutions.
- The site-specific natural and cultural structure should be determined on the basis of traditional, local, and scientific knowledge.
- With a transparent and participatory system, fair societal benefits should be produced.
- Biological and cultural diversity and the ability of ecosystems to develop over time should be preserved.
- It should be applied in landscape scale.
- The trade-offs between the production of economic benefits for development, and future options for the production of the full range of ecosystems services must be recognized and addressed.
- A specific challenge should be addressed with the overall design of policies, and measures or actions.

These approaches can be placed into five main categories, as shown in the Tab. 1. These approaches actually emerged before NbS. They exhibit similarities in terms of their functioning. Although it contributes to the development of an operational framework for NbS, it should not be considered an inclusive concept under NbS [13].

It is very important to integrate NbS into ecological processes to achieve the desired success [19]. Because the basis of sustainability in nature depends on the healthy functioning of ecological processes [20]. Ecosystems have integrity when their natural components are intact including: abiotic components (the physical elements, e.g., water, rocks), biodiversity (the composition and abundance of species and communities in an ecosystem), and ecosystem processes (the engines that make ecosystem work, e.g., fire, flooding, predation) [21]. In other words, ecological integrity refers to the ecological processes and the ability to support and maintain a diverse community of organisms [22]. From this definition, it is understood that, as De Leo and Levin have stated, if an ecosystem has ecological integrity, a set of ecological qualities in spatial and temporal scale must be intact [23]. For these reasons, the solutions that will take the nature as an example should have a function to include the preservation and continuity of the integrity in nature. In addition, the solutions should have systems that evaluate the components and processes that form the integrity.

The ecological processes in a landscape affect the quality of water and the way it moves through a system (in the hydrologic cycle). These processes also affect soil formation, erosion, sediment transport and deposition, which have a great impact on hydrology [10]. Therefore, it is very important to understand the concept of ecological process in water resource issues.

Tab. 1: Category of NbS Approaches [13]

Category of NbS Approaches	Examples
Ecosystem restoration approaches	• Ecological restoration • Ecological engineering • Forest landscape restoration
Issue-specific ecosystem-related approaches	• Ecosystem-based adaptation • Ecosystem-based mitigation • Climate adaptation services • Ecosystem-based disaster risk reduction
Infrastructure-related approaches	• Natural infrastructure • Green infrastructure
Ecosystem-based management approaches	• Integrated coastal zone management • Integrated water resources management
Ecosystem protection approaches	• Area-based conservation approaches, including protected area management

Ecological processes reveal the function of landscape [24]. In other words, the landscape function is the potential and capacity of the landscape [25]. In this context, first the landscape function must be well understood. The landscape function is the expression of ecological, cultural, and visual processes that form a landscape and ensure its continuity [19]. Landscape functions are ecological processes formed by the interaction of natural and cultural landscape elements (e.g., topography, geology, geomorphology, hydrology, climate, soil, wildlife, human activities) in terms of natural life and landscape continuity [26]. Landscape function deals with ecosystems in terms of processes related to transport, utilization, and cycling of scarce and limiting resources (e.g., water, topsoil, organic matter, and propagules) in space and time scale [27]. Processes occur with the interaction of these landscape elements [28]. The landscape function is the process of shaping a landscape, but also this process is affected from shaped landscape [29]. Even all these processes interact with each other [30].

Ecological processes involve changes in the flow and quality of events in landscape over time [19]. Ecological processes determine and affect species composition, habitat structure, and ecological health of landscape [30]. The four fundamental ecological processes are the hydrologic cycle, nutrient (mineral) cycling, energy flow, and community dynamics (plant and animal succession) [2]. However, these processes can be diversified such as climatic process, natural disturbance regime, structural complexity, purification services, biotic interactions, population dynamics, genetic diversity, movements of organisms-biodiversity-habitats, predator/prey cycles [19,20,30,31] and grouped according to the purpose of the study [19]. In fact, these perceived landscape features are the result of ecological processes defined as the rhythm of nature. Also, ecological processes will not be equally important for each landscape. The key processes that create the landscape may be different [2].

The analysis of these processes is one of the best ways to understand the landscape. In landscape function analysis, these processes are analyzed. Landscape function analysis is used to investigate and determine how the ecological processes in question are in the landscape and whether there are some problems in this process [24]. Within the analysis, first, the descriptive and distinctive features and formation process of the landscape are classified and then these classes are evaluated [26].

Ecological processes can be analyzed. The analysis of processes is important in understanding and protecting the landscape. Analyses can be made by evaluating the field studies and observations and/or maps that reveal the natural and cultural characteristics of the landscape [24,26,29,32].

3 Hydrologic Cycle

Water is stored in the oceans, in the atmosphere, as well as on and under the land surface. The water is stored and moved in the form of surface water in rivers, lakes, ponds, ice caps, and below the surface as groundwater. All these water bodies are called as water reservoirs. Oceans form the biggest and largest water reservoirs. The residence time and storage of water in each reservoir varies depending on geological, environmental, and other conditions. Water moves from one reservoir to the other. The sun's radiant energy plays a very significant role in this movement and atmospheric pressure, temperature, and the amount of water vapor present in the air, all play a dominant role in these processes. There is an endless circulation of water among different spheres of the Earth. It is popularly known as the hydrologic cycle (water cycle) [33–36]. If the cycle of water is examined in detail, we observe that the water that evaporates from the ocean with the sun's energy starts to move by the circulation of winds around the planet. Upon rising over the contours of the mountains, it cools and becomes rain, providing moisture to forests, grasslands, scrub, agricultural lands, cities, and others. It feeds streams, rivers, lakes, and groundwater before eventually returning to the sea. On this long journey, it is absorbed by plants and consumed by animals and humans [1,2,37].

It is of two kinds as terrestrial hydrologic cycle and global hydrologic cycle. The terrestrial hydrologic cycle is the water resources formation mechanism on a given area of the land, like a river basin or a watershed. The global hydrologic cycle affects global climate and other geological and physical processes. In addition, the oceans that have the largest mass of water in the world play a very crucial role in the global hydrologic cycle [34].

Hydrologic cycle involves several causative factors and components. The major components (or) elements of the hydrologic cycle are: Precipitation, evaporation, transpiration, evapotranspiration, interception, surface runoff, condensation, infiltration, groundwater base flow, sublimation [38]. Precipitation is water being released from clouds as rain, sleet, snow, or hail. Rainfall is the most common form of precipitation occurring in almost all parts of the world. The snowfall is another form of precipitation. Evaporation is the conversion of a solid or liquid into gas by the heat energy. Transpiration is the process of releasing the water absorbed by the plants through their root system after utilizing the nutrients for building their tissues in a specified time. Evapotranspiration is water evaporating from the ground and transpiration by plants. Interception is the process of retaining water on the leaves of vegetation. A small amount of rainfall is intercepted by vegetation. Runoff is the flow of precipitation reaching the Earth, not infiltrated in saturated soil. Condensation is the conversion of water from a vapor to liquid. Infiltration is the process by which water on the ground surface enters the soil. Percolation refers to filtration of water through soil and permeable rocks. The rainwater that is falling over the ground surface percolates down through the soil and reaches the groundwater zone. Depending upon the slope of the groundwater system, the groundwater moves towards the rivers, lakes, or the oceans. This flow of groundwater is known as base flow. The process of direct evaporation of snow into water vapour is known as sublimation [34,37,38]. In this context, it should be said, hydrological cycling is one of stochastic processes in nature. Also most of the ecological processes affect hydrological cycling in different ways over the short-term and long-term. These processes are dynamic and change over spatial and temporal scale [39].

Water process analysis forms a part of the hydrologic cycle. It defines the movement of water in the space.

4 Water Process Analysis

In order to determine the reach path (trace) of the precipitation water to groundwater, a number of analyses and drilling studies on rock and soil are required. In the determination of the infiltration zones, field observations, instrumental measurements, soil mechanics measurements, and geophysical measurements are required[1] [40]. However, it is important to determine the precipitation water path in ground in order to make spatial planning and land use decisions regarding water resources [41]. In this context, water process analysis was developed by Şahin (1996).

The basic idea of the water process analysis is to evaluate the affecting factors reaching process of precipitation water to the underground (within soil and rock). Because while a part of the rain falling on the Earth passes directly to the stream, the other part is filtered into the ground with the effect of gravity-capillary and molecules stresses. Leakage water first feeds the soil moisture. When the maximum

1 However, hydrography analysis and infiltrometer studies are among the techniques that are generally used to find out the infiltration capacity of the soil. Infiltrometers are generally called as rain simulators. In this system, a small testing ground is splashed with rain and parameters like infiltration ratio, penetration, and surface flow are observed.

subsurface storage is exceeded, some of the water forms the underwater flow, the rest supplying ground water storage. In other words, the rainwater flows into underground with the influence of gravity and is collected in the pores of the formation on a certain impermeable layer [42]. Factors affecting infiltration and percolation are effective in the progression of precipitation water in soil [40]. In this movement of water, land slope, area use, vegetation cover, soil and rock layer properties have priority effect [27,40,43,44]. Some of these properties are given below.

- The slope of the topography affects the flow of precipitation in the surface, soil, and rock layers [23].
- The presence of vegetation (forests, meadows, planting areas) above the soil has a significant impact on the underground journey of precipitation water. Presence of vegetation increases infiltration. Plants should not be seen as consumers of water, but actually as water recyclers [10,27].
- Infiltration increases in the area where canopy density (vegetation density) is too high. Due to various factors, infiltration capacity values up to 25 mm/h can be achieved in an empty area. Vegetation density can increase this by five times [42].
- The amount of interception (as a percentage of rainfall) varies according to tree types. Conifers have higher interception values relative to broadleaf trees, because the amount of water that is intercepted by plants is substantially reduced due to the loss of leaves in the winter [45]. In addition, evaporation on plant surfaces increases from short grass to trees (increase direction: short grasses, ball-shaped plants, shrubs and trees). From this point of view, it can be said that when a field covered with bushes or forest is turned into pasture, more of the falling rainfall is transferred to the soil and more useful water for herbaceous plants accumulates in the soil. Infiltration is mostly in forests, then in bushes and lumpy formations, and at least in areas covered with grass. The most infiltration is in forest area, because there are more alive and dead cover under forests and shrubs. Then, the increase in infiltration is seen in lumpy formations, short grasses and at least bare areas, respectively [46].
- Patches are important in the surface drainage network. For example, dense patches of perennial grasses cause drainage network to have a tortuous path. Also, as the perennial grasses patch density decreases, the surface flow rate of drainage water will increase [27].
- The infiltration capacity of the soil is strongly influenced by the soil texture, that is, the relative proportions of sand, silt, and clay [47]. As the amount of clay and silt in the soil increases, the infiltration decreases. However, as the amount of sand and gravel in the soil increases, infiltration increases [42].
- As stated in Singer and Munns (1999), the average soil pore size, distribution of soil pore sizes, and connectivity of soil pores are of greater importance for infiltration. The soil pores must be large enough and offer sufficient continuity in order for infiltration to occur. Soil pores are classified by size into macropores (> 0.075 mm), mesopores, and micropores (< 0.03 mm) [47].

- The effect of soil texture and vegetation on the infiltration rate is great. The infiltration capacity according to whether the same soil type is covered by vegetation, that is, the amount of precipitation water that goes down to the soil within a unit time, is very different. This difference is in the form of increased infiltration of 100 % in the soil which is vegetation [48].
- The amount of moisture in the soil is important for infiltration. As the moisture in the soil increases, the infiltration rate decreases [47].
- Erosion, sediment transport, and deposition affect infiltration [10].
- The movement of precipitation water in the rock layers depends on the porosity and permeability of the rocks. Permeability is rock layer's ability to allow fluids. Porosity is the percentage of void space in a rock layer. These features vary according to the properties of rocks forming rock layers. For example, magmatic and metamorphic rocks have low porosity. Soluble rocks, such as limestone, are too porous due to melting and fracturing. Sand and gravel deposits are the most permeable layers [44].

Water Process Analysis was first developed by Şahin. Şahin evaluated permeability status of soil and rocks structure (two variables) with holistic approach and then mapped the results. In this context, numerical slope, soil and rock structure permeability maps were analyzed by Geographic Information Systems (GIS). In the analyses, 1/25,000 scale digital maps were used. In addition, the variation of permeability rate was evaluated. Evaluations were made based on literature review and expert opinion. This method has three stages (Fig. 1). In the first stage, soil structure permeability data has been created on the basis of soil classes and soil texture. The variation of soil structure permeability was divided into five value ranges (very high, high, moderate, low, very low). The second step of this stage is the overlay analysis (GIS) of soil structure permeability and slope because permeability rate decreases with increasing slope angle. The result of the overlay analysis was evaluated according to Tab. 2. In the second stage, rock structure permeability map has been created on the basis of rock type. The variation of rock structure permeability was divided into three value ranges (high, moderate, low). The second step of this phase is the overlay analysis (GIS) of rock structure permeability and slopebecause permeability rate decreases with increasing slope angle. The result of the overlay analysis was evaluated according to Tab. 3. In the third stage, soil and rock structure permeability data, which were evaluated according to the slope, were subjected to overlay analysis with GIS. The new data is considered as the soil+rock infiltration map according to the soil and rocks. Evaluation of the new data was made according to Tab. 4[49].

Uzun et al. (2015) had developed the method by adding the criteria of vegetation cover to the water process analysis (Fig. 2). Because vegetation cover has a guiding effect in the infiltration of precipitation water [10,27,42,45,46,50]. This new method is applied in fourstages. The first threestages were applied as in the method developed by Şahin [49]. Soil+rock structure permeability data obtained from this process was subjected to overlay analysis by vegetation cover infiltration data. In the

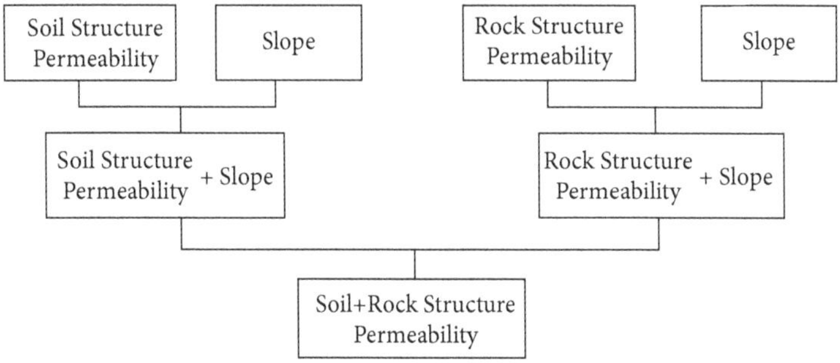

Fig. 1: Water process analysis, stages [49].

Tab. 2: Structure Permeability According to the Slope Value [49]

Soil Infiltration Value	**Slope Rate (%)**					
	0–2	**2–6**	**6–12**	**12–20**	**20–30**	**>30**
Very high	vh	vh	vh	h	h	h
High	vh	vh	h	h	m	m
Moderate	vh	h	h	m	m	m
Low	l	l	l	l	l	l
Very low	vl	vl	vl	vl	vl	Vl

Very high (vh), high (h), moderate (m), low (l), very low (vl)

Tab. 3: Rock Structure Permeability According to the Slope Value [49]

Rock Ppermeability value	**Slope Rate (%)**					
	0–2	**2–6**	**6–12**	**12–20**	**20–30**	**>30**
High	vh	vh	vh	h	h	m
Moderate	vh	h	h	h	m	m
Low	l	vl	vl	vl	vl	Vl

High (h), moderate (m), low (l)

Tab. 4: Soil+Rock Structure Permeability Value [49]

Rock Permeability Value	Soil Infiltration Value				
	Very High	**High**	**Moderate**	**Low**	**Very Low**
Very high	vh	vh	h	m	l
High	vh	h	m	l	vl
Moderate	h	h	m	l	vl
Low	h	m	l	vl	vl
Very low	h	m	l	vl	Vl

Very high (vh), high (h), moderate (m), low (l), very low (vl)

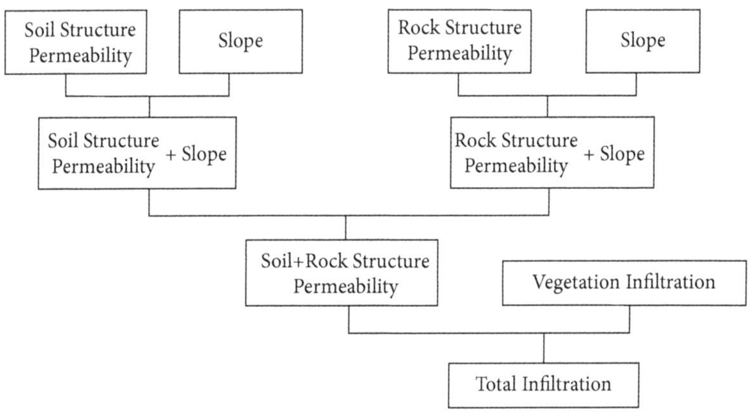

Fig. 2: Water process analysis, stages [50].

method, satellite imagery was used to determine vegetation and land cover. The result of the overlay analysis was evaluated according to the Tab. 5.

1: *Forest areas: coniferous, mixed, leafy*
2: *Shrub areas, irrigated areas, sparse vegetation*
3: *Herbaceous plants, irrigated agricultural areas*
4: *Bare areas, cliffs*
5: *Artificial areas: built areas, hard pen.*

Şahin [49] and Uzun et al. [31] have mapped precipitation water journey in soil and rock layer. They used digital data and overlay analysis in their methods. The

Tab. 5: Total Infiltration Value (Rock+Soil+Vegetation) [50]

Soil+Rock Structure Permeability Value	Vegetation Infiltration Value				
	Very High[1]	High[2]	Moderate[3]	Low[4]	Very Low[5]
Very high	vh	vh	h	m	l
High	vh	h	h	m	l
Moderate	H	h	m	m	l
Low	M	m	m	l	vl
Very low	M	l	l	vl	Vl

very high (vh), high (h), moderate (m), low (l), very low (vl)

total infiltration map was produced with the holistic analysis. This methods were used by some researchers [26,29,31,32,49,51].

In Fig. 3, a total infiltration map of the Hendek district determined based on soil infiltration and rock permeability is given.

5 Conclusion

Nature-based solutions (NBS) are inspired and supported by nature and use, or mimic natural processes to contribute to the improved management of water. An NBS can involve conserving or rehabilitating natural ecosystems and/or the enhancement or creation of natural processes in modified or artificial ecosystems [10]. At this point, it can be said that it is quite important to integrate nature-based solutions with ecological processes. Therefore, ecological processes are a key for water management and planning.

In this chapter, water process analysis method was developed on the basis of water cycle, which is one of the ecological processes. In the method, the infiltration of the precipitation water through the soil-rock layer or the vegetation-soil-rock layer was evaluated with a holistic approach. Evaluation of the analyses was made on the basis of expert opinion. As a result, total infiltration map was formed.

The development of the method will ensure that the total infiltration map is more accurate. There are two ways to improve the method. The first is to increase the criteria used in the method. In other words, other factors affecting infiltration can be mapped and included in the method. The second way is to increase the expert opinions and to create the infiltration evaluation tables which are formed with these views by statistical methods (e.g., analytical hierarchy, delphi technique). Also these values can be give numerical form.

The total infiltration map can contribute to many water resources studies (e.g., water management, watershed planning, green infrastructure planning).

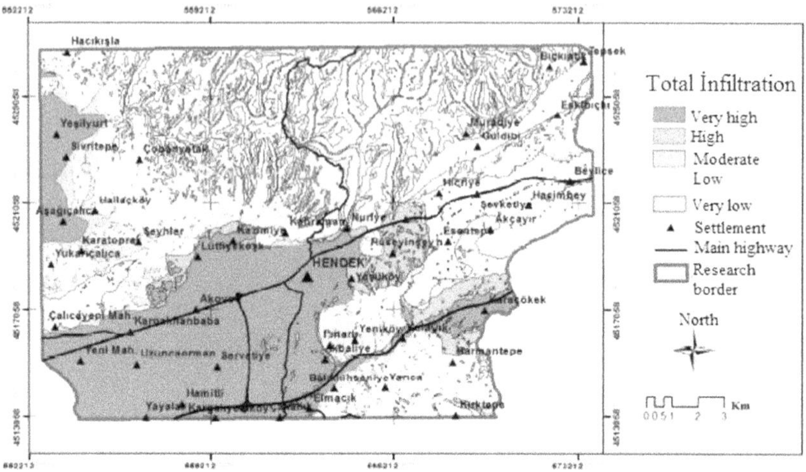

Fig. 3: Total infiltration map of Hendek district (Sakarya, Turkey) determined according to soil infiltration and rock permeability [29].

Especially, it is practical data in terms of upper-scale planning. The total infiltration map is also important in the development of spatial planning and land use decisions. Because precipitation water breaks down pollutants as it passes through an area. Contaminated water reaches to soil, rocks, and groundwater. Therefore the best way to reduce pollution is to prevent it in the first place. In this context, high infiltration areas can be preserved. For example, industry, agriculture, cemetery, etc., in these areas may be prevented from intensive land use. This map can also be used to assess landscape character and landscape sensitivity. Landscape character map and landscape sensitivity map are important for space planning because this map is a tool to help understand what the landscape is. If the landscape is understood, ecological processes will be understood. [29,31,40,49–51].

As a result, water process analysis contributes to the production of important data in terms of water conservation and spatial planning. The total infiltration map is a practical spatial data especially in terms of upper-scale planning. The method is a nature-based approach and has been developed on the basis of ecological processes.

References

1 Mitchell, V. G., Mein, R. G., & Mc Mahon, T. A. *Modelling the Urban Water Cycle*, 2001. Environmental Modelling & Software, 16, 615–629.

2 Muir, J. *Ecological Processes*, 2009. Mexian Biodiversity. Available from: www.biodiversidad.gob.mx (Date of access:16.11.2018).

3 Aksungur, N., & Firidin, Ş. *Use of Water Resources and Sustainability*, 2008. Yunus Research Bulletin, 8 (2), 9–11.

4 Turan, T., & Bayhan, Y. K. *Comparing the Protection of Water Resources Politics in European Union and Turkey*, 2009. Atatürk University Journal of the Institute of Social Sciences, 13 (2), 331–340.

5 World Resources Instıtute Available from: www.wri.org/our-work/topics/water (Date of access: 02.11.2018)

6 Akkaya, C., Efeoğlu, A., & Yeşil, N. *The European Union Water Framework Directive and Applicability in Turkey*, 2008. The Union of Chambers of Turkish Engineers and Architects (UCTEA). II. Water Policy Congress, Ankara.

7 Aküzüm, T. Çakmak, B., & Gökalp, Z. *Evaluation of WaterResources Management in Turkey*, 2010. Journal of Agricultural Sciences Research, 3 (1), 67–74.

8 Flörke, M., Kynast, E., Bärlund, I., Eisner, S., Wimmer, F., & Alcamo, J. *Domestic and Industrial Water Uses of the Past 60 Years as a Mirror of Socio-Economic Development: A Global Simulation Study*, 2013. Global Environ Change, 23, 144–156. Available from: www.doi.org/10.1016/j.gloenvcha.2012.10.018

9 Liu, J., Yang, H., Gosling, S. N., Kummu, M., Flörke, M., Pfister, S., Hanasaki, N., Wada, Y., Zhang, X., Zheng, C., Alcamo, J., Oki, T. *Earth's Future Water Scarcity Assessments in the Past, Present, and Future*, 2017. Earth's Future, 5 (6). ISSN: 2328–4277, DOI num: 10.1002/2016EF00518

10 Connor, R., Coates, D., Uhlenbrook, S., & Koncagül, E. *The United Nations World Water Development Report*, 2018. Available from:www.unesdoc.unesco.org/images/0026/002615/261594e.pdf (Date of access: 09.11.2018).

11 Water Resources Management Available from: www.worldbank.org (Date of access: 02.11.2018).

12 Nature-Based Solutions. Available from: www.ec.europa.eu/research/environment/index.cfm?pg=nbs (Date of access: 20.11.2018)

13 Cohen-Shacham, E., Maginnis, S., Walters, G. M., & Janzen, C. *Nature-Based Solutions to Address Global Societal Challenges*, 2016. Publisher: IUCN, Gland, Switzerland. DOI 10.2305/IUCN.CH.2016.13.en ISBN 978-2-8317-1812-5

14 Nature-Based Solutions Available from: www.iucn.org (Date of access: 20.11.2018).

15 Nature-Based Solutions & Re-Naturing Cities. European Commission. Available from: www.greengrowthknowledge.org/ (Date of access: 20.11.2018).

16 Nesshöver, C., Assmuth T., Irvine, K. N., Rusch, G. M., Waylen, K. A., Delbaere, B., Haase, D., Walters, L. J., Keune, H., Kovacs, E., Krauze, K., Külvik, M., Rey, F., Dijk, J., Vistad, O. I., Wilkinson, M. E., & Wittmer, H. *The*

Science, Policy and Practice of Nature-Based Solutions: An Interdisciplinary Perspective, 2017. Science of the Total Environment, 579: 1215–1227.www.doi.org/10.1016/j.scitotenv.2016.11.106

17 Artman, M., & Sartison, K. *The Role of Urban Agriculture as a Nature-Based Solution: A Review for Developing a Systemic Assessment Framework*, 2018. Sustainability, 10, 1937. DOI:10.3390/su10061937

18 Europe's Ecosystem Research Network Available from: www.alter-net.info/outputs/conf-2015 (Date of access: 20.11.2018)

19 Bennett, A. F., Haslem, A., Cheal, D. C., Clarke, M. F., Jones, R. N., Koehn, J. D., Lake, A. S., Lumsden, L. F., Lunt, I. D., Mackey, B. G., Nally, R. M., Menkhorst, P. W., New,T. R., Newell, G. R., O'Hara, T., Quinn,G. P., Radford,Q., Robinson, D., Watson, J. E. M., & Yen, A. L. *Ecological Processes: A Key Element in Strategies for Nature Conservation*, 2009. Ecologıcal Management & Restoratıon, 10 (3). DOİ:10.1111/j.1442-8903.2009.00489.x

20 What Is Ecological Integrity? www.bcauditor.com/sites/default/files/publications/2010/report3/files/ (Date of access: 22.11.2018).

21 What Is Ecological Integrity? Available from: www.pc.gc.ca/en/nature/science/conservation/ie-ei (Date of access: 22.11.2018).

22 Ecological Integrity Available from: www.oceanhealthindex.org (Date of access: 22.11.2018).

23 Vogiatzakis, I. N., Manolaki1, P., Zomeni, M., Trigkas, V., Papatheodoulou, A., Christou, O., Michael, K., & Victor, M. *Assessing the Ecological Integrity of Landscapes in Cyprus*, 2015. Journal of Mediterranean Ecology, Vol. 13, 2015: 29–35 R. D. Pieper (ed.). Society for Range Management.

24 Çetinkaya, G., & Uzun, O. *Landscape Planning*, 2014. Birsen Publisher. Page: 219. ISBN: 978-975-511-608-2, İstanbul, Turkey.

25 Bolliger, J., & Kienast, F. *Landscape Functions in a Changing Environment*, 2010. Landscape Functions in a Changing Environment, Landscape Online, 21, 1–5. DOI:10.3097/LO.201021

26 Aytaş, İ., Bilgili, B. C. & Şahin, Ş. *Analysis of Landscape Character in Çankırı*, 2016. International Geography Symposium, 696–713, Ankara, Turkey.

27 Tongway, D. J., & Hindley, N. L. *Landscape Function Analysis: Procedures for Monitoring and Assessing Landscapes-with Special Reference to Minesites and Rangelands.* 2004. Published by CSIRO Sustainable Ecosystems, Australia.

28 Şahin, Ş. Perçin, H., Kurum, E., Uzun, O., & Bilgili, B. C. *National Technical Guideline for Landscape Character Analysis and Assessment at theRegional and Sub-Regional (Provincial) Levels*, 2014. Republic of Turkey Ministry of Interior, General Directorate of Local Administrations, Republic of TurkeyMinistry of Environment and Urbanization, General Directorate of Spatial Planning Republic of Turkey, Ministry of Forestry and Water Affairs, General Directorate of Nature Conservation and National Parks, Ankara, Turkey.

29 Karadağ, A.A., & Yıldız, K. *Evaluation of Landscape Functions in the Sample of Hendek District*, 2013. Düzce University Journal of Forestry, 9 (1), 77 – 96.

30 Considering Ecological Processes in Environmental Impact Assessments. Available from: www.epa.gov/sites/production/files/2014-08/documents/ (Date of access: 22.11.2018).

31 Uzun, O., İlke, E.F., Çetinkaya, G., Erduran, F., & Açıksöz, S. *Landscape Planning: Bozkir - Seydişehir - Ahirli - Yalihüyük District of Konya and Suğla Lake Village Landscape Management Conservation and Planning*, 2012. Project. ISBN: 978-605-4610-09-9. Ankara, Turkey.

32 Uzun, O. Gültekin, P. *Process analysis in Landscape Planning, The Example of Sakarya/Kocaali, Turkey*, 2011. Scientific Research and Essays, 6 (2), 313–331

33 Pagano, T., & Sorooshian, S. *Hydrologic Cycle* (in Encyclopedia of Global Environmental Change), 2002. Volume:1, ISBN 0-471-97796-9

34 Balasubramanian, A. & Nagaraju, D. *The Hydrologıc Cycle*, 2015. Technical Report, University of Mysore, Centre for Advanced Studies in Earth Science. Available from: www.researchgate.net/publication/315125743_THE_HYDROLOGIC_CYCLE (Date of access: 27.11.2018).

35 Understanding Ecosystem Processes. Available from: www.for.gov.bc.ca/hra/publications/pamphlets/Rangeland_Health_Pamphlet3.pdf) (Date of access: 22.11.2018).

36 Biswas, A. K., Braga, B. P.F, Tortajada, C., & Palermo, M. *Integrated Water Resources Management in Latin America*, 2009. Routledge, New York.

37 The Water Cycle. Available from: www.gw.govt.nz/assets/our-environment/water-supply/pdfs/turning-on-the-tap/turning-on-the-tap-section-2.pdf (Date of access: 26.11.2018).

38 Hydrologic Cycle. Available from: www.nasa.gov/audience/forstudents/5–8/features/Observatorium_Feat_5-8.html (Date of access: 26.11.2018).

39 Schreier, H., Kurian, M., & Ardakanian, R., *Integrated Water Resources Management: A Practical Solution to Address Complexity by Employing the Nexus Approach*, 2014. Institute for Integrated Management of Material Fluxesand of Resources, Germany.

40 Yolcubal, İ. 2018. Hydrogeology Course Notes. Kocaeli University, Geology Engineering Department. Available from: www.jeoloji.kocaeli.edu.tr/dosyalar/dersNotlari/hidrojeoloji_ders%20notlari_prof_dr_irfan_yolcubal.pdf (Date of access: 21.02.2019)

41 Uzun, O. Landscape Assessment Development of Management Model for Düzce, Asarsuyu Watershed, 2013. Ph.D. Thesis. Ankara University Graduate School of Natural and Applied Sciences, Ankara, Turkey.

42 Okkan, Ü. Hydrology Course Notes, 2018. Balıkesir University Civil Engineering Department, Balıkesir, Turkey. Available from: www.drumutokkan.com/lisans-dersleri/hidroloji/ (Date of access: 21.02.2019)

43 Öztekin, T., Öztekin, S. Effects of Bulk Density and Drainage System on Infiltration, 2007. Journal of Agricultural Faculty of Gaziosmanpaşa University (JAFAG), 24 (2), 67–75.

44 Dirik, K. *Grounwater Course Notes*, 2018. Hacettepe University, Faculty of Engineering, Ankara, Turkey.

45 Yurtseven, İ., Zengin, M. *Neural Network Modelling of Rainfall İnterception in Four Different Forest Stands*, 2013. Annals of Forest Research, 56 (2), 351–362.

46 Fidan, C., Duran, C., Kırış, R. *The Effect of Plant Formations on Water Adjustments*, 2008. Unıon of Chambers of Turkısh Engıneers and Archıtects. 2. Water Policy Congress, Ankara, Turkey.

47 Haghnazari, F., Shahgholi, H., Feizi, M. *Factors Affecting the İnfiltration of Agricultural Soils: Review*, 2015. International Journal of Agronomy and Agricultural Research (IJAAR), 6 (5), 25–35. University of Mysore

48 Çepel N. *Rainfall Catchments of the Dams up the Ecological Principles for Land Use Planning*, 1986. Istanbul University Journal of Faculty of Forestry, 36 (20), 17–27.

49 Şahin, Ş. *A Research on Determination and Assessment of Landscape Potential for Dikmen Valley*, 1996. PhD Thesis. Ankara University Graduate School of Natural and Applied Sciences, Ankara, Turkey.

50 Uzun, O., Müderrisoğlu, H., Demir, Z., Kaya, L., G., Gultekin, P., Gunduz, S. *Yeşilırmak Basin Landscape Atlas*, 2015. T.C. Ministry of Forestry and Water Affairs, General Directorate of Nature Conservation and National Parks, AKS Planning Engineering Limited Company, Ankara.

51 Demiroğlu, D., Yücekaya, M., Günaydın, A. S., Taşçıoğlu, S. 2017. *Ecological Approach to Urban Parks: The Case of Urban Parks in Kilis*, Turkey. Fresenius Environmental Bulletin, 26 (12), 7142–7149.

Assist. Prof. Dr. Aylin Aras and Assoc. Prof. Dr. Egemen Aras

3 A New Solution Proposal for Pedestrian–Vehicle Traffic: Uzun Street Sample in Turkey

Abstract: This chapter was made in order to propose a solution to ease the traffic flow of Uzun Street where the pedestrian and motor vehicle traffic is dense. In the scope of the study, a structural walking platform at the levels of the second floors of the buildings lined up along the axis of Uzun Street will be built. Motor vehicle traffic will be made possible under the platform and passages to the side streets will be built. Thus, both the pedestrian density of the street will be eased and the second-floors will be used as stores.

Keywords: Uzun Street, Urban design, Pedestrian Crossings, Double-story road

1 General Information

Trade areas are important for the development of the city, and they have played an active role in the formation of the city for ages. Particularly, from 19th century onwards city centers and trade areas have been in interaction with each other in the development and organization of the cities, and this turned the city centers into trade centers. City centers today do not only meet people's daily needs but also they serve for other needs such as socio-cultural and entertainment activities, being the place where people meet each other and talk. Trade areas in the city centers are not only the places where people go for shopping but also the places where people gather in order to carry out social activities.

There is a need for a better transportation system in the city centers that will support trade. There is no doubt that there is a need for building non-stop pedestrian axes that support the pedestrian traffic, or a need for building pedestrian crossings and vehicle roads that hardly intersect each other.

Many researchers have sought solutions for pedestrian and vehicle traffic in the busy streets. They explore how to design both pedestrian and vehicle traffic simultaneously [1,2,3,4,5,6]. The ease of window-shopping for the people and a safe shopping atmosphere will all create a positive effect for the trade center. Free from the negative effects such as gas and noise, shoppers want spaces where they can relax and reach the shops.

Particularly, in the city of Trabzon, many city dwellers as well as those who come from the rural places constitute a majority of the population that uses the city center. Those people who come from rural as well as city dwellers use the city center mostly for cultural and educational facilities (private courses, hobby courses, etc.), fun activities (cinema, theatre), for shopping, and to meet each other.

Gocer has outlined the aims of the pedestrian trade routes as following:

- To create spaces where city people meet or come across each other in the city.
- To create points of attraction in the community life.

- To create a forum for the city dwellers, and provide places where they can take part in the planning and discuss with each other on various topics.
- To help create good relations among the people without force or organizations.
- To provide a safe atmosphere in which pedestrians may walk safely, free from the traffic.
- To help solve the problems between the pedestrians and motor vehicle traffic in favor of pedestrians.
- To separate the pedestrians and motor vehicle traffic.
- To limit vehicle traffic and foster pedestrian traffic.
- To develop planning decisions to avoid such operations as knocking down a building or road extensions, and to use the buildings in their actual forms [7].

This chapter was made in order to propose a solution to ease the traffic flow of Uzun Street where the pedestrian and motor vehicle traffic is dense. Yalcinkaya Erol and Sancar have done research on the urban identity and image of Trabzon. In this study, 56.7% of the surveyed people stated that Uzun Street was a symbol for Trabzon. This shows that Uzun Street's Trabzon is an important transportation axis for urban identity and image [8].

It is possible to use it together with the urban city, to age and to gain more value for the local people as long as it functions, by conserving the buildings with a re-functioning approach at the single level [9]. However, maintaining the city center should also be considered in urban scale. The city center should be revitalized by mobilizing the economic, social, and cultural mechanisms. Thus, it is thought that the people of the region will be relocated to important axes in the city center and the urban image will be restored. Therefore, a design proposal has been prepared in order to make this transportation axis comfortable and effective. In the scope of the study, a structural walking platform at the levels of the second floors of the buildings lined up along the axis of Uzun Street will be built. Motor vehicle traffic will be made possible under the platform and passages to the side streets will be built. Thus, both the pedestrian density of the street will be eased and the second floors will be used as stores.

2 The Geography and Cultural History Development of the City of Trabzon

Trabzon is a city that was established in the north-east of Eastern Anatolia, on the shore of a natural harbor of Black Sea, and on the way to historical transit road between Asia and Middle East. It is on the 41 degree of northern latitude and 39 and 43 degrees of eastern longitude. It covers an area of 4686 square km. and it is surrounded by Rize in the east, Bayburt in the south-east, Gümüşhane in the south, Giresun in the west, and Black Sea in the north.

Trabzon is one the oldest cities of the Black Sea coasts. There are two main reasons for the establishment and importance of the city. The first one is that there are few connections between the coast and the inner parts. In order to reach Harşit and Çoruh Valleys, one has to go beyond an altitude of 3000 km. There is nowhere a valley which is large enough to be a road to inner parts other than the Zigana in Trabzon. Zigana Passage starts from Trabzon and reaches to inner regions. This road (the passage) starts from Değirmendere valley, the south of harbor, and rises slowly for 30 km to the sides of the Zigana mountains. After the distance of 66 km is covered, one can reach the Zigana tunnel at an altitude of 2030 meters. This makes it possible to reach the Harşit Valley and then to the Mid-Anatolian plateaus. The first reason why the city of Trabzon was established goes to the geographical location of the city as mentioned above. The other reasons why the city of Trabzon was established can be given to the fact that it is a naturally protected area by Yoroz Burnu and Güzel Hisar hills which prevent winds, and it has a natural bay that serves like a harbor for the small ships and that the city has a land that is suitable to establishment and development.

Trabzon whose known history dates back to 4000 years is one of the rare cities that has always attracted world attention due to its geographical location. Its geographical significance, being a cross road, and home to many civilizations throughout the ages are among the most important factors that make it important. The cultural life of the city is also very colorful. It reflects, first of all, the "urban culture" for ages. The existence of education-culture-and trade centers for ages in the city is the indication of the fact that the city has a rich and cultural accumulation.

In many historical documents it is possible to come across the existence of cultural institutions in Trabzon that were developed further after the conquer of Trabzon in 1461 by Sultan Mehmet, the Conquerer. After the Turkish Republic was proclaimed, M.Kemal Atatürk called the people of Trabzon as "clever, productive, initiative and hardworking" and many investments in terms of education and culture were made in the city. World famous travelers visited the city and thus it became an attraction for many people in the world. Hundreds of travelers from Ksenophon to Evliya Çelebi and from Fallmerayer to Frunze visited Trabzon and they recorded their thoughts, ideas, and observations in the form of "seyehatname" or "travels book." The focal point in those books is that "Trabzon is an indispensible city since it has an mystical nature and geography, and since it constitutes an important link between Middle Asia, Caucasia, Far East, Middle East, İstanbul, and Europe [10].

Trabzon has a number of tourist attractions, some of them dating back to the times of the ancient empires that once existed in the region. In the city itself, one can find a hub of shops, stalls, and restaurants surrounding the "Meydan," a square in the center of the city, which includes a tea garden.

3 Uzun Street and Trabzon

The city center of Trabzon, one of the oldest cities in the world, has four main axes that are parallel to each other. The streets that have dense pedestrian and vehicle traffic are Uzun Street, Maraş and Kunduracılar Streets, and Yavuz Selim

Fig. 1: Uzun Street before the renovation works.

boulevard. With the construction of Tanjant road, there has been a relief but pedestrian traffic still remains problematic as ever.

Speaking about the social life in Trabzon today, the first place to be remembered is Uzun Street, the busiest place of the city. It is a long street with stores and passages and a walking area where many people meet each other. In fact, Uzun Street is the mostly perceived shopping area of people in Trabzon and this perception dates back to a distant past (Fig. 1).

Uzun Street is a fully equipped place that can meet any needs any time. At the one end, there is Meydan Park and the other end extends to the end of the city as the name refers to.

Uzun Street today looks like a rather narrow street, whose ground floors are used as stores, that is surrounded by stone walls and patios. It is possible to see clothing stores, shoe stores, cinemas, restaurants, book sellers, doctors, pharmacy, and patisseries which all are necessary in shopping centers. There is a continuous rush in there and the noise can be regarded as a moderate hum of voices. The city dwellers use and meet their needs continuously, and it is no less crowded after the dark [11].

The side streets along Uzun Street contribute a lot to the texture of the city. Some important ones are Zeytinlik Avenue that is home to Kostaki House on the way to Boztepe, the City Club Avenue that has many stone buildings, and Hacı Kasım slope that links Uzun Street to south.

Uzun Street is an identity for Trabzon. Every morning the rush hours start in there, become denser during the work hours, and slowly calms down after the dark.

4 The Analysis of City Renovation

While cities are growing and developing, a city model must be designed according to the projected population increase and the change of its functions. The process of urban renovation and restructure gain importance as a special tool of the internal developmental impetus of the city in line with the new land use regulations that are recently developed in the world [12]. Particularly the physical and economic restructuring of a city depends on the development and change of its population and economic structure. The analysis of the economic structure of the city can give us the future sizes of the city, as well as the location of the center business areas in the land use [13].

Particularly, the failures felt strongly in big cities and metropolitans are not only urban and physical ones but also related to the social structure of the society. The way to deal with them is to renovate the city. One definition of renovation says that "renovation is to fix the problems in the present cities and centers, and re-planing and application in order to modernize" [14]. Another definition says that it is the cleansing of the poor neighborhoods with a public initiative, fixing and protection of the old buildings, better accommodation and commerce opportunities, and the modernization of the city in order to meet the changing conditions and in order to create more public buildings [15]. This definition ensures the importance of the public in the renovation practices. As is seen in the essence of the city renovation concept, there lies dynamism in terms of change, transformation, and revitalization. Based on the definitions above the concept of city renovation can best be described as "the changing, developing, revitalizing and recreating the city texture and infrastructure that is worn out in the course time, and that lost its potential value and holds no more value other than the land itself in the framework of a strategic approach designed within the social and economic programs and according to the most recent socio-economic and physical conditions" [16,17,18].

City renovation is a comprehensive application concerning many cycles such as the population, social-economic and cultural values, the economic frame for the work done, and the local and administrative problems. This shows that city renovation is a multidimensional application. Particularly the social, economic, planning, legal, and administrative dimensions among others are the ones that directly affect the renovation [19,20,21].

Today, due to changing living conditions, people are almost indoors in daylight. They spend most of the day without daylight [22]. Therefore, users in the city thought that spatial arrangements that can meet the socio-cultural needs in the shopping center and where they can receive daylight can be useful. With this study, one of the important axes of the city has been chosen as Uzun Street study area. It is thought that the users can benefit from the socio-cultural opportunities of the city in the open air.

4.1 Renovation Works in Uzun Street

The project encompassed all aspects of urban planning, both in the center of Trabzon and in the surrounding districts: streets and boulevards, regulations imposed on facades of buildings, public parks and water works, city facilities, and public monuments. The planning was influenced by many factors, not the least of which was the city's history of street revolutions.

Fig. 2: The closure of Uzun Street to the traffic.

After the start of the motor vehicle traffic, it was felt that the streets were very narrow in Trabzon where the historical texture is felt at a high level. Moreover, with the increasing population Uzun Street became too small to meet the shopping and sightseeing needs of the city people. The shop owners in Uzun Street filed a suit because of the fast drivers, and lack of space for the pedestrians, and the Trabzon Municipality closed the street to the traffic and opened it to the pedestrians only (Fig. 2 and 3).

The project will ease the pedestrian traffic there and it caused some reactions in the city. The minivan traffic became all the more difficult after it was forbidden in Uzun Street. The load of the vehicle traffic was moved to Maraş Street to a great extent, and this caused a chaos in there. Also the owners of the auto-park on the streets who were subjected to economic losses asked for specific solutions. After the pressures of those people, some diversions from the project were made.

5 Discussions

The ever-lasting traffic problem in Uzun Street was not solved after the street was closed to traffic and a new proposal for solution is being sought by the designers in the city.

Fig. 3: Views from Uzun Street after the renovation.

In this study, a solution proposal will be made, and it claims to ease both the pedestrian and vehicle traffic on the street. This proposal claims to bear no resemblance with another project and is to be the first in Turkey.

In the scope of the project the intersection of Cumhuriyet Street and Uzun Street, where the traffic is very dense, will be used as a study area (Fig. 4). The construction of a platform that covers an area of 380 square meters is planned, which is at the second floor level of all other buildings on the site. The upper side of the platform will be open to pedestrians and the below will be open to pedestrians and vehicles (Fig. 5). Thus, Uzun Street, which is one of the densest axes of the city, will be transformed into an open shopping center.

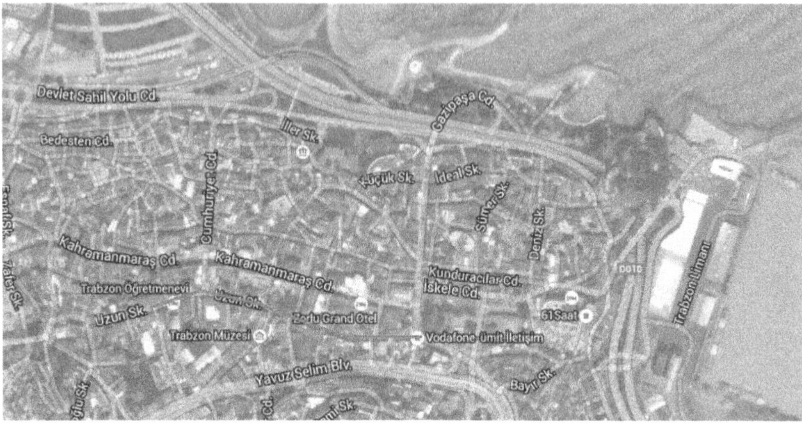

Fig. 4: The area (red line) where the platform will be built.

The platform is planned to be built on a steel platform. Steel columns will be 20x20 cm long and be directed one in every three meters and thus the structure will be made. Main beams will be 20x10 cm and secondary beams 15x5 cm. Since only the pedestrians will use the platform, there is no need for a separate structure system (Fig. 5).

In the platform that is planned, there will be four exits. Two secondary exits were chosen other than the ones at both ends. Automatic stairs will serve at these points (Fig. 5). Moreover, considering the handicapped, two lifts will be made on two points.

The floor of the platform will be a semi-transparent material (laminated glass) and thus sun light will be reflected on the below sections, and in this way the stores below will be able to take the advantage of natural light. People who are on the platform will feel psychologically better thanks to laminated glass to be used, and feel physically safer thanks to the multi-labelled glass to be used in the structure (Fig. 5). Moreover, adequate illumination system on the upper sides of the tunnel sections will be made by considering the aesthetical and technical concerns and that will fit into the beams. Moreover, the system will be a smart system and will adjust itself according to the density of the daylight and thus energy will be saved. In the tunnel section, ventilating shafts in certain distances will be made in order to prevent the accumulation of the exhaust gases from the cars, and these ventilating shafts will be covered by the columns and beams and the aesthetical perception and technical suitability will be sustained.

Laminated glass resists glass fallout in earthquakes. The level of protection afforded is a function of the thickness of the interlayer and the framing detail, the thickness and number of plies of glass, the size and strength of the glass, the temperature of the glass, and the method of installation [23].

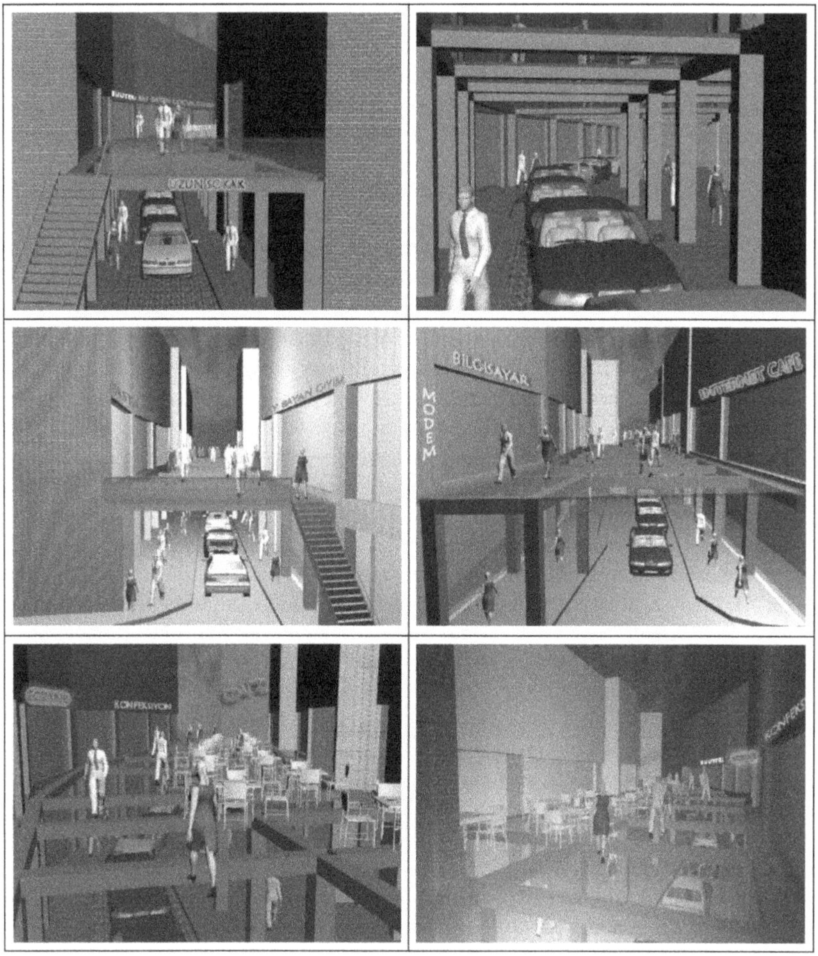

Fig. 5: The structure of the platform.

In rainy weather, in order to prevent the accumulation of water on the upper sides and the leak o fit to the down sides, water channels along the platform will be built and they will be led to the ground at certain distances and to the sewage system.

6 Conclusions

The pedestrian traffic in Uzun Street, which is the most crowded street of the city and symbolizes the character of the city, became all the more too difficult with the increasing population and vehicles, and this density has finally reached to an intolerable point. Since the city square was established in a relatively narrow area, it is impossible to move the stores and many other units all designed for different functions to anywhere else. The Tanjant road that has been opened to traffic in recent years is unable to bring some relief to the city in the respect.

Along with the city renovation works done by the Trabzon Municipality, Uzun Street was closed to vehicle traffic and reserved only for pedestrians. The motor vehicle traffic flows on Maraş Street. With this project trade life in Uzun Street became much better but the vehicle traffic became tighter. The worse is that it became virtually impossible to solve it anymore. The auto garages on the Uzun Street became useless. Upon the pressures of the auto garage owners a partial use of the garages were allowed, thus making many people again upset and the safety of the pedestrians were again ignored. All these led to the conclusion that the aim of the Uzun Street renovation project was diverted.

The traffic of Uzun Street and the city square which now seems unlikely to be solved will be solved with this study. The pedestrian and vehicle traffic on Uzun Street prevents each other. The separation of both on Uzun Street was not enough to provide complete comfort. In the new proposal to be made with this aim some part of the pedestrian traffic was isolated from the vehicle traffic. Through a 400-meter long platform to be made on the street, it will be possible to pedestrians and vehicles to function healthily on the street. Another aim of the project is prevent a further loss of the street in terms of trade.

Along with the platform the offices in the first floors of the nearby buildings will be more valuable and new stores will be opened there. Thus, in many narrow spaces many stores will be opened and the trade will increase (Fig. 5). In the below parts, there will be vehicle and pedestrian traffic together. Since the upper side will be closed, the tunnel section will be like a shopping center. Especially in winter times the decreasing amount of shoppings and trade outside will increase again with this application and thus the place with an easy access will again be revitalized. An additional platform will be built on the back of the Zorlu Grand Hotel that occupies some part of Uzun Street and the first floor of the hotel will be reorganized as a service space and then can be turned into a beautiful open cafeteria. Thus a large area there will be used. Considering that the present hotel does not have such a service space, it is clear that this area is an important advantage for the hotel (Fig. 5).

With the project proposed it is aimed to ease the city traffic and trade and to systemize it for future use, and to provide symbolic Uzun Street with its former identity.

The concept of a double-story street has not yet been introduced into the world literature and for this reason a comparative study was not done. However, this study is still a theory and when the time comes to implement it, it will be possible to discuss it by all parties involving scientists and academicians alike.

References

1 Trpkovic, A., and Stanic, B., *Modern Concept of Streets and Space Design - Shared Space*, Proceedings of the Second International Conference on Traffic and Transport Engineering (ICTTE), 2014, pp. 652–658.

2 LeeJungHyung, *A Study on the Street Management System for Improving the Pedestrian Environment - Focus on Complete Streets Policy and Design Guidelines in U.S.A.*, Journal of The Urban Design Insitute of Korea, 2014, pp. 75–91.

3 Allen, M., *Design and construction of Shoreditch High Street Bridge, UK*, Proceedings of the Institution of Civil Engineers-Transport, 2013, 166(1), pp. 19–25.

4 Kaparias, I., Bell, M. G. H., Dong, W. L., Sastrawinata, A., Singh, A., Wang, X. X., and Mount, B., *Analysis of Pedestrian-Vehicle Traffic Conflicts in Street Designs with Elements of Shared Space*, Transportatıon Research Record, 2013, 2393, pp. 21–30.

5 Biddulph, M., *Radical Streets? The Impact of Innovative Street Designs on Liveability and Activity in Residential Areas*, Urban Design International, 2012, 17(3), pp. 178–205

6 Kim, Sei-Yong, *A Street Design Proposal, Utilizing Historical and Cultural Resources*, Journal of The Urban Design Insitute of Korea, 2012, 13(1), pp. 33–50.

7 Gocer, O., *Trade Areas in the Cities* (In Turkish), 1984, Gumussuyu: ITU Publication.

8 Yalcinkaya Erol, S., and Sancar Ozyavuz, K., *Today and Tomorrow of Trabzon City: Urban Identity and Image* (In Turkish), 6th International Sinan Symposium, Edirne, TURKEY, 29–30 Nisan 2010, pp.110–116.

9 Kurak Acici F., and Ozdemir I., Landmarks in The Formation of Trbzon City Center: Old and New Structures, IAPS-CSBE & Housing Network International Symposium, Istanbul, TURKEY, 12–16 October 2009, 1(1), pp.58–58.

10 Kayaoglu I. G., Ciravoglu O., and Akalın, C., *Trabzon Is a Passion* (In Turkish), 1997, Istanbul: Yapı Kredi Publication.

11 Capa, M., and Cicek, R., *Life in Trabzon at the Beginning of the Twentieth Century* (In Turkish), 2004, Trabzon: Serander Publication.

12 Kohl, C., and Krier, R., *Principles of Urban Design or How to Build a Town*, Proceedings of the International Conference on Research by Design, 2001, Delft, The Netherlands.

13 Menn, C., *The Place of Aesthetics in Bridge Design*, Structural Engineering International, 1996, 6(2), pp.93.

14 Hasol, D., *Encyclopedic Dictionary of Architecture* (In Turkish), 1998, Istanbul: Y.E.M. Publication.

15 Keles, R., *Dictionary of Science City* (In Turkish), 1980, Ankara: Turk Dil Kurumu Publication.

16 Bednar, M. J., Interior Pedestrian Systems, 1990, B.T.Batsford Ltd., pp.145–180, London.

17 Demirtas, Y., and Esgin, I., *Experience of an Urban Renewal: Barcelona* (In Turkish), Journal of the Chamber of City Planners, 2006, 36(2), pp. 155–162.

18 Ozden P. P., and Kubat, A. S., *Proposals on Adaptability of Urban Renewal in Turkey*, ITUdergisi/a Mimarlık, Planlama, Tasarım, 2003, 2(1), pp. 77–88.

19 Woodruff, S., and Billington, D. P., *Aesthetics and Economy in Pedestrian Bridge Design*, International Journal of Space Structures, 2007, 22(1): pp.81–89.

20 Robertson, K. A., *Pedestrian Malls and Skywalks: Traffic Separation Strategies in American Downtowns*, 1994, Newcastle: Athenaeum Press Ltd.

21 Ugur, H., and Giritlioglu, C., *Traditional Urban Patterns and the Characteristics of Their User*, ITUdergisi/a Mimarlık, Planlama, Tasarım, 2006, 5(1): pp.29–34.

22 Kurak Acici F., and Islamoglu O., *Traditional Architecture-Modern Architecture Readings in Architectural Education* (In Turkish), Mimaran, Mimarlar Odası Konya Şubesi Mimarlık ve Kültür Dergisi, 2011, 3, pp.7–11.

23 www.glasswebsite.com

Assist. Prof. Dr. Aylin Salıcı

4 Smart Cities and Their Criticism from the Perspective of Landscape Architecture

Abstract: Many planning approaches have appeared in order to minimize the negative effects of urbanization on nature and people, eliminate population and urbanization pressure, and create more efficient, livable cities. Some of them are green city, eco city, livable city, extended urbanism, digital city, smart city, and so on. According to the International Organization for Standardization, smart cities are a new concept and a new model, which applies the new generation of information technologies, such as the Internet of Things, cloud computing, Big Data, and space/geographical information integration, to facilitate the planning, construction, management and smart services of cities (ISO, 2014). The developments in especially the information technology in the last two decades have brought to mind the possibility of using technology in solving the problems that are seen especially in existing cities. Among the concepts listed above, "Smart city" is a relatively recent one. While it still lacks a certain definition, it is considered within the scope of different urban development scenarios, and it mainly aims at meeting the requirements of cities through solutions integrated with technology. However, cities naturally involve a continuous change and transformation. If this change and transformation are not managed in a controlled manner within the framework of some basic principles, it can result in unqualified and ordinary environments that have lost their original values. In conclusion, integrating the features of all concepts would be a more solution-based and befitting approach for a landscape architect, rather than handling the cities, which are different from each other with all their natural characteristics, through a single concept and trying to attach an identity or describing them.

Keywords: Smart cities, Smart applications, Sustainable cities

1 The Concept of Smart City

Current literature involves numerous definitions of "Smart City" and therefore, there are different applications regarding this concept.

This term, which has been used since 1990, defines the role of communication infrastructure in urban production and growth. In addition, its definition has differentiated due to the developments in information and communication technologies since it was first coined and has been turned by politicians into a strategic instrument frequently used for increasing the competitive profiles of cities [1]. Some prominent definitions related to this concept in literature are as follows:

The idea of smart city emerged as an attempt to mitigate current problems related to urban areas and create better livable cities. Smart city concept is commonly associated with information and communications technology (ICT) systems. It is a concept which is enriched by high-technology to provide a connection between people, information, and city so as to increase life quality and build a sustainable, attractive greener city with competitive and pioneering commerce [2].

According to the International Organization for Standardization, smart cities are a new concept and a new model, which applies the new generation of information technologies, such as the Internet of Things, cloud computing, Big Data, and space/geographical information integration, to facilitate the planning, construction, management, and smart services of cities [3].

A smart city may be defined as smart when the investments in socio-human capital, urban infrastructure, and rational management of natural resources encourage a sustainable economic development and a high quality of life through participatory actions and commitment from community members [4].

2 The Components of Smart Cities

There are several approaches in literature and practice regarding the planning of transforming into smart cities and monitoring this transformation process. Among these approaches, Cohen's "Smart Cities Wheel (SCW)" methodology is one of the most prominent. According to this approach, which is also accepted by the European Union (EU), a city should meet six basic criteria in order to be regarded as a smart city: [5].

1. Smart mobility: This component, with its broadest definition, involves integrated transportation systems supported with information and communication technologies. Primarily, it aims at bringing environment-friendly transportation solutions that embrace the disadvantaged groups especially. Within the scope of this component, producing real-time traffic information and sharing this with passengers, drivers, and operators is one of the primary issues [6].
2. Smart living: People's lives are facilitated with information and communication technologies and a calmer and more secure environment is ensured for the city residents.
3. Smart governance is described as an important characteristic of a smart city that is based on citizen participation and private/public partnerships. Smart governance depends on the implementation of a smart governance infrastructure that should be accountable, responsive, and transparent. This infrastructure allows collaboration, data exchange, service integration, and communication [7].
4. Smart environment: It involves solutions like renewable energy, smart systems, micro systems, smart utility meters, advanced air pollution monitoring systems, green buildings and city planning, energy efficient smart street lights, solid waste management, smart water management, and drainage systems.
5. Smart economy: Applications like increasing efficiency through information and communication technologies, e-commerce, advanced production and supply systems, smart clustering and business ecosystems, and living laboratories are handled within the scope of this component.
6. Smart people: The purpose within the scope of this component is to form an inclusive society that encourages creativity and innovation through developing people's ability to use and produce information and communication technologies.

Since the components listed above are applications that feed and develop each other, handling these components holistically with a system approach during smart city design shall increase the efficiency in practice.

3 Applications of Smart City with New Technologies and İnnovative Approaches

- In the United States, disability activists in cities like Berkeley, California, and Urbana, Champaign, Illinois have prepared digital-accessibility maps with the help of smart city technologies for disabled people living there within the scope of "Smart Cities for Everyone" project [8].
- With the intelligent waste management system in San Francisco, the United States, 80% of the garbage is supplied to the economy. The 2020 target of the city is "zero loss" in the garbage [9].
- In Atlanta, The Atlanta Smart Corridor project resulted in an adaptive traffic control system that uses real-time data to best manage traffic flows in a previously congested city area. This "smart" transportation system provides clear efficiency benefits for the city and reduced driving time by as much as 40% in some instances. Environmental sensors also captured emissions and have also shown that there are significant health benefits from smart technology: NOx and VOC vehicle emissions have reduced by 1% and 8.5%, respectively [10].
- In Barcelona, The city administrators built the "Smart City Campus" with the support of private organizations [11].

In Hong Kong, where more than 12 million passengers travel per day, 630 thousand vehicles are being used. The city has one of the most widely used transportation infrastructures in the world with its roads which are 2086 km in length. In the city, road traffic information service, route finding service, smart road network are prepared with the help of smart city technologies [12].

- The Japan Intelligent Community Alliance, established by major industry partners in Japan, plays an important role in making Japan's smart grid roadmap. In 2010, in four cities smart city pilot applications within the context of the smart grid have been started [13].
- The smart city model implemented in Paris is integrated with the sustainable city, connected cities, and open cities. In this context, in Paris, both practices covering the examples of "common multi-model mobility" between urban culture and technology have been applied, and also, a transformation has been realized protecting the valued elements of urban history and cultural heritage [14].
- In addition to reducing the urban space dedicated to motorized vehicles, the Seine river banks are becoming a pedestrian way and the ambitious Paris Bicycle Plan is optimizing biking trips. For a better quality of urban life, the city municipality supports shared mobility initiatives, such as Autolib (electric car) and electric Velib (electric bike), to decrease air pollution [15].

- One of the smart cities in North America is Toronto. In city not only the public but also the private sector are encouraging the promotion of low carbon technologies. In this city, garbage trucks are operated with natural gas generated from garbage [16].

Conclusion

Many planning approaches have appeared in order to minimize the negative effects of urbanization on nature and people, eliminate population and urbanization pressure, and create more efficient, livable cities. Some of them are green city, eco city, livable city, extended urbanism, digital city, smart city, and so on.

While the principles of these approaches differentiate, the purpose of these approaches, all of which fundamentally aim at creating awareness, is to plan livable cities that preserve natural ecosystems with low carbon emission, low energy and water consumption, and thus ensure sustainable development. Here, the critical issue is to try to reach the goals of such approaches through a series of planning/design decisions during planning of a new settlement or improving an existing city. It is difficult, however, to say that current policy, planning, design, and application tools have not been sufficiently successful in solving sustainability and life quality–related problems in existing cities.

The developments in especially the information technology in the last two decades have brought to minds the possibility of using technology in solving the problems that are seen especially in existing cities. Among the concepts listed above, "Smart city" is a relatively recent one. While it still lacks a certain definition, it is considered within the scope of different urban development scenarios, and it mainly aims at meeting the requirements of cities through solutions integrated with technology.

However, cities naturally involve a continuous change and transformation. If this change and transformation is not managed in a controlled manner within the framework of some basic principles, it can result in unqualified and ordinary environments that have lost their original values. It is certainly unavoidable and necessary to utilize the advantages of technology. But the basic problems in the existence and running of a city could only be solved as long as they are associated with the sustainability of all natural and cultural systems (social, political, historical, legal, administrative, and economic) of the city. In this regard, technology should be considered as a factor that could be included in the solution, rather than being a fundamental solution itself. In other words, technology should be the instrument, not the purpose.

In conclusion, integrating the features of all concepts would be a more solution-based and befitting approach for a landscape architect, rather than handling the cities, which are different from each other with all their natural characteristics, through a single concept and trying to attach an identity or describing them.

References

1 Akgün, A.A., kentler, A. Kentsel Planlama Ansiklopedik Sözlük, Ninova Yayınları. İstanbul, Türkiye. 2012. s 7.

2 Akaraci S., Usmanb M.A., Usmanb M.R. and Ahn D.J., From Smart to Smarter Cities: Bridging the Dimensions of Technology and Urban Planning, IEEE ICSGTEIS, International Conference on Smart Green Technology in Electrical and Information Systems. 2016: Bali, Indonesia.

3 ISO. Smart cities. ISO/IEC JTC 1 Information Technology Preliminary Report. 2014 [cited 2015]; Available from: https://www.iso.org/files/live/sites/isoorg/files/developing_standards/docs/en/smart_cities_report-jtc1.pdf.

4 Tîrlă, M.L., Manea, G., Vijulie, I., Matei, E., Cocoş, O. Green Cities–Urban planning models of the future. Cities in the Globalizing World and Turkey: A Theoretical and Empirical Perspective, St. Kliment Ohridski University Press, 2014, p. 462. Sofia.

5 Anthopoulos, L.G., Vakali, A. Urban Planning and Smart Cities: Interrelations and Reciprocities. In: The Future Internet Assembly. Springer, Berlin, Heidelberg, 2012. pp. 178-–189.

6 Gül, A., Çobanoğlu, Ş.A. The Assessment of the Smart City Implementations in Europe and the Analysis of the Transformation of Çanakkale as Smart City. Suleyman Demirel University Journal of Faculty of Economics & Administrative Sciences, 2017, 22: pp. 1543–1565.

7 Chourabi, H., Nam, T., Walker, S., Gil-Garcia, J. R., Mellouli, S., Nahon, K., Pardo, T.A., Scholl, H.J. Understanding Smart Cities: An Integrative Framework. In System Science (HICSS), 2012 45th Hawaii International Conference on (pp. 2289–2297). IEEE.

8 Hamraie, A. A Smart City is an Accessible City. [cited Nov 6, 2018]; Available from: https://www.theatlantic.com/technology/archive/2018/11/city-apps-help-and-hinder-disability/574963.

9 https://geturkiyeblog.com/bir-bilim-kurgudan-fazlasi-akilli-sehirler. Bir bilim kurgudan fazlası: Akıllı Şehirler. [cited 08.01.2016].

10 Resendez J. Smart Cities for All: A Vision for an Inclusive, Accessible Urban Future. 2017 [cited 06.2017]; Available from: http://smartcities4all.org/wp-content/uploads/2017/06/Smart-Cities-for-All-A-Vision-for-an-Inclusive-Accessible-Urban-Futur...-min.pdf.

11 March, H., R.R. Fumaz. Smart Contradictions: The Politics of Making Barcelona a Self-Sufficient City. European Urban and Regional Studies, 2016, 23(4): pp. 816–830.

12 http://www.ebelediye.info/dosya/akilli-sehir-cozumlerinde-dunya-ornekleri. Akıllı şehir Çözümlerinde dünya örnekleri. 28.02.2018.

13 Yenilmez, M. Smart Grid Developments in Distribution System Automation. Karabük University Faculty of Engineering Department of Electrical-Electronics Engineering, Master's Thesis. 2016.

14 Carriat, J. Paris Ends Autolib Electric Car Sharing Contract with Bollore. Paris Smart City 2050. For a Sustainable, Dense and Connected City. 2017. [cited June 21, 2018] Available from: http://vincent.callebaut.org/object/150105_parissmartcity2050/parissmartcity2050/projects/user.

15 https://www.reuters.com/article/us-china-autos-trucks-electric-focus/a-slew-of-electric-truck-plans-may-deliver-the-goods-for-chinas-ev-ambitions-idUSKCN1NK35I. 2018.

16 https://www.toronto.ca/city-government/accountability-operations-customer-service/city-administration/staff-directory-divisions-and-customer-service/solid-waste-management-services/, City of Toronto, Solid Waste Management Services, cited: 14.03.2018.

Assist. Prof. Dr. Aysun Tuna and Graduate Student Ayşe Gülan Çelebi

5 Site Management in the Protection of Archeological Landscapes: A Review on the Archeological Site Management Plans of World Heritage Sites Located in Turkey

Abstract: Archeological sites are non-renewable resources, attributed to the past via anthropogenic effects within cultural landscape. Protection is of vital importance in terms of ensuring the sustainability of these sites and handing down to the next generations. Aiming at sustainable protection of archeological sites, the Archeological Site Management Plan comprises of classifying the archeological heritage, defining its characteristics, making an archeological inventory of the region afterwards, using the archeological sites with interactive sites, evaluating up of the new usage manner to be brought into the site for meeting the potential needs of society, as well as the administrative structuring to designate the preventive actions and action plans in this scope and the administrative activities undertaken by this structure. In this chapter, the archeological heritages in UNESCO's World Heritage List after 2005 (incl. the year 2005) are examined within the scope of preparation process for site management plan following the releasing of "*Regulation on Principles and Procedures concerning the establishment of Site Management & Monumental Works Board, as well as identifying their duties and management sites*" in the year 2005 in Turkey. The management plan contents of Catalhoyuk, Neolithic Site, Bergama Multi-Layered Cultural Landscape, Ephesus, Archaeological Site of Ani, Aphrodisias Ancient City, Archaeological Site of Gobeklitepe, which can be seen in the official website of Ministry of Culture, are examined and subjected to comparison.

Keywords: Archaeological landscape, Site management, WHL (World Heritage List), UNESCO, Turkey

1 Protection of Archeological Sites and Site Management

Archeological sites are non-renewable resources, attributed to the past via anthropogenic effects within cultural landscape. Also, archeological sites, which are the common property of the human kind, are the data memories of the society reflecting the emotions and thoughts of civilizations from past to date. Protection is of vital importance in terms of ensuring the sustainability of these sites and handing down to the next generations. Athens Charter, which was accepted in 1931, is recognized as the first serious attempt in protection of archeological sites. Within the scope of fundamental principles as prescribed by this charter, the requirement on respecting the historical identities of monumental structures and the idea of examining the monumental works along with its immediate vicinity were underlined (1). In the first article of Venice Charter, it can be understood that the protection scope is extended based on the idea to apply it not only to great works of art but also to other works of the past which have acquired cultural significance with the passing of time. After publishing the Venice Charter, ICOMOS was established in Warsaw with the objective to support and direct any research concerning the methods and techniques on the protection, protection and evaluation up of ancient monuments and sites, in 1965. As an efficient organization in the protection of historic areas, UNESCO was effective in laying the conception of protection onto an international basis by publishing the World Heritage List within the same period. It was decided to include the matter within an international agreement, in the 16th General Conference of UNESCO between 17th October to 21st November 1972 in Paris, gathering on the purpose of introducing the cultural and natural sites with universal values to the world, which are recognized as the common heritage of the entire human kind, raising the awareness of society for protecting this universal heritage, as well as ensuring the cooperation required for sustaining the cultural and natural values that are being deformed due to various reasons, and the Convention concerning the Protection of the World Cultural and Natural Heritage was entered into on 16th November 1972 (2). Raising awareness on protection led the extension of approaches for this matter. It was understood that protection must be addressed via multiple aspects, including the physical properties of the area, social and economical aspects, etc. Additionally, certain topics like participation and transparency gained more importance within the scope of protection sustainability. In this context, site managements comes to the fore as a policy of protection. In the emergence of the concept "site management," protection of cultural heritages, which are recognized as protection element, in its own environment, minimizing the confronted risks and handing down to the future by being integrated into our daily lives, were effective in the process. Historical (ancient) and archeological sites were determinant factors in turning the protection from a single structure-scale to an integrative and extensive perspective, in the development of site management. Aiming at sustainable protection of archeological sites, the Archeological Site Management Plan comprises of classifying the archeological heritage, defining its characteristics, making an archeological inventory of the region afterwards, using the archeological sites with interactive sites, evaluating up of the new usage manner to be brought into the site for meeting the potential needs of society, as well as the administrative structuring to designate the preventive actions and action plans in this scope and the administrative activities undertaken by this structure. Requiring the "site management plan" from countries applying for being recognized within the "World Heritage List" as specified by UNESCO World Heritage Center within the scope of immovable cultural assets, involving the archeological sites, highlights the significance of site management.

2 UNESCO's Management Guidelines for World Heritage Sites

In this section of the chapter, the site management plan content is examined, which is required to be prepared for candidate or already listed sites in UNESCO's World Heritage List. Management Guidelines for World Heritage Site is recognized as the fundamental guide for Management of World Heritage Sites, published by the International Center for the Study of Preservation and Restoration of Cultural Property (ICCROM), as one of the consultancy organs of World Heritage Committee. In this guide, which was drawn up by Feiden and Jokilehto in 1993, it is particularly underlined that cultural heritage can be rendered usable for various objectives like education, research, tourism, and employment after being protected and re-functioned. The management plan, which is required for site management, comprises of the following sections in the guide: a preface for identifying the general framework of the current status of the site, description of the site, assessment and aims, policies, bibliography (3).The first section, named as "Site Description," comprises of the location, definition, and privileges of the site, along with related generic information and supporting materials (maps, charts, photos, etc.). Then, natural and anthropogenic analyses are included, comprising of anthropological, ethnographical, archeological, historical, art history, architecture, technical, scientific topics, including cultural information and climate, hydrology, geology, geomorphology, seismology, soil and human-borne dangers. Additionally, such supporting information on land use, resource utilization, ownership, public interest and economical benefits, and tourism is included. In the "Assessment and Aims" section, identifying the protection status of the site, as well as an assessment of characteristics & potentials of the site are included. Description of projects concerning the site, headline and classification, project preparation process details are included under the third section of "Policies for Overall Site Management." This section is followed by annual operation plan, the relation between annual plan and mid- and long-term plans, costs and staging up of operations, under the headline "Work Programme." The appendices, references, amendments within the management plan are included in the bibliography. The preparation process of management plan must be as follows according to the specified content:

- The minimum appropriate level of maintenance should be defined that is compatible with respecting the significance of the cultural resource, the needs of users, and the local climatic constraints.
- An outline of the planning regulations affecting the whole site should be prepared in line with the decision taken by centralized and regional administration.
- Other plans and legislation that affect the site should be considered by the "Site Commission" and integrated into the overall schemes of management, maintenance, development or research.
- The general management plan should identify personnel required to operate and maintain the site, and define their tasks.
- Proposals for the development of facilities should be included in the management plan, but under a separate budget heading. Projects can be brought forward for approval by stages:
- Description of the site and specifying the borders, along with the interaction areas (buffer zones),
- Identifying the values in the site,
- Assessment of resources,

- Defining the targets and threats,
- Describing the project,
- Preparing the work schedule and annual plans,
- Implementation process,
- Documentation, reporting, and controlling/reviewing,
- Storing information and data,
- Revision and review of the site description,
- Reviewing the revised targets and threats,
- Identifying the future projects,
- Preparing the revised working schedule and annual plans (3).

Demas underlines the importance of participative management planning approach rather than the hierarchical management mentality, during the preparation of archeological site management plan, in addition to the management guidelines (4). This approach is based on the identification of interests' community (partners and shareholders). For the archeological site management plan to be successful, the common shareholders of each and every community of interest must actively participate in the planning process. The decisions on the archeological site management to be taken by the centralized and local administrative organizations, employees of the protection site, international consultancy organs, local community, non-governmental organizations, and other shareholders, in line with the planning principles of the planning process, along with the applicable regulations and legislation, will primarily refer to the people living in the respective area. For determining how the residences would be affected by these decisions during the planning process, participation is a must in the decision-making (4).

Orbaşlı lists the communities of interest that bear the potential to contribute in the planning process of archeological site management: research; archeology and science communities; decision-making mechanisms on local, regional, and national scale; local people, associations, and foundations that may provide support, sponsors, visitors, academicians, international organizations (i.e., ICOMOS, UNESCO) (5).

3 Archeological Site Management in Turkey

The "management" phenomenon to gain importance in the protection and development procedures of cultural heritage in Turkey, and including it within the national protection legislation, is prescribed by law no.: 5226. This law directly focuses on "Site Management." The existence and necessity of Site Management is included in the definition of "Site Management" within the respective law.

Upon the amendments made in 2004, the preparation of management plans concerning the archeological sites and ruins are provided with a legal ground within the scope of Protection of Cultural and Natural Properties Code no.: 2863, and the works and studies were kicked of de facto after publishing *Regulation on Principles and Procedures Concerning the Establishment of Site Management & Monumental Works Board, as Well as Identifying Their Duties and Management Sites* (6). The planning process, as identified within the scope of this regulation, can be seen on Tab. 1.

Tab. 1: Site Management Planning Process in Turkey (7)

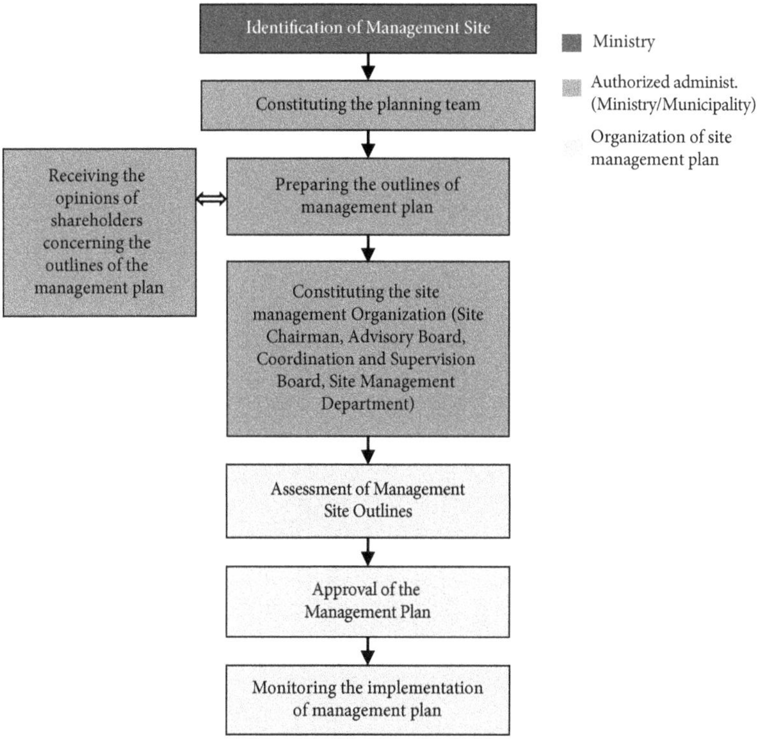

4 Assessment of Archeological Site Management Plans on the World Heritage Sites in Turkey

In this section, the archeological heritages in UNESCO's World Heritage List after 2005 (incl. the year 2005) are examined within the scope of preparation process for site management plan following the releasing of *Regulation on Principles and Procedures Concerning the Establishment of Site Management & Monumental Works Board, as Well as Identifying Their Duties and Management Sites*" in the year 2005 in Turkey. The management plan contents of Catalhoyuk, Neolithic Site, Bergama Multi-Layered Cultural Landscape, Ephesus, Archaeological Site of Ani, Aphrodisias Ancient City, Archaeological Site of Gobeklitepe, which can be seen in the official website of Ministry of Culture, are examined and subjected to comparison (8) (Tab. 2–3).

Tab. 2: The Authority Share for Management Plans and Their Relation to World Heritage List Studies (Prepared Based on the Data & Information from Ulusan and Website of Ministry of Culture and Tourism.) (7) (8)

Management Plans	Purpose of Preparation	Legal Responsible	Method for Conducting	Budget Source	Period of Preparation	Relation to World Heritage List Studies
Catalhoyuk, Neolithic Site	Upon the World Heritage Committee decision	Ministry of Culture and Tourism	By the planning team established within the Ministry	Ministry of Culture and Tourism	2010–2013	Inscribed on the List in 2012
Bergama Multi-Layered Cultural Landscape	Nomination to the World Heritage List	Bergama Municipality	By the planning team established within the municipality	Bergama Municipality	2010 on going	Inscribed on the List in 2014
Ephesus	Nomination to the World Heritage List	Selcuk Municipality	Procurement (Ege Plan Planning Office and Assoc. Dr. Savaş Zafer ŞAHİN)	Selcuk Municipality	2010–2014	Inscribed on the list in 2015
Archaeological Site of Ani	Within the scope of UN Joint Program of Alliances for Culture Tourism in Eastern Anatolia	Ministry of Culture and Tourism	By the planning team established within the Ministry	UN Millennium Development Goals Fund and Ministry of Culture and Tourism	2009–2015	Inscribed on the List in 2016
Aphrodisias Ancient City	Nomination to the World Heritage List	Ministry of Culture and Tourism	Procurement (MSGSÜ Research Center for Urbanism)	Geyre Foundation	2010–2013	Inscribed on the List in 2017
Archaeological Site of Gobeklitepe	Nomination to the World Heritage List	Ministry of Culture and Tourism	DAI, Project of Gobekli Tepe, Bradenburg University of Technology, Ministry of Culture and Tourism	Historical Revival Project for Sanliurfa (Supported by EU), DAI	2017-	Inscribed on the List in 2018

Tab. 3: Comparison of the Management Plans of Archeological Sites, Listed in UNESCO World Heritage List (Prepared Based on the Data National Site Management Plan from Website of Ministry of Culture and Tourism) (8)

	Main headings of Site Management Plans	**Çatalhöyük (Konya) 2012**	**Bergama (İzmir) 2014**	**Efes (İzmir) 2015**	**Ani (Kars) 2016**	**Afrodisias (İzmir) 2017**	**Göbeklitepe (Şanlıurfa) 2018**
INTRODUCTION	Objective of management plan	✓	✓	✓	✓	✓	✓
	Site borders for management plan	✓	✓	✓	✓	✓	✓
	Completed works to date	-	✓	✓	✓	-	✓
	International partnerships	-	-	-	✓	-	✓
STAKEHOLDER ENGAGEMENT	Identification and analysis of shareholders	✓	✓	✓	✓	✓	✓
	Identifying shareholder participation strategies	-	-	✓	✓	✓	✓
SITE DESCRIPTION	General environmental properties, physical structure	✓	-	✓	✓	✓	✓
	Relation with the urban area	-	-	-	-	✓	✓
	The current land use	-	-	-	-	✓	✓
	Transportation	✓	✓	✓	✓	✓	✓
	Infrastructure	-	-	-	-	✓	-
	Historical development process	✓	✓	✓	✓	✓	✓
	Data on excavation works and findings at the site	✓	✓	✓	✓	✓	✓
	Protection works	✓	✓	✓	✓	✓	✓
	Administrative structure	✓	✓	✓	✓	✓	✓

(continued on next page)

Tab. 3: Continued

	Main headings of Site Management Plans	Çatalhöyük (Konya) 2012	Bergama (İzmir) 2014	Efes (İzmir) 2015	Ani (Kars) 2016	Afrodisias (İzmir) 2017	Göbeklitepe (Şanlıurfa) 2018
	Administrative distribution	-	-	-	✓	✓	✓
	Site borders	-	-	✓	✓	✓	✓
	Ownership/property	-	-	-	-	✓	✓
	Socio-economic structure	-	-	✓	✓	✓	✓
	Tourism	✓	✓	✓	✓	✓	✓
	Visitor profile/statistics/impacts of visitors	-	-	-	-	✓	-
	Researches on current plan, project and protection practices	✓	✓	✓	✓	✓	✓
IMPORTANCE VALUES OF THE SITE	Cultural importance (UNESCO Criteria)	✓	✓	✓	✓	✓	✓
	Intangible cultural heritage	-	✓	-	-	✓	✓
ASSESSMENT AND REVIEW	SWOT Analysis	✓	✓	✓	✓	✓	✓
	Identification of needs and priorities	-	-	✓	✓	✓	✓
	Identifying the carrying capacity	-	-	-	-	✓	-
	Risk analyses	-	-	-	-	✓	-
	Perception and media analyses	-	-	✓	-	-	-

VISION, OBJECTIVE AND STRATEGIES	**Action Plans, Organization Model, Distribution of Roles, Staging and Timing**	Management and organization	✓	✓	✓	✓	✓	✓
		Protection and planning	✓	✓	✓	✓	✓	✓
		Accessibility	✓	✓	✓	✓	✓	✓
		Environmental planning, visitor management, and presentation	-	-	✓	✓	✓	✓
		Raising awareness on the importance and its value	✓	✓	✓	✓	✓	✓
		Training, raising awareness, and participation	✓	✓	✓	✓	✓	✓
		Risk management	✓	-	✓	-	✓	✓
MONITORING, ASSESSMENT/PERFORMANCE INDICATORS			✓	✓	✓	✓	✓	✓

Conclusion

Within the scope of this chapter, the site management plans of the heritage sites in Turkey, which are listed in the World Heritage List that comprise of archeological heritages, are examined.

Until 2015, it was deemed sufficient for an asset to have a matured management plan despite not being approved for being nominated in World Heritage List, thus taking binding decisions on the approval procedure for the plan within a certain period of time in the decision to be taken by the World Heritage Committee upon registering the asset to the List. Catalhoyuk Neolithic City and Bergama Multi-layered Cultural Landscape's site management plans were approved within the scope of this binding decision based on the date it is accepted in the List. However, certain articles of Operational Guidelines for the Implementation of World Heritage Convention, which can be considered as the Regulation of World Heritage Agreement, according to the decision no.: 39.COM.11 taken in 39th cycle meeting of the World Heritage Committee in 2015 at Bonn. According to these amendments, which also involve matters on the management plan, it became a must for the management plans to be approved while presenting the nomination files for World Heritage List as of 2015.

Examining the main topics and contents of the archeological site management plans, the following have been inferred:

The objective and planning restrictions of all the heritage sites' management plan are identified in the introduction; information on the completed works & studies to date can be seen in the site management of Efes, Ani and Gobeklitepe, while the international partnership details are included in Ani and Gobeklitepe site management plan;

While the general environmental characteristics and physical structure are defined under "Site Description," the urban relations, current land use, infrastructure, ownership, and visitor profile/statistics & impacts were not addressed, in general;

SWOT (Strengths, Weaknesses, Opportunities, Threats) Analyses are applied for each and every heritage sites under "Assessment and Review," the risk analyses and perception & media analyses of the carrying capacity were not carried out, but the necessity to do so was stated in the action plans;

Under "Vision, Objective and Strategies," the subheadings of "action plans, organization model, staging and timing" were included in the Afrodisias Site Management Plan, while the vision, objective, and strategies in these subheadings were mentioned in different subheadings within other site management plans;

The monitoring, assessment, and performance indicators were included in all the site management plans.

Achieving an integrative and overall protection perception for the urban landscape, from the mentality of conserving the protection manner based on a single structure scale, was a determinant factor in developing the conception of site management. The management plans, drawn up based on the sense of site management, were the guiding documents for sustaining the originality and integrity of the protection sites, without being disrupted. It is a must for these documents to be the product of a dynamic process that is continuously monitored, controlled, as well as bearing the ability to be subjected to changes based on emerging needs. Another characteristic on this matter is that the process should be maintained with the participation of respective shareholders. In this sense, it can be said that the management plans are addressed with the strategic plan approach. This practice increases the achievability of ensuring the plans to be actualized, drawing up of projects in line with realistic needs, and accordingly owning an urban and plan management that is directed via local dynamics. It is thought that the site management plans, which are discussed based on the strategic planning approach as an alternative to the heavily criticized centralist and nontransparent, uncontrolled plans and projects, have played/will play a critical role in the sustainable protection of cultural heritages.

References

1 European Council of Town Planners, *The new charter of Athens 2003: the European Council of Town Planners' vision for cities in the 21st century: Lisbon, 20 November 2003 = La nouvelle charte d'Athènes* 2003, Citta di Castello: Alinea Editrice.

2 UNESCO World Heritage Centre. *Operational guidelines for the implementation of the world heritage convention*, 2015, Paris: UNESCO.

3 Feilden, Jokilehto, M., Management Guidelines for World Cultural Heritage Sites, 1993, ICCROM, Rome, p:122

4 Demas, M., *Planning for Conservation and Management of Archaeological Sites: A Value-Based Approach, Management Planning for Archaeological Sites,* Proceedings of the Corinth Workshop Edited by Gaetano Palumbo and Jeanne Marie Teutonico, The Getty Conservation Institute, 2002, Los Angeles, p:27–54

5 Orbaslı, A., *Guidelines for Preparing Management Plans for Prehistoric Sites, Mediterranean Prehistoric Heritage*, Training, Education and Management, 2007

6 Ministry of Culture and Tourism. https://en.unesco.org/news/unesco-database-national-cultural-heritage-laws-updated/ Access date: 24.02.2019.

7 Ulusan, E. Türkiye'de Kültürel Miras Alanlarında Yönetim Planlaması Deneyimi, Journal of Urban Studies, 19 (7), ISSN:1307-9905, 2016 p: 372–401.

8 Ministry of Culture and Tourism. http://www.kulturvarliklari.gov.tr/TR-204390/ulusal-yonetim-planlari.html Access Date:18.10.2018.

Assoc. Prof. Dr. Ayşe Durukan Kopuz

6 Planning Experience of Hermann Jansen in Gaziantep City during 1930s

Abstract: The establishment of the Republican regime in 1923 in the history of Turkish architecture indicates an important starting/turning point in our history consciousness. In the mentioned period, a new nation-state were formed according to examples in the West, instead of the Ottoman Empire, which had a feudal government form for almost 500 years. The architecture of the period will be re-established as the beginning of a new way of thinking, based on the differences between the old and the new, the traditional and the contemporary, the reactionary and the progressive. So, the new architecture was imported as the new face of Turkey. In this chapter, architectural planning and urban development efforts as to extend modernity project throughout Anatolia by German architect and city planner Hermann Jansen will be studied, especially to highlight the early planning efforts of Hermann Jansen made for Gaziantep which is in the south east of Turkey along the Syrian border. The research will focus on both the public and town planning activities held by Jansen in Gaziantep. The urban designs for new state purposes, as well as their reception in the highly new developing context of Turkish Republic, require the study to emphasize the principles as well as the professional perspective and background of the architect of the contemporary architectural context. Meanwhile, the urban plans produced during this period, land use decisions, and transportation system design reflecting the transition to the industrial city model and a dual structure consisting of modern and traditional texture in urban space would be observed.

Keywords: Hermann Jansen, Gaziantep, Urban planning, Modernity project

1 Introduction

The establishment of the Republican regime in 1923 in the history of Turkish architecture indicates an important starting/turning point in our history consciousness. In the mentioned period, a new nation-state were formed according to examples in the West, instead of the Ottoman Empire, which had a feudal government form for almost 500 years. The architecture of the period will be re-established as the beginning of a new way of thinking, based on the differences between the old and the new, the traditional and the contemporary, the reactionary and the progressive (1). So, the new architecture was imported as the new face of Turkey.

Since the transition from Ottoman Empire to Turkish Republic caused many necessities in several ways, numerous European architects and urban planners were invited to Turkey officially to perform the new aspect of the State. Not only does their work constitute many studies, but it also provided the formation of new cities through the planning policy in the government program. Although the complex interactions of different factors played a significant role in the development of Modern Turkish Architecture after the foundation of the Turkish Republic, the prominent European architects with their unique background transformed personal characteristics, and this constitutes one of the key aspects in this process. More than 20 German, Austrian, and Swiss architects have worked in Turkey between the years 1925 and 1955 (2).

It was the unique case of republic from modern/industrialized cities because of their reflection on architectural designs and community (3). So, the declaration of Ankara as a capital city merged new urban development design strategies in order to constitute a model for other cities in Turkey. One of the prominent German planners Hermann Jansen (1869–1847) who was the winner of the Ankara master plan competition among the other urban planners, designed Ankara city master plan (4). After approval of the Jansen plan in 1932 for Ankara, he had the chance to prepare new urban plans, parks, and public areas in Anatolia.

Yet, architectural planning and urban development efforts as to extend modernity project throughout Anatolia by Hermann Jansen were researched in detail. Except big cities such as Ankara, Adana, Mersin, and İzmit, what is particularly evident from this historical review is that Gaziantep reflects all the characteristics which the cities have undergone through modernity project of the new state by the architectural efforts of Jansen's planning for this city. He has not only planned the first urban project of the city but also new cemetery area, republic square, entrance gate of city, and Kavaklık coffee house.

So, the aim of this research is to highlight the early planning efforts of Hermann Jansen made for Gaziantep which is in the south east of Turkey along the Syrian border. The research will focus on the projects of both public and town planning activities held by Jansen in Gaziantep. The urban designs for new state purposes, as well as their reception in the highly new developing context of Turkish Republic, requires the study to emphasize the principles as well as the professional perspective and background of the architect of the contemporary architectural context. Meanwhile, the urban plans produced during this period, land use decisions and transportation system design reflecting the transition to the industrial city model, and a dual structure consisting of modern and traditional texture in urban space can be observed.

Although the Republican Period corresponds to a very interesting period in terms of Turkish Architecture, the studies on urban/public space arrangements of this period are quite limited. Except Ankara and İstanbul, planning history in Turkey have been researched by some architects and urban planners such as Burak Beyhan and Selçuk Uğuz on Mersin plans, Duygu Saban Ökesli on Adana city, Emel Karakaya on İzmit urban plan, Suat Çabuk and Kemal Demir on Kayseri ,and more (5,6,7,8). This research is trying to make a contribution to this necessity, as an another secondary city Gaziantep during 1930s in Turkey.

2 The Evolution of Urban Plan of Jansen

Gaziantep has been household to different cultures for centuries because of the strategic location of its being on an important place. It is located among the two regions where the first civilizations, such as Mesopotamia and the Mediterranean, were born and also at the crossroads from the Mediterranean to the west, north, and east, and an important commercial axis "Ipek Yolu" was passing through this region. Every new civilization established in Gaziantep has led to the transformation of the urban structure of the city with the social and cultural dynamics that it carries with it. The most effective focal point of the city was Gaziantep Castle. Topography and cultural data have been influential in the physical formation of the city with the beginning of the castle which constitutes the center of the first settlements and neighborhoods (9).

When we look at the development of Gaziantep at the beginning of 20th century, we see a stable mode for a certain time. The reason why there was no change in morphology in this period was the fact that public buildings and civil architecture structures were destroyed during the two-year Antep War. However, the economic collapse had been created by the war and the Armenians had to leave the city; this prevented the physical development of the city. This stagnation, which continued throughout the early Republican period, gained momentum with the implementation of the development plans of Jansen in 1935 and Kemal Ahmet Aru in 1950 (10).

In the context of 1930s urban planning strategies to perform a regional focus, Gaziantep city was one of cities planned to be an industrial peripheral in the gridal system. The traditional morphology of the city shows that the parallel lines to slopes is mostly evident where the grid system is distinctive for flat spaces. This tissue is researched by Kemal Ahmet Aru (who prepared the second plan of the city in 1950). According to the tissue analysis of him, the traditional morphology of the city of Gaziantep shows the F and G group texture style (Fig. 1). According to this, the characteristics of the F-type Turkish urban texture is

- On the plans, settlement suitable for hill slope, but self-developing organic ways.
- Narrow roads for running,
- Urban services in low slopes, which people can easily access, the G group urban texture characteristics,
- Geometric and symmetrical texture layout,
- Montaneously free organic form roads in non-metric style, on the large building islands,
- Pedestrian paths in large islands,
- Less intensive building islands,

can be sorted in the form (11).

The road system again, in the traditional city texture of Gaziantep, shows parallel lines to the slope in the hill settlements and the grid system. Empty areas within the texture are the areas used for agricultural purposes, and they show a different feature than the structuring in urban texture (12).

Meanwhile during the Republican period, the urban structure that emerged has increased on the legacy of the new state. The oldest plan showing the urban fabric of the 20th century is the development plan of Hermann Jansen of 1935. This plan

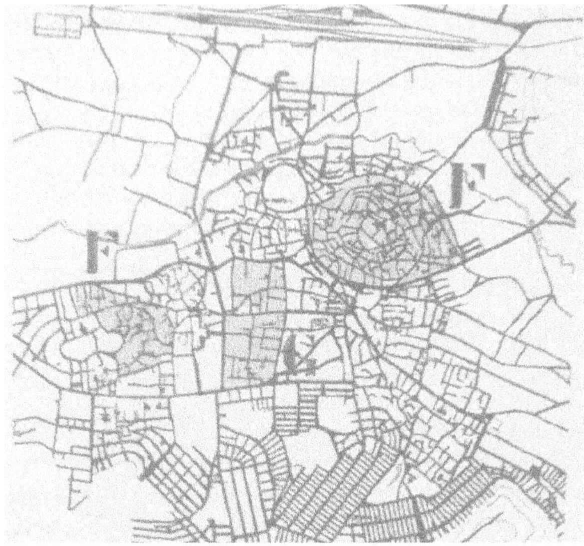

Fig. 1: Traditional morphology texture F and G type style of Kemal Ahmet Aru for Gaziantep (Aru, 1998).

is an important document drawn with scientific data by using the latest state of the city as a base for the Republic (Fig. 2). Naturally, the plan reflects the main decisions that meet the current needs, but not underestimating the future development of the city.

In this plan, the city was to be composed of five parts: industrial district, central existing old district, new residential district, labor house district, and small industrial district (Fig. 2). Airport is put on the north part, outside the city close to new residential areas Jansen designed in all cities he planned. But an interesting space "race square" is unique for Gaziantep. He placed a huge area for it, near to sport areas and animal houses (Figs. 3, 4). This is because according to Jansen a modern city which considers the health of residents should have a big sports complex, tennis courts, a public bath, a swimming pool, and a restaurant located in a good garden (13). Within this context three tennis courts were proposed and the race square is located near the residential area.

Another fascinating point is the interconnections between railways, highways, and airport which were well arranged here. The existence of railway facilities in close proximity to the proposed industrial district was also another supporting factor. The northern end of city for the industrial activities was in line with the connection with Anatolia, while the small industrial district is in line with Halep/Syrian border.

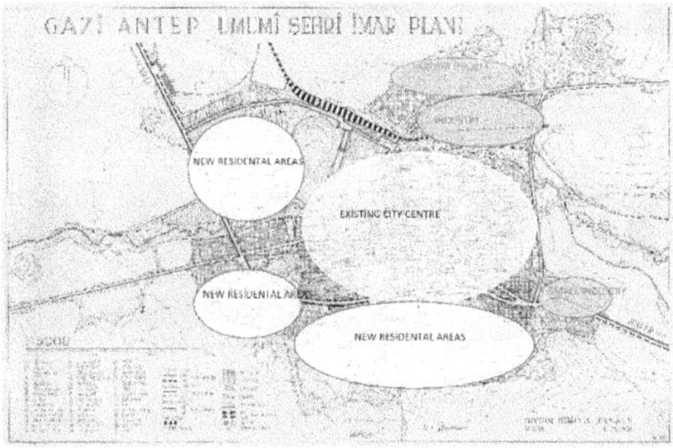

Fig. 2: The planning districts for the development plan of Gaziantep dated 1938 (revised for the paper from the copy at TU Berlin Architecture Museum Inv. Nr. 23414).

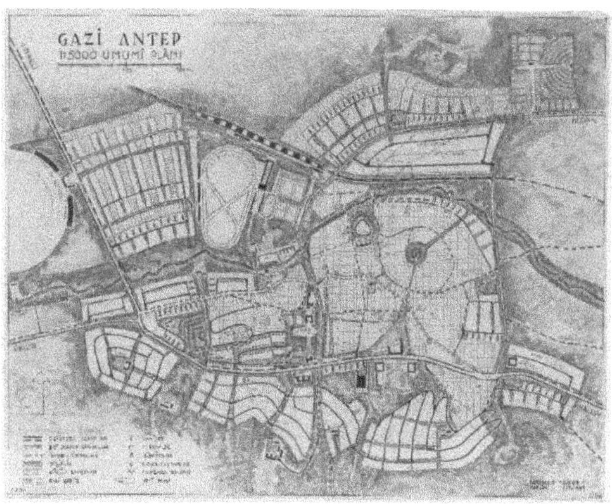

Fig. 3: The general development plan of Gaziantep dated 8.9.1938 (TU Berlin Architecturemuseum Inv. Nr.23413).

Besides them, the urban development principles of Gaziantep have similar parts with Ankara city plan, because Hermann Jansen, who designed the first zoning plan of Gaziantep, drew the zoning plan of capital Ankara, as a competition project in 1928. Some of the principles outlined in the disclosure report of the preliminary

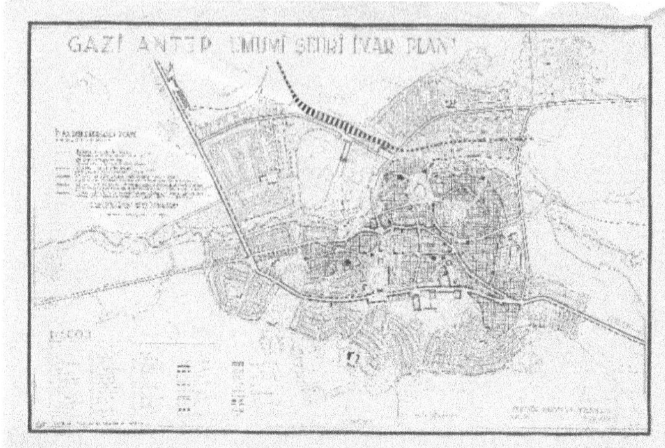

Fig. 4: The planning strategy for the development plan of Gaziantep dated 10.12.1938 (TU Berlin Architecturemuseum Inv. Nr.23415).

project that emerged as a result of the competition show similarities with the principles of the project that the planner had drawn for Gaziantep. The first of these principles is that the castle is seen as the source of urban aesthetics. The second one is the goal of providing urban health with green areas, sports fields, children's gardens, parks, and empty spaces. The third of the principles that are similar in both projects is that Jansen does not have an area to serve as a trade center. Jansen considered the center of the city as a castle, and with a physical and visual center, it excluded a functional city center. Accordingly, the commercial need of the new city will be covered by the commercial center in the old city. Although Jansen has designated areas for the commercial center in his project for Ankara, this park was replaced by Youth Park in 1932. The last one is the decisions about transportation. Jansen maintained the north-south and east-west connections in Ankara (14). This transportation decision, which was given as historical data before the competition, was also applied in Gaziantep. In the north-south direction, the Hürriyet Caddesi positioned the Suburcu Street-Atatürk Boulevard as the main transportation route (Fig. 5).

In the zoning plan for Gaziantep, Jansen aimed to enlarge the city in all directions, and divided the south of the city into residential areas and the north to the official usage areas. In this way, both sides of Alleben Stream are planned as a big green band. This green band is preserved until today. Preservation of large openings around Alleben Stream as green gardens is one of the successes of the Jansen's plan. Although Jansen's plan was to expand the city in every direction, after the Republic, the city evolved from north to west, the urban settlement emerged from its existing organic form and changed sharply with wide road lines. The zoning plan prepared by Jansen in 1935 preserved the organic

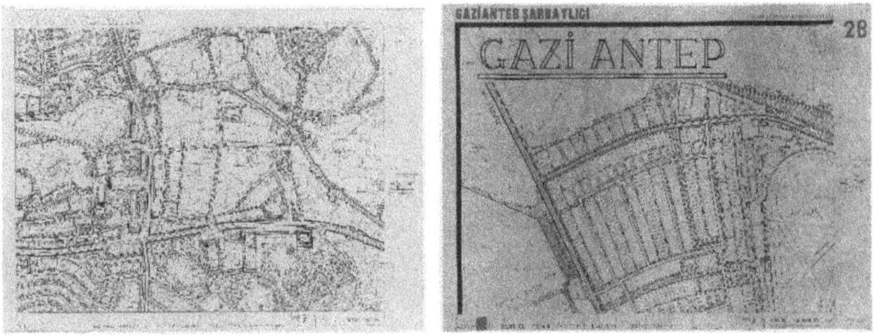

Fig. 5: The detailed planning of Gaziantep dated 1938. Area of old part city, new residental area in north-west part (Source: TU Berlin Architecturemuseum Inv. Nr.23421, 23417).

texture. However, it started to settle in the grid plan system in west and north directions (15). In the following years, the zoning plan was prepared by Aru and Söylemezoğlu in 1950, Jansen Plan underwent partial changes and continued to maintain the grid plan system. In this grid system, besides the buildings constructed with new materials and techniques, there are also buildings using traditional materials and techniques made after the Republican era (16) (Fig. 6). So, when the proposed general schema for Gaziantep is analyzed in comparison with Jansen's other urban designs, one can easily categorize matches between Gaziantep plan and the ones in Turkey and Europe. Essentially he employed a template for all the cities he planned. According to this template the city is divided into certain zones including old city, new city, industrial area, sports area, and port.

3 Other Projects of Jansen for Gaziantep

Since the cities have entered a planned and modern process in the republic period, new public spaces have been designed. Especially after the first five years of the period, "Republic Square, the town hall, the government house and the community house" started to be constructed as the symbol of the state in the Anatolian cities with the decrease of the negative effects of the war and the strengthening of the new system.

Within this concept Jansen designed many projects for Gaziantep. Jansen planned new cemetery project for Gaziantep in 1936. The castle was pointed on the hills particularly. The project showed a combination of semi-circular and grid shape together on a flat area (Fig. 7). It was located on the north part outside of city, near to industrial zone and labour neighbourhood area on the urban plan.

Later, in 1936 29th August), he designed Gaziantep Martyrs Monument (Şehitler Abidesi) surrounded by community center and park through Atatürk boulevard.

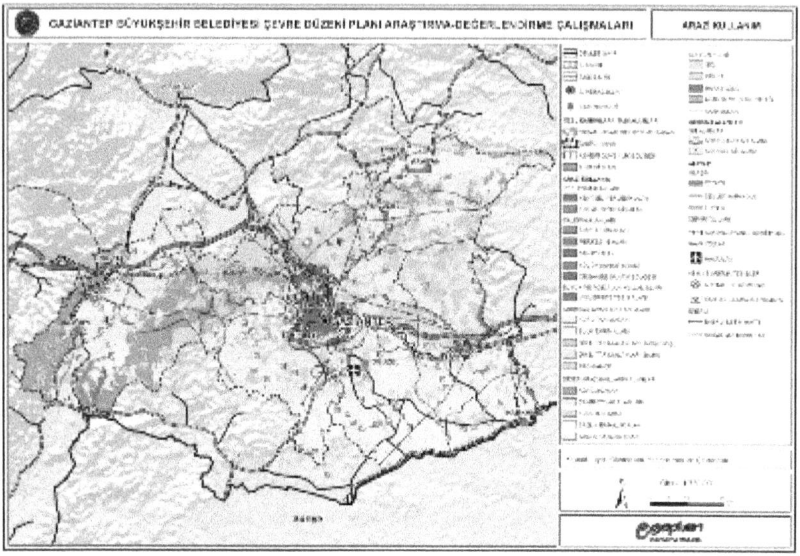

Fig. 6: Usage of Land Of Gaziantep, 2011 (https://gantep.bel.tr/files/yeniduyuru/plani-aciklama-raporu6846.pdf.

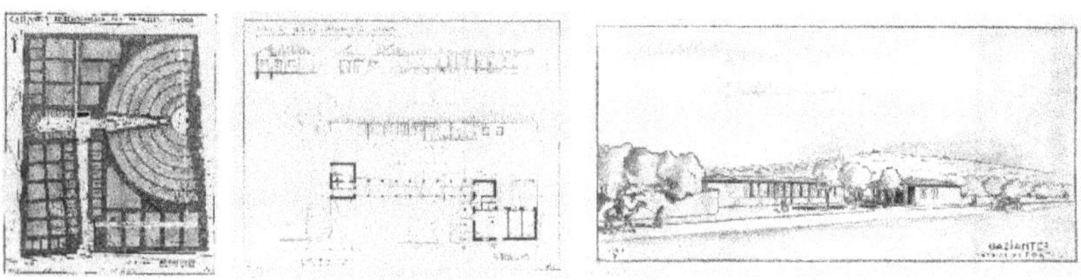

Fig. 7: Gaziantep new cemetery design of Jansen in 1936 (yeni mezarlık-friedhofanlage) (Source: TU Berlin Architecture museum, Inv. Nr. 23406, 23407, 23408).

In Turkey, after establishment of Republic regime, modernization project which comprises many actions, especially public works, showed its reality. During the period a dense building program was noticed. Community Center buildings were built near to Republic square in all cities and urban places (17) (Fig. 8).

In 1937 he planned Gaziantep Republic square behind the new province building. He put the government building, province building, and municipality building in the same axis, which were surrounded with huge green areas (Figs. 9,10).

In 1939, he designed a tower for entering the city and Kavaklık coffee house along a green bend through Alleben Stream (Fig. 11).

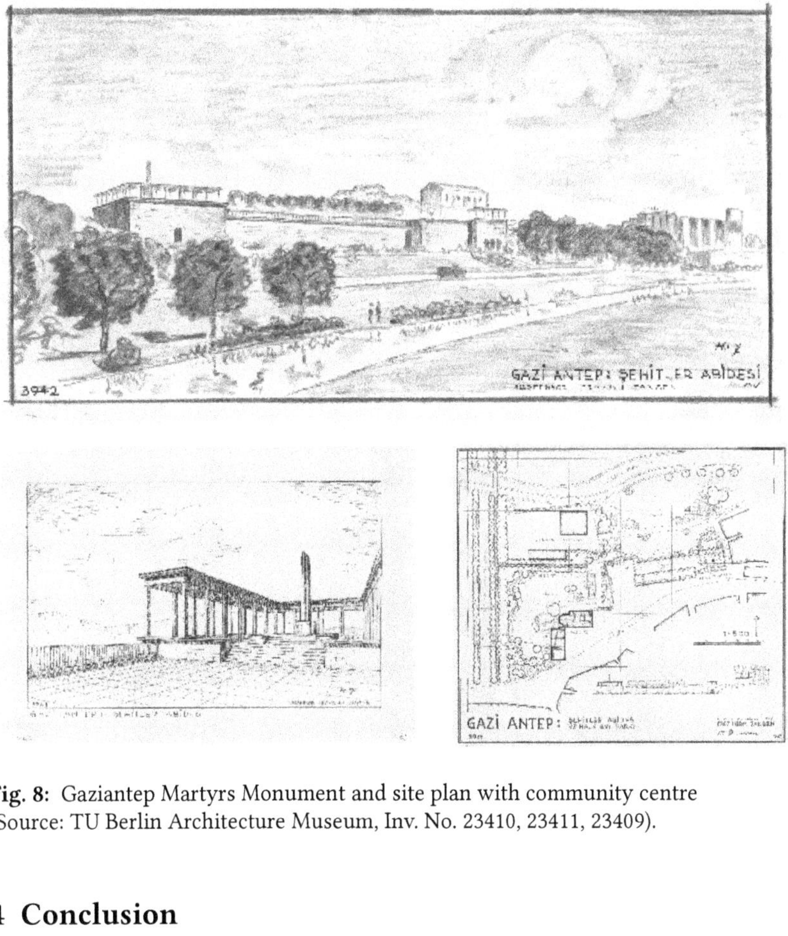

Fig. 8: Gaziantep Martyrs Monument and site plan with community centre (Source: TU Berlin Architecture Museum, Inv. No. 23410, 23411, 23409).

4 Conclusion

After the proclamation of the Republic, the physical change in the urban scale was manifested by the first zoning plan. This zoning plan, which was built in 1935 to contribute to the targeted restructuring of the city after the Republic, served to expand the main arteries to adapt to the traffic, to determine the growth direction of the residential areas, to provide space for the fairground, the sports area, and the airport to be constructed in the future. With the decisions taken in this zoning plan, significant progress has been made towards defining the identity of the new regime and starting the structural establishment. By this aim, Hermann Jansen made a general development plan besides partial planning as cemetery area, administrative district, entrance gate, and coffee house for Gaziantep.

Fig. 9: Gaziantep republic square, regional council, city hall (Source: TU Berlin Architecture Museum Inv. No. 23412, 23427).

Fig. 10: Gaziantep republic square (Source:http://www.eskiturkiye.net/2466/gaziantep-cumhuriyet-meydani).

As a result, the plans of Jansen for Gaziantep shows that his planning approach included functional planning principles, financially successful new industrial spaces with aesthetic concerns, conserving historical center of the city. Without a doubt, he preserved this existing historical district, as long as possible planning the street system and providing green spaces. What is further evident from the research is that all the projects for Gaziantep were designed in Berlin. For that reason, the plans for other cities would have similarities in common.

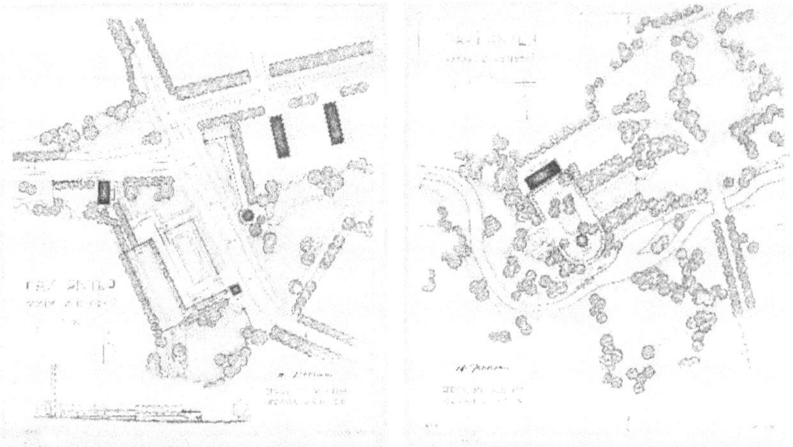

Fig. 11: Partial planning of the western city entrance and a coffee house on Kavaklık Site (Source: TU Berlin Architecture Museum Inv. No. 23429, 23430).

References

1 Bozdoğan, S., Modernism and Nation Building: Turkish Architectural Culture in the Early Republic, University of Washington Press, USA, 2001: p. 34.

2 Zander, Y,P., A Comparative Study on the Works of German Expatriate Architects in Their Home-land and in Turkey during the Period of 1927–1950, Unpublished Phd thesis, İzmir İnstitute of Technology, 2007: p. 8.

3 Durukan Kopuz, A., Spatial Evaluation of Primary Sugar Factories in early Republican Period in Turkey, A/Z ITU Journal of the Faculty of Architecture **14** (3), 2017: p. 129

4 Tankut, G., The Reconstruction of the Capital Ankara: (1929–1939), Anahtar Books, İstanbul, 1993.

5 Beyhan,B., Uğuz, S, Planning as a Tool for Modernization in Turkey: The Case of Hermann Jansen's Plan for Mersin, Journal of Faculty of Architecture, **29** (2), 2012: pp. 1–34

6 Saban Ökesli, D, Hermann Jansen's Planning Principles and His Urban Legacy in Adana, *METU JFA*, **26** (2), 2009: pp. 45–67

7 Karakaya, E., Policy-Oriented Planning in 1930s in Turkey: İzmit Urban Plan, A/Z ITU Journal of the Faculty of Architecture **14** (2), 2017: pp. 9–20

8 Çabuk, S., Demir K., Urban Planning Experience in Kayseri in the 1940s: 1945 Oelsner-Aru City Plan, A/Z ITU Journal of the Faculty of Architecture **10** (1): 96–116, 2013.

9 Yüce, B., The Change of Architectural Tissue of Gaziantep between Ottoman Period and Early Republican Period (1839–1950), Istanbul Technical University, Science Knowledge Program, Unpublished Master Thesis, İstanbul, 2010: p. 1.

10 Yüce, B., The Change of Architectural Tissue of Gaziantep between Ottoman Period and Early Republican Period (1839–1950), Istanbul Technical University, Science Knowledge Program, Unpublished Master Thesis, İstanbul, 2010: p. 16.

11 Aru, K.A., Turkish City, Construction Industry Center Publications, İstanbul, 1998.

12 Aru, K.A., Turkish City, Construction Industry Center Publications, İstanbul, 1998.

13 5. Beyhan,B., Uğuz, S, Planning as a Tool for Modernization in Turkey: The Case of Hermann Jansen's Plan for Mersin, Journal of Faculty of Architecture, **29** (2), 2012: p. 22

14 Tankut, G., The Reconstruction of the Capital Ankara: (1929–1939), Anahtar Books, İstanbul, 1993.

15 Gül, G., Traditional Texture Preservation and Development Proposal in the District of Bey (Kayacık), Gaziantep, Unpublished Master Thesis, Gazi University, 2005.

16 Gül, G., Traditional Texture Preservation and Development Proposal in the District of Bey (Kayacık), Gaziantep, Unpublished Master Thesis, Gazi University, 2005.

17 Durukan A., Ulusu Uraz, Community centre buildings as cultural institutions of the young Republic of Turkey, ITU Journal A, **7**, (1), 2008.

Gaziantep-2040 City Perimeter Explanation Report, p. 52 (cited 10.10.2018 available https://gantep.bel.tr/files/yeniduyuru/plani-aciklama-raporu6846.pdf)

Gaziantep Republic Area (cited 10.10.2018 available http://www.eskiturkiye.net/2466/gaziantep-cumhuriyet-meydani)

TU Berlin Architecturemuseum

Assoc. Prof. Dr. Barış Ergen and Assoc. Prof. Dr. Kemal Demir

7 Evaluation of Erciyes University Eastern Campus Social Center Project from Lefebvrian Perspective

Can social space be defined in terms of the projection of an ideology into a neutral space? No. Ideologies dictate the locations of particular activities, determining that such and such a place should be sacred, for example, while some other should not, or that a temple, a palace or a church must be here, and not there. But ideologies do not produce space: rather, they are in space, and of it. (Henri Lefebvre [1], 1991, Production of Space, page 210)

Abstract: University campuses can be thought of as urban spaces, in the sense that they feature a great degree of social diversity due to the multicultural nature that tends to define university settings and the international profile of the teaching staff and students. Otherwise stated, universities, because of their social diversity, are social spaces that reflect or even idealize the image of urban life. In this research, the proposed socio-cultural and religious facility to be built on the Erciyes University Eastern Campus will be looked at by contrasting the design needs of a university setting with the prevailing ideology, the political discourse that inspired it, and the impact it has had on Turkey over the past 20 years. The architectural design itself will be discussed on the basis of the spatial triad formation, a spatial analysis proposed by Lefebvre, using the interpretation method. A multicultural dynamic can be given to the structure of campuses by applying architectural designs that function to counteract the idea of creating religious facilities that accord with the governing Islamic political discourse that has arisen in the last 20 years in Turkey. With the rise of globalization, there has been an increasing number of students and instructors from all around the world descending on university campuses, and this has generated the need for universities to accommodate an array of different religious practices. It is in this area that design professions, such as architecture and urban planning, have an important role in producing the proper effect or perception of the intellectual design of campus areas.

Key Words: University campuses, Social center, Architectural and urban design, Spatial ideology, Lefebvrian approach.

1 Introduction

The Republic of Turkey was established in 1923. The constitutions of 1921 and 1924 defined the religion of the state, but this practice was abolished in 1928 and eventually replaced by the secularism principle, which was adopted in 1937 [2,3]. The idea of secularism is equated with a variety of concepts, including religious sciences, social sciences, state and public administration, and modernity [4–9]. Nonetheless, secularism can be defined in basic terms as the separation of politics and religion, a definition emphasized by Lægaard [10] in his research. After the 1994 local elections in the Republic of Turkey, the political discourse began to undergo a change, which particularly took hold as an Islamic political approach once the Welfare Party came into power in 1997 as part of a coalition. Following the 2002 general elections and the coming to power of the Justice and Development Party, this change became a major turning point for Turkey [11–14]. The effects of the Islamic political ideology have steadily begun to be reflected in urban and public spaces [15,16]. As Lefebvre [1] mentioned, ideologies do not produce urban space, but rather, ideology exists within the space relatively and it is the space itself. Therefore, the question is, how has this transformation, based on Islamic political ideology, unfolded in the process of creating the social spaces of universities? Political Islam has been reflected in universities in two major ways. First, there has been a trend to establish Faculties of Theology or Islamic Studies in the majority of universities. Secondly, there has been a tendency towards constructing mosques on university campuses as a reflection of the Islamic political ideology on urban space. In this latter context, Lefebvre's argument that ideologies dictate certain activities on urban space is validated, in the sense that the mosques are seen as reflections of sacred elements on space. Prior to the 2000s, with the secular state ideology in place, religious symbols and structures were never seen in public spaces or public state buildings; however, since then, Islamic symbols and related structures have started to be seen increasingly in public spaces.

According to data from the Council of Higher Education (COHE) in Turkey, as of January 2018 there are 128 state universities [17]. Of these 128 state universities, 60 of them have a Faculty of Theology and 26 a Faculty of Islamic Sciences, with 40 of the 60 theology faculties and the 26 Islamic Science faculties being established after 2002. These figures are a clear indication of the extent to which Islamic political ideology has been reflected on universities and campus sites.

Religion is just one of the elements of spiritual life. As stated by Kazanjian [18], spiritual life includes more than humanitarian, religious, and spiritual beliefs and experiences. Universities view student satisfaction as a significant measure of success, and spirituality is one of the most important indicators for determining the satisfaction of the students [19–21]. In today's world, there is a marked increase in religious diversity among university students, and therefore the need to have new and inclusive religious places has increased [22]. As Cragun et al. [23] emphasize, very little information is known about the religious life of the students on university campuses. Barnet et al. [24], in their study, noted that active participation in groups occurs in the form of attitudes, knowledge, behavior, and/or symbols. Therefore, the role that symbols play is as equally effective in social life as in urban spaces.

In Lefebvre's [1] book *The Production of Space*, space is conceptualized as a triad, where society is hidden within a social space and where the spatial experience of the community is realized and deciphered. Spatial scientists, urban planners, and social engineers have described the concept of space in terms of "representation of space," where they seek to discover what is to be understood from this space, how life is to be lived in this space, and how best to design this space. "Representational spaces" are spaces where life, along with its symbols and images, is carried out, and as such, is directly related to the inhabitants and users of the space, which can include artists, and to a lesser degree, writers and philosophers.

This study deals with the social requirements of the Erciyes University East Campus Social Center Project, the reflection of the political Islamic ideology on the space and the way the designer conceptualizes places related to the project.

2 Erciyes University Eastern Campus Social Center Project

Erciyes University was founded in 1978. The Gevher Nesibe Faculty of Medicine, established in 1969 under Hacettepe University, and the Kayseri Business Administration Faculty, founded in 1977, are the two main faculties of Erciyes University. Erciyes University also incorporated two higher education institutions, the Kayseri Higher Islam Institute in 1967 and the Kayseri State Architecture and Engineering Academy in 1977, re-designating them as the Faculty of Theology and the Faculty of Architecture, respectively [25]. Based on the premise that Erciyes University had a Faculty of Theology and a Faculty of Medicine that served the city of Kayseri and its surrounding provinces, a mosque was built on the western campus in 1987.

As of the 2016–2017 academic year, there were 61,961 students enrolled at Erciyes University, with 13,601 of the enrolled students being in the associate degree program, 40,222 being undergraduates, 6698 being masters students, and 1440 being doctoral students. The distribution of students by units and gender is shown in Tab. 1 below. Erciyes University has a prestigious status, not only in terms of number of students but also in terms of scientific research. Erciyes University was recognized as one of the top ten research universities in Turkey in 2017, along with Ankara University, Boğaziçi University (Istanbul), Gazi University (Ankara), Gebze Technical University (Kocaeli), Hacettepe University (Ankara), Istanbul University (Istanbul), Istanbul Technical University (Istanbul), İzmir High Technology University, and Middle East Technical University (Ankara). Therefore,

Tab. 1: The Total Number and Distribution of Students Enrolled at Erciyes University. Source: COHE, https://istatistik.yok.gov.tr/ [17]

Foundation Degree Students			Undergraduate Students			Graduate Students			PhD Students			Total Students		
M	F	T	M	F	T	M	F	T	M	F	T	M	F	T
9158	4443	13601	20893	19329	40222	3939	2759	6698	873	567	1440	34863	27098	61961

Erciyes University is the only university outside of the three major metropolitan areas of Ankara, Istanbul, and Izmir, to have achieved this honor of being a top research university. This prestige holds Erciyes University to a higher standard and mission, particularly in terms of hosting foreign students. Tab. 1 shows the total number of registered students at Erciyes University, which as can be seen is 61,961.

Erciyes University has 1663 foreign students from 103 different countries. The hosting of students from different societies and countries has brought with it the need to accommodate different religious requirements and places of worship. In addition to the students, Erciyes University also hosts teaching staff from different countries. Tab. 2 shows the distribution and number of foreign teaching staff at Erciyes University by country.

Tab. 2: Distribution of Foreign Teaching Staff of Erciyes University by Countries. Source: COHE, https://istatistik.yok.gov.tr/ [17]

Country	Prof.	Assoc. Prof	Assist. Prof.	Lecturer	Instructor
USA	0	0	0	1	1
Azerbaijan Democratic Republic	3	6	0	9	0
People's Republic of Bangladesh	0	0	1	1	0
United Kingdom	1	0	0	0	0
People's Republic of China	1	0	0	0	0
Republic of Estonia	0	0	0	0	1
French Republic	1	0	0	0	0
Republic of Georgia	0	1	0	0	0
Republic of India	0	0	1	2	0
Republic of Iraq	0	0	1	0	0
Islamic Republic of Iran	1	0	0	0	0
Japan	0	0	0	2	0
Republic of Kazakhstan	0	0	0	0	1
Kyrgyz Republic	0	1	0	0	0
Republic of Korea	1	0	0	0	0
Libyan Republic	0	0	1	0	0
Arab Republic of Egypt	0	0	0	1	0
Romania	0	1	0	0	0
Russian Federation	0	0	0	2	0
Republic of the Sudan	0	0	1	0	0

Fig. 1: Eastern and western campuses and the campus project area.

A recognized need emerged to erect a religious facility that would serve public housing units, presidency and administrative units, the faculty of pharmacy, the construction and security unit, the faculty of architecture, the faculty of fine arts, the Mustafa Çıkrıkçıoğlu vocational high school, the faculty of tourism, the permanent education center, the faculty of communication, the faculty of law and justice vocational high school, the physical education and sports vocational high school, the veterinarian faculty, the faculty of education, the nanotechnology application and research center, and the drug research center. As a reflection of the governing political Islamic ideology, the idea was introduced to build a social center, one that would function as a symbol of political Islam and therefore include a mosque and related structures. An instructor from the Faculty of Architecture was appointed by the presidency to carry out the mosque and social center project. The Eastern and Western campuses of Erciyes University, as well as the project area proposed for the Eastern campus, are shown in Fig. 1.

The architectural designer recognized that a mosque was not enough to accommodate the need for worship, but instead, a social center was necessary, one that would serve different religious groups and at the same time include shopping spaces, museums, and the like, which students could benefit from. The goal is to create a living center whose units are designed to consistently address the needs and desires of different users. The design therefore aimed to establish a faith center consisting of a mosque, synagogue, church, and museums, as well as recreation, shopping, dining, meeting, and sharing areas. In the planning stages, there are certain key elements to the design that an architect considers, including the topographic structure of the area, which serves as one of the main pieces of data

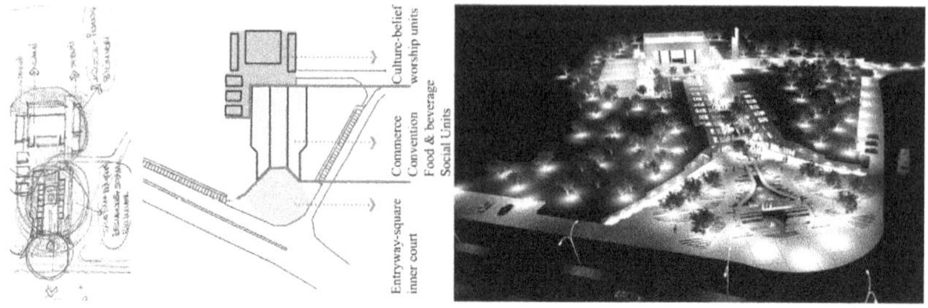

Fig. 2: Draft drawing and night view of the entrance square.

helping to orient a designer's plans, and the incline of the land, which can be taken advantage of in raising the visibility of important structures—in this case, the houses of worship—to a higher spatial perception. In the design of the campus project, the designer hid the social gathering and sharing units under the topographic pattern to give greater visibility to the religious buildings, which would be accessible through the main thoroughfare. The design basically involves three main uses of space, with the first part featuring an entrance square; the second part featuring shopping, dining, meeting, and social units; and the third part featuring cultural and religious units. The three spaces that make up the design will be analyzed using Henri Lefebvre's mental interpretation of the space approach, or as he referred to it, the "Representation of Space" (conceived space).

2.1 Entrance Square

The designer used the ornamental fountain as a design component in this space. Specifically, the pool design is supposed to hold allegorical meaning, as it is in the form of the tree of life to represent the dawn of humankind. This pool is designed to be perceived from the front courtyard of religious units in the upper elevation. The roots of the tree of life are based on the culture and faith center in the spatial scheme. When viewed from the square at the ground level, the spatial perceptibility of the religious units is held in the foreground in terms of scale, proportion, and form. Regarding the two-dimensional plan view of the square, it can be noticed that the square is designed as eye abstraction. The emphasis of the design here is to make Erciyes University's uniqueness prominent. Fig. 2 shows the appearance of the entrance section and the design stages.

2.2 Commercial and Socio-Cultural Units

These spaces have four basic functions: commercial, socio-cultural, meeting and sharing, and event hosting. The commercial units include a stationery store, market, pharmacy, and bank; the socio-cultural units feature student clubs, meeting spaces, and hobby workshops; the meeting and sharing units have cafes, restaurants, and

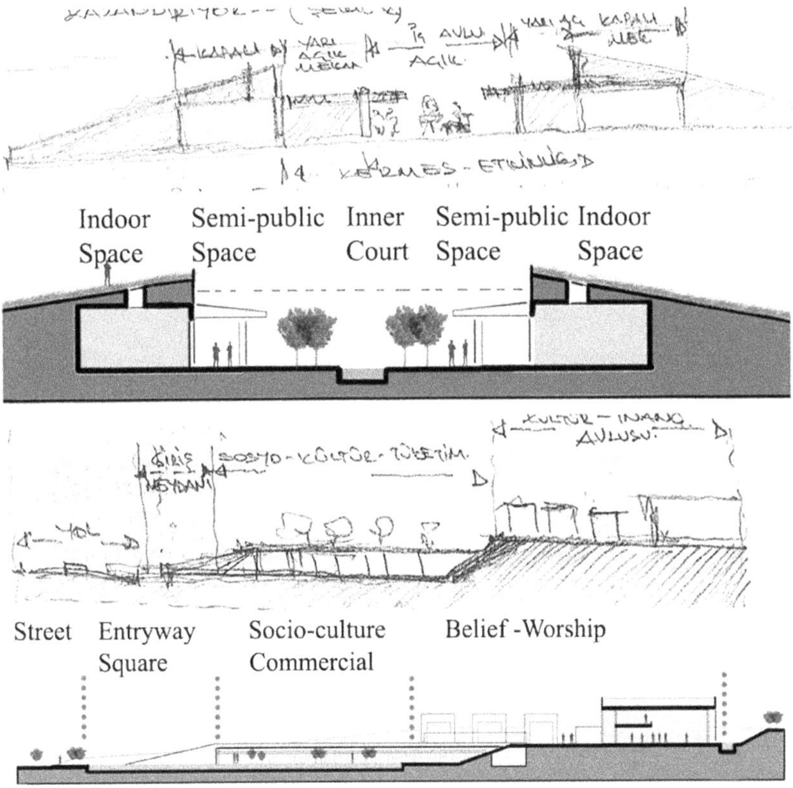

Fig. 3: Draft drawings of commercial and socio-cultural activity areas.

game and entertainment venues; and finally, the event hosting units consist of multipurpose halls with small-scale conference rooms and event and meeting spaces. The commercial and socio-cultural units are designed to support each other in two directions on the main axis. The goal here was to create a seamless living and entertainment environment, full of activities, where more user access would be provided to the space. In this way, the design of a lived and used space would be fulfilled. Fig. 3 shows the spatial relationship between the commercial and socio-cultural units, along with design stages, draft drawings, and topography.

2.3 Cultural and Religious Facilities

This space includes religious places and a museum. The mosque, church, and synagogue, which serve as the places of worship of the three Abrahamic religions,

Fig. 4: View of religious facilities in relation to other units.

constitute the spatial elements. A simple language was chosen in the spatial design, with no dome or minaret used for the mosque, nor bell tower for the church. The aim was to develop a timeless approach to integrate the religions. In the design of the facades for the places of worship, only the most natural and basic materials would be used. The facades are made to be transparent in some places, to allow for the integration of the interior and exterior spaces and to give a greater feel of spaciousness in the places of worship. In the spatial design, a large courtyard is proposed in front of the church, synagogue, and the mosque. The tower, which is placed to represent the center of faith, serves as the central defining feature of the space. A significant issue in the design is to place emphasis on time. Contemporary materials and a contemporary design approach were used as the design basis, taking special care to avoid emulating the Mimar Sinan mosque architectural design (replica) so popularly applied in Turkey. The fourth building is designed as a museum and with the flexible nature of its space, it is made to be adaptable for use by other religions. Fig. 4 shows the spatial relationship between the religious units, the topography, and the commercial, socio-cultural units.

In the design of the Erciyes University Eastern Campus Social Center, the Erciyes Campus Faith Center was modelled after the Acropolis in Ancient Greek Civilization, including the visual sophistication [26] provided by the location of the Acropolis in a high position.

A similar approach can be seen in the Imperial Forum and the spatial relationship that exists between it and the city of ancient Rome [26]. This relationship seen

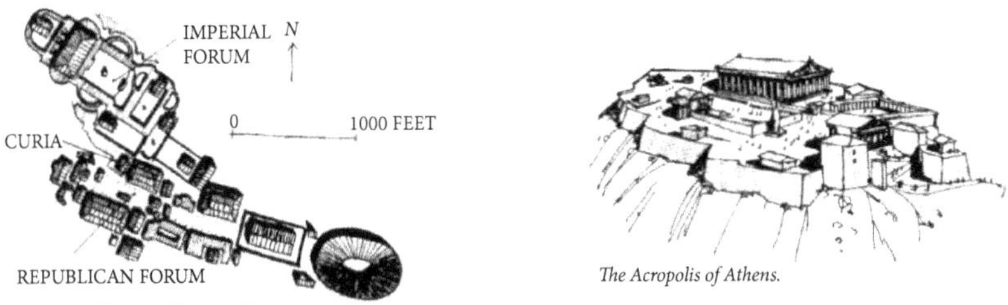

The Republican and Imperial Forums together.

The Acropolis of Athens.

Fig. 5: Positions of Imperial Forum and Acropolis. (Source: Spreiregen [26]).

in the spatial diagram of the Imperial Forum can also be observed in the design of the Erciyes University Eastern Campus Social Center. Furthermore, based on Rob Krier's [27] urban space book, in the Erciyes University Eastern Campus Social Center Project, an approach that reflects the elements of a market area, ceremonial area, church and administration building, all of which were characteristic elements of the squares in the medieval cities, was adopted. Fig. 5 shows the relationship and location of the commercial and religious units in the ancient Greek and Roman civilizations. Just as in the case of the Acropolis, the dominant position of the religious units at higher areas is one of the significant features of the socio-cultural center project design.

In the Ottoman cities, the mosque courtyards served as the city squares and also functioned as large gathering places, shopping areas, and bazaars. Since the center of social life was the mosque, the planned gathering places were designed to be part of the mosque courtyard and its surroundings [28–29]. The designer for the Social Center Project was inspired by the Ottoman cities, as well as the ancient Greek and medieval cities. By positioning commercial and cultural units on the axis that opens out to places of worship, he adapted the gathering function in the Ottoman cities to the Social Center Project. Fig. 6 shows the location of the socio-cultural center project planned for the Erciyes University Eastern Campus through 3D visualization.

It is meant to be reinterpreted with contemporary material and design to meet the needs of the contemporary age, which carries traces from the past to the present in the form of a traditional space.

3 Discussion

This chapter can be seen as a response to Lefebvre's question "Can social space be defined in the projection of an ideology into a neutral space?" Considering that the space itself is the epitome of the ideology, the Erciyes University Eastern Campus Social Center Project has been designed with a secular understanding, where all beliefs share equal space and places of worship are available for all beliefs,

Fig. 6: The location of the Erciyes University Eastern Campus socio-cultural project proposal in the field.

avoiding a design that is based on an Islamic political approach alone. The need to have religious places of worship in university settings has become increasingly important due to the rising mobility of students and instructors in the globalized world. Since universities are idealized places of social life, they should contribute to building relationships between societies. The word *university* derives from the Latin word "universus," meaning "corporation, society," or "the whole, the aggregate and entire" [30]. Given that this term embodies the ideas of integration, social cooperation, and gathering together, to focus on just one single religion in university spaces would be incompatible with the today's mainstream understanding of multiculturalism.

The idea behind the Erciyes University Eastern Campus Social Center Project emerged out of the gradually rising Islamic political discourse and the implementations this discourse engendered in Turkey. As Lefebvre mentioned, given that the idea of space includes social values, social interaction affects the space, and the space cannot be thought independent from ideology. This understanding of social interaction affecting the space is also applicable for the universities in Turkey. However, the designer for the Erciyes University Eastern Campus Social Center project does not simply reflect the current political ideology of Turkey onto the space; rather, the integrative meaning inherent to being a university can be seen in the worship places designed for the project.

The design of the Erciyes University Eastern Campus Social Center project was inspired in part by the architectural principles used in ancient civilizations, where the emphasis was on their own spatial experience, and as such, the layout is filled with "appropriated" space, as Lefebvre mentioned in his book, *Production of Space*. Lefebvre [1] summarizes the idea that spatial formation is created by being articulated over the traces of ancient civilizations as follows:

> The fact is that the shortage of space is a distinctly socio-economic phenomenon, one which can only be observed, and which only occurs, in quite specific areas, namely in or near urban centers. These may have grown up from historically established centers, from the old cities, or they may have evolved out of new towns.

The architect created the social center project using the design principles of ancient civilizations. However, it is worth noting that old cities evolve, and in the process, new ones are created. In examining the design, it was seen that although the design principles of the ancient civilizations were taken into account, the use of contemporary technology, materials, and design demonstrates the evolution of these design principles.

As Lefebvre emphasizes, ideology as an architectural approach is within and integrated with the space. In Turkey, the traces of two ideologies can be seen in urban spaces. This first one is the general political ideology, which has been operating in Turkey for the past 15 years and can be described as a growing conservatism based on Islamic political discourse and its applications. This ideology has been reflected as an ideological approach on urban space through the construction of mosques on university campuses. The second ideological approach is based on the multiculturalism and increased mobility that has emerged out of globalization, bringing people from different societies and beliefs together. The designer had this ideological approach in mind when designing the Erciyes University ("Universus" in Latin) Eastern Campus Social Center Project, as it is reflected in the different places of worship and the emphasis on promoting a secular ideology, which maintains an equal distance from all societies and beliefs.

> Urban space gathers crowds, products in the markets, acts and symbols. It concentrates all these, and accumulates them. (From Lefebvre's (1991) *Production of Space*)

The Erciyes University Eastern Campus Social Center embodies all these elements of an urban space, as defined by Lefebvre. The center gathers together a crowd of multicultural students and faculty members; it includes commercial units offering market products, or, from a different perspective, the scientific productivity can be interpreted as products as well; and its places of worship feature symbols as a simple language used in the act of worship. If scientific productivity is considered as the highest level of satisfaction, then the religious worship places that will enable all the faculty members and students from different nations to establish connections with the space will serve to further increase this level of satisfaction.

4 Conclusion

University campuses do not consist of only research units. Students and instructors constitute the basis of universities and are at the core of the meaning of universities and society. Universities should include traces of the community and areas of religious worship as spaces reflecting the ideal form of society. These are indispensable features in today's understanding of globalization, not only for Turkey, but also for the entire world.

Both the ideological and symbolic design principles governing the dialect of the conceived space of the proposed social center project for the university campus are emphasized, along with the concepts of perceived space (spatial practices), conceived space (representations of space), and lived space (representational spaces), as described by Lefebvre in the production of the urban space. Urban and architectural design play a significant role in shaping the space. It is unarguable that political ideology affects the space. However, the designer had different symbolic approaches to the identity of the space. Urban planning and architecture professions play an important role in spatialization and the conceptualization of space. The designer took a critical approach against conservatism and political Islam, which have recently been on the rise in Turkey, by considering the secularism principle, which was instituted by Atatürk for Turkey. On the basis of this principle, the goal was the unification of the religious and sectarian beliefs of all societies living in Turkey. When looking at the mosque typology used in Turkey, it can be seen that it is largely a replica of the architect Sinan's works. However, the Erciyes University Eastern Campus Social Center boasts a unique design, without including a dome or a minaret, one that reflects the requirements of the contemporary era in terms of materials used and the design.

Following the Welfare Party's rise to power after the 1994 local elections in Turkey and the ushering in of Islamic political ideologies, the Justice and Development Party rose to become the sole power in 2002, sealing the Islamic political ideology. Since 2002, the Justice and Development Party has continued to be the single ruling party. The influence of political Islam directly affects universities because of the appointment of rectors by the President of Turkey. This effect also influences the structure and number of faculties of the universities (as in the case of the establishment of Faculties of Theology), and leaves spatial symbolic traces (such as the construction of mosques in universities). However, the aim of the designer was to create places of worship that served all religious views, not focusing only on a single religion, to live up to the concept of secularism, which was added to the constitution on February 5, 1937, with the words, "No one can be questioned regarding religion and sectarian or philosophical thought; it is certainly allowed to celebrate any religious rite."

Although this project has been appreciated by the university administration in terms of design concept, it has not been implemented yet. The assertive meaning, as well as the parts of the project that are not compatible with the Islamic political ideology of today's Turkey, has prevented it from being implemented.

Acknowledgement:

This work is part of the research project numbered 050313 which has been approved by Erciyes Technopark.

References

1 Lefebvre, H., *The Production of Space.* 1991, Translated by D. Nicholson-Smith. UK: Oxford and USA: Cambridge, Blackwell. 454.

2 Ertan, T.F., *Türkiye Cumhuriyeti Anayasalarında Laiklik.* Ankara Üniversitesi Türk İnkılâp Tarihi Enstitüsü Atatürk Yolu Dergisi, 2007. **39**: pp. 409–423.

3 Tayhani, İ., *Türkiye Cumhuriyeti'nin Temeli Laiklik.* Ankara Üniversitesi Türk İnkılâp Tarihi Enstitüsü Atatürk Yolu Dergisi, 2009. **43**: pp. 517–529.

4 Bhargava, R., *How Secular Is European Secularism?* European Societies, 2014. **16**(3): pp. 329–336.

5 Hart, J.G., *Deep Secularism Faith and Spirit.* International Journal of Philosophical Studies, 2016. **24**(5): pp. 639–662.

6 Grigoriadis, I.N., *Islam and Democratization in Turkey: Secularism and Trust in a Divided Society.* Democratization, 2009. **16**(6): pp. 1194–1213.

7 McLennan, G., *Is Secularism History?* Thesis Eleven, 2015. **128**(1): pp. 126–140.

8 Possamai, A., *Post-Secularism in Multiple Modernities.* Journal of Sociology, 2017. **53**(4): pp. 822–835.

9 Weir, T.H., *Germany and the New Global History of Secularism: Questioning the Postcolonial Genealogy.* The Germanic Review, 2015. **90**: pp. 6–20.

10 Lægaard, S., *Multiculturalism and Secularism: Theoretical Understandings and Possible Conflicts.* Ethnicities, 2017. **17**(2): pp. 154–171.

11 Gurses, M., *Islamists, Democracy and Turkey: A Test of the Inclusion-Moderation Hypothesis.* Party Politics, 2014. **20**(4): pp. 646–653.

12 Kaya, I., *Contemporary Turkey: An Islamic-capitalist variety of modernity?* Social Science Information, 2014. **53**(2): pp. 197–212.

13 Keyman, E.F., *Modernity, Secularism and Islam The Case of Turkey.* Theory Culture & Society, 2007. **24**(2), pp. 215–234.

14 Taniyici, S., *Transformation of Political Islam in Turkey.* Party Politics, 2003. **9**(4), pp. 463–483.

15 Batuman, B., *Minarets without Mosques: Limits to the Urban Politics of Neo-liberal Islamism.* Urban Studies, 2013. **50**(6): pp. 1097–1113.

16 Karaman, O., *Urban Neoliberalism with Islamic Characteristics.* Urban Studies, 2013. **50**(16): 3412–3427.

17 COHE, *Data Management System of Council of Higher Education.* [cited 2018, 3rd March]; Available from: https://istatistik.yok.gov.tr/.

18 Kazanjian, V., *Spiritual Practices on College and University Campuses: Understanding the Concepts - Broadening the Context,* Journal of College and Character, 2013. **14**(2), pp. 97–104.

19 Mohebpour, I., et al., *Religiosity, Religious Acceptance, Social Interaction, and Satisfaction with University Experience*, International Journal of Christianity & Education, 2017. **21**(3): pp. 197–212.

20 Kuh, G.D., and R.M. Gonyea, *Spirituality, liberal learning, and college student engagement*. Liberal Education, 2006. **92**: pp. 40–47.

21 Mooney, M., *Religion, college grades, and satisfaction among students at elite colleges and universities*. Sociology of Religion, 2010. **71**: pp. 197–215.

22 Karlin-Neumann, P., and J. Sanders, *Bringing Faith to Campus: Religious and Spiritual Space Time and Practice at Stanford University*. Journal of College and Character, 2013. **14**(2): 125–132.

23 Cragun, R.T., et al., *Chapel Use on College and University Campuses*, Journal of College and Character, 2014. **15**(2): pp. 103–118.

24 Barnett, K., Ropers-Huilman, R., and L. Aaron, *A Planning-Process Perspective on Student Activists' Upward Influence Attempts to Effect Campus Change*. Southern Communication Journal, 2008. **73**(4): pp. 332–346.

25 Erciyes University, *History of Erciyes University*. [cited 2018, 3rd March]; Available from: https://www.erciyes.edu.tr/kategori/ERU-HAKKINDA/Tarihce/2/58

26 Spreiregen, P.D., *Urban Design: The Architecture of Towns and Cities*. 1965, USA: McGraw-Hill Book Company. 243.

27 Krier, Rob., *Urban Space*. 1991, Hong Kong: Academy Editions, Fifth Impression. 173.

28 Önder, S., and F. *Aklanoğlu, Kentsel Açık Mekan Olarak Meydanların Irdelenmesi*, S. Ü. Ziraat Fakültesi Dergisi, 2002. **16**(29): pp. 96–106.

29 Ersağdıç, Y., *Kent Meydanlarının Rekreasyonel İşlevleri Açısından İncelenmesi ve Bu İşlevlerin Konak Meydanı Örneğinde İrdelenmesi*. 1998, İzmir: E.Ü. Fen Bilimleri Enstitüsü Peyzaj Mimarlığı Anabilim Dalı Yüksek Lisans Tezi. 160.

30 Dictionary, *University*. [cited 2018, 3rd March]; Available from: http://www.dictionary.com/browse/university

Lecturer Dr. Betül Atakan Öznam and Assist. Prof. Dr. Ayça Yeşim Çağlayan

8 Recreational Area Quality in Residential Environment

Abstract: Common goal for all physical, social, managerial and economic circles is "the creation of a livable space" that can respond to individual and social needs such as shelter, security, and recreation and that is egalitarian, and that means something to individuals. In the modern urban system fictionalized over home-dwelling life, there is an increasing need for spaces where town-dwellers can make use of their leisure time. Urban green spaces are the main components of the cities because they offer areas where inhabitants can benefit from their leisure time, as well as high life experience.Although there are different groups of individuals having the ability to translate the opportunities offered into recreational opportunities or groups of different users having various demographic structure, age, and sex characteristics, some common physical characteristics increase the recreational value of open spaces in the immediate vicinity of the housing. The areas that create an environment for the individual's experience in both natural and cultural environments in terms of the sources of recreation that allow for various uses, and that are rich with regard to the presence of sustainable equipment and facilities define the areas with high recreational value.In this process, which requires evaluation in terms of physical, social, individual, managerial, and economic different variables; with the planned distribution of adequate space intensities in urban scale, creation of public spaces where the housing and its immediate vicinity is designed with a holistic approach, which are reliable, where easy accessibility is provided through the walking distances, which allow for socialization, which offer a rich variety of recreational opportunities, which are multi-functional, which are designed according to their function, and most importantly, which in terms of both topography and living environments protect the nature in the region in which they are applied, and where health and comfort requirements fulfilled should be determined as the main goal. Addressing the residential areas together with their environment through a holistic and participatory planning approach is of great importance with regards to urban sustainability and quality oflife.

Keywords: Residential environment, Non-Residential areas, Recreation, Quality, Urban green spaces, Recreational areas

I Urban Life, Recreation, and Green Areas in Residential Environment

Due to the ease of access to employment and infrastructure facilities in a lot many countries of the world, the majority of the population prefer to live in cities. Gül and Küçük (2001) state that architectural structures, open-green spaces ,and their relations with each other and the integrity they form constitute the general character of a city [1]. Open-green spaces have an important position in balancing the disrupted relationships between human beings and the nature, and improving urban living conditions. Therefore, in developed countries, the quality and quantity of open-green spaces are considered to be the indicators of civilization and life quality. What is important for ensuring the quality of urban life is to establish a healthy urban structure. For this purpose, also, economic, cultural, environmental, social, and administrative needs should be met and those should individually be regarded as a right, and they should be implemented, preserved and improved [2].

In the modern urban system fictionalized over residential life, there is an increasing need for spaces where town-dwellers can make use of their leisure time. Urban green spaces are the main components of the cities because they offer areas where inhabitants can benefit from their leisure time, as well as high life experience. Housing, one of the basic requirements for the individual and family as a shelter, is a phenomenon with economical and spatial content for the society. It is impossible to think of the housings as a single structure. It is not only a physical shelter with its environment, land, infrastructure, social relations, but also a whole covering all social facilities and services required for the welfare and happiness of the human beings [3]. Türkoğlu states that the housing and its surrounding is a concept that encompasses the social environment that includes both the dwelling and the neighborhood relations, and expresses that ambience defines the individual's relations with the environment in both social and physical ways [4]. In many studies conducted on this subject, it is emphasized that the non-residential common areas (such as gardens) are effective in terms of bringing together individuals from different age groups and forming a shared life style; and that those venues are crucial for social actions and establishing friendships; and that the dwellings and their surroundings are determined and organized by the traditions, preferences, social status, and cultural level of the people living in that area. The demographic structure, socio-economic characteristics, education and culture level, working and labor potential of the population constituting that city emerge as an important factor in the physical configuration of the quality of the residential immediate surroundings, and the formation of social structures [5]. Studies carried out show that people hierarchically tend to meet their shelter, safety, comfort, socialization, self-expression, and aesthetic needs, respectively around the dwelling [6]. Rapoport [7], who states that the surrounding of the housing contains much more than the housing itself, in a sense, indicates that non-residential spaces are indicative of the socio-cultural status of people. In today's conditions, dwellings, far beyond providing the individuals with the possibility of sheltering, have significant impacts on the formation of social structure with the vital quality standards they offer in both interior and exterior spaces. Yet, open spaces around the residential environment are featured areas where families spend their social lives, where they get into social interaction with the individuals around them, where healthy growth and self-renewal environments for individuals of different age groups are provided. It should be kept in mind that healthy, high-quality environments to be provided by function and user diversity in these areas are the indicators of the quality of life of the city. In housing layouts, the areas other than the buildings are defined as non-residential areas. Basis of the arrangement of non-residential areas can be listed as follows:

a. Creating a healthy environment that increases the usage value of the dwelling
b. Formation of a certain quality in that environment created
c. Arranging areas differentiated according to usages
d. Ensuring the integrity of the landscaped environment within the city.

Non-residential areas have the characteristics of continuity of daily life, but they must address the biological, social, and psychological needs of people. It is also expected to have a positive impact on their mental activities such as comprehension, perception, thinking, and interpretation. Researches have shown that people's perception and experiencing of the surroundings of the housing vary depending on traditions, preferences, social status, and culture level they have. In the researches done before, it is revealed that the individual and the surroundings of the housing mutually affect each other. In this context, it may be possible to talk about a positive environment-human interaction only if the environment becomes compatible with the perceptual, sensory, and semantic dimensions of the user as a result of individuals' reflecting their personal characteristics to the environment, and so landscaping their environment in their relation with the housing and its surroundings. Otherwise, individuals have a desire to change environmental conditions or to move to other places, and in cases where this cannot be achieved, environmental discrepancy or psychological discontent occurs [5].

Particularly, besides being a transition area between private property and public ownership, the surrounding of the housing, which covers the most common public outdoor uses of the city and forms the main living space of people, is of great importance in terms of meeting the quality criteria such as access to cultural, sportive, recreational, and commercial functions and bringing together different social groups that are integrated with human scale [8].

As well as shaping the lifestyles of individuals, the physical and social characteristics of the residential environment lead to an important potential for the development of the type of behaviors encouraging sports and effecting the health in a positive way [9]. Actions taken in daily life such as getting together in the vicinity of the housing complex, walking together, sitting in open spaces, playing sports, using common social areas, etc., should be thought to be increasing the social interaction between people; and landscaping works should be carried out in accordance with these principles [10].

Akşit emphasizes that the open and green spaces around the housing are important in terms of bringing people together for social activities, meeting the mental and physical needs of the individuals, providing the transition between the places by softening the masses in the building spaces, forming a scale between human and environment, and providing comfort for life [11].

The mass housing practices, which are based on the principles of providing the population living in the cities with a structure suitable for urban life and establishing a quality and healthy living environment in the settlement areas, have started to be accepted as widespread housing production in our country since the 90s.

Mass housing areas are important residential areas that we can assess at a neighborhood scale. In the areas around the housing complex, they confront us as user expectations of the multi-functional spatial structure, which simultaneously incorporates sports, shopping, dining activities in leisure times.

As a matter of fact, mass housing becomes more of an issue as "in a designated area, together with a large number of dwellings, the structures having quality and quantity to meet the needs of those living in those dwellings; including, social, cultural, educational, health units and recreational facilities" [12]. One can talk about two different types of outdoor features feeding the activities that can be carried out in the vicinity of mass housing complexes. The first of these are the open spaces, which are within the boundaries of the mass housing area, and which are in other words under private ownership but can be defined as common open spaces because they are reserved for common use. The other one encompasses spaces that are open to the public use in the immediate vicinity in the quarter where the mass housing is located [8].

In this context, on one hand, the mass housing areas offer their own recreational potential but on the other hand they benefit from the recreation opportunities offered by the environment they are located in. Also, the level of benefiting from each other increases at the rate of socio-demographic structure in urban environment, and the socio-demographic structure which constitutes the mass housing area, and urban environment land use types overlap.

Recreation activities are of great importance for a better organization of the society and attaining a modern structure. Recreation provides a great number of social benefits such as purpose-oriented activities, maintenance of public and social relations, sense of responsibility that is developed through the relations with the society by leaving individuality behind, sustaining the phenomena of groups and friendship that are important in social life and ensuring the sustainability of sociality etc. [3].

In short, recreation defined as the use of free time (leisure) contributes to the uniting of the society's members under common values and their having a sense of belonging. According to Ardahan, recreational activities are the tools that make the distances between the individuals shorter that eliminate the prejudices about differences in society, and that affect the physical and mental health of the society and the level of social capital positively as well. Although it may vary by content and type, participation in recreational activities satisfying important needs such as belonging to a group, physical and mental development, and self-expression plays an important role in the socialization of individuals and in the development of their social environment, in their getting rid of stress that monotonousness work and daily life bring and in increase in their life satisfaction level [13].

According to Karaküçük [14], leisure activities are classified according to the goals, wishes and tastes of individuals in participating in activities. He groups the recreation according to purposes such as relaxation, culture, art, establishing communal/social relations, sporting, and touristic purpose leisure activities. Other recreation classifications are by age factor (youth, children, and adult recreation, etc.), by number of participants (individual, group, mass and family recreation, etc.), by time (seasonal, weekend, and retirement period recreation, etc.), according to social structure (by luxury, traditional, etc.), and according to the space where the activities are performed [14].

Open and green spaces, which are of great importance in terms of healthy development of cities, are, within built-up areas, common use areas which are organized in order to realize various recreational activities and to get closer to nature [15]. Considering that the quality of urban life is composed of components for physical, social, and economic environmental quality, it is concluded that the presence of open and green spaces together with recreation areas are among the determinants of the physical environment quality [16].

Green spaces can also be characterized as cultural focal points where social communication takes place in terms of meeting social needs such as encounter of people from every walk of life, meeting, talking and sharing urban life, and providing socio-cultural continuity and development [17].

Today's modern people prefer recreation areas, to lead a healthy life both physically and mentally, and green spaces and parks where they could perform recreation and various activities [3].

Also, another important feature of urban green spaces in terms of health is their ensuring physical activity possibilities through their recreational functions. It is known that physical activity is very important with regards to physical health problems such as cardiovascular diseases, diabetes, and obesity, and it is known to be effective in eliminating mental health problems such as stress and depression [9].

In the researches about the use of green spaces around the housing, it is revealed that individuals prefer to use the parks in the area of their residence. In line with this trend, neighborhood parks emerge as common use areas where individuals meet their basic recreation needs. These areas help in socializing, and satisfying children's play needs, especially for those in low income groups.

The spread of urban green areas to the whole city in terms of quality and quantity starting from the residential environment is of great importance for the healthy development of cities.

II Factors Affecting the Recreational Area Quality in Residential Environment

Common goal for all physical, social, managerial, and economic circles is "the creation of a livable space" that can respond to individual and social needs such as shelter, security, and recreation and that is egalitarian, and that means something to individuals. For open area usage performance, overlapping of planning approach of the space, that is, the activity area, with intended use is the essential condition. [18].

Recreational use in the immediate vicinity of the residence occurs in the form of an outdoor recreation, that is, land-source use [19]. For this reason, generally being considered in a simple description, recreation approaches are scrutinized on the basis of recreational resources [20].

However, the service/product (as a benefit) produced in the outdoor recreation is being provided directly by the user/individual. In other words, production and consumption are simultaneously carried out by the user/individual [19]. This process is a complex, multi-variable equation that covers individual characteristics such as personal interest, awareness, fund of knowledge, and ability to translate the facilities offered into recreational opportunity. Different perspectives in different scales diversify that process [21]. For example, while a small puddle of pavement or a mud area in nature is a factor that needs to be taken care of for grownups, it means a playground for children.

Due to its structure mentioned, recreational sources exhibit a character that is not static and can be varied according to time (frequency of recreation use, time, duration, etc.) and space (scale, position, form, etc.) [20]. Briefly, the source of recreation has a personalized nature based on the individual evaluation of users [19].

In the quality of life and housing area satisfaction studies and for the assessment of environmental quality, the individual assessment criteria such as a risk-free living environment, residential environment that creates feeling of life not alienated from the nature, the general condition of the structures in the surrounding environment (aesthetics, form, color, intensity, etc.) are considered as indicators of spatial quality [2]. At the same time, having environmental identity, the presence of the housing immediate vicinity areas that create a sense of location, having an arrangement made aiming at socialization of individuals and development of neighborly relations, or giving place to quiet recreation and amusement areas allowing for spending time away from other users are also perceived as qualified spatial arrangements as an indicator of demand sensitivity [22,23].

On the basis of correct spatial arrangements there is demand sensitivity [23]. According to Gültekin, the user's perception of the quality of social life in the housing environment is focused on the common areas that will support and develop social life and in this perception social spaces are mostly evaluated for spending spare time [24].

According to Rapoport, the components that determine the environmental quality should be considered in two dimensions as "Social Dimension" that encompasses qualities such as social characteristics, social composition, social homogeneity, the diversity and quality of the services provided in the environment, the prestige of the environment, and neighborhood relations; and "Physical Dimension" that covers the design approach [25].

PPS (Projects for Public Space) notes that the impact of the **socialization opportunities**, in another words of the area, on socialization includes the interaction of the users, the diversity of the users, and the intensity of the friendship relations; It emphasizes that the impact level can be evaluated by street use, utilization efficiency of the areas, existence of different user groups, and studies on social connections [22].

Open space arrangements and social facilities, which allow users to participate in different levels of social life, are important in terms of being planned and designed together with the dwellings in order to create a holistic environment [26].

Although there are different groups of individuals having the ability to translate the opportunities offered into recreational opportunities or groups of different users having various demographic structure, age and sex characteristics, some common physical characteristics increase the recreational value of open spaces in the immediate vicinity of the housing. The areas that create an environment for the individual's experience in both natural and cultural environments in terms of the sources of recreation, that allow for various uses, and that are rich with regard to the presence of sustainable equipment and facilities define the areas with high recreational value [8,22,27–32]. For example, studies have shown that the spaces where sports, entertainment, and group meetings can be held for the users between the ages of 20 and 40 are especially in demand and problem-free accessibility of these places is stated as the primary criterion [24,33].

Inadequacy of sitting-resting areas, gathering areas, and multi-function recreational areas which allow for socialization is considered among the reasons for failure of open and green areas in the immediate vicinity of the dwelling [22].

Existence of the recreational equipment and facilities that we can consider among the design criteria for their intended use should be examined not only quantitatively but also qualitatively, and its suitability for its function should be questioned. Conformity to its function is the state that the function is suitable for its position and place in terms of cultural behavior patterns. Considering that the function may change over time, it will also need to be flexible for the sustainability of functional convenience [34]. Otherwise, designs that are not suitable for ergonomics, weak park entrances, visually inaccessible areas, non-functional spaces, aimless paths that do not reach the desired function area, empty and unproductive areas are formed; consequently the spatial quality value decreases [22]. At the same time, in order for the open and green areas to be functional, the amount of green space per capita, the amount and footprint amount and the capacity of the facilities and equipment should be planned to be large enough to serve the purpose of use [35].

Space quality can also be measured with **the existence of the environmental factors that threaten human health**, recreational comfort, and **environmental aesthetic values,** as well as recreational possibilities such as facility and equipment diversity [28]. In the researches, dust and smoke are considered as the most harmful environmental factors in terms of recreational comfort, and it is emphasized that measures should be taken to prevent this situation [38]. Priority should be given to wind-related spatial solutions, taking into account the health and comfort when meeting the climatic requirements. In the designs of various scales, the shade/sun conditions of the structures and the surrounding areas should be taken into consideration, and open space uses should be arranged within this framework [36].

The comfort in urban open and green spaces is evaluated with respect to security, care, and attractiveness by The project for Public Spaces group (PPS) [22]. In order to ensure a safe and livable environment around residential areas, open-green spaces and public spaces serving there must have the quality to encourage use at various times and especially at night [8,36].

When a formal classification of urban open spaces is made, areal recreation sources that are mostly in the vicinity of the housing come to the fore, just like neighborhood parks. However, connections such as linear roads, water-stream-river banks, crossings are not only for transportation purposes, but they create sources that are often used for physical activity and whose recreational values are high. In particular, to create an environment for active recreation and to encourage the community to active recreation, in an environment whose open-closed balance is offset, it is obligatory **to create a circulation that ensures continuity of movement**, and to plan and design the roads in a shared road status by reducing the traffic density in the vicinity of the housing [9,37].

Tekeli (1994) [34], defined aesthetic value (visual satisfaction) as one of the basic spatial quality criteria in a viable urban design, and emphasized the importance of the **space having been assigned a meaning, not only aesthetically** but also **by means of associative perception**, and also location quality of a design, more importantly, and its gaining identity in terms of spatial quality [34].

In his approach Rapoport deals with the aesthetics in the context of physical components that determine the quality of the environment, and notes that structural diversity and wealth, structural design styles, symbolic qualities are the factors affecting the identity and overall appearance of the environment, and environmental quality. At the same time, he evaluates landscape, housing-nature relationship, maintenance and repair, transportation, density, topography, and noise levels as elements that affect the quality within physical components as well [25].

In this context, Velioğlu and Öztürk, in general, summarize some of the criteria related to the structural design in the housing and its surroundings as follows: harmony in aesthetic values and physical items; value of use; adapting to changing requirements; strength value; reflecting structure-environment harmony; providing image continuity; value of authenticity; buildings providing a continuity and creating space (street, square, etc.); their specifying the settlement form (in terms of street-building layout, street elements, etc.); sample creation; symbolic values; and creating awareness, diversity around itself and thus being an effective item or element [38].

It is essential for the open and green areas to be positioned in accordance with the intended use to be able to take on the task of "recreative space" effectively and to be able to determine the spatial quality values [39], so they should be accessible. Accessibility of all of the individuals of the community should be ensured by safe pedestrian-friendly roads where connections to urban sub-centers by public transport systems are made, and where walking distance is observed.

III Conclusion and Recommendations

Open-green spaces in cities preferred for living due to employment and infrastructure facilities emerge as an important factor that balances human-nature relationship and improves urban living conditions thanks to recreation opportunities, etc., they offer.

When it is taken into account that the modern city system is fictionalized over residential life, it is concluded that the housing environment plays an active role in meeting the social and physical needs of individuals and shaping their manner of life.

In general, the criteria that increase the quality level of recreational areas in the immediate vicinity of the housing, in fact, constitute the basic criteria of the livability of a space.

In this process, which requires evaluation in terms of physical, social, individual, managerial, and economic different variables, with the planned distribution of adequate space intensities in urban scale, creation of public spaces where the housing and its immediate vicinity is designed with a holistic approach, which are reliable, where easy accessibility is provided through the walking distances, which allow for socialization, which offer a rich variety of recreational opportunities, which are multi-functional, which are designed according to their function, and most importantly, which in terms of both topography and living environments protect the nature in the region in which they are applied, and where health and comfort requirements fulfilled should be determined as the main goal. Addressing the residential areas together with their environment through a holistic and participatory planning approach is of great importance with regards to urban sustainability and quality of life.

Especially, in our globalized world where daily transportation takes place with increasing kilometers in most of the countries, where we are in the same environment with the people we have never met before, where we communicate in our virtual environments with the help of developing technology, dwelling's immediate vicinity and neighborhood scale increase their importance with their human values as a "social privacy zone" that slows down life for us that makes us live the moment in which individuals realize themselves, and that we need to protect sensitively.

References

1 Gül, A., Küçük, V., 2001. *Kentsel Açık-Yeşil Alanlar ve Isparta Kenti Örneğinde İrdelenmesi*, Süleyman Demirel Üniversitesi Orman Fakültesi Dergisi Seri: A, Sayı: 2, Yıl: 2001, ISSN: 1302–7085, Sayfa: 27–48.

2 Karadayı, H., 2006. *Yaşam Kalitesi ve Kullanıcı Memnuniyetinin Kentsel Tasarımdaki Etkisine Çok BoyutluYaklaşım*, İ.T.Ü.Fen Bilimleri Enstitüsü, Yüksek Lisans Tezi, İstanbul.

3 Uyanık, H.N., 2016. *Yeni Kent Kurgusunda Rekreatif Yeşil Alanlar ve Parklar Üzerine Sosyolojik Bir Araştırma*, T.C. Selçuk Üniversitesi Sosyal Bilimler Enstitüsü, Yüksek Lisans Tezi, Konya.

4 Türkoğlu, H., 1993. *İstanbul'da Konut Bölgelerinde Kullanıcıların Konut ve Konut Çevresinden Memnuniyet Derecesinin Belirlenmesi*, İstanbul Teknik Üniversitesi Araştırma Fonu, İstanbul.

5 Çerçi, S., 1997. *Konut Yakın Çevresinin Kullanıcı Bilişsel, Duygusal ve Davranışsal Parametrelere Bağlı Olarak Değerlendirilmesi*, İ.T.Ü. Fen Bilimleri Enstitüsü, Doktora Tezi, İstanbul.

6 Kellekçi, Ö.L. ve Berköz, L. 2006a. *Konut ve Çevresel Kalite Memnuniyetini Yükselten Faktörler*, İTÜ Dergisi/a, mimarlık, planlama, tasarım 5 (2): 167–178.

7 Rapoport, A., Wohlwill, J., 1985. *Thinking about Home Environments: A Conceptual Framework, Home Environments Advance in Theory and Research.* Altman, I and Werner, CM. S.255–286. Plenum Pres, New York, USA.

8 Kalyoncuoğlu B.B, Arın, Ö., Uzun, K., 2017. *Toplu Konut Çevresi Açık Alanlarda Genç ve Yetişkin Mekansal Kalite Tercihleri ve Yenilikçi Yaklaşımlar*, Tasarım+Kuram, Sayı:24, sf.11–30.

9 Kısar Koramaz, E., Türkoğlu H., 2010. *Konut Çevresi Tasarımının Sağlıklı Yaşama Etkisi*, [cited 2018, 10 October] Avaliable from: http://www.skb.gov.tr/konut-cevresi-tasarimlarinin-sagli-s326k/

10 Yüksel, D.Y., Aydınlı, S., Pulat, G., 1996. *Toplu Konut Uygulamalarında Niteliksel Değerleri Araştırmaya Yönelik Bir Çevre Analizi*, Cilt-1, Konut Araştırmaları Dizisi 4, Toplu Konut İdaresi Başkanlığı, ODTÜ Basım İşliği, Ankara. ISBN 975-19-1330-6.

11 Akşit, N., 1994. *Toplu Konut Planlamasında Konut Dışı Alan Kullanımı ve Standartları Üzerine Bir Araştırma.* Basılmamış Yüksek Lisans Tezi, MSGSÜ, Fen Bilimleri Enstitüsü, İstanbul.

12 Erdoğan, Ö., 2009. *Toplu Konut Uygulamalarının Dış Mekan Tasarım İlkeleri Açısından İrdelenmesi: Adana Seyhan ve Yüreğir İlçeleri Örneği*, Çukurova Üniversitesi Fen Bilimleri Enstitüsü, Adana.

13 Ardahan, F., 2013. *Komşuluk, farklılığa hoşgörü ve yaşam doyumu arasındaki ilişki, ilişkinin güçlendirilmesinde rekreatif etkinliklerin önemi*, International Journal of Human Sciences, Volume:10, Issue:1, ISSN: 1303-5134.

14 Karaküçük, S., 2014. *Rekreasyon: Boş Zaman Değerlendirme*, Gazi Üniversitesi, Ankara.

15 Şahin, Ş., M., Barış, 1998. *Kentsel Doku İçerisinde Açık ve Yeşil Alan Standartlarını Belirleyen Etmenler*, Peyzaj Mimarlığı Dergisi, s.10, İstanbul.

16 Kemp, I., K., Leidelmeijer, G., Marsman, A. Hollander, 2003. *Urban Environmental Quality and Human Well-Being Towards a Conceptual Framework and Demarcation of Concepts*, A Literature Study, Landscape and Urban Planning, 65, s.5–18.

17 Onsekiz, D., 2003. *Kentsel Ölçekte Eğlence Mekanları İle Konut Alanları İlişkisi – Ankara Kenti Örneği*, Gazi Üniversitesi, Fen Bilimleri Enstitüsü, Ankara.

18 Gökmen, P.G., V. Ok, A. Özsoy, N. Esin, 2005. *Toplu Konutlarda Açık Mekan Yaşantısı ve Kalite: Ataköy İçin Bir Yeniden Değerlendirme*, Konut Değerlendirme Sempozyumu, İ.T.Ü. Mimarlık Fakültesi Baskı Atölyesi, İstanbul. ISBN: 975 – 561- 263 – 7. S: 61–69.

19 Akesen, A. 1978: *Türkiye'de Ulusal Parkların Açıkhava Yönünden Nitelikleri ve Sorunları*, İstanbul Üniversitesi Orman Fakültesi Yayınları, İ.Ü. Yayın No: 2484

20 Hall, C.M. & Page, S.J., 2014. *The Geography of Tourism and Recreation, Environment, Place and Space*, Fourth edition, ISBN: 978-0-415-83398-1, NewYork.

21 Bell, S. 1997. *Design For Outdoor Recreation*, ISBN:0419 20350 8, London.

22 Projects for Public Spaces Inc., 2000. *How to Turn a Place Around.* A Handbook for Creating Successful Public Spaces, New York.

23 Yücel, A., 1996. *İstanbul'da 19.yüzyılın Konut Biçimleri, Tarihten Günümüze Anadolu'da Konut ve Yerleşme*, Tarih Vakfı Yayınları, Türkiye

24 Gültekin, A.T. 2015. *Konut Kullanıcılarının Kentsel ve Sosyal Yaşam Kalitesi Farkındalığı: Ankara/Çayyolu-Prof. Dr. Ahmet Taner Kışlalı Mahallesi Örneği*, Ankara Araştırmaları Dergisi (Journal of Ankara Studies), 3 (2): 165–182.

25 Rapoport, A., 1977. *Human Aspects of Urban Form*, Pergamon Press, Oxford, England.

26 Bakan, K., Konuk, G., 1987. *Türkiye'de Kentsel Dış Mekanların Düzenlenmesi*, Tübitak Fotoğraf Klişe Laboratuarı ve Ofset Tesisleri.

27 Ergen, Y.B.,1989. *Yüksek Bina ile Açık Mekan İlişkisinin Kent Planlamasına Yansıyışı*, Yüksek Binalar I.Ulusal Sempozyumu, İ.T.Ü. Mimarlık Fakültesi.

28 Sadıkoğlu, H., Özsoy, A., 2017. *Modern Dönem Konut Stokunda Mekansal Kalitenin İncelenmesi Üzerine Bir Araştırma*, Megaron 2017; 12 (2): 203–212 DOI: 10.5505/megaron.2017.93823

29 Ibes, D.C., 2015. *A Multi-Dimensional Classification and Equity Analysis of an Urban Park System: A Novel Methodology and Case Study Application*, Landscape and Urban Planning, 122–137.

30 Bedimo-Rung, A.L., et al., 2005. *The Significance of Parks to Physical Activity and Public Health: A Conceptual Model*, American Journal of Preventine Medicine; 28 (2S2): 159–168, Elsevier.

31 Koohsari, M.J., et al., 2015. *Public Open Space, Physical Activity, Urban Design and Public Health: Concepts, Methods and Research Agenda*, Health & Place Volume 33, pp.75–82.

32 Kaczynski, A.T., et al. 2008. *Association of Park Size, Distance, and Features with Physical Activity in Neighborhood Parks*, American Journal of Public Health, 98(8).

33 White, R. 2001. *Youth Participation in Designing Public Spaces*, Youth Studies Australia, 20 (1): 18.

34 Tekeli, İ., 1994. *Bir Kentsel Tasarım Kuramının Geliştirilmesi Üzerine Düşünceler*, Kent Planlama, Politika ve Sanat, ODTÜ Mimarlık Fakültesi Yayını, s.591–617.

35 Etli, B., 2002. *Edirne İli Merkez İlçe Yeşil Alan Sisteminin Peyzaj Mimarlığı İlkeleri Yönünden İrdelenmesi*, Trakya Üniversitesi, Bilimsel Araştırmalar Dergisi, B Serisi, Cilt 3, No.1, s.47–59.

36 Gökmen, P.G., V. Ok, A. Özsoy, N. Esin, 2005. *Toplu Konutlarda Açık Mekan Yaşantısı ve Kalite: Ataköy İçin Bir Yeniden Değerlendirme*, Konut Değerlendirme Sempozyumu, İ.T.Ü. Mimarlık Fakültesi Baskı Atölyesi, İstanbul. ISBN: 975 – 561- 263 – 7. S: 61–69.

37 Kocabay, Z. ve Bilgen, P., 1995. *Nitelikli Çevreleri Elde Edilme Sürecinde Bir Model: Toplu Konut İdaresi Uygulamaları*, Mimari ve Kentsel Çevrede Kalite Arayışları Sempozyumu, s.216–222, Taşkışla, İstanbul.

38 Velioğlu, A., Öztürk, K., 1993. *Tarihi Çevrelerde Mimari Değerler ve Tasarım Üzerine Bazı Ölçütler*, Yapı Dergisi, Mayıs, 138, S:38.

39 Aydemir, S.E., 2004. *Kentsel Donatılar – Açık Alanlar ve Rekreasyon, Kentsel Yeşil Alanlar*, Kentsel Alanların Planlanması ve Tasarımı, Bölüm:12, sf.285–337, Akademi Yayınevi, Trabzon.

Prof. Dr. Bülent Cengiz and Oğuz Erdi Yakan

9 Visual Landscape Assessment in Güzelcehisar Village-Bartin (Western Black Sea Region of Turkey)

Abstract: Coastal areas are sensitive landscapes due to their cultural values and unique ecological characteristics. Güzelcehisar, a rural settlement along the Black Sea coast, which is part of the central district of the Bartın province has 80-million-year-old unique geomorphological formations known as "Lava Columns" along its coastline due to its volcanic structure. The diameters of these Lava Columns vary between 50–100 cm and their heights exceed 30 meters. The Lava Columns at Giant's Causeway and Causeway Coast World Heritage Site in Northern Ireland and Fingal's Cave in Scotland and Devils Postpile National Monument in California, USA, are international centers of tourism in the world. The Lava Columns at Güzelcehisar in Turkey are among the rare natural formations with national and international tourism potential. A total of six different visual landscape areas were defined as coastal, archaeologic, forest, agricultural, rural and lava columns. A total of nine parameters were selected as visual quality parameters, which are naturalness, coherence, upkeep, clarity, vista characteristic, diversity, complexity, landform and safety. Visual landscape assessment was carried out with 35 experts. A total of 48 photos from Güzelcehisar Village were used in the survey with 8 for each of the six visual landscape areas. The experts carried out assessments according to the visual quality parameters for each visual landscape area. SPSS 16.0 software was used for survey assessment. The relationships between the parameters used for evaluating the visual landscape area were put forth via Spearman correlation analysis. As a result, suggestions were developed for contributing to the sustainable development of rural landscapes in coastal settlement areas.

Keywords: Visual landscape, Güzelcehisar, Lava columns, Coastal village, Survey

1 Introduction

Coastlines as the intersection points of land and water ecosystems have significant ecological importance [1–4]. In addition to their ecological characteristics, coastal areas also have high landscape values that enable multidimensional spatial analyses for tourism and recreational use [2,5] and cultural use [2,4,6,7]. Coastal areas are dynamic formations that vary subject to the quality and intensity of the human activities that they are involved in. The expectations of humans as one of the most important factors shaping the landscape with an ever increasing population on coastal resources have varied rapidly which has led to an increased utilization pressure on unit area. In this regard, different area use types resulting from cultural activities at coastal areas make up the coastal landscape [4,7]. Landscape characteristic is put forth by whether that specific landscape with its various elements or units differs either in a good or bad way from the other landscapes or whether it has a perceivably varying or dominant characteristic [8]. According to Kaplan and Kaplan [9] and Müderrisoğlu and Eroğlu [10] visual landscape is an aesthetic product generated in the mind by human perception shaped via human psychology when faced with a natural and cultural landscape. At this point, the concept of aesthetics for the individual emerges as a manifestation of the mental and spiritual structures [4, 11]. The concept of visual quality plays an important role in the planning and design stages for defining the elements that make up the landscape or their structure. Hence, determining the visual quality in an area is very important for determining the potential related with the already existing landscape structure, as well as the potential of the elements and components that make up the landscape [8]. Measurement of visual quality takes place via appraisal of a source. The aesthetic and visual characteristics of a place or route emerge when it is evaluated in a functional relationship with all sensory organs and primarily the eyes [12].

The geological and cultural importance of coastal areas, their geomorphology and landscape have led to the emergence of various national and international references/symbols subject to their relevant habitats. In this regard, it is important for Heritage Coasts to understand the concepts of geodiversity and geoconservation [4, 13]. According to Williams et al. [14]; 34% of the already existing coastline at England and Wales is protected under the Heritage Coast Programme with the main principles of:

i. Preserving natural beauty dependent on the natural landscape instead of man-made activities and encouraging entertainment activities, enabling the sustainable use of coastlines for public entertainment and recreation,
ii. Protecting/improving the coastline and encouraging a preservation consciousness,
iii. Enabling and improving public participation,
iv. Defining the coastlines [4, 13].

Güzelcehisar, selected as the study area, has 80-million-year-old unique geomorphological formations called "Lava Columns" due to its volcanic structure. The diameters of these lava columns vary between 50–100 cm and their heights exceed 30 meters. The lava columns at Giant's Causeway and Causeway Coast World Heritage Site in Northern Ireland and Fingal's Cave in Scotland and Devils Postpile National Monument in California, USA, are international centers of tourism in the world. The lava columns at Güzelcehisar in Turkey are among the rare natural formations with national and international tourism potential. Visual landscape assessments via experts were carried out in this study in accordance with the visual quality criteria for the visual landscape areas determined in Güzelcehisar. As a result, suggestions were developed for contributing to the sustainable development of rural landscapes in coastal settlement areas.

2 Material

The study area is located in the Güzelcehisar Village part of the central district of Bartın province. Güzelcehisar Village is located to the west of the Bartın city center and is 17 km away from the city center.

Reasons for selecting Güzelcehisar Village as the study area were as follows:

- Its unique natural and cultural landscape values,
- Its coastal settlement with rural characteristics which has significant tourism potential,
- Unique 80-million-year-old geomorphological formations which can be considered as natural monuments,
- The fact that it is a preservation priority site, since it includes strictly protected areas and first degree archaeological conservation sites,
- It has different landscape areas,
- As of 14.08.2017, it has been declared as the 112nd natural monument in our country as Bartın Güzelcehisar Basalt Columns,
- Its flora and fauna,
- It is one of the rare areas in the Bartın in the Black Sea climate where citrus fruits can be cultivated due to its microclimatic characteristics,
- Important recreational activities (tent camping, trekking, swimming, photography, and fishing activities, etc.).

3 Method

The purpose of this chapter was to contribute to the sustainable development of rural landscapes via visual landscape assessment method in coastal areas and to put forth the visual landscape potential of Güzelcehisar on the basis of its various landscape value. For this purpose, the study method is comprised of two main stages. These are:

1st stage: Defining the visual quality parameters and visual landscape areas: Studies on visual quality and assessment [15–20] were used for determining the parameters and a total of nine parameters were selected as naturalness, coherence, upkeep, clarity, vista characteristic, diversity, complexity, landform and safety.

Scientific studies by Cengiz et al. [16], Cengiz and Cengiz [13] and Cengiz et al. [4] were used for the visual landscape assessment of Güzelcehisar and a total of six different landscape areas were determined in the light of field studies. These are: Coastal Landscape, Archeological Landscape, Forest Landscape, Agricultural Landscape, Rural Landscape and Lava Columns Landscape.

2nd stage: Expert Opinion Evaluation: A total of about 1000 photographs were obtained during field studies for the six different visual landscapes of the Güzelcehisar coastal settlement. The photos used in the survey were determined by a panel of five experts. These photos were selected to reflect the characteristics of each visual landscape area. A total of 48 photos were used in the survey form with 8 photos for each of the six visual landscape areas. The surveys were applied via e-mail on 35 experts (landscape architect, urban and regional planning, architect, civil engineer, survey engineer, environmental engineer, forest engineer, agriculture engineer, archeologist, and economist). The survey was comprised of questions on the demographic characteristics of the experts such as gender, age, level of education, profession, occupation, level of income, and was designed so as to question nine theoretical parameters of naturalness, coherence, upkeep, clarity, vista characteristic, diversity, complexity, landform and safety from the photos of each visual landscape area. The experts carried out an evaluation at a 5-point Likert scale taking into consideration all parameters of the visual landscape area. The experts were asked to sort the 5 choices from 1 to 5 according to the level of importance for each landscape area. The experts were asked to interpret the adjective pairs determined via semantic differential method for each of the areas by way of the photographs presented. SPSS 16.0 software was used for survey evaluation. The relationship between the parameters used for evaluating the visual landscape area was put forth via Spearman correlation analysis.

4 Results and Discussions

The demographic characteristics of the experts who took part in the survey have been given below (Tab. 1). Of the experts who participated in the survey, 37.1% (13 people) were female, 62.9 % (22 people) were male.

The experts who took part in the survey were asked, "What are the recreational activities that may stand out in Güzelcehisar?" to which they responded as "Visiting Natural/cultural values" as the first preference with a ratio of 54.3 %, "Swimming" as the second preference with a ratio of 31.4 %and "Nature observation" as the third preference with a ratio of 20% (Tab. 2).

The experts were also asked, "What are the most important landscape characteristics of Güzelcehisar?" to which they responded as "Lava columns landscape" as the first preference with a ratio of 51.4 % (Tab. 3).

- **Expert Evaluations on Coastal Landscape**

According to the experts who took part in the survey, the "Vista Characteristic" parameter stands out in the evaluation of Güzelcehisar Coastal Landscape according to the parameters provided (Tab. 4).

Evaluations were made for the top three photos of the coastal landscape in the study area (Fig. 1).

Spearman Correlation Analysis Results for the Coastal Landscape Parameters

It was determined that there is a positive and moderate relation between naturalness and coherence ($p<0.01$, $r=0.558$). It was observed that the natural vista, the vista characteristics and the landscape characteristics it encompasses were included as a whole. Visual coherence increases as the naturalness parameter of the vista increases.

Positive and moderate relations were determined between coherence and clarity ($p<0.01$, $r=0.433$), vista characteristic ($p<0.01$, $r=0.515$) and diversity ($p<0.01$,

Tab. 1: Demographic Characteristics of the Experts

Individual characteristics		Number (N)	Percent (%)
Gender	Female	13	37,1
	Male	22	62,9
Age	21–29	7	20
	30–39	12	34,3
	40–49	12	34,3
	50–59	4	11,4
Level of Education	Undergraduate	8	22,8
	Graduate	8	22,8
	Doctorate	19	54,4
Profession	Landscape Architect	15	42,9
	Urban and Regional Planner	2	5,7
	Architect	3	8,6
	Civil Engineer	2	5,7
	Survey Engineer	2	5,7
	Environmental Engineer	1	2,9
	Forest Engineer	2	5,7
	Agricultural Engineer	2	5,7
	Archeologist	4	11,4
	Economist	2	5,7
Occupation	State Employee	7	20
	Private	9	25,7
	Academician	19	54,3
Monthly Income	1500–3000 TL	5	14,3
	3000–5000 TL	7	20
	5000–10000 TL	18	51,4
	10000+ TL	5	14,3
Knowledge of Güzelcehisar	Yes	35	100
	No	-	-

Tab. 2: Recreational Activities That May Stand Out in Güzelcehisar Based on Expert Opinion

What are the recreational activities that may stand out in Güzelcehisar?	1st Preference		2nd Preference		3rd Preference	
	Number (N)	Percentage (%)	Number (N)	Percentage (%)	Number (N)	Percentage (%)
Visiting natural/ cultural values	19	54.3	5	14.3	5	14.3
Scientific trips	1	2.9	3	8.6	2	5.7
Camping	-	-	4	11.4	3	8.6
Nature observation	-	-	4	11.4	7	20.00
Fishing activities	-	-	2	5.7	2	5.7
Photography	3	8.6	3	8.6	4	11.4
Swimming	10	28.5	11	31.4	3	8.6
Trekking	2	5.7	-	-	4	11.4
Picnic	-	-	3	8.6	5	14.3
Total	**35**	**100**	**35**	**100**	**35**	**100**

Tab. 3: The Most Important Landscape Characteristics of Güzelcehisar Based on Expert Opinion

What are the most important landscape characteristics of Güzelcehisar?	1st Preference		2nd Preference		3rd Preference	
	Number (N)	Percentage (%)	Number (N)	Number (N)	Percentage (%)	Number (N)
Coastal landscape	16	45.7	**15**	**42.8**	4	11.4
Archeological landscape	1	2.9	5	14.3	**13**	**37.1**
Forest landscape	-	-	3	8.6	8	22.9
Agricultural landscape	-	-	1	2.9	1	2.9
Rural landscape	-	-	1	2.9	5	14.3
Lava columns landscape	**18**	**51.4**	10	28.5	4	11.4
Total	**35**	**100**	**35**	**100**	**35**	**100**

Tab. 4: Results for the Evaluation of the Landscape Parameters of the Coastal Landscape

Coastal Landscape			
Parameters	**Average Score**	**Percent (%)**	**Parameter Characteristics**
Naturalness	3.9	11.2	Natural
Coherence	4.0	11.5	Coherent
Upkeep	3.7	10.6	Well-kept
Clarity	3.7	10.6	Clear
Vista Characteristic	4.6	13.1	Very good
Diversity	3.8	10.9	Various
Complexity	3.8	10.9	Less complex
Landform	3.7	10.6	Partly uneven
Safety	3.7	10.6	Safe
Total	**34.9**	**100**	

Fig. 1: Top 3 photos of the coastal landscape.

r=0.472). Level of visual coherence increases when the images of the site are not complex, when the vista is beautiful and when it contains many different landscape elements.

A positive and moderate relationship was determined between clarity and complexity ($p<0.01$, $r=0.437$) and a positive and weak relationship was determined between clarity and landform ($p<0.05$, $r=0.361$). It was observed that there are more than one different landscape elements in the images of the study area and that the landscape elements are perceived in a regular manner in the three-dimensional site. The facts that the landscape elements and three-dimensional structure are easily perceived in the images increases the clarity parameter level.

A positive and moderate relationship was determined between vista characteristic and; diversity ($p<0.01$, $r=0.555$). The fact that there are many different elements in the images increases the vista characteristic (Tab. 5).

Tab. 5: Correlation between the Parameters Related with the Coastal Landscape

	Naturalness	**Coherence**	**Clarity**	**Vista Characteristic**
Coherence	0.558**			
Clarity		0.433**		
Vista characteristic		0.515**		
Diversity		0.472**		0.555**
Complexity			0.437**	
Landform			0.361*	

**p<0.01 * p<0.05

➢ Expert Evaluations on Archeological Landscape

It was determined that the landscape characteristic is "very natural" with a highest score of 4.5 for the "Naturalness" parameter as a result of the evaluations of Güzelcehisar Archeological landscape by the experts who participated in the survey based on the provided parameters (Tab. 6).

Evaluations were made regarding the top three photos of the archeological landscape in the study area (Fig. 2).

Spearman Correlation Analysis for the Parameters Related with Archeological Landscape

A positive and moderate relationship was observed between naturalness and coherence ($p<0.01$, $r=0.530$), a positive and moderate correlation was determined between naturalness and vista characteristic ($p<0.01$, $r=0.525$) and a negative and weak relationship was determined between naturalness and landform ($p<0.05$, $r=-0.378$). The fact that the remaining landscape elements are included as a whole in the images and that the images have a beautiful vista increase the naturalness of the area.

A positive and strong relationship was observed between coherence and clarity ($p<0.01$, $r=0.671$), a positive and moderate relationship was observed between coherence and vista characteristic ($p<0.01$, $r=0.565$), a positive and weak relationship was observed between coherence and complexity ($p<0.05$, $r=0.394$) and a negative and moderate relationship was observed between coherence and landform ($p<0.01$, $r=-0.585$). The fact that there is a high level of foreign elements in the image which are integrated with the landscape elements in the background has resulted in an increased correlation between coherence and clarity, since it makes it easier to interpret the image. Vista characteristic of the images and low complexity increases the level of coherence. The fact that the landscape elements in

Tab. 6: Evaluations for the Landscape Parameters Related with the Archeological Landscape

Archeological Landscape			
Parameters	**Average (1–5 points)**	**Percent (%)**	**Parameter Characteristic**
Naturalness	4.5	13.9	Very natural
Coherence	4.1	12.7	Coherent
Upkeep	2.7	8.4	Moderate
Clarity	3.7	11.5	Clear
Vista Characteristic	4.4	13.6	Good
Diversity	3.6	11.1	Various
Complexity	3.5	10.8	Less complex
Landform	2.7	8.4	Uneven
Safety	3.1	9.6	Moderate
Total	**32.3**	**100**	

Fig. 2: Top 3 photos of the archeological landscape and preference ratios.

the background are irregular and the existence of foreign elements resulted in a converse relationship between coherence and landform. Irregular perception of landform decreases the level of coherence.

Positive and moderate correlations were determined between clarity and vista characteristic ($p<0.01$, $r=0.499$), as well as between diversity ($p<0.01$, $r=0.579$) and complexity ($p<0.01$, $r=0.434$). The fact that the images have a vista characteristic, that they include many different elements and the lack of a complex three-dimensional structure make it easier to interpret the images.

Positive and moderate correlations were determined between vista characteristic and diversity ($p<0.01$, $r=0.440$) and between vista characteristic and complexity ($p<0.01$, $r=0.478$), whereas a negative and moderate correlation was

Tab. 7: Correlation between Archeological Landscape-Related Parameters

	Naturalness	**Coherence**	**Upkeep**	**Clarity**	**Vista Characteristic**	**Diversity**
Coherence	0.530**					
Clarity		0.671**	0.427*			
Vista Characteristic	0.525**	0.565**	0.428*	0.499**		
Diversity			0.451**	0.579**	0.440**	
Complexity		0.394*		0.434**	0.478**	
Landform	-0.378*	-0.585**			-0.428*	-0.336*
Safety			0.398**			0.404*

**p<0.01 * p<0.05

determined between vista characteristic and landform ($p<0.05$, $r=-0.428$). The fact that the images contain many different elements and the lack of a three-dimensional complex structure play a role on the beauty of the vista. The level of vista characteristic increases with increasing diversity and complexity. Irregular structure of landscape elements in the images has resulted in a converse relationship between vista characteristic and landform. Failure to perceive the landform properly decreases the vista characteristic.

A negative and weak correlation was observed between diversity and landform ($p<0.05$, $r=-0.336$), whereas positive and moderate correlations were determined between diversity and safety ($p<0.05$, $r=0.404$). Irregular landscape elements have resulted in the failure to perceive many different elements in the images. Failure to properly perceive the landform decreases the diversity parameter. Proper perception of the diversity parameter in the images increases the safety parameter as well (Tab. 7).

➢ Expert Evaluations on Forest Landscape

"Naturalness" parameter stood out in the evaluation of the forest landscape subject to the parameters. It was determined that the landscape characteristic is "natural" (Tab. 8).

Evaluations were made regarding the top three photos of the forest landscape (Fig. 3).

Spearman Correlation Analysis for the Parameters Related with Forest Landscape

Positive and moderate correlations were determined between naturalness and coherence ($p<0.01$, $r=0.457$), as well as between naturalness and diversity ($p<0.01$,

Tab. 8: Evaluation of the Landscape Parameters Related with Forest Landscape

Forest Landscape			
Parameters	**Average (1–5 p.)**	**Percent (%)**	**Parameter Characteristics**
Naturalness	4.3	13.4	Natural
Coherence	4.0	12.5	Coherent
Upkeep	3.4	10.7	Moderate
Clarity	3.1	9.7	Balanced
Vista Characteristic	4.2	13.1	Good
Diversity	4.1	12.8	Various
Complexity	3.3	10.3	Balanced
Landform	2.5	7.8	Rough
Safety	3.1	9.7	Moderate
Total	**32**	**100**	

Fig. 3: Top 3 photos of the forest landscape and preference ratios.

r=0.449). The naturalness of the area increases when the landscape elements in the background are located as a whole and when there are many different elements in the images.

Positive and moderate correlations were determined between coherence and upkeep (p<0.01, r=0.445), as well as between coherence and vista characteristic (p<0.01, r=0.499). The level of coherence increases when the elements that make up the image are well kept, when the images have a vista characteristic and when these two parameters are well integrated.

Positive and weak correlations were determined between upkeep and vista characteristic (p<0.05, r=0.387), as well as between upkeep and complexity (p<0.05, r=0.355), while a positive and moderate correlation (p<0.01, r=0.431) was determined between upkeep and safety. The level of upkeep increases when the images have a vista characteristic, when the three-dimensional structure is not complex and when the images present a safe appearance.

A positive and moderate correlation was determined between clarity and safety ($p<0.01$, $r=0.500$). Since there are no confusing elements in the images, they pose no threats with regard to safety. The clarity parameter level increases with increasing safety parameter level.

A positive and weak correlation was determined between the vista characteristic and diversity ($p<0.05$, $r=0.398$). The vista characteristic is affected when there are many different elements in the images. The vista characteristic parameter level increases with increasing diversity parameter level.

A negative and moderate correlation was determined between diversity and landform ($p<0.01$, $r=-0.468$). While diversity is low in flat areas, it was observed that diversity increases at mountainous areas.

A positive and weak correlation was determined between complexity and safety ($p<0.05$, $r=0.362$). The fact that the three-dimensional structure in the images is simple and clear indicates that the area poses no threats (Tab. 9).

➢ Expert Evaluations on Agricultural Landscape

"Diversity" stood out in the evaluation of the Güzelcehisar Agricultural landscape subject to the parameters. It was determined that the landscape has a "diverse" characteristic (Tab. 10).

Evaluations were made regarding the top three photos of the agricultural landscape in the study area (Fig. 4).

Spearman Correlation Analysis for the Parameters Related with Agricultural Landscape

A positive and high correlation was observed between naturalness and coherence ($p<0.01$, $r=0.635$) and a positive and moderate correlation was observed between naturalness and vista characteristic ($p<0.05$, $r=0.403$). It was determined that the already existing scenery was complete with vista characteristics, as well as landscape elements. The increase in the level of the coherence parameter highly increases the level of the naturalness parameter. The increase in the level of the vista characteristic parameter moderately increases the naturalness parameter.

Positive and moderate correlations were observed between coherence and upkeep ($p<0.01$, $r=0.421$), clarity ($p<0.01$, $r=0.452$), vista characteristic ($p<0.01$, $r=0.539$) and complexity ($p<0.01$, $r=0.432$); whereas positive and high correlations were determined between coherence and safety ($p<0.01$, $r=0.680$). The facts that the elements which comprise the visual are well-kept, that the images are not complex, that there is a scenic beauty and that the three-dimensional structure is understandable without any threat affect the coherence parameter.

Positive and moderate correlations were observed between upkeep and clarity ($p<0.01$, $r=0.490$), vista characteristic ($p<0.,01$, $r=0.448$) and safety ($p<0.01$, $r=0.506$), whereas a positive and weak correlation was observed between upkeep

Tab. 9: Correlation between the Parameters Related with Forest Landscape

	Naturalness	**Coherence**	**Upkeep**	**Clarity**	**Vista Characteristic**	**Diversity**	**Complexity**
Coherence	0.457**						
Upkeep		0.445**					
Vista Characteristic		0.499**	0.387*				
Diversity	0.449**				0.398*		
Complexity			0.355*				
Landform						-0.468**	
Safety			0.431**	0.500**			0.362*

**p<0.01 *p<0.05

Tab. 10: Evaluation of the Landscape Parameters Related with Agricultural Landscape

Agricultural Landscape			
Parameters	**Average (1–5 p.)**	**Percent (%)**	**Parameter Characteristics**
Naturalness	3.5	10.9	Natural
Coherence	3.7	11.5	Coherent
Upkeep	3.5	10.9	Moderate
Clarity	3.6	11.2	Balanced
Vista Characteristic	3.7	11.5	Good
Diversity	3.8	11.8	Various
Complexity	3.4	10.5	Balanced
Landform	3.3	10.2	Rough
Safety	3.7	11.5	Moderate
Total	**32.2**	**100**	

Fig. 4: Top 3 photos of the agricultural landscape and preference ratios.

and complexity ($p<0.05$, $r=0.357$). The facts that the images do not have complex elements, that there is a scenic beauty and that the three-dimensional structure is not complex affect the upkeep parameter.

Positive and moderate correlations were observed between clarity and the vista characteristic ($p<0.01$, $r=0.533$) and complexity ($p<0.01$, $r=0.520$), whereas positive and weak correlations were observed between clarity and landform ($p<0.05$, $r=0.353$) and safety ($p<0.05$, $r=0.350$). The facts that the images have a scenic beauty, that the three-dimensional structure is not complex, that the landscape elements are perceived regularly and that they do not pose any threats affect the clarity parameter.

Positive and weak correlations were observed between the vista characteristic and; diversity ($p<0.05$, $r=0.357$), complexity ($p<0.05$, $r=0.354$) and safety ($p<0.05$,

Tab. 11: Correlation between Agricultural Landscape-Related Parameters

	Naturalness	**Coherence**	**Upkeep**	**Clarity**	**Vista Characteristic**	**Diversity**
Coherence	0.635**					
Upkeep		0.421**				
Vista characteristic		0.452**	0.490**			
Diversity	0.403*	0.559**	0.448**	0.533**		
Complexity					0.357*	
Landform		0.432**	0.357*	0.520**	0.354*	
Safety				0.353*		0.360*
Coherence		0.680**	0.506**	0.350*	0.428*	

**p<0.01 * p<0.05

r=0.428). The facts that the images have many different elements, that the three-dimensional structure is not complex and that it does not pose any threats affect vista characteristic.

A positive and weak correlation was observed between complexity and landform (p<0.05, r=0.360). The facts that the landscape elements are observed regularly and that the three-dimensional structure is perceived comfortably do not cause any complexities (Tab. 11).

➢ **Expert Evaluations on Rural Landscape**

"Diversity" stood out in the evaluation of the Güzelcehisar Rural landscape subject to the parameters (Tab. 12).

Evaluations were made regarding the top three photos of the rural landscape in the study area (Fig. 5).

Spearman Correlation Analysis for the Parameters Related with Rural Landscape

A positive and high correlation was observed between naturalness and coherence (p<0.01, r=0.604); a positive and moderate correlation was observed between naturalness and upkeep (p<0.01, r=0.485), whereas a positive and weak correlation was observed between naturalness and clarity (p<0.05, r=0.392). The facts that there are foreign elements integrated with the landscape elements in the background, that it is well-kept and that there are no complex elements when evaluating the images affect the natural vista characteristics. An increase in the coherence parameter level highly increases the naturalness parameter. An increase in the upkeep

Tab. 12: Evaluation of the Landscape Parameters Related with Rural Landscape

Rural Landscape			
Parameters	**Parameters**	**Parameters**	**Parameter Characteristics**
Naturalness	3.3	10.4	Natural
Coherence	3.5	11	Coherent
Upkeep	3.4	10.7	Moderate
Clarity	3.5	11	Balanced
Vista Characteristic	3.8	12	Good
Diversity	3.9	12.4	Various
Complexity	3.3	10.4	Balanced
Landform	3.3	10.4	Rough
Safety	3.7	11.7	Moderate
Total	**31.7**	**100**	

Fig. 5: Top 3 photos of the rural landscape and preference ratios.

parameter level moderately increases the naturalness parameter. An increase in the clarity parameter weakly increases the naturalness parameter.

Positive and moderate correlations were observed between coherence and upkeep ($p<0.01$, $r=0.507$), complexity ($p<0.01$, $r=0.511$) and safety ($p<0.01$, $r=0.469$), whereas a positive and weak correlation was observed between coherence and clarity ($p<0.05$, $r=0.375$). The facts that the elements that make up the image are well-kept, that the three-dimensional structure can be easily perceived, that it poses no threats and that there are no complex elements when evaluating the images constitute a whole thus affecting the coherence parameter. An increase in the upkeep, complexity and safety parameter levels moderately increases the coherence parameter. An increase in the clarity parameter level weakly increases the coherence parameter.

Tab. 13: Correlation between the Parameters Related with Rural Landscape

	Naturalness	**Coherence**	**Upkeep**	**Clarity**	**Vista Characteristic**	**Diversity**
Coherence	0.604**					
Upkeep	0.485**	0.507**				
Clarity	0.392*	0.375*				
Vista Characteristic			0.462**	0.442**		
Diversity					0.418*	
Complexity		0.511**		0.359*	0.339*	
Land Form				0.364*		-0.367*
Safety		0.469**	0.502**			

**p<0,01 * p<0,05

A positive and moderate correlation was observed between clarity and vista characteristic ($p<0.01$, $r=0.442$); whereas positive and weak correlations were observed between clarity and complexity ($p<0.05$, $r=0.359$) and landform ($p<0.05$, $r=0.364$). The facts that the images have a scenic beauty, that the three-dimensional structure is comprehensible and that the landscape elements are perceived in an orderly manner do not result in any complexities when evaluating the image.

A positive and moderate correlation was observed between the vista characteristic and diversity ($p<0.05$, $r=0.418$), whereas a positive and weak correlation was observed between the vista characteristic and complexity ($p<0.05$, $r=0.339$). The facts that the images have many different elements and that the three-dimensional structure is easily perceived affects the vista characteristic. An increase in the diversity parameter level moderately increases the vista characteristic parameter. An increase in the complexity parameter level weakly increases the vista characteristic parameter.

A negative and weak correlation was observed between diversity and landform ($p<0.05$, $r=0.367$). The facts that the landscape elements are perceived in an orderly manner play a role in easily perceiving the diversity of the different elements in the images (Tab. 13).

➤ Expert Evaluations on Lava Columns Landscape

"Naturalness" stood out in the evaluation of the Güzelcehisar rural landscape subject to the lava column landscape parameters and it was determined that the landscape characteristic is "very natural" (Tab. 14).

Evaluations were made regarding the top three photos of the lava columns landscape in the study area (Fig. 6).

Tab. 14: Evaluation of the Landscape Parameters Related with Lava Columns Landscape

Lava Columns Landscape			
Parameters	**Parameters**	**Parameters**	**Parameter Characteristics**
Naturalness	4.9	14.5	Natural
Coherence	4.5	13.3	Coherent
Upkeep	3.4	10	Moderate
Clarity	3.9	11.5	Balanced
Vista Characteristic	4.7	13.9	Good
Diversity	4.0	11.8	Various
Complexity	4.0	11.8	Balanced
Landform	2.0	5.9	Rough
Safety	2.5	7.3	Moderate
Total	**33.9**	**100**	

Fig. 6: Top 3 photos of the lava columns landscape and preference ratios.

Spearman Correlation Analysis for the Parameters Related with Lava Columns Landscape

Positive and moderate correlations were observed between coherence and upkeep ($p<0.01$, $r=0.478$), vista characteristic ($p<0.01$, $r=0.432$) and complexity ($p<0.01$, $r=0.458$), whereas a positive and weak correlation was observed between coherence and clarity ($p<0,05$, $r=0,384$). The facts that the elements which make up the visual are well-kept, that they have a scenic beauty, that the three dimensional structure can be easily perceived and that there are no complex elements when evaluating the images play a role on the coherence between all these landscape elements.

A positive and high correlation was observed between upkeep and complexity ($p<0,01$, $r=0,630$); whereas positive and weak correlations were observed between

Tab. 15: Evaluation of the Landscape Parameters Related with Lava Columns Landscape

	Coherence	**Upkeep**	**Clarity**	**Vista Characteristic**	**Landform**
Coherence					
Upkeep	0.478**				
Clarity	0.384*	0.377*			
Vista Characteristic	0.432**	0.339*	0.454**		
Diversity			0.472**	0.521**	
Complexity	0.458**	0.630**	0.384*	0.547**	
Land Form					
Safety		0.361*			0.592**

**$p<0.01$ *$p<0.05$

upkeep and clarity ($p<0.05$, $r=0.377$), vista characteristic ($p<0.05$, $r=0.339$) and safety ($p<0.05$, $r=0.361$). The facts that the three-dimensional structure can be perceived comfortably, that there are no complex elements when evaluating the images, that there is a scenic beauty and that there is no danger involved play a role on the upkeep of the elements that make up the image. An increase in the complexity parameter level highly increases the upkeep parameter. An increase in the clarity, vista characteristic and safety parameter levels weakly increases the upkeep parameter.

Positive and moderate correlations were observed between clarity and vista characteristic ($p<0.01$, $r=0.454$) and diversity ($p<0.01$, $r=0.472$), whereas a positive and weak correlation was observed between clarity and complexity ($p<0.05$, $r=0.384$). The facts that the images have a scenic beauty, that they have many different elements, that the three-dimensional structure is perceived comfortably and that there are no complex parameters involved when evaluating the images affect the clarity parameter.

Positive and moderate correlations were observed between vista characteristic and; diversity ($p<0.01$, $r=0.521$) and complexity ($p<0.01$, $r=0.547$). The facts that there are many different elements in the images and that the three-dimensional structure can be perceived comfortably affect the vista characteristic.

A positive and moderate correlation was observed between landform and safety ($p<0.01$, $r=0.592$). Landscape elements may be perceived in an orderly manner, since the images are not associated with any risk or danger (Tab. 15).

5 Conclusion and Suggestions

It was determined in this chapter that the parameters of naturalness, coherence, upkeep, clarity, vista characteristic, diversity, complexity and safety contribute positively to the visual assessment of the landscape, whereas the landform characteristic contributes negatively.

When the visual quality parameter scores of photos which received the highest scores were examined, it was observed that the parameters of naturalness, coherence, vista characteristic and diversity had the highest average.

A relationship was determined between the vista characteristic score and naturalness score of each area when the landscape parameter results are evaluated for each area. Photos with the highest vista characteristic scores also had the highest naturalness scores.

Among all parameters evaluated for each area, naturalness had the highest score among all other parameters with 4.9 and vista characteristic had a score of 4.7, vista characteristic had the highest score in coastal landscape with a score of 4.6 and naturalness had the highest score in archaeological landscape with a score of 4.5.When the responses of the experts to the question, "What are the various recreational activities that may stand out at Güzelcehisar?" were examined, the selected parameters support the fact that the Güzelcehisar Castle, Lava Columns and the natural beach have significant impact on why "Seeing Natural/Cultural assets" was ranked number 1 with 54.3%, swimming was ranked number 2 with 31.4% and nature observation was ranked number 3 with 20%. Moreover, the fact that Lava Columns Landscape was ranked number 1 with 51.4 %, Coastal Landscape was ranked number 2 with 42.8% and Archaeological Landscape was ranked number 3 with 37.1% as the responses of the experts to the question "What are the most important landscape characteristics of Güzelcehisar?" further emphasize the importance of these areas. It was put forth as a result of this study carried out for visual landscape assessment that the variation of tourism which is important for regional development within the context of the sustainability of the rural landscape characteristic of Güzelcehisar Village shall contribute to the local public. It is of special importance that a natural formation such as the lava columns is put forward for tourism purposes under the concept of geotourism and heritage coast by way of master plans which should be supported with long-term plans. In addition, Güzelcehisar as a destination for geotourism may be further supported and inventories may be developed by carrying out detailed studies on the natural-cultural values, biological diversities and geological characteristics that make up the geosources. Village design guide should be developed which encompasses various structural characteristics of Güzelcehisar such as its topographical structure, climate, vegetation, water sources, geology in addition to emphasizing their importance in the landscape further contributing to the sustainability of the rural characteristic and landscape quality of Güzelcehisar.

Acknowledgements

This study was prepared within the scope of the project coded 2018-FEN-CY-009 entitled "Visual Landscape Assessment Within the Context of the Sustainable Development of Rural Landscapes at Coastal Areas: Güzelcehisar Village Example" and we hereby wish to acknowledge the support of the Bartın University Scientific Studies Projects Commission.

References

1 Lindgren, D., *Assessment of Ecological Value of Coastal Areas Using Morphometry and Secchi Depth: A Case Study with Data from the Swedish Coast.* Journal of Coastal Research, 2010. **26**(3): 429–435.

2 Scialabba, N., *Integrated Coastal Area Management and Agriculture, Forestry and Fisheries.* Food and Agriculture Organization of the United Nations. Rome, 1998.

3 Cicin-Sain, B. & Knetch, R.W., *Integrated Coastal and Ocean Management.* Concepts and Practices. Island, Washington, 1998. p. 517.

4 Cengiz, B., Cengiz, C. Karakoç, H., *Bartın İli Güzelcehisar Lav Sütunları Ve Sahilinin Turizm Ve Rekreasyon Amaçlı Peyzaj Uygulama Projesi.* Batı Karadeniz Kalkınma Ajansı 2016 Yılı Küçük Ölçekli Altyapı Mali Destek Programı TR81/16/KÖA/0042 Kodlu Proje, ISBN: 978-975-17-4073-1, Bartın. 2018.

5 Marin, V., Palmisani, F., Ivaldi, R., Dursi, R., & Fabiano, M., *Users Perception Analysis for Sustainable Beach Management ın Italy.* Ocean & Coastal Management, 2009. **52**, 268–277.

6 Kayır, Z.G., *Çevre Korumaya Ekolojik Yaklaşım: Dalyan Örneği.* Doktora Tezi (basılmamış). Mimar Sinan Üniversitesi Sosyal Bilimleri Enstitüsü Sosyoloji Anabilim Dalı, Kurumlar Sosyolojisi Programı, İstanbul. 1994.

7 Cengiz, C., *Kıyı Alanlarında Ekolojik Planlama: Yalova-Armutlu Örneği.* Doktora Tezi, Ankara Üniversitesi Fen Bilimleri Enstitüsü, Ankara. 2009.

8 Ak, M.K., *Akçakoca Kıyı Bandı Örneğinde Görsel Kalitenin Belirlenmesi ve Değerlendirilmesi Üzerine Bir Araştırma.* Doktora Tezi, Ankara Üniversitesi Fen Bilimleri Enstitüsü, Peyzaj Mimarlığı Anabilim Dalı, Ankara. 2010.

9 Kaplan, S. & Kaplan R., *The Experience Of Nature: A Psychological Perspective.*Cambridge University Press, Cambridge. 1998.

10 Müderrisoğlu, H. ve Eroğlu, E., *Bazı İbreli Ağaçların Kar Yükü Altında Görsel Algılanmasındaki Farklılıklar.* Süleyman Demirel Üniversitesi Orman Fakültesi Dergisi, 2006. **A**(1): 136–146.

11 Düzgüneş, E., ve Demirel, Ö., *Evaluation of Rural Areas in Terms of Landscape Quality: Salacik Village (Trabzon/Turkey) Example.* Environmental Monitoring and Assessment, 2015. 187: 310.

12 Erdönmez, İ. M. Ö. ve Kaptanoğlu, A. Y. Ç., *Peyzaj Estetiği ve Görsel Kalite Değerlendirmesi.* İstanbul Üniversitesi Orman Fakültesi Dergisi, 2008. 1(58): 39–51.

13 Cengiz, C. & Cengiz, B., *The Güzelcehisar-Bartın Coastal Landscape Heritage Project.* International Conference on Sustainable Development, 2016. 411–419. Skopje, Macedonia.

14 Williams, A.T., Sellers, V. & Philips, M.R., *An Assessment of UK Heritage Coasts in South Wales: J A Steers Revisited,* Journal of Coastal Research, Special Issue 50 (Proceedings of the 9th International Coastal Symposium), 2007. 453–458. Gold Coast, Australia.

15 Tveit, M., Ode, A. and Fry, G., *Key Concepts in a Framework for Analysing Visual Landscape Character,* Landscape Research, 2006. **31**(3): 229–255.

16 Cengiz, C., Cengiz, B., Bekçi, B. ve Tekdamar, D., *Bartın Güzelcehisar Kıyı Yerleşiminin Ekolojik Planlaması.* Bartın Üniversitesi Bilimsel Araştırma Projesi, 2015. BAP-2011- 2-26, Bartın.

17 Gonzalo, F. and Hermann, M., *Visual quality: An examination of a South American Mediterranean Landscape, Andean Foothills East of Santiago (Chile).* Urban Forestry & Urban Greening, 2014. **13**(2): 261–271.

18 Kalivoda, O., Vojar, J., Skrivanová, Z. and Zahradník, D., *Consensus in Landscape Preference Judgments: The Effects Of Landscape Visual Aesthetic Quality and Respondents Characteristics.* Journal of Environmental Management, 2014. 137: 36–44.

19 Kıroğlu, E., *Erzurum Kenti ve Yakın Çevresindeki Bazı Rekreasyon Alanlarının Görsel Peyzaj Kalitesi Yönünden Değerlendirilmesi.* Yüksek Lisans Tezi, Atatürk Üniversitesi Fen Bilimleri Enstitüsü, Peyzaj Mimarlığı Anabilim Dalı, Erzurum. 2007.

20 Kiper, T., Korkut, A., ve Topal, T. Ü. (2017). Görsel Peyzaj Kalite Değerlendirmesi: Kıyıköy Örneği. KSÜ Doğa Bil. Derg., 20(3): 258–269.

Burcu Özdemir, Ph.D.

10 Cybernetics' Twin of Urban Environments: Urban Informatics

Abstract: The desire for clearly understanding the urban environment with its dynamic, complex and vibrant structures, also their patterns and processes evolving in space-time is becoming a more achievable target with the increasing volume and variety of spatial data, developments and advancements in related fields (informatics, statistics, computing) and ICT in the last decades. The urban informatics field is establishing its place to utilize the development and tools of informatics in the domain of urban environments with integration methodologies from urban science, information and computing science. Particularly, cognitive informatics are promising for intelligence computing in informatics to deal with complex urban systems. This study attempts to broadly explore an emerging discipline of urban informatics and make a prediction for an urban ecosystem in the future digital world. As a result of all the progress made in urban and informatics science and technology, cybernetics' twin of urban environment will be born in a virtual world of urban science as another dimension of urban ecosystems. This cybernetics' twin will probably be part machine and part live organism in a self-organizing system, by learning from the knowledge of urban science and all processing of human brains in real-world cities, and by utilizing the development of informatics science.

Keywords: Urban environment, Urban informatics, Cognitive informatics, Urban systems, Cybernetics city

1 Introduction

Towards the end of the 20th century with the development of the Internet, computer and communication technologies became more involved in everyday lives. Computers, with the recent advancement of information and communication technologies (ICT), have become an essential part in all fields.

ICT and computers have rapidly entered urban environments where data production is the highest [1]. Because city scales are expanding with urban dynamics, the components of the urban system have become further complex in detail, the amount of urban data has prompted a need for an integrated urban information system to manage data, perform analytics and drive decisions. Especially in urban environments, there are wide solution sets that need to be scrutinized to evaluate the numerous possibilities in the problem-solving and decision-making processes [2]. These solution sets can have an elaborate and massive structure based on the city scale. In addition, cities are consistently evolving within a dynamic and integrated structure. Thus, the structure of cities may develop in unexpected ways, regardless of what was planned. It is necessary to have an effective urban-based integrated information system with real-time data to conduct accurate and efficient interventions or revise decisions, otherwise relying upon classical methods or the foresight of decision makers can take more time and be limited [3]. These needs related to urban informatics have ensued from necessity of not only the public sector in urban contexts, but the private sector as well.

Today, many public and private sector enterprises have digitized their existing data, established their own information systems, from which they produced their own data in order to effectively use it to improve their capabilities. Meanwhile, these trends have lead to a big market in the domain of urban technology [4]. Therefore, expanding amounts of data, advancements in technology and developments in informatics science have created their own ecosystem of the digital transformation. Data publishers, data consumers and technology suppliers have been three main actors of the digital ecosystem and cities have become one of the important hosts for this ecosystem. Moreover, location-time-based data has become a necessity for many enterprises. Widespread smart devices, including those with Global Positioning System (GPS) access, have provided low-cost and fast spatial data collection and production; in this way, location-based data have been produced by numerous sources such as enterprise, smartphones, social media and sensors [5]. As a result, the extensive availability of information in urban areas enhances the overall data. Ever expanding volumes of urban data and the dynamic structure of cities have given rise to topics related to the informatics field focused on urban environments with considerations for urban data such as storage, management, classifying, computing, sharing, transforming, visualizing, analytics, accessibility, reusing, as well as people and technology. Thus, the urban informatics field has been establishing its place to utilize the development and tools of informatics in the domain of urban environments [6].

This study attempts to broadly explore an emerging discipline of urban informatics and make a prediction for an urban ecosystem in the future digital world. The following sections expand upon the topics and components in Fig. 1, demonstrating a proposal formation of urban informatics.

2 Informatics

The modeling of the natural world contains three essences: matter, energy and information. Information is the supplementing essence of matter and energy. Informatics is the science of information, the ways of processing and transferring data between matter, energy and information [8]. Informatics is a broad field and is generally defined as an admixture of computer science, information systems, software engineering and computer engineering. It also may be seen as computing, in a general sense [9]. As Wang points out, informatics has expanded from the classical information theory [10–13], and contemporary informatics [14–16], to cognitive informatics [17–19] in the past half century [8]. Mulder and Weert proposed a working draft definition for informatics: "The discipline of informatics deals with the manipulation of objects (data, entities, i.e.) with the aim to solve a problem, specified by a requirements analysis under a given set of constraints, resulting in

Informatics
Tools, Methods and Advancements

- Approaches
  - Classical information theory
  - Contemporary informatics
  - Cognitive informatics
- Tools
  - Data capture / collection
  - Data storage
  - Data management
  - Data analytics/ mining
  - Data reuse / sharing
  - Modelling / simulation / visualization
  - Multi-criteria decision support
- Methods
  - Artificial intelligence / Machine learning / Deep Learning
  - Neural networks
  - Fuzzy logic
  - Pattern recognition
  - Agent technologies
  - Genetic algorithms
- *ICT*, Information and Communication Technologies

\+

Urban Environment
Components, Systems and Approaches

- Components 3 environments and their systems
  - Natural environments
  Climate, forest, water, natural habitat-ecosystem.
  - Built environments
  Infrastructure, transportation, housing, urban energy use,
  - Socio-economic environments
  Public health and wellness, safety and security, communities, municipal management-finance, economic development
- (3E) Optimization for decision making
  - Ecology
  - Equality
  - Economy
- Application Scale / Borders
  - Nation, region, metropolitan area, city, town, neighbourhood, street, building or special domain.
- Dynamic / Spatiotemporal Structures
  - People- living- mobility
  - Governance – Economy
  - Networks (relations between the objects)

Urban Informatics

- Urban Big Data
  - Spatial geometrics data
  - Spatial thematic (attribute) data
- Urban Computing
  - Urban sensing and data acquisition
  - Data bases / Urban data clouds
  - Urban data management
  - Urban data analytics
  - Service providing
- Urban Ubiquity Systems
- Cybernetics Twin of Urban Environment

Fig. 1: Informatics, urban environment and urban informatics. Fig. 1 is expanding version of the urban informatics diagrams [7] with approaches [22] and methods [25] in informatics, optimization [34] in decision making and urban computing [42] in urban informatics.

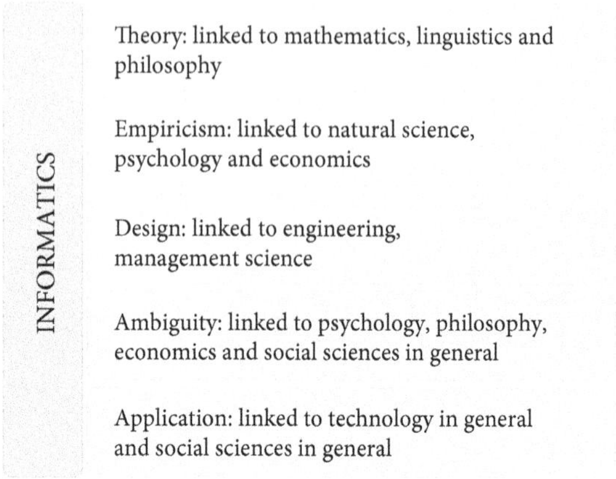

Fig. 2: Five paradigms of informatics discipline.

a maintainable working system with an automated component, in which man and machine interact adequately" [p. 9] and put forward theory, empiricism, design, ambiguity and application of the informatics discipline (Fig. 2) by extending theory, abstraction and design of the computing discipline [20].

Informatics terms are generally used with a prefixed application field such as clinical/medical/health informatics, community/organizational/social informatics, urban/environmental informatics, tourism informatics, security informatics, educational informatics and so on. Recently, such informatics research fields are ever-growing and by term of informatics they mean either the study of information content with utilizing representation, technology, methods and strategies or a variety of computing research [21].

Through necessity, informatics has ensued despite the challenges of using complex and large-scale data in conjunction with data capture or collection, storage, classification or organization, management, mining, observation and sharing, the needs of data analytics, modeling, simulating, visualization with the aim to enhance efficiency automate some work, reduce errors, support accelerated and reliable decision making, manage complex tasks, extract knowledge and even wisdom from information or data within special domains or fields. Moreover, information science with big data issues, trends of an intelligent system/trends of intelligent systems, and advancements in ICT has encountered its own problems and challenges independent from application fields. Therefore, cognitive informatics has risen as a profound interdisciplinary research to deal with the problems of modern informatics, computation, software engineering, artificial intelligence, neural psychology and cognitive science [8]. As a multidisciplinary branch of cognitive informatics studies the internal information processing and mechanism of human brains and mind-based data, another studies the external information

The Subject of AI

- Learning Systems
- Knowledge Representation and Reasoning
- Planning
- Knowledge Acquisition
- Intelligent Search
- Logic Programming Soft Computing - Fuzzy Logic /Artificial Neural Nets (ANN) / Genetic Algorithms
- Management of Imprecision and Uncertainty

Applications of AI Techniques

- Expert Systems
- Image Understanding and Computer Vision
- Navigational Planning for Mobile Robots
- Speech and Natural Language Understanding
- Scheduling
- Intelligent Control

Fig. 3: The subjects and applications of AI.

processing such as contemporary informatics studies are computer based, classical informatics studies are channel based [22].

Particularly, the four meta types of brain models in cognitive informatics: object, attribute, relation and action HAVE connections between them [23] that are promising for intelligence computing in informatics to deal with complex systems. Recently, the application of artificial intelligence (AI), machine learning and deep learning have become common in many fields due to the availability of computation and data, technological developments, rising demand and limits of time and human capability. Clarifying these three terms: artificial intelligence as a general category is the external circle, machine learning is the inner circle in the middle that contains deep learning at its core. Actually, the most common form of learning in the artificial intelligence community is *supervised learning* to train the machine or agent [24]. Fig. 3 demonstrates briefly the subject and application of AI techniques with the aim to guide for reasoning, learning, planning, perception, knowledge acquisition, intelligent search and uncertainty management in informatics research fields [25].

Consequently, this section broadly indicates approaches, methods, techniques, tools, skills and advancements of the informatics science in order to utilize them to fulfill the needs and requirements of a complex system such as an urban environment.

3 Urban Environment

Urban environment is a term distinguishing a discipline that originated from both the natural and the social sciences and applies to a place described by an intensity of social interactions, various organizations and systems, a dense pattern of development such as any area scaled from metropolis to neighborhoods [26]. Shortly, "urban environments are places where most of the people in our contemporary society live" [27]. Moreover, considering the projected future population by

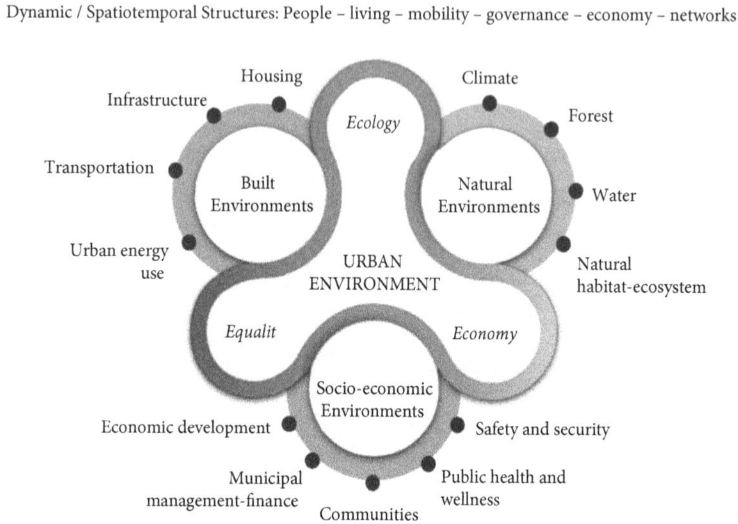

Fig. 4: Urban environments and structures.

the year 2050, two-thirds of the world's people will be living in urban areas, and urban population will increase by 31 million annually[1] [28] so it is obvious that urban areas (cities) will determine the quality of life in the world. Therefore, urban environment takes into consideration social relations, effects of human activities, global and local issues and solutions for problems with holistic viewpoints [26].

Urban environment emerged from three environments, natural, built and social-economic environments. Each environment has subsystems with an interaction between the other environments subsystems. Fig. 4 depicts urban environment with components with a holistic view.

Natural environments in urban areas have subsystems such as climate, forest, water and natural habitat which make up the ecology subset network of the world. Increasing threats to humanity (public health and wellness) and the natural environment have resulted from the rapid urbanization and the decisions on built environment (housing, infrastructures, transportations and land use). Moreover, the complexities of urban environments have created a need to consider a holistic view of physical structures in urban areas such as its intangible cultural, economic, ecological and social factors [29]. All environments in the urban system are tightly integrated to each other and any decision related to built environments

1 As reported in the paper of *Population and The Urban Future* by permission of The Population Institute [28].

or socio-economic environments is a reflection on the situ of the natural environment. Obviously, there is a dynamic interaction between people, cities and ecology. People are shaping cities and nature, then cities are shaping people's lives, then ecology is also shaping quality of life and cities. Hence artificial divides between nature and cities in ecological science shifted to integrate urban places into components of long-term resilience [30] and sustainability. As a result, urban environments embodying integrated urban sustainability—which is consisting of economic sustainability and its substrates (growth, development, productivity, trickle down), social sustainability and its substrates (equity, empowerment, accessibility, participation, sharing, cultural identity, institutional stability criteria), ecological sustainability and its substrates (ecosystem integrity, carrying capacity and biodiversity)—in urban planning practice to mitigate or prevent the negative urbanization impacts. [31,32]. Indeed, urban planning has always been challenging because of its dynamic [29] and spatial temporal structure such as the interplay of people, living, mobility, government regulations, entrepreneurial activities in economy and all system networks.

As mentioned in the US EPA technical report: "Decisions about how and where we build our communities have significant impacts on the natural environment and on human health" [33]. So, especially multi-criteria–based decision making in the complex system of urban environment should be based on a holistic approach "(3E) economy, ecology and (social) equality" [34] to enhance quality of life and environments, vibrant livelihoods, urban durability to natural disasters, government efficiency, economic development, public safety and security and to briefly generate urban sustainability. Real-time, location, information-based decision making in urban environments have determined the requirement of urban informatics, to provide the knowledge for a timely action by the six stages of decision in a life cycle [35]. These stages are (1) preparing for urban problems and trends by learning analysis of past situations, (2) predicting/determining the likelihood of situations by defining where, when, what, who and possibly why, (3) prevention to keep urban sustainability up (4) detection, early on, of any upcoming issues to mitigate impact of negative situations, (5) response in a timely manner by analyzing data and models, and (6) recovery to minimize negative outcomes or facilitate effort in unplanned issues. Decision or action process in urban environment needs data, analysis, information, model and systems [35] to manage complex, large-scale urban areas, to clearly understand the integration of subsystems, to reduce errors and side effects in process and application, to become more efficient and effective in planning and administration, to automate some work, to share information or reuse output data as input in other processes, to visualize results and so on, under the circumstances of continuous changes in urban environments and its subcomponents.

The desire remains to clearly understand the urban environment with its dynamic, complex, vibrant structures, and its components' functions, relations, integration with each other, also their patterns and processes evolving in space-time is becoming a more achievable target with the increasing volume and variety of spatial data, developments and advancements in related fields (informatics, statistics, computing) and ICT in the last decades [27], also with integration methodologies from urban science, information and computing science. As a consequence of these, urban informatics appeared as a scientific research, eventually.

4 Urban Informatics

Broadly, urban informatics is "the application of computers to the functioning of cities" to control the city system and "the ways in which computers are being embedded into cities as hardware and as software" to enable the interaction between populations and the system [1]. Urban informatics is also defined when "a disciplinary domain is situated at the intersection of notions, trends and considerations for place, technology, and people in urban environments" [36]. Urban informatics has a multidisciplinary aspect like urban planning; also it fuses together the computing science with the aspect of different traditional sciences [7]. Urban informatics takes a trans-disciplinary approach to understand the nature of the urban ecosystem with its constituents that are technological, social, physical (architectural) [36], ecological and economical in nature. Urban informatics is also an interdisciplinary research field in the context of urban environment. By extension, urban informatics is comprised of both studying and comprehension of socio-cultural characteristics of people [37], space–time scale of city evolution and dynamics [38], city functions within and across urban domains and systems [7], interaction between these systems and city administration such as public policy, operations, services, management and controlling [39]. Therefore, urban informatics gathers the varying aspects of urban planning, computer sciences, operations, management and decision making, [7] which leads to cross-training within the field. As pointed out by Townsend, the urban planning discipline can easily prepare its own students to systematically cross-train with expanding the connection to informatics then other disciplines, "because planning is already connected to a hodgepodge of disciplines that offer insights on the city: engineering, economics, sociology, geography, political science, law and public finance" [4, pp. 303–304].

Consequently, urban informatics evolved from information science and urban science, and potentially belongs to both scientific epistemologies of them. Fig. 5 depicts the general framework of urban informatics formation with the main purposes.

The significant increase in the production of urban data (also with developments and trends of smart cities) makes it more challenging to analyze cities with the complexity and high volume of spatial data, but advanced computational methodologies and tools strengthen comprehension [27]. Cities are an enormous data source for a wide array of data types. Urban data consists of different data types, spatial geometric data and thematic (attribute) data, collected from sensory and the Internet of thing (IoT) in smart cities or by devices, vehicles, buildings and humans. Divided into two broad categories, cities are physically machine and socially organism; the physical part provides measurable data, while the social part

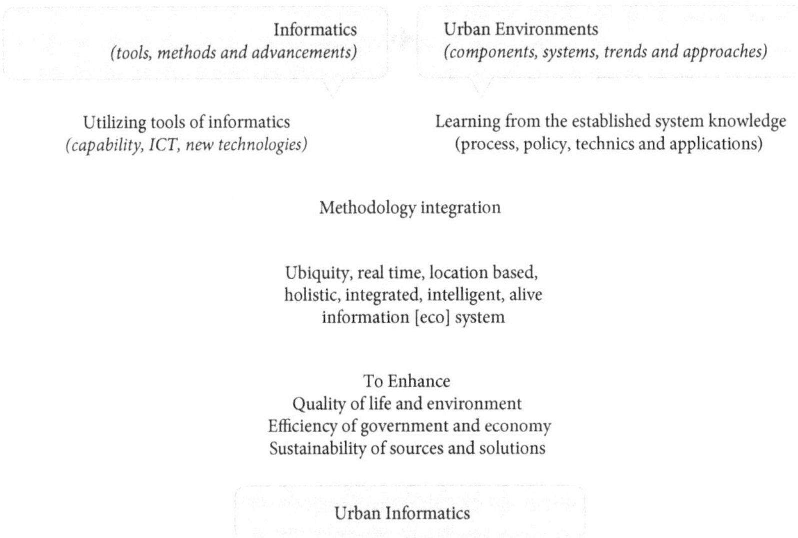

Fig. 5: General framework of urban informatics formation.

provides some unmeasurable or latent data. Furthermore, classifying and ordering urban data of complex structures to interpret their meaning is a big challenge because cities are dynamic (a space of flows), functioning in space and time, an ecosystem consisting of interactions or relations between all urban environments networks, patterns, processes and people [40]. In a word, urban data like big data faces compounded issues. For instance, considering a data-linked human activity, such as visiting a place, attending a concert or using a mobile phone, creates a high amount of data and only a part of them can manage the geographic information system, the rest just (have) some sort of geographic references [41]. It is necessary to know (in which urban phenomena this data must be used or what its role is in urban systems. In that point, urban computing was defined as "a process of acquisition, integration, and analysis of big and heterogeneous data generated by diverse sources in urban spaces, to tackle the major issues that cities face," to understand urban phenomena and predict the future [42]. Hence, urban computing is seemingly a subtopic of urban informatics as a method or process for dealing with urban data. Fig. 6 puts forward the general framework of urban computing [42].

Zheng et al. pointed out three key challenges "urban sensing and data acquisition, computing with heterogeneous data and hybrid systems blending the physical and virtual worlds" resulted from the goals in seven categories– "urban planning, transportation systems, environment, urban energy consumption, social application, economy, and public safety-security"–and the framework [42]. On the

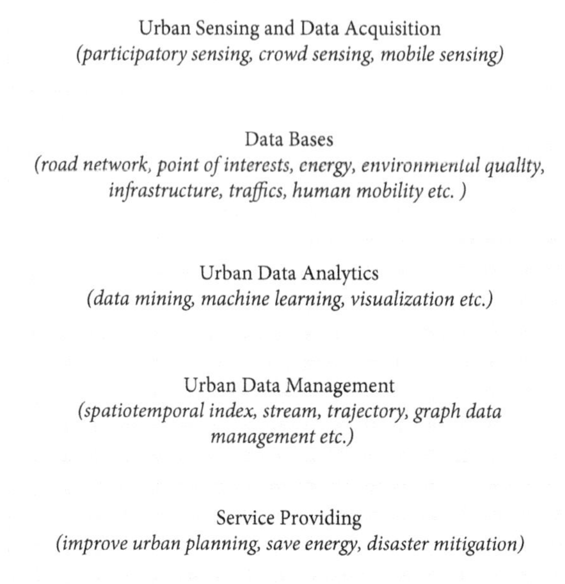

Fig. 6: General framework of urban computing [42].

other hand, Valle et al. determined that information integration is a core problem in urban computing and claimed that Semantic Web technologies are a valid solution for it.

With further advancements in ontology and real-word semantics of an object (data) is promising to deal with heterogeneous data problems in urban areas which are due to technological and communicational differences such as in hardware, software, and operating systems [43]. When considering the Web 4.0 ability, development of a ubiquitous system in geographical information, urban informatics studies will attempt to bring more holistic urban ubiquity computing systems into urban environments independently from types of data sources, hardware, software and operating systems while keeping track of opportunities (such as artificial intelligence with it is similarity to the human brain) provided by cognitive informatics in the future.

5 Concluding Remark

As a result of all the progress made in urban and informatics science and technology, cybernetics' twin of urban environment will be born in the virtual world of urban science as another dimension of urban ecosystems. This twin of cybernetics will probably be part machine and part live organism in a self-organizing system, by learning from the knowledge of urban science and all processing of human brains in real-world cities, and by utilizing the development of informatics science.

For now, it can be said that urban informatics is at the initial stage of research and development; there are many challenges such as urban data computation management, heterogeneous geospatiality, queries, integration, training and other issues. Like urban science, urban informatics will evolve with the progression of time. This evolution in urban informatics will enable the reconfiguring of administrative duties, participation of citizens, the mentality of all related processes and systems and new scientific interdisciplinary research topics in urban domains as an irresistible development for urban areas.

Urban informatics with ubiquitous computing does not have to trap humanity in a virtual world, ever-changing innovations in the field allow for a future with less time spent computing and more efficient technology that yields to the needs of people.

References

1 Batty, M., *Urban Informatics and Big Data.* A Report to the ESRC Cities Expert Group. 2013, Available online: http://www. smartcitiesappg.com/wp-content/uploads/2014/10/Urban-Informatics-and-Big-Data.pdf (accessed on 5 January 2018)

2 Li, X., and Yeh, A.G.O., *Integration of Genetic Algorithms and GIS for Optimal Location Search.* International Journal of Geographical Information Science, 2005. **19**(5): pp. 581–601.

3 Ozdemir, B., and Tokyay, O., *Multi Criteria Decision Support for the Best Allocation of the Primary School Area.* SGEM2013Conference Proceedings, June 16–22, 2013. **1**: pp. 457–464.

4 Townsend, A. M., *Smart Cities: Big Data, Civic Hackers, and the Quest for a New Utopia.* 2013, WW Norton & Company, New York.

5 Zhang, C., Zhao, T., and Li, W., *Current and Future Challenges of Geospatial Semantic Web.* 2015, Geospatial Semantic Web Springer, Cham, pp. 167–189.

6 Kontokosta, C.E., *Urban Informatics in the Science and Practice of Planning.* Journal of Planning Education and Research, 2018. https://doi.org/10.1177/0739456X18793716

7 *Urban Informatics Diagram Available Online* http://web.cs.toronto.edu/research/areas/si.htm (accessed on 10 November 2018)

8 Wang, Y., *On Cognitive Informatics.* Brain and Mind, 2003. **4**(2): pp.151–167.

9 Lidtke, D., Myers, P., Cassel, L., Gupta, G., Hacquebard, A., Mulder, F., and Waligorski, S., *A Common Core of Concepts for Informatics Majors.* 1998. In Informatics in Higher Education Springer, Boston, MA., pp. 17–20.

10 Bell, D.A., *Information Theory*, 1953, Pitman, London.

11 Hartley, R.V.L. *Transmission of Information*, Bell System Technical Journal, July, 1928, p. 535.

12 Shannon, C.E., *A Mathematical Theory of Communication*, Bell System Technical Journal, 1948, 27, pp. 379–423 and 623–656.

13 Shannon, C.E., and Weaver, W., *The Mathematical Theory of Communication*, 1949, Illinois University Press, Urbana, IL, USA.

14 Wang, Y., S. Patel, and R. Johnston, eds. *Proceedings of the 7th International Conference on Object-Oriented Information Systems* (OOIS'01), 2001, Calgary, Canada, August, Springer, London, p. 552.

15 Wilhelm, R., ed. *Informatics: 10 Years Back, 10 Years Ahead*, 2001, LNCS 2000, Springer Verlag, Berlin.

16 Zhong, Y., *Principles of Information Science* (in Chinese), 1996, BUPT Press, Beijing.

17 Wang, Y., *On a New Frontier: Cognitive Informatics*, Invited Talk, Proceedings of the 7th International Conference on Object-Oriented Information Systems (OOIS'01), 2001, Calgary, Canada, Springer, August.

18 Wang, Y., *On the Informatics Laws of Software*, Keynote Lecture, Proceedings of the 1st IEEE International Conference on Cognitive Informatics (ICCI'02), 2002, Calgary, Canada, IEEE CS Press, August.

19 Wang, Y., *A New Mathematics for Describing Notions and Thoughts in Software Design*, Proceedings of the 1st IEEE International Conference on Cognitive Informatics (ICCI'02), 2002, Calgary, Canada, IEEE CS Press, August.

20 Mulder, F., and Van Weert, T., *Towards Informatics as a Discipline: Search for Identity*. In Informatics in Higher Education, 1998, Springer, Boston, MA., pp. 3–10.

21 Sawyer, S., and Eschenfelder, K.R., *Social Informatics: Perspectives, Examples, and Trends*. Annual Review of Information Science and Technology, 2002. **36**(1): pp. 427–465.

22 Wang, Y., *On Cognitive Informatics*. In Proc.of the 1st IEEE International Conference on Cognitive Informatics (ICCI'02), IEEE, 2002, Calgary, Canada, IEEE CS Press, August, pp. 34–42.

23 Wang, Y., and Wang, Y., *Cognitive Informatics Models of the Brain*. IEEE Transactions on Systems, Man, and Cybernetics, Part C (Applications and Reviews), 2006. **36**(2): pp. 203–207.

24 LeCun, Y., Bengio, Y., and Hinton, G., *Deep Learning*. Nature, 2015. **521**(7553): pp. 436–444.

25 Konar, A., *Artificial Intelligence and Soft Computing: Behavioral and Cognitive Modeling of the Human Brain*. 1999. CRC Press, Boca Raton, FL.

26 Sénécal, G., *Urban Environment: Mapping a Concept.* Introductory Note. Environnement Urbain/Urban Environment, 2007. **1**.

27 Helbich, M., Arsanjani, J.J., and Leitner, M., (Eds.). *Computational Approaches for Urban Environments*, 2015. **13**. Springer, Switzerland.

28 Fornos, W., Thirunarayanapuram, D., and Burdett, H.N., *Population and the Urban Future*, Time-Saver Standards for Urban Design, ed: Watson, D., 2003, McGraw-Hill, New York.

29 Hamilton, A., Wang, H., Tanyer, A.M., Arayici, Y., Zhang, X., and Song, Y., *Urban Information Model for City Planning.* Journal of Information Technology in Construction (ITCon), 2005. **10**(6): pp. 55–67.

30 Seto, K.C., Parnell, S., and Elmqvist, T., A *Global Outlook on Urbanization.* Urbanization, Biodiversity and Ecosystem Services: Challenges and Opportunities, 2013. Springer, Dordrecht. pp. 1–12.

31 Basiago, A.D., *Economic, Social, and Environmental Sustainability,* Development Theory and Urban Planning Practice. Environmentalist, 1998. **19**(2): pp.145–161.

32 Khan, M.A., *Concepts, Definitions and Key Issues in Sustainable Development: The Outlook for the Future.* Proceedings of the 1995 International Sustainable Development Research Conference, Manchester, England, Mar. 27–28, 1995, Keynote Paper, pp. 2–13.

33 US Environmental Protection Agency (US EPA). *Our Built and Natural Environments: A Technical Review of the Interactions Between Land Use, Transportation, and Environmental Quality.* 2013. US EPA, Washington, DC.

34 Edgeman, R., and Eskildsen, J., *An Excellence Equation:(E3 Governance+ 3E Strategy)-> 3P Performance.* Journal of Positive Management, 2012. **3**(1): pp. 50–66.

35 Tien, J.M., *Viewing Urban Disruptions from A Decision Informatics Perspective.* Journal of Systems Science and Systems Engineering, 2005. **14**(3): pp. 257–288.

36 Foth, M., Choi, J.H.J., and Satchell, C., *Urban Informatics.* Proceedings of the ACM 2011 Conference on Computer Supported Cooperative Work, 2011, March. ACM, p. 1–8.

37 Bilandzic, M., and Venable, J., *Towards Participatory Action Design Research: Adapting Action Research and Design Science Research Methods for Urban Informatics.* Journal of Community Informatics, 2011. **7**(3). Retrieved from

http://ci-journal.net/index.php/ciej/article/view/786

38 Portugali, J., *Toward a Cognitive Approach to Urban Dynamics.* Environment and Planning B: Planning and Design (2004). **31**(4): p. 589–613.

39 Kitchin, R., Coletta, C., and McArdle, G., *Urban Informatics*, Governmentality and the Logics of Urban Control: The Programmable City Working Paper 25. 2017.

40 Batty, M. ., *The New Science of Cities.* 2013, MIT Press, Cambridge, MA.

41 Valle, E.D., Celino, I., and Dell'Aglio, D., *The Experience of Realizing a Semantic Web Urban Computing Application.* Transactions in GIS, 2010. **14**(2): pp. 163–181.

42 Zheng, Y., Capra, L., Wolfson, O., and Yang, H., *Urban Computing: Concepts, Methodologies, and Applications.* ACM Transactions on Intelligent Systems and Technology (TIST), 2014. **5**(3): p. 38.

43 Sheth, A.P., *Changing Focus on Interoperability in Information Systems: From System, Syntax, Structure to Semantics.* Interoperating Geographic Information Systems, 1999, Springer, Boston, MA., pp. 5–29.

Dr. Burcu Salgın, Dr. Kofi Agyekum, and Dr. Barbara Simons

11 Designing Out Damp: Exploring Basic Design Principles to Control Excessive Moisture in Buildings

Abstract: Moisture can enter buildings through many ways. It is a threat to buildings and the health of occupants. It can ultimately damage the structure and its associated materials. Occupants of damp buildings are at increased risks of experiencing health problems like respiratory disorders, allergic rhinitis and asthma. The first measure to be taken to prevent any moisture-related problems that may occur during a building's lifespan is related to design decisions. In this context "Designing out Damp" refers to architectural design decisions that are explored to deal with moisture problems in buildings. Moisture and its related activities in buildings have been reviewed, and basic design principles to control moisture in buildings have further been identified in this chapter. The basic design principles explained should help designers in designing out damp during the design stage of buildings.

Keywords: Designing out damp, Dampness, Moisture in buildings, Moisture control, Rising damp

1 Introduction

Buildings are constantly evolving as a result of the changes around them. The demands placed on a building or an element of its construction by users relates to its location and siting, climatic and environmental conditions, the manner in which it is used, current and past levels of damage, deterioration and decay, etc. [1]. Oftentimes, these changes involve water which leads to a fast deterioration of the structure, properties and the health of occupants. Processes such as condensation and capillarity all affect buildings negatively. Research has shown that of all the defects associated with buildings, moisture is the most frequent and dangerous and contributes more than 50% of all known building failures [2–4]. The first measure to put in place to prevent any moisture-related problems that may occur during the building lifespan is related to design decisions. In this context, "designing out damp" refers design decisions to eliminate or minimise the risks of moisture throughout the life of a building. This study looks at the problems of moisture in buildings and some basic design principles as control mechanisms.

2 Overview of Moisture and Its Activities in Buildings

2.1 Problems of Moisture in Buildings

Moisture in buildings, if not properly controlled, can create serious problems for the building, its occupants and furnishings [5]. As the single most important agent of building deterioration, moisture has been an issue of great concern to building professionals [6]. Moisture from any source may lead to staining of wall surfaces, damage to paints, corrosion of metal surfaces, increase in the deterioration of building furnishings and structural materials [5], deterioration and disbanding of common building materials like ceiling tiles, wood products, etc., and the development of biological agents like moulds, dust mites and the likes [7].

A study by Tsongas [8] on most residential buildings in the United States of America and Canada showed that most of these buildings experienced a number of moisture problems. These instances were demonstrated when 98% of residents living in 344 buildings in Iowa state reported of at least one type of moisture problem experienced in their buildings [8]. Rousseau [7] also reported on the severity of moisture problems on a number of buildings in Canada. In similar surveys in countries with different climates such as Finland, Mauritius and Portugal many similar issues were identified [9–11]. Recently in Ghana, this problem is vastly reported [2, 12–14].

According to Straube [15], biological growth such as fungus associated with moisture can have severe impacts on buildings and the health of occupants. For moisture to be present in buildings, there should be a source of moisture available, a means or route for the moisture to travel and a driving force to enhance the moisture movement [15–17]. The elimination of any one of these conditions should avoid moisture problems in buildings. However, in reality, it is difficult to manage or remove all the sources of moisture to completely create moisture-free buildings [15–17].

2.2 Sources of Moisture in Buildings

There are several ways by which moisture migrates into building envelopes. According to Brook [18], the sources of moisture can be categorized into external and internal. Ways in which moisture is generated internally include the activities of building occupants such as bathing and cooking. The external sources of moisture can also include rainwater penetration, foundation drainage, among others [18]. Sources of moisture in buildings can be from many areas. The major classifications include liquid water sources, leakages through roofs and walls, diffusion of moisture through walls, and capillary rise of groundwater into basements or crawl spaces [19]. All these sources of moisture have direct influences on the moisture balance of houses.

2.3 Mechanisms of Moisture Transport in Buildings

There are several ways by which moisture travels to and from buildings. The movement of moisture in building materials follows the thermodynamics and structural engineering laws of physics [20]. The major ways through which moisture is transported in buildings are rain and pipe leakages, capillary action, diffusion and convection [15,21,22]. The key processes involved in the wetting and drying and the moisture transport mechanisms involved in the movement of moisture to and from buildings is proposed by Straube [15].

2.3.1 Vapour Convection

This is the key mode of transport used to move and dispense water vapour along ducts and within spaces. This mode of transport is the major cause of interstitial condensation. It is also a major source of moisture inside building enclosures after rain penetration and should be controlled [15,23].

2.3.2 Capillary Suction

This moves liquid moisture slowly and progressively through porous materials. It results from a negative pressure formed in the pores of materials [15,22]. The smaller the pores, the larger the negative pressures and the higher the moisture rise [15]. The rate at which moisture is transported by this mechanism is very slow but can act for many years. Capillary suction can be managed by installing barriers to these flows [15,22].

2.3.3 Gravity Flow

According to Straube [15], it is the most powerful means by which moisture travels in building materials. Due to gravitational flow water travels downwards on the vertical and inclined surfaces of a building [24]. It requires relatively large openings since capillary forces tend to engulf the forces of gravity in small pores. As a result, water may not flow out of a saturated brick but can flow through a screw hole in a plastic window sill [15].

According to Straube [15], all modes of moisture transport that cause moisture problems can also assist in the drying of building materials and surfaces. As a result, attempts to block transport mechanisms may not always be the best approach for managing moisture [15]. However, "a sensible combination of reducing the source strength, controlling and directing transport mechanisms and encouraging drying should be used" [15].

3 Basic Design Principles to Control Moisture in Buildings

Moisture control is very important to protect occupants' health and the structural and material performance of the building—during its lifespan—from any physical or chemical damages. Literature reports of the several problems that cause moisture to include poor design decisions [25–27], poor workmanship [26,28] and ventilation factors [9, 29, 30]. For the purpose of this study, solutions for the moisture problems are handled in terms of design decisions of the building. According to Pereira et al. [25], design errors of the building can adversely affect the building's performance towards humidity in the future. Chew [31] reported that design decisions affect the building, especially when it is sited in a wet area. Al-Hammad et al. [27] emphasized that architects should avoid designing irregular shapes to avoid collection of moisture, dust or water. As it is commonly known that future behaviour and endurance of the building depends on the design decisions in the early stage of the building's lifespan [25], it is important to know basic design principles to control dampness. According to EPA [32], moisture control which is based on design decisions consists of controlling liquid water; managing indoor humidity; and using moisture tolerant materials. This section of the study adopted a similar categorisation used by EPA [32].

3.1 Controlling Liquid Water

Liquid entry and liquid production are one of the common problems in buildings. It can be categorised as follows: liquid water from rain and melting snow; liquid water from plumbing leaks; liquid or vapour from the ground; and water production from indoor activities [15]. The relevant precautions can be as simple as ensuring adequate site drainage, accurate foundation and wall construction details, accurate roof assembly details and proper plumbing installation details [32].

The first step in controlling liquid water is to properly drain the site. That is locating the building in dry or well-drained site. Properly designed site drainage avoids building damage and the need for remediation during the building's life. With this goal, the building needs a rainwater runoff system to divert rain from the roof into the site drainage system. Ring drainage is the basic technique used to surround the building and protect it by channelling away water. However, another separated drainage system should be installed for sewer drainage from sanitary fixtures.

Accurating the foundation and wall insulation details in terms of water during architectural detailing stage is the second step in controlling liquid water. The basic prevention method for basement walls is named as coating on the exterior of a foundation wall. A coating on the exterior of a foundation wall prevents water and water vapour from moving through the wall. Proper waterproofing materials should be selected for the basement walls taking into account climate and topographic characteristics where the building is constructed.

External walls should be carefully designed in terms of water like foundation walls. Wall assemblies should be detailed in guidance of the climate to manage water vapour and condensation. But in any climate, the most important criteria are keeping the wall dry and preventing the condensation of water vapour on cool surfaces within the dry portion of the exterior wall assembly [32]. It is important to detail the walls that have rainwater protection behind the cladding in the form of air barrier materials (i.e., the drain plane). The breadth of this space differs based on the cladding type and purpose. Sometimes the drainage space also acts as a "ventilation space" [33]. Special design consideration should be given to the openings of the wall to prevent seepage problems. With this goal, the window and door design are crucial. Particular attention should be drawn to the design of joints in window/door frames and sections. The detail for the gap between the window/door frame and wall opening should be improved and a non-shrink waterproof grout should be included to fill this gap. Achieving the properly constructed window sill with sufficient fall can eliminate seepage due to ponding of water [34].

Designing properly the roof assembly details is another important issue to acheive controlling the liquid seepage. Providing an effective path for water to flow in a controlled direction is one of the most important principles in design although the recommended minimum slopes vary from climate to climate. Choosing the proper roof slope (flat, low-slope or steep-slope) is related to the local rainfall rates. The NRCA [35] states: "The criterion for judging proper slope for drainage is that there be no ponding water on the roof 48 hours after a rain during conditions conducive to drying." Adequate rainwater runoff system and drainage outlets should be provided to avoid water ponding. Architectural team provide the roofing documents that should indicate roof slope and drainage elements, such as roof drains, scuppers or gutters [36]. Downpipes along exterior walls are more vulnerable to condensation and for that reason designing them with a precision that allows easy access for periodic inspection and repair is important [32]. All the product which is to be mounted on the roof should be decided at the design stage and the details are prepared for them not to harm waterproofing material. At the design stage, conscious efforts should be made for the selection of suitable materials to shield or shade openings in the external wall from possible entry of water with due regard being given to the exposure to the elements of such areas [34].

Water and moisture problems associated with poorly designed plumbing system can cause damages that can affect nearly any location in the building. Not only the places that can be easily diagnosed but also the places that are difficult to see and difficult to access. For that reason, plumbing system must be designed not only to prevent primary problems but also to allow easy maintenance to avoid future problems. With this goal, plumbing should be located where the leaks will be diagnosed quickly, and plumbing should be insulated to protect water inside the plumbing from freezing in cold weather [32]. Proper maintenance accesses for plumbing system should be provided such as ceiling floor, etc. Plumbing pipes should not be designed to be embedded in common partition walls separating units. Moreover, the selection criteria of plumbing system are also important. Materials for plumbing and drainage pipes should comply with the local building code requirements as the building was being constructed. It is important to select long-lasting and durable products with capability of withstanding high water pressure.

3.2 Managing Indoor Relative Humidity

Effective management of indoor humidity is mostly related to ventilating the indoor air. Ishak et al. [37] stated that poor ventilation will lead to mould growth.

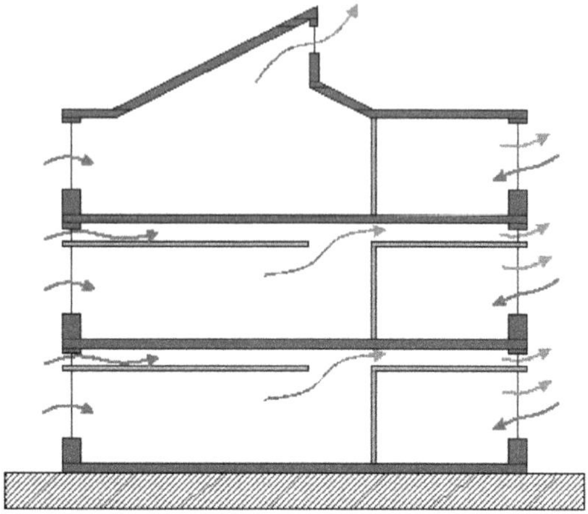

Fig. 1: Cross-ventilated building schematic section [40].

According to WHO [30], ventilation can be categorised as (1) natural ventilation (depends on openings in the buildings and air movement); (2) mechanical ventilation; and (3) hybrid ventilation (that uses an auxiliary low energy extract fan or natural ventilation which is combined with a fully independent mechanical system). When these three categories are evaluated, the architectural design stage decisions are mostly related to natural ventilation principles. Natural ventilation is carried out by the movement of the fresh air from the outside to the inside, and by expelling the same amount of used air [38]. Natural ventilation methods are divided into three groups as single-sided, cross and stack ventilation [39]. Considering the vertical air movement, it is believed that the most effective ventilation methods are cross and stack effect ventilation (Fig. 1 and Fig. 2). It is thought that single-sided ventilation is not as good as cross and stack effect ventilation because for any air to exit, there must be the same amount of air coming. That means there must be at least two openings for a good exchange in between inside and outside environment. To reach the target, the most important no-cost decision is related to position the openings in plans and sections.

When the architectural team place the windows preferably on opposite sides in floor plans (Fig. 3/a, b, c, e, f) and place the windows in different levels in sections (Fig. 1) to maintain cross air flow, it is the most effective way to remove internal moisture. Additionally, wing walls help to improve effective natural ventilation as an external projection [41] (Fig. 3/d, e, f). Properly positioned wing walls produce positive and negative pressure regions on the inlet and outlet windows, respectively. For that reason, they are capable of producing cross-ventilation even during low external wind velocity conditions [41].

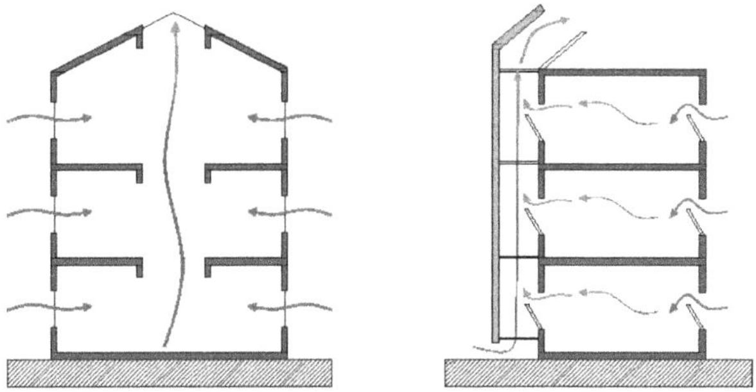

Fig. 2: Stack-ventilated building schematic sections [44,45].

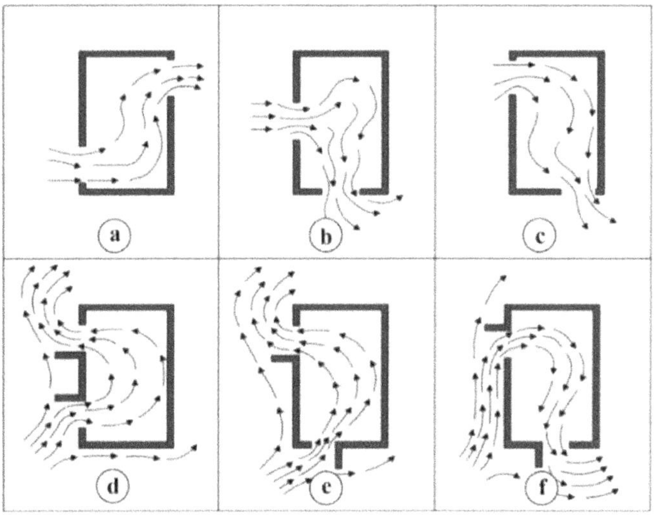

Fig. 3: Naturally ventilated building schematic plans [41,46].

Floor plans and sections that allow breeze to flow through the building result in natural ventilated building. Once architectural designers are made aware of this problem, these simple solutions could be applied at an early stage of the design.

Another natural ventilation method, the stack effect uses air pressure differences due to height to pull air through the building. The statement "hot air rises and cold air comes in to takes its place" is used to ventilate the building naturally. Towers and chimneys create a stack effect inside the building, and they work to move the air up and out. Multistorey buildings should have a vertical atrium or a shaft to connect the airflows of different floors (Fig. 2). Ismail and Rahman [42] mentioned that there are stack ventilation strategies as natural, advanced and fan-induced. From these three types of stack effect, fan-induced strategy is supported by mechanical systems that are defined by ASHRAE [43] as "intentional movement of air into and out of a building using fans and intake exhaust vents." When natural strategies are compared to fan-induced and advanced strategies, natural strategies can be a part of design stage with no or low-cost by positioning the openings easily.

3.3 Using Moisture Tolerant Materials

The last designing out damp criterion is to use moisture tolerant materials that can withstand wetting in areas like bathroom toilet, kitchen, etc. These areas are prone to getting wet easily by design and so should be assembled using only durable, moisture tolerant and mould resistant materials. The use of waterproofing system on the floor and walls is recommended. The waterproofing membrane/material should cover the entire floor and turn up every wall [34].

4 Conclusion

Moisture can enter and be seen in buildings through many different processes ranging from rainwater penetration, foundation drainage, leakages through roofs and walls, diffusion of moisture through walls, and capillary rise of groundwater into basements or crawl spaces. The issue of moisture in buildings is known to cause a myriad of problems for buildings and its occupants. In extreme cases, because of the moisture problems; building products should be changed and sections of residential homes have to be reconstructed. These are expensive and time-consuming works during the building lifespan. It is believed that basic design principles which are explained in this study can help to prevent moisture-related problems in buildings. Additionally, controlling water with admixtures, managing internal relative humidity and the use of moisture tolerant materials will go a long way to providing comfort for occupants as far as moisture is concerned.

Acknowledgement

The authors would like to acknowledge Miss Kübra Nur Eminel who assisted us in redrawing the figures.

References

1 Watts. D., B. Colston., and A. Bulow, *Predicting Damp-Related Problems in Historic Buildings and Monuments Used for Library and Archive Purposes*, in RICS Foundation Construction and Building Research Conference, 2001: London. p. 690–700.

2 Agyekum, K., K. Blay, and A. Opoku, *Mechanisms for Preventing Rising Damp in New Building Infrastructure.* International Journal of Building Pathology and Adaptation, 2018. https://doi.org/10.1108/IJBPA-06-2018-0048.

3 Blay, K., K. Agyekum, and A. Opoku, *Actions, Attitudes and Beliefs of Occupants in Managing Dampness in Buildings.* International Journal of Building Pathology and Adaptation, 2018. https://doi.org/10.1108/IJBPA-06-2018-0044.

4 Halim, A.A., S.N. Harun, and Hamid, Y., *Diagnosis of Dampness in Conservation of Historic Buildings.* Journal Design+Built, 2012. **5**(1): p. 1–14.

5 Stanke, D., et al., *Managing building moisture.* 1998, American Standard Inc.: The United States of America.

6 Dacquisto, D.J., J.H. Crandell, and J. Lyons, *Building, Moisture and Durability: Past, Present and Future Work.* 2004, US Department of Housing and Urban Development, Office of Policy Development and Research: Washington, D.C.

7 Rousseau, D. *Adapting to Climate Change: Some Observations,* Building Research & Information, 2004. **32**(1): p. 58–60.

8 Tsongas, G. *Case Studies of Moisture Problems in Residences.* in *the 2nd Annual Conference on Durability and Disaster Mitigation in Wood-Frame Housing,* November 2000. 2000: Madison, WI.

9 Lourenco, P.B., E. Luso, and M.G. Almeida, *Defects and moisture problems in buildings from historical city centres: A case study in Portugal.* Building and Environment, 2006. **41**(2): p. 223–234.

10 Bholah, R. and A.H. Subratty, *Humid Building Problems in Mauritius.* Indoor and Built Environment, 2003. **12**: p. 221–225.

11 Lappalainen, S., et al., *Evaluation of Priorities for Repairing in Moisture-Damaged School Buildings in Finland.* Build Environment, 2001. **36**: p. 981–986.

12 Agyekum, K., B. Salgin, A.K. Danso, *The Health Impacts of Damp Housing Conditions: Lessons for Inhabitants Living in Damp Tropical Buildings,* in *6th International Conference on Infrastructure Development in Africa-ICIDA 2017, 14–16 April 2017.* 2017: Kumasi, Ghana. p. 257–269.

13 Agyekum, K., J. Ayarkwa, and E. Adinyira, *A Holistic Survey of Dampness in a Three-Bedroom Residential Bungalow in Kumasi, Ghana.* Case Studies in Construction Materials, 2014. **1**: p. 125–137.

14 Agyekum, K., et al., *Preliminary assessment of dampness in walls of residential buildings in four climatic zones in Ghana.* Journal of Sustainable Development, 2013. **6**(9): p. 51–61.

15 Straube, J. *Moisture in Buildings,* ASHRAE Journal, 2002.

16 Straube, J. *Review and Hygrothermal Simulations of Planned Insulation Retrofit to Hitchcock Hall, Dartmouth College.* 2007, Building Science Consulting Inc.: Waterloo, ON.

17 Straube, J. and C. Schumacher, *Interior Insulation Retrofits of Load-Bearing Masonry Walls in Cold Climates.* 2007 [cited 2018 20 October]; Available from: buildingscience.com.

18 Brook, D.M. *Home moisture problems.* 2008 [cited 2018 10 September]; Available from: https://catalog.extension.oregonstate.edu/sites/catalog/files/project/pdf/ec1437.pdf.

19 Canadian Wood Council, *Moisture and Wood-Frame Buildings.* 2000.

20 Graham, C.W. *A Building Moisture Load Theory for Improvement of Whole House Durability.* [cited 2013 08 May]; Available from: www.pathnet.org.

21 MacDonald, R. *Moisture Effects on Facility Life: Sources and Corrections.* Advances in Pork Production, 2005. **16**: p. 197–203.

22 Wilson, A. *Moisture Control in Buildings: Putting Building Science in Green Building.* 2003 [cited 2013 06 August]; Available from: http://www.shelterecology.com/articles/art_Moisture.html.

23 US Department of Energy Webpage. *Building energy software tools directory.* [cited 2018 18 September]; Available from: https://www.buildingenergysoftwaretools.com/

24 Voutilainen, J. *Methods and Instrumentation for Measuring Moisture in Building Structures.* A PhD Thesis submitted to the Department of Electrical Engineering, Helsinki University of Technology, Espoo, Finland, 2005: p. 1–157.

25 Pereira, C., J. Britob, and D.J. Silvestre, *Contribution of Humidity to the Degradation of Façade Claddings in Current Buildings.* Engineering Failure Analysis, 2018. **90**: p. 103–115.

26 Kubba, S.A.A. *Architectural Forensics.* 2008, McGraw-Hill: The United States of America.

27 Al-Hammad, A., S. Assaf, and M. Al-Shihah, *The Effect of Faulty Design on Building Maintenance.* Journal of Quality in Maintenance Engineering, 1997. **3**(1): p. 29–39.

28 Asamoah, R.O., et al., *Construction Practices Contributing to Rising Damp in Kumasi Metropolitan and Ejisu Municipal Assemblies in Ghana.* Journal of Construction Engineering. **2017**: p. 1–6.

29 Ge, G., F. Xiao, and S. Wang, *Neural Network Based Prediction Method for Preventing Condensation in Chilled Ceiling System.* Energy and Building, 2012. **45**: p. 290–298.

30 World Health Organization, *WHO Guidelines for Indoor Air Quality: Dampness and Mould.* 2009.

31 Chew, M.Y.L. *Defect Analysis in Wet Areas of Buildings.* Construction and Building Materials, 2005. **19**(3): p. 165–173.

32 EPA, *Moisture Control Guidance for Building Design, Construction and Maintenance.* 2013.

33 Lstiburek, J. *Understanding Drainage Planes.* 2006 [cited 2018, 18 September]; Available from: https://buildingscience.com/documents/digests/bsd-105-understanding-drainage-planes.

34 Buildings Department. *Guidelines on Prevention of Water Seepage in New Buildings.* [cited 2018 19 September]; Available from: https://www.bd.gov.hk/english/documents/circular/GWS.pdf.

35 NRCA. *The NRCA Roofing Manual: Membrane Roof Systems 2011.* 2011, National Roofing Contractors Association: Rosemont, IL.

36 Wilen, J.P. *Proper roof slope and drainage are important to prevent excessive water accumulation.* 2012 [cited 2018 19 October]; Available from: http://docserver.nrca.net/technical/9862.pdf.

37 Ishak, S.N.H., A.H. Chohan, and A. Ramly, *Implications of Design Deficiency on Building Maintenance at Post Occupational Stage.* Journal of Building Appraisal, 2007. **3**(2): p. 115–124.

38 Sev, A. and A. Özgen, *Yüksek Binalarda Sürdürülebilirlik ve Doğal Havalandırma*, Yapı Dergisi, 2003. **262**: p. 92–99.

39 Wood, A. and R. Salib, *Natural Ventilation in High-Rise Office Buildings, CTBUH Technical Guides.* 2012, Taylor & Francis LTD: London.

40 BRE. *The Environmental Building: A model for the 21st century.* [cited 2018 19 September]; Available from: http://projects.bre.co.uk/envbuild/envirbui.pdf.

41 Sushil Kumar, T.V.K., et al., *An Analysis of Effect of Wing Walls on Natural Ventilation.* Indian Journal of Science and Technology, 2015. **8**(17): p. 1–4.

42 Ismail, M. and M.A. Abdul Rahman, *Stack Ventilation Strategies in Architectural Context: A Brief Review of Historical Development, Current Trends and Future Possibilities.* IJRRAS, 2012. **11** (2): p. 291–301.

43 ASHRAE, *ASHRAE Handbook: Fundamentals.* 1997, American Society of Heating, Refrigerating, and Air-Conditioning Engineers: Atlanta, GA.

44 Aflaki, A., et al., *Relevant Indoor Ventilation by Windows and Apertures in Tropical Climate: A Review Study*, in *Emerging Technology for Sustainable Development Congress*, 2014, Bangi, Malaysia.

45 Kelly, N. *Ventilation for Low Energy Buildings.* [cited 2018 19 September]; Available from: https://www.slideserve.com/yukio/ventilation-for-low-energy-buildings

46 Watson, D. and K. Labs, *Climatic Building Design: Energy Efficient Building Principles and Practices.* 1993, McGraw Hill: New York.

Assoc. Prof. Dr. Canan Cengiz and Res. Asst. Aybüke Özge Boz

12 Urban Greenway Systems within the Context of Sustainable Landscapes

Abstract: The use of greenways is encouraged in both rural and urban areas for land conservation. Urban greenways are mostly used for recreational greenways in cities with paths and different kinds of trails based on natural corridors in addition to public rights of way, abandoned rail beds and canals . Greenways are networks of land containing linear elements that are planned, designed and managed for multiple purposes including ecological, recreational, cultural, aesthetic or other purposes compatible with the concept of sustainable land use. In addition to facilitating the planning and shaping of communities, they also improve the quality of life of neighborhoods and communities while protecting the cultural and historic resources in cities. Cultural resources may be connected via greenways into a type of network or system with greater value and higher use than the sum of the constituent parts as a kind of landscape synergy. Greenway systems are important for improving urban quality of life, contributing to urban ecology and urban aesthetics, protecting natural and cultural landscape values, providing urban biodiversity, presenting urban recreational area, emphasizing urban identity, contributing to social sustainability and evaluating sustainable transportation options in an integrative and multidisciplinary framework within the context of the planning and design of sustainable landscapes.

Keywords: Urban greenway, Sustainable landscapes, Greenway benefits, Greenway planning, Urban ecology, Quality of city life, Sustainability

1 Introduction

Greenways are linear green spaces that are generally formed along channels, rivers, historical routes, valleys, routes, ridges and railroads with goals such as recreation, protection and economic growth. According to a certain concession, greenways support sustainable growth due to their intrinsic structure. It has been indicated by recent studies on greenways that greenways comprise an inseparable whole with sustainability and the principles set forth by the World Commission on Environment and Development; that these areas should be planned in a well-balanced manner with their economic, environmental and social characteristics [1,2]. Beatley and Manning [3], Fábos and Ahern [4], Flink and Searns [5] have suggested greenways frequently during urban planning stage for greener, healthier and more livable urban spaces [1].

Greenways are directly related with landscape ecology within the context of sustainability. Landscape ecology emphasizes the interaction between spatial pattern and ecological process, that is, the causes and consequences of spatial heterogeneity across a range of scales (Tab. 1) [6].

Tab. 1: Definition of Commonly Uused Terms in Landscape Ecology [Modified 7]

Term	Definition
Biodiversity	Different variety of life forms, especially number of species including number of ecosystem types and genetic variation within species
Connectivity	A measure of how connected or spatially continuous a corridor, network or matrix is.
Corridor	A strip of a particular type that differs from the adjacent land on both sides.
Ecosystem	A relatively homogeneous area of organisms interacting with their environment.
Fragmentation	The breaking up of a habitat, ecosystem or land-use type into smaller parcels.
Habitat	The ecosystem where a species lives, or the conditions within that ecosystem.
Heterogeneity	The uneven, non-random distribution of objects. (Contrasts with homogeneity, and also with arrangement where objects are spatially configured in a particular way)
Network	An interconnected system of corridors.
Road corridor	A linear surface used by vehicles plus any associated, usually vegetated parallel strips.
Stream (or river) corridor	A strip vegetation enclosing, or on one side of, a water channel
Sustainability	The condition of maintaining ecological integrity and basic human needs over human generations

The use of greenways is encouraged in both rural and urban areas for land conservation. However insufficient attention has been paid to their potential for serving urban biodiversity conservation goals [8]. The increasing number of urban centers makes it necessary to study the interaction between living organisms and urban environment defined as the environment surrounded by man-made structures, such as residential and commercial buildings, paved surfaces etc. [9, 10]. Within this scope, urban ecology developed as a branch of ecology in the last few decades [9].

Weiland and Richter [11] have put forth that urban ecology is inherently an interdisciplinary field of study. It examines the complex interactions between natural and social sciences. Thereby, urban ecology can be used for improving the conditions of urban environments in addition to understanding the urban systems as a result of this interdisciplinary nature. For example, it is required to comprehend how urban systems function and to what extent they are affected from global and local processes so that we can analyze how to maintain the water cycle in a region as well as the affecting factors such as the use of landscape, the effect of green spaces, climate conditions, the coexistence of species, etc. [9]. It is through the integration of several disciplines which examine the ecological and human dimensions of urban ecosystems that urban ecology evolved as a unique field of study. Urban ecology is a primary discipline which creates the knowledge base, conceptual frameworks and tools crucial for building and maintaining sustainable and resilient cities and towns for future generations [12]. There are distinctive patterns of change for cities, as well as unique spatial organizations which result in patterns of species behaviour, population dynamics and the formation of communities that are specific to the urban environment. Urban ecology is synonymous with "sustainable cities" in fields of policy and planning [13].

2 Greenway History

Greenway movement in Europe has developed differently in comparison with its counterparts in America due to social, economic and cultural differences. Greenways in Europe have followed a non-continuous and piecewise approach which has spread out to many countries [1, 14]. Greenways provide significant contributions with regard to sustainability in both cases [1]. Majority of early efforts in landscape planning put forth by visionary landscape architects and planners would today be called greenways. Frederick Law Olmsted is credited by many greenway planners as the founder of the greenway movement [15,16]. The Boston Park System, or "Emerald Necklace" of Olmsted is often referenced to as the first significant greenway in the United States [15].

Together with Frederick Law Olmsted, one of the pioneers of landscape architecture in USA, Fábos [17] has classified greenways in five different periods according to the evolution that they have undergone since their scientific emergence. These are:

- 1867–1900: first landscape architects and greenway planning period,
- 1900–1945: the period during which landscape architects carried out studies on greenway planning,
- 1960–1970: the influence of the environmental decades on landscape and greenway planning,
- 1980–1990: the period encompassing the start of the greenway movement nationally and internationally, as well as its applications.

The history of greenways date back to the ancient world. The ceremonial avenues used in ancient times are known as the first examples of greenway use. Greenways are among the most appealing approaches in the latter part of the 20th century for integrating open space types in cities. Footpaths, bridges, cycleways, bridlepaths, stream valleys, linear parks, waterfront reservations, covered arcades, elegant streets and greenways of every type may be used for interlinking public open

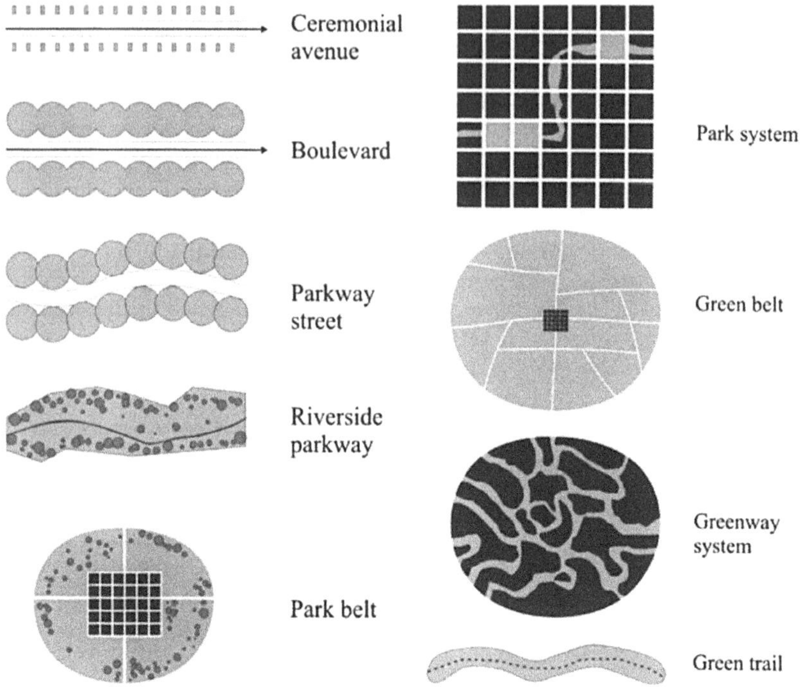

Fig. 1: Ancestors of the greenway concept, shown diagrammatically [18,19].

spaces [18,19]. The "red infrastructure" of roads, railways and airports can be complemented by the "green infrastructure" as the system extends into the urban hinterland. Open spaces are among the most frequent "Alexander patterns" with regard to such a planning [18]. Turner [18] has explained the historical development of green corridors using a diagram that depicts the evolution from the ceremonial avenues of the past to the green corridor systems of our day.

As shown in Fig. 1, the ancestors of the greenway concept include [18]:

- the ceremonial avenue, as in Ancient Egypt;
- the boulevard, originally a walk on the bulwark of a fortified town but later a tree-lined street;
- the parkway, planned for recreational transport; a riverside parkway, for urban recreation;
- the park belt, on the perimeter of a settlement for the recreational needs of a whole town; a park system, planned to inter-connect urban parks;
- the green belt, planned to control urban sprawl;
- the greenway system a settlement for the recreational needs of the entire city;

- the green trail, planned as a recreational routeway in an urban or rural area.

Perhaps the most important changes since Olmsted's Boston Park System has been the change in how greenways are defined today. The primary use of Olmsted's Park System was recreation. The greenways of today have at least three major functions or definitions. Firstly, greenways are: ecologically significant corridors and natural systems; secondly, users have access to a network of trails and water link land and water-based recreational sites and areas by way of recreational greenways and thirdly greenways provide historical heritage and cultural values. All three functions and benefits are often provided by greenways as multipurpose corridors [5,20].

3 Green Infrastructure

Urban green infrastructure is an essential part of the urban network that serves the interests of both people and nature. Green infrastructure is defined as the combined structure, position, connectivity and types of green spaces in and around cities that together provide improved quality of life and ecosystem services [21–23]. According to Benedict and McMahon [24], green infrastructure is comprised of parks, gardens, greenways, forests, wetlands, agricultural fields, green roofs and green walls regardless of its ownership, condition or size [21]. According to Orr et al. [25] urban green spaces (UGS) are an essential part of urban green infrastructure system. UGS come in many forms and the greenway is one of them [21]. The importance of greenways is widely accepted with regard to the development of urban ecological networks, and, together with greenbelts, they provide a comprehensive green infrastructure [26]—or ecostructure—on which cities are built [27].

4 Greenway Planning

Greenways are defined as "a linear open space established along either a natural corridor, such as a riverfront, stream, valley, or ridgeline, or overland along a railroad right-of-way converted to recreational use, a canal, scenic road, or other route," "Any natural or landscaped course for pedestrian or bicycle passage," "An open-space connector linking parks, nature reserves, cultural features, or historic sites with each other and with populated areas," "Locally, certain strip or linear parks designated as parkway or greenbelt" [28,16]. Urban greenways are defined as places for nature in the city where people can fulfill recreational needs and achieve solitude and retreat without leaving the public realm [21,29] and they are generally designed with multi-use trails providing opportunities for physical activity, recreation and transportation [21,30]. Urban greenways are mostly used for recreational greenways in cities with paths and different kinds of trails based on natural corridors in addition to public rights of way, abandoned rail beds, and canals [21,16]. They also have various important functions and provide significant benefits improving the quality of life through the use of land for multiple purposes [21,31].

Various studies have been carried out which put forth the ecological, environmental, social, recreational, aesthetic, educational, and economic benefits of urban greenways as well as their importance. As an example, urban greenways sustain and improve hydrological functions, provide a place for remediation of contaminants by way of natural processes in addition to a location for significant levels of carbon entrapment. In addition to facilitating the planning and shaping of communities, they also improve the quality of life of neighborhoods and communities while protecting the cultural and historic resources in cities. Park components are also interconnected by urban greenways to form a recreation, cohesive park, and open space system in cities thereby allowing for safe and uninterrupted pedestrian movement between parks throughout the community. The aesthetic and natural beauty of neighborhoods and communities are also improved by urban greenways, and they also serve to protect public infrastructure investment as well as enhance property values [8,16,17,21,32–35].

Greenways are networks of land containing linear elements that are planned, designed and managed for multiple purposes including ecological, recreational, cultural, aesthetic or other purposes compatible with the concept of sustainable land use. The aforementioned definition includes five key concepts which require further discussion. First of all, greenways are primarily linear with regard to spatial configuration. Greenways are based on particular characteristics and opportunities inherent in linear systems that provide distinct advantages with regard to movement and transport of materials, species or nutrients. Secondly, greenways are defined by the linkage characteristic which is a key concept that defines the greenway in addition to relating it often at multiple scale levels to the larger landscape context. Greenways as integrated systems attempt to realize a synergy based on the advantages of linkages across spatial scales. Thirdly, greenways are multifunctional, based on an assumed or negotiated spatial and functional compatibility of certain uses. The process of establishing goals in greenway planning is particularly important due to this characteristic, since all goals cannot be optimized, trade-offs and compromises must be made to reflect the ecological, cultural, social and aesthetic goals associated with greenways. As an example, in case it is not possible to achieve compatibility, the needs of recreation and wildlife habitat protection often conflict with each other and may require spatial segregation, specific management or elimination of one of the uses [35]. According to IUCN [36], WCED [37] and Lyle [38] fourthly, greenways strategy is consistent with the concept of sustainable development, in that it is based on an assumed complementarity between nature protection and economic development. In addition to protecting nature, other human uses of the landscape are also recognized and legitimized while a balance is tried to be established between resource use and protection. Sustainability should be regarded as a special globally accepted goal and paradigm for the future [35]. According to Checkland [39] and Zonneveld [40] lastly, greenways represent a distinct spatial strategy based on the particular characteristics and

Tab. 2: Ecological Network and Greenway Terms [35]

Term	**Term usage**	**Functions: Biotic, Cultural, Multifunctional**	**Scale: Continental, National, Regional, Local**
Ecological Networks	Europe	B	C,N,R,L
Habitat Networks	Europe/America	B	N,R,L
Ecological Infrastructure	Europe	B	C,N,R,L
Greenways	America	B, C, M	N,R,L
Wildlife Corridors	America	B	R,L
Riparian Buffer	Europe/America	B, M	R,L
Ecological Corridors	America	B	R,L
Environmental Corridors	America	M	R,L
Greenbelts	Europe/America	C	R,L
Landscape Linkages	America	B	R,L

advantages of integrated linear systems [35]. Greenways should be considered not as a replacement to but comprehensive landscape and physical planning but as a complementary concept to them [35].

The increasing use of green infrastructures in different situations and heterogeneous projects have created a wide range of design styles from the first experiments by Ebenezer Howard in the United Kingdom and Frederick Law Olmsted in the US Llausàs and Roe [41], which differ according to context, scale of the project, objectives and strategies. According to Ahern [35], Ecological Network for EU and Green Systems in the Balkans, Greenways in the USA, Biodiversity Corridors in the Philippines and Urban Greening in Singapore are just a few examples. According to Fábos and Ryan [15] (Tab. 2) [42], European ecological networks and the American greenways are among the two most popular green infrastructure solutions. Source Ahern [35] the context can be urban, fringe, rural or natural; the extension of the project can vary from a local level to the macro-continental order structures Jongman [43], Jones-Walters; its objectives can be ecological, anthropogenic, cultural, recreational or mixed [42].

It is important when designing a greenway to consider its impact on natural processes. Like environmental corridors which are their natural counterparts, there are six basic functions of greenways [44]:

- *habitat* for plant and animal communities,
- *a conduit* for plants, animals, water, sediment and chemicals,
- *a barrier* preventing movement,
- *a filter* allowing some things to pass while inhibiting others,

- *a source* for animals or seeds which move to other parts of the landscape, and
- *a sink* for trapping sediment, toxins or nutrients.

5 Greenway Typologies

An attribute of greenways according to Ahern [45] is that they provide an approach to open space planning which responds better to the growing awareness of and concern for the interconnectedness of ecological systems. Thereby significant improvement is put forth over approaches which determine open space on a residual basis of areas that cannot be developed. According to Forman and Godron [46], this approach generally includes floodplains, wetlands, steep slopes, and water resources, as well as agricultural, visual, and historical resources into an open space plan, but often fails to consider the configuration, juxtaposition, and functional relationships between these as an essential component of an ecologically sound approach to development [47].

Greenways are sometimes referred to as environmental corridors, landscape linkages, wildlife corridors, or riparian buffers which provide an important means for protecting natural areas and providing opportunities for recreation [16,35,48]. According to Fábos and Ahern [4], greenways make significant contributions to many ecological and societal values [48]. According to Forman [7,49] and Forman and Godron [46] the greenways help maintain biological diversity, protect water resources while conserving soils, supporting recreation, enhancing the community and cultural cohesion in addition to providing migration routes for different species during climate or seasonal change [48]. Smith and Hellmund [50] have put forth that greenways can be used to create interconnected networks of open space that may include more traditional non-linear parks and natural areas. They can help to maintain ecological integrity in human-dominated landscapes, especially with regard to sustaining high-quality water reserves and preserving biological diversity [48].

Little [16] defined the following five types of greenways:

1. Urban riverside (or lakeside) greenways, which are usually created as part of a redevelopment program along neglected and often run-down city waterfronts.
2. Recreational greenways, featuring paths and trails of various kinds, often a relatively long distance, based on natural corridors as well as canals, abandoned rail beds and other public rights of way.
3. Ecologically significant natural corridors, usually along rivers and streams and sometimes ridgelines, to provide for wildlife migration and species interchange, nature study and hiking.
4. Scenic and historic routes, usually along a road or highway (or less often a waterway), the most representative of them making an effort to provide pedestrian access along the route or at least places to alight from the car.
5. Comprehensive greenway systems or networks, usually based on natural land forms such as valleys and ridges but sometimes simply an opportunistic

assemblage of greenways and open spaces creating an alternative municipal or regional green infrastructure.

The English landscape architect Tom Turner [51] argued for special purpose greenways: parkways, blueways, paveways, glazeways, skyways, ecoways and cycleways [27].

Tab. 3 shows the main components and main benefits of greenways. When we rank the components based on their benefits, riverfronts/streams and parks are the most important. They are followed by natural/landscaped courses, nature reserves, valleys, streets, scenic roads/highways, railroads, canals and ridgelines, respectively. It is well known that greenways provide the connections linking these components to one another. When designing a greenway, taking advantage of as many of these components is an important criterion for that greenway to be identified as high performing. For example, if a greenway uses riverfronts/streams, parks, natural/landscaped courses, scenic roads, railroads, and streets then its effects and benefits to the society would be much more advanced than if it only provided for riverfronts and streams [31].

6 Ecological Benefits of Greenways

The term *environmental corridor* is sometimes used instead of greenways which refers to landscape linkages, wildlife corridors or riparian buffers that provide an important means to protect natural areas and to provide recreation opportunities [16,35].

According to Peterjohn and Correll [52] and Binford and Buchenau [53] these flows can be moderated by greenways to maintain more sustainable levels of landscape function through buffering. There are two main types of buffering that relate to greenways: riparian corridor/wetland filtering and patch edge buffering. Riparian systems and wetlands are generally situated downhill from terrestrial environments and are thus the recipients of materials moved by the mass flow processes of erosion and mass wasting. The nature of these materials generally has an impact on the ecological integrity of riparian ecosystems. Filtering sediments, controlling erosion, and regulating water temperature are among means with which such impacts are mitigated by greenway buffers located between disturbances and riparian corridors and wetlands. Protecting interior habitat from external disturbances is another function of greenway buffers [35].

It has been indicated by Brown et al. [54] and World Resource Institute et al. [55] that greenways provide an opportunity to reduce the impacts of habitat fragmentation. Habitat fragmentation is among the most severe threats for biological diversity and is a primary cause of the extinction crisis [47]. It has been indicated by Wilcove et al. [56] and Opdam [57] that loss of habitat and habitat isolation are among the two main effects of fragmentation. Population size is decreased and extinction rates are increased as a result of habitat loss, while isolation decreases the likelihood of recolonisation of otherwise productive habitats [47].

Tab. 3: Greenways Main Components and Benefits/Functions [31]

Main Benefits Main Components	Aesthetic	Recreation	Conservation	Connectivity	Biological diversity	Wildlife Migration	Economic benefit	Preservation	Social interaction	Water discharge
Riverfronts/ Streams	x	x	x	x	x	x	x	x	x	x
Valleys		x	x	x	x	x		x		x
Ridgelines		x	x			x				
Railroads	x	x		x					x	
Canals	x	x		x						x
Scenic roads/ Highways	x	x		x					x	
streets	x	x		x			x		x	
Natural/ Landscaped courses	x	x	x	x	x	x		x	x	x
Parks	x	x	x	x	x	x	x	x	x	x
Nature reserves		x	x	x	x	x		x		x
Cultural features/ Historical sites	x	x	x				x	x	x	

One of the most important ecological functions of greenways is that they act as a buffer for protecting surface waters from pollution due to non-point sources along the river corridors. When used in urban areas, they enable the protection of the areas that are under threat from the developing sections of the city [58]. They also have other ecological functions such as: improving air quality thanks to the air flow provided due to their corridor effect, filtering gases and dusts, cleaning the air, reducing noise and absorption of radioactive and poisonous gases.

Public Health Benefits of Greenways

It has been put forth in recent studies that there is a "positive relationship between well-being, health, and green space" [59,60]. Payne [61] has asserted that older park users perceive themselves to be in better health, feel less stressed and are less likely to be overweight in comparison with non-park users; that they visit doctors less frequently [59]. There is a positive association between levels of physical activity and closeness to green spaces in the neighborhood [59,60]. There are other studies which show that exposure to nature improves physical and psychological health as well as evidence indicating the relationship between green space and physical activity. Therefore, minority populations are less likely to engage in physical activity which can lead to negative health effects including obesity, diabetes, heart disease, respiratory disease, hypertension and mental illness due to the impact of the physical environment on the behaviour of individuals [59,62].

Social Benefits of Greenways

Greenways contribute to many ecological and societal values [4]. Greenways or open spaces provide means for social interaction while strengthening the sense of community in addition to improving physical and emotional health.

Greenway systems provide alternative travel choices for transit, schools, workplaces and shopping areas. These alternative choices for travel help in reducing congestion, improving air quality and providing safe alternatives to residents who wish to bike or walk to nearby destinations [63]. Greenways connect residential areas with schools, parks, workplaces, urban centers, public activity areas, transfer areas thereby enabling the residents to prefer non-motorized vehicles. Therefore, paths are transformed into greenways that can be used for bicycle rides and for short walks. Moreover, they also reduce air pollution while improving air quality and reducing energy costs. Greenways provide alternative access routes that connect people, societies and rural areas. Greenways do not only protect good lands, natural plant and animal species; they also connect people with open air recreational facilities and natural life. They provide possibilities for sharing by way of fiber optic lines, infrastructure and railway lines [64].

Thanks to their natural characteristics, scenic value and similar attributes, greenways improve the quality of life of societies. According to the study by Lee [65], many green corridors in Chicago Oakland affect the local public due to their scenic characteristics. The friendship and neighboring relationships between the locals who use the corridor are improved and thereby the corridors are transformed into locations of focus where certain activities are organized [66,67].

Economic Benefits of Greenways

According to Marples [68], the economic benefits of greenways have been emphasized during 1960–1970s. These are as follows: greenways provide new and increased work opportunities, the value of lands close to the greenways are increased, tourism is increased by greenways [64]. According to Jongman [69], corridors generate a wind line for agriculture areas and animal rearing, prevent soil erosion and desertation while providing forest products which are resources that the local public may use for supporting themselves [70]. According to Flink and Searns [5], greenways increase the economic value of the surrounding areas while generating recreation-based work opportunities and attractive areas for tourists [1]. In addition, greenway users also provide economic contributions to businesses along the route for their food, beverage, accommodation, etc., needs.

Aesthetic Benefits of Greenways

Aesthetic quality and public perception may be increased by greenway systems. While forming a "green infrastructure" that connects people with spaces, greenways also work as mechanisms for the protection of open spaces [66,71]. In this scope, greenways present opportunities to planners and designers for developing new norms in urban planning and design [66]. Silva et al. [72] put forth that plants constitute a significant corner stone for urban biodiversity. However, the importance of green spaces in urban areas is due to their role not only in ensuring ecological sustainability but also in social and aesthetic aspects of urban life [73]. It is possible via greenways to provide a visible structure and legibility to the landscape [35].

7 Discussion and Conclusion

Urban quality of life is related with providing sufficient technical, economic and social infrastructure in addition to the environment that supports urban life as well as the urban landscape developed by way of ecologic and aesthetic approaches. Greenway applications provide important opportunities in designing integrated development plans which make use of social, economic, ecologic and cultural criteria. Greenways are applications with significant importance as multi-functional linear systems which connect urban and rural areas.

Greenway systems are important for improving urban quality of life, contributing to urban ecology and urban aesthetics, protecting natural and cultural landscape values, providing urban biodiversity, presenting urban recreational area, emphasizing urban identity, contributing to social sustainability and evaluating sustainable transportation options in an integrative and multidisciplinary framework within the context of the planning and design of sustainable landscapes.

Cultural resources may be connected via greenways into a type of network or system with greater value and higher use than the sum of the constituent parts as a kind of landscape synergy [35]. The greenway concept as a planning and design has been embraced by other professions dealing with planning and design and comprises a tool that is beyond landscape architecture. Thus, the concept of greenway has become a tool not only for protecting resources and connecting places but also a means of improving communication between different professions dealing with planning and design, thus contributing to a better global landscape [74].

References

1 Kurtaslan B. Ö., *Kentsel yeşil yollar ve sürdürülebilirlik: Boston Rose Fitzgerald Kennedy Yeşil Yolu örneği.* Uluslararası Hakemli Tasarım ve Mimarlık Dergisi, 2014. **1**(2): pp. 86–97.

2 Lindsey, G., *Sustainability and Urban Greenways: Indicators in Indianapolis.* Journal of the American Planning Association, 2010. **69**(2): pp. 165–180.

3 Beatley, T. and K. Manning, *The Ecology of Place: Planning for Environment, Economy, and Community.* USA: Island Press, 1997.

4 Fábos, J. G. and J. F. Ahern, *Greenways: The Beginning of an International Movement.* Amsterdam: Elsevier, 1995.

5 Flink, C. and R. Searns, *Greenways: A Guide to Planning Design and Development.* USA, 1993.

6 Turner, M. G. and R. H. Gardner, *Landscape Ecology in Theory and Practice: Pattern and Process.* Second Edition. New York: Springer-Verlag, 2015. 482.

7 Forman, R.T.T., *Land Mosaics: The Ecology of Landscapes and Regions.* Cambridge: Cambridge University Press, 1995.

8 Bryant, M. M., *Urban Landscape Conservation and the Role of Ecological Greenways at Local andMmetropolitanSscales.* Landscape and Urban Planning, 2006. **76**: p. 23–44.

9 Cengiz, C. *Urban Ecology.* Advances in Landscape Architecture, Murat Özyavuz, 2013 [cited 2018 22, November]; Available from: https://www.intechopen.com/books/advances-in-landscape-architecture/urban-ecology

10 Niemelä, J., *Ecology and Urban Planning.* Biodiversity and Conservation, 1999. **8**: pp.119–131.

11 Weiland, U. and M. Richter, *Applied Urban Ecology: A Global Framework.* First Edition. Edited by Matthias Richter and Ulrike Weiland. Blackwell Publishing Ltd, 2012.

12 McDonnell, M.J., *The History of Urban Ecology: An Ecologist's Perspective.* Urban Ecology Patterns, Processes, and Applications, 2011. pp: 5–13. DOI: 10.1093/acprof:oso/9780199563562.001.0001

13 Sukopp, H., *On the Early History of Urban Ecology in Europe.* Preslia, 2002. **74**: pp. 373–393.

14 Toccolini, A., N. Fumagalli, and G. Senes, *Greenways Planning in Italy: The Lambro River Valley Greenways System.* Landscape and Urban Planning, 2006. **76**: pp. 98–111.

15 Fábos, J.G. and R.L. Ryan, *International Greenway Planning: An Introduction.* Landscape and Urban Planning, 2004. **68**: pp. 143–146.

16 Little, C., *Greenways for America.* Baltimore and London: The John Hopkins University Press, 1990.

17 Fábos, J. G., *Greenway Planning in the United States: Its Origins and Recent Case Studies.* Landscape and Urban Planning, 2004. **68**: pp. 321–342.

18 Turner, T., *Greenway Planning in Britain: Recent Work and Future Plans.* Landscape and Urban Planning, 2006. **76**: pp. 240–251.

19 Turner, T., *Landscape Planning and Environmental Impact Design.* London: UCL, 1998.

20 Fábos, J. G., *Introduction and Overview: The Greenway Movement, Uses and Potentials of Greenways.* Landscape and Urban Planning, 1995. **33**: pp. 1–13.

21 Akpınar, A., *Factors Influencing the Use of Urban Greenways:A Case Study of Aydın, Turkey.*Urban Forestry & Urban Greening, 2016. **16**: pp. 123–131.

22 Foster, J., A. Lowe, and S. Winkelman, *The Value of Green Infrastructure for Urban Climate Adaptation.* Center for Clean Air Policy, 2011.

23 Forest Research, *Benefits of Green Infrastructure.* Report by Forest Research. Farnham: Forest Research, 2010.

24 Benedict, M. A. and E. T. McMahon, *Green Infrastructure: Smart Conservation for the 21st Century.* Sprawl Watch Clearinghouse Monograph Series, 2002.

25 Orr, S., J. Paskins, and S. Chaytor, *Valuing Urban Green Space: Challenges and Opportunities.* UCL public policy, 2014.

26 Walmsley, A., *Greenways and the Making of Urban Form.* Landscape Urban Planning, 1995. **33**: pp. 81–127.

27 Ignatieva, M., G. H. Stewart, and C. Meurk, *Planning and Design of Ecological Networks in Urban Areas.* Landscape and Ecological Engineering, 2011.**7**: pp. 17–25.

28 Ahern, J., *Greenways in the USA: Theory, Trends and Prospects.* Cambridge: Cambridge University Press, 2003, 34–55.

29 Luymes, D. T. and K. Tamminga, *Integrating Public Safety and Use into Planning Urban Greenways.* Landscape and Urban Planning, 1995. **33**: pp. 391–400.

30 Lindsey, G., et al., *Urban Greenways, Trail Characteristicsand Trail Use: Implications for Design.* Journal of Urban Design, 2008. **13**(1): pp. 53–79.

31 Akpınar, A., *High Performing Greenways Design: A Case Study of Gainesville, GA.* Artium, 2014. **2**(2): pp. 102–121.

32 Cengiz, B., *Urban River Landscapes.* Advances in Landscape Architecture, Murat Özyavuz, 2013 [cited 2018, 22 November]; Available from: https://www.intechopen.com/books/advances-in-landscape-architecture/urban-river-landscapes

33 Cengiz, B., R. C. Smardon, and Y. Memlük, *Assessment of River Landscapes in Terms of Preservation and Usage Balance: A Case Study of the Bartın River Floodplain Corridor (Western Black Sea Region, Turkey).* Fresenius Environmental Bulletin, 2011. **20**(7): pp. 1673–1684.

34 Giordano, L.C. and P.S. Riedel, *Multi-criteria Spatial Decision Analysis for Demarcation of Greenway: A Case Study of the City of Rio Claro, Sao Paulo, Brazil.* Landscape and Urban Planning, 2008. **84**: pp. 301–311.

35 Ahern, J., *Greenways as a Planning Strategy.* Landscape and Urban Planning, 1995. **33**: pp. 131–155.

36 IUCN, *The World Conservation Strategy.* IUCN, UNEP, WWF, Gland, 1980.

37 World Commission on Environment and Development, *Our Common Future.* USA: Oxford University Press, 1987.

38 Lyle, J. T., *Regenemtive Design for Sustainable Development.* New York: John Wiley, 1994.

39 Checkland, P., *Systems Thinking, Systems Practice.* Chichester: John Wiley, 1989.

40 Zonneveld, W., *Conceptvorming in de Ruimtelijke Planning: Patronen en Processen (Vol. 1) and Encyclopedic van Planconcepten (Vol. 2).* Planogische Studies 9, Universiteit van Amsterdam: Planologisch en Demografisch Instituut, 1989, 256–292.

41 Llausàs, A. and M. Roe, *Green Infrastructure Planning: Cross-National Analysis between the North East of England (UK) and Catalonia (Spain).* European Planning Studies, 2012. **20**(4): pp. 641–663.

42 Balestrieri, M. and A. Ganciu, *Greenways and Ecological Networks: Concepts, Differences, Similarities.* Agricultural Research & Technology: Open Access Journal, 2017. **12**(1): pp. 1–3.

43 Jongman, R. H. G., *Nature Conservation Planning in Europe: Developing Ecological Networks.* Landscape and Urban Planning, 1995. **32**(3): pp. 169–183.

44 Labaree, J. M., *How Greenways Work a Handbook on Ecology.* National Park Service and Atlantic Center for the Environment, 1992.

45 Ahern, J., *Planning and Design for Sustainability in a Changing New England Landscape.* 1989, Proc.Landscape/Land Use Planning Committee of the

American Society of Landscape Architecture's 1989 Annual Meeting, Orlando, FL. American Society of LandscapeArchitecture, Washington, DC. 1–12.

46 Forman, R. T., and M. Godron, *Landscape Ecology.* New York: Wiley, 1986, 619.

47 Linehan, J., M. Gross, and J. Finn, *Greenway Planning: Developing a Landscape Ecological Network Approach.* Landscape and Urban Planning, 1995. **33**: pp. 179–193.

48 Miller, W., et al., *An Approach for Greenway SuitabilityAanalysis.* Landscape and Urban Planning, 1998. **42**: pp. 91–105.

49 Forman, R. T., *Corridors in a Landscape: Their Ecological Structure and Function.* Ekologiya, 1983. **2**: pp. 375–385.

50 Smith, D. S. and P. C. Hellmund, *Ecology of Greenways.* Minneapolis, Minnesota: University of Minnesota Press, 1993.

51 Turner, T., *Greenways, Blueways, Skyways and Other Ways to a Better London.* Landscape Urban Planning, 1995. **33**: pp. 269–282.

52 Peterjohn, W. T. and D. L. Correll, *Nutrient Dynamics in an Agricultural Watershed: Observations on the Role of a Riparian Forest.* Ecology, 1984. **65**: pp. 1466–1475.

53 Binford, M. W. and M. J. Buchenau, *Riparian Greenways and Water Resources.* In: D. H. Smith and P. C. Hellmund (Editors), Ecology of Greenways. Minneapolis: University of Minnesota Press, 1993, 69–104.

54 Brown, L., C. Flavin, and S. Pastel, *Vision of a Sustainable World.* In: L. Brown (Editor), The Worldwatch Reader on Global Environmental Issues. New York: Norton, 1991, 299–316.

55 World Resource Institute, The World Conservation Union (IUCN), United Nations Environmental Programme, *Global Biodiversity Strategy.* 1992.

56 Wilcove, D. S., C. H. McLellan, and P. Dobson, *Habitat Fragmentation in the Temperate Zone.* In: M. E. Soule (Editor), Conservation Biology: The Science of Scarcity and Diversity. Sunderland, MA: Sinauer, 1986, 237–256.

57 Opdam, P., 1991. Metapopulation Theory and Habitat Fragmentation. Landscape Ecology, 1991. **5**(2): pp. 93–106.

58 Conine, A., et al., *Planning for Multi-Purpose Greenways in Concord, North Carolina.* Landscape and Urban Planning, 2004. **68**(2–3): pp. 271–287.

59 Kappler, C. and L. Miller,*Re-imagining the Urban Greenway: An Alternative Transportation Strategy and Vacant Land Use Plan for the Woodbridge Neighborhood of Detroit.* A practicum submitted in partial fulfillment of the requirements for the degree of Master of Landscape Architecture (Natural Resources and Environment). USA: University of Michigan, 2009.

60 Tzoulas, K., et al., *Promoting ecosystem and human health in urban areas using Green Infrastructure: A Literature Review.* Landscape and Urban Planning, 2007. **81**: pp. 167–178.

61 Payne, L., et. al., *Local Parks and the Health of Older Adults: Results of an Exploratory study.* Parks and Recreation,1998. 33(10): pp. 64–71.

62 Sherer, P. M., *The Benefits of Parks: Why America Needs More City Parks and Open Space.* The Trust for Public Land, 2003.

63 Pennsylvania Greenways Partnership Commission with Committee Assistance, Benefits of Greenways: A Pennsylvania Study. 2002 [cited 2018, 10 November]; Available from: https://www.americantrails.org/images/documents/BenefitsPA2002.pdf.

64 Hoşgör, Z., *Kentsel yeşil yollar ve yeşil yolların planlama stratejisi: İstanbul-Haliç için bir öneri.* Yüksek Lisans Tezi, İstanbul: İstanbul Teknik Üniversitesi Fen Bilimleri Enstitüsü, 2005.

65 Lee, B.K., *The Dynamic Nature of Emotions during a Trail-Based Leisure Experience: An Application of Affect Control Theory (ACT).* Unpublished Doctoral Dissertation, Texas A&M University, College Station, TX: Department of Recreation, Park and Tourism Sciences, 1999.

66 Salıcı, A., *Çatalan Baraj Gölü-Deli Burun Aksında Seyhan Nehrinin yeşil koridor potansiyelinin araştırılması.* Doktora Tezi, Adana: Çukurova Üniversitesi Fen Bilimleri Enstitüsü, 2009.

67 Shafer, C. S., B. K. Lee, and S. Turner, *A Tale of Three Greenway Trails: User Perceptions Related to Quality of Life.* Landscape and Urban Planning, 2000. **49**: pp. 163–178.

68 Marples, P. J. *Sumas Way Recreational Greenway Design: A Case Study of Sumas Way Abbotsford, B.C.* Master's thesis, Aberta, Canada: The University of Calgary, 2002.

69 Jongman, R. H. G. *The Context and Concept of Ecological Networks* .2004, R.H.G. Jongman and G. Pungetti (eds.), Ecological Networks and Greenways Concept, Design and Implementation, (Sf. 7–33), Cambridge, UK: Cambridge University Press, 2004, 345.

70 Tokuş, M., *Kentsel yeşil ağlar: İstanbul Sarıyer örneği.* Yüksek Lisans Tezi, İstanbul: İstanbul Teknik Üniversitesi Fen Bilimleri Enstitüsü, 2012.

71 Fábos, G. J., *Introduction and Overview: The Greenway Movement, Uses And Potentials Of Greenways.* Landscape and Urban Planning, 1995. **33**:pp. 1–13.

72 Silva, A. I., E. Isidro, and A. L. Soares, *Lisbon's public gardens: Landscape heritage assessment.* 27th ECLAS (European Council of Landscape Architecture Schools) Conference Proceeding, 29 September–2 October. 2010: Istanbul.

73 Bekci, B., C. Cengiz, and B. Cengiz, *Evaluating Urban Biodiversity in Terms of User Preferences: Urban Residential Landscapes in Bartın (Turkey).* Fresenius Environmental Bulletin, 2012. **21**(6): pp. 1626–1634.

74 Riberio, L. and B. Terasa, *Greenways for Recreation and Maintenance of Landscape Quality: Five Case studies in Portugal.* Landscape and Urban Planning, 2006. **76**: pp. 79–97.

Assoc. Prof. Dr. Demet Demiroğlu and Assoc. Prof. Dr. Aybike Ayfer Karadağ

13 A Sustainable and Integrated Framework for Local Governments: Smart City

Abstract: Urban areas, which cover only 2% of the world's surface and which are the most important regions where people live, produce 80% of the greenhouse gas emissions and consume 75% of the natural resources in the world, and these rates are increasing in parallel with the urbanization rate of population. For this reason, various planning and urban concepts/approaches have been developed in recent years, which include the efforts to make urban resources more efficient and economical, and to make cities more livable such as sustainable cities, ecological cities, green cities, slow cities, low carbon cities. Though they have different names, what these approaches have in common is to create awareness of different ecosystems and to plan sustainable cities that consume low energy and water, preserve natural ecosystems, and thus ensure sustainable development. These approaches, which have developed with the increasing awareness of environmental consciousness and climate change, are dealt together with the developments in information and communication technologies in the 21st century; and, they entail the emergence of the concept of "Smart Cities" which bring great innovations and conveniences to local governments in many areas such as transportation, communication, health, safety, energy efficiency, water use and waste management. This chapter defines the concept of smart city; provides information about its components; and examines its role in achieving the sustainability targets of local governments.

Keywords: Smart city, Sustainability, Sustainable local government, Targets of smart city, Component of smart city

1 Introduction

In recent years, the concept of sustainable development has become the key word of many academic studies, international conferences and projects and has attracted much public debate. Though the concept was first defined in 1987 in Our Mutual Future report prepared by UN Environment and Development Commission[1] [1]—it appeared in 1972 in "Rome Club Report" known as the "Limits of Growth." The study in question was prepared due to the concern that the environmental problems would destroy and consume natural resources and prevent development [2]

Since then, sustainability and sustainable development have been and continue to be among the important topics of urban planning. As a matter of fact, urban areas, which cover only 2% of the world's surface and which are the most important

1 Sustainable development "meets the needs of the present without compromising the ability of future generations to meet their own needs" [1].

regions where people live produce 80% of the greenhouse gas emissions and consume 75% of the natural resources in the world, and these rates are increasing in parallel with the urbanization rate[2] of population [3]. For this reason, various planning and urban concepts/approaches have been developed in recent years, which include the efforts to make urban resources more efficient and economical, and to make cities more livable. Various planning and design approaches such as "Sustainable Cities" [5] "Ecological Cities" [6], Green Cities [7], Slow Cities [8], Low Carbon Cities [9] are some examples that start with "New Urbanism" (Congress for the New Urbanism) movement in early 1990s [10]. Though they have different names, what these approaches have in common is to create awareness of different ecosystems and to plan sustainable cities that consume low energy and water, preserve natural ecosystems and thus ensure sustainable development [5–10].

These approaches, which have developed with the increasing awareness of environmental consciousness and climate change, are dealt together with the developments in information and communication technologies in the 21st century; and, they entail the emergence of the concept of "Smart Cities" which bring great innovations and conveniences to local governments in many areas such as transportation, communication, health, safety, energy efficiency, water use and waste management.

This chapter defines the concept of smart city; provides information about its components; and examines its role in achieving the sustainability targets of local governments.

2 Smart City Approach and Its Targets

The first studies on smart city began to be published in 1992. These are the articles written by Laterasse and the book written by Gibson and others. The first example of the smart city is, according to Hollands, a smart laboratory which was established in 2008 in Southampton, England. The smart laboratory was created to develop a portal to support smart card applications [11].

Although the concept of smart city is the most common and preferred definition, it is possible to come across various other definitions. Among these are Intelligent City, Information City, Knowledge City, The Talented City, Wired City, Digital City [12,13].

Washburn and Sindhu [14] define the smart city as a system that includes city management, education, health, public security, transportation and public services in order to provide the city with more intelligent, interdependent and efficient critical infrastructure by using smart information and communication technologies (ICTs).

---

2 The urbanization rate, which was 30% in 1950, has reached 54.5% today. This rate is estimated to reach 66% in 2050 [4].

City economist Andrea Caragliu defined the smart city as "a city that invests in human and social capital, has established its traditional (transportation) and modern (Information and communication technologies–ICT) communication infrastructure, provides sustainable economic growth and high-quality life, and employs participatory governance in the management of natural resources." In general, three basic qualities put forward by Caragliu play key roles in explaining the term *smart city*. These are; forming friendship with environment, using information technologies for smart management and having sustainable development goals [15]. Chourabi et al. [16] describe smart cities as sustainable and livable cities[3]. According to Cohen's [17] definition, smart cities are the cities where resources are used more effectively and wisely; cost, energy savings, service delivery and quality of life are improved; and environmental pollution and carbon emissions are low. According to Dameri [18], the smart city is the place where digital data, services and communication take place via Internet connections in order to improve the quality of life and reduce the environmental footprint. According to Anthopoulos [19] the smart city is the innovation and use of ICT for the sustainability of cities in the economic, social and environmental areas by contributing to the overcome of some challenges encountered.

According to the United Cities and Local Administrations Organization (UCLG), "smart cities" as new city models refer to more livable, functional, competitive, innovative cities which are able in information management by using the highest level of information and communication technologies in cities [21]. The smart city concept in the EU goes beyond the use of information communication technologies for more efficient resource utilization and less emissions. This concept means more efficient urban transport networks, refurbished water networks and waste disposal facilities, and illumination and heating of buildings. In addition, this concept includes more interactive and responsive city management, safer public areas and meeting the needs of the aging population [22].

Despite the rising importance of smart cities, there is no generally accepted definition. One of the most important problems in developing a common definition is the dynamic nature of technological developments and the different meanings attributed to three areas (academic world, industry and management) on which the smart cities are based and the differences of their objectives [23] In the light of these definitions, it is necessary to make a broad definition of the smart cities in an operational understanding: smart city is "a high-quality life through human and social capital investment, traditional (transportation) and modern

3 Characteristic features of sustainable cities include sustainable development, mixed land use, use of public transport and bicycle paths, intensive use of wind and solar energy, protection of natural hydrological systems, protection of wetlands, reproduction of forests, use of natural sewage treatment methods, reduction and recovery of wastes and recycling. These cities also encourage citizen participation to improve the relationship between business and local governments [20].

(ICT) communication infrastructure, sustainable economic growth and intelligent management of natural resources and a logical management of natural resources through governance" [24].

All these definitions inform us that the smart city is a concept in which

- resources are used more effectively and wisely,
- Cost and energy saving applications are present,
- the quality of life conditions is high,
- environmental pollution is low,
- carbon emission is low,
- citizens have high participation in decisions about the city.

When the smart city definitions are examined, three different approaches are identified. These are [25]:

Sustainable Smart City: One of the most popular approaches for smart cities is the consideration of energy consumption. In this approach, great emphasis is placed on energy savings, alternative energy resources and more efficient means of transportation. The biggest advantage of the approach is to turn investments into easy economic savings.

Intelligent City with Sensors: This approach emphasizes how the city is perceived by citizens. The city manages thousands of different sensors dispersed inside the city such as traffic sensors, air pollution sensors, sound sensors, humidity sensors and camera sensors. These sensors provide critical information to solve some of the city's most complex problems. In addition, it uses the existing technology to manage the impressive amount of data generated. When using these sensors, they provide solutions by providing a sensible view of the problems.

Collaborative Smart City: It is the third and a popular approach of smart city. This approach is based on the ability of its citizens to participate in daily operations of the city. It contributes to the development of participatory policies in city management by operating digital participation mechanisms with data provided to citizens.

While it is known that smart cities are basically derived from the idea of information and communication technologies, which is the focal point for the cities of the future, new approaches require those services to be treated as a whole. The adaptation of technological developments to city life is not enough on its own to define cities as smart cities. A smart city is expected to provide residents with sustainable, standardized and participatory life and effectively integrated services. A field having information transfer infrastructure is not enough for it to be defined as "smart" [26].

Apart from their functionality for employing modern technologies, smart cities aim to provide good governance, economic development, educational opportunities and social equality in order to improve the society and life quality [27]. In this context, Caragliu et al. [24] emphasize that cities will only be considered smart when human and social capital, and traditional and modern communication infrastructure, together with participatory governance, provide sustainable economic development and quality life. In this case, the concept of smart city is not only related to the dissemination of information and communication technologies, but also directly related to addressing the needs of people and society and improving the life quality [28].

Dameri [29] argues that four key target sets can be defined for smart cities:

- **Environmental sustainability**: This target is linked to the most important urban problems (traffic, pollution, waste) and technologies such as energy production, mobility, logistics in smart city applications.
- **Quality of life and welfare**: It is a comprehensive goal that can be directly related to the policies of local governments.
- **Participation:** It is a goal linked to e-management initiatives in which citizens will play an active role by participating in city administration.
- **Information and intellectual capital**: It is the abstract aspect of the city and it is a goal that states that the smart cities need abstract resources as well as physical one for a better economic and social development.

The vision of the smart city can be explained as the integration of communication and automation systems (outdoor sensors, mobile terminals, etc.) with private and public spaces. However, smart technologies should also aim for an eco-friendly and energy efficient environment by being integrated to the city. It aims to reduce the city's carbon footprint by providing green energy infrastructure (heat and electricity generating solar panels, home wind turbines, electrified vehicle charging stations, efficient energy distribution and control through smart grids, etc.) [30].

3 Components of Smart City Approach

Another important dimension of the concept of smart city is "Smart City Components" which stand out for the implementation of the concept and its adaptation to urban evaluation. As one of the first examples of this situation, Nijkamp defined six typologies of urban capital to analyze the level of urban sustainability in his study conducted for United Nations. The researcher revealed the "Nijkamp Hexagon" with components such as "Ecoware: nature/environment component," "Hardware: concrete material component," "Finware: economic and financial component," "Orgware: organizational component," "Software: human component," "Civicware: social component" [31].

Today, in the center of smart cities lies the creation of ICT and social and human center and their mutual integration in order to create a greater and more sustainable economic growth and a better quality of life. In this context, while different components are defined for planning of the transformation process of the smart cities and monitoring of the implementation phase, the components that come to fore are the ones expressed in Cohen's Smart City Wheel (Fig. 1) Smart cities are expressed as cities performing forward-looking performance in the context of these six components, which are considered as "smart" combinations of citizens with high awareness and ability in these areas, and urban layers [32–35]. These six

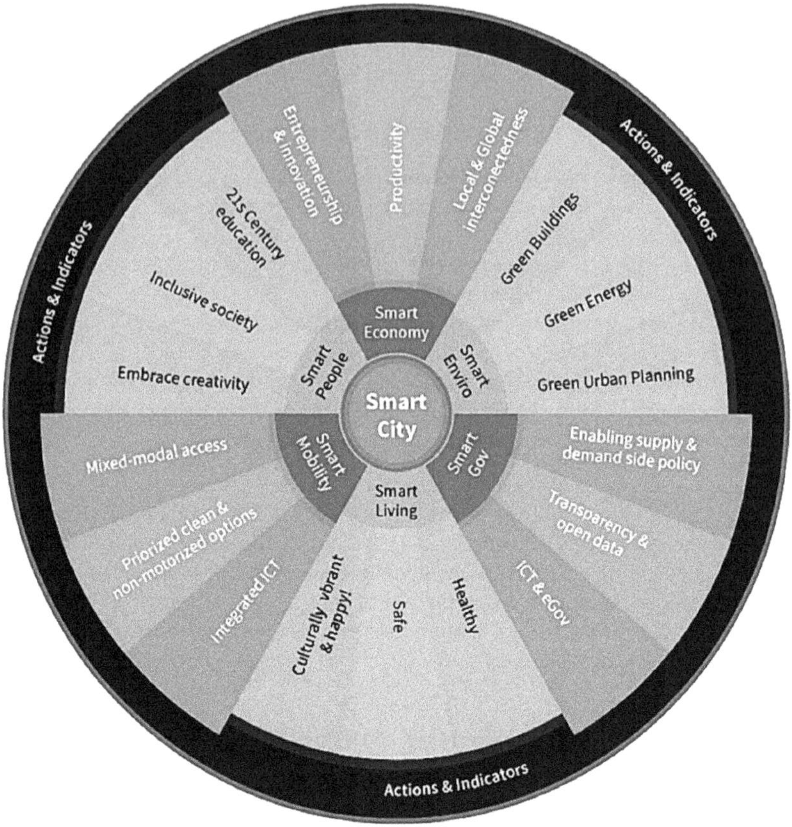

Fig. 1: Cohen's Smart City wheel (smart city components) [17].

components are: smart economy, smart human, smart transportation, smart governance, smart life, smart environment.

Smart Economy: The important point here is to benefit from the emerging new technologies in a way that will contribute to the city's entrepreneurship and production capacity [36]. While advocating innovation and entrepreneurship, the development of new and high technologies are emphasized and innovation to strengthen ties between the domestic economy and the world economy is emphasized [33]. By developing smart applications by local governments, it is possible to implement applications such as meeting consumers and local businesses or increasing the city's general appeal with the brand value created through smart city. The smart economy should direct strategic resources to strategic investments, support the development of local and national brands and ensure the development of urban tourism [37].

Smart Human: One of the most important elements in the construction and maintenance of a smart city is the human capital of that city. In a city where technology usage practice and technology interest are not widespread, smart applications will not flourish. In this context, in terms of human capital [37], some factors in smart cities come to fore such as adequacy of human and social capital, its education, level of quality, lifelong learning tendency, creativity, flexibility and desire to participate in public life [38].

Smart Transportation: This component includes integrated solutions supported with information and communication technologies, mixed access models and environment-friendly non-motorized transportation options. Indicators of smart transport are that it includes human mobility along with vehicle mobility; it supports cycling and walking; it provides suitable roads for pedestrians and cyclists; and it provides integration between transportation types [37]. In addition, thanks to sensors placed in various places in a city, the information about occupancy rate in car parks, the icing conditions on the road, traffic flow according to the vehicle density at the stop points, the place of public transportation vehicles are transferred to drivers and people via mobile applications [33].

Smart Governance: There is a need for a multilateral approach in the process of developing new applications based on technology, as well as implementing the applications and turning them into benefits. Within the scope of this component, efficient and effective communication between different level stakeholders, transparency in public administration and participatory decision-making mechanisms are established through interoperable information and communication technologies solutions. In this sense, the governance of the smart city has an inclusive effect, including other components. Smart cities are cities that use technology to promote overall sustainability. If sustainability is defined as the right balance between environmental, economic and capital stock targets, governance plays an important role in finding this balance [39].

Smart Life: It is classified under the headings such as housing, health, education, security which shape people's life in the city by using ICTs. Characteristics of smart life in smart cities are; sufficient availability of cultural facilities, good health facilities, ensuring individual security, quality of housing, attractive educational and tourist facilities and ensuring social cohesion [37,40].

Smart Environment: It is an element that protects and conserves water resources, and supports the renewable energy resources, and in which natural resources are protected, pollution is reduced, sustainable resource management is provided, biodiversity and urban ecosystem are preserved, and recreational opportunities are vast [37]. It contains solutions such as renewable energy, smart grids, micro grids, smart meters, advanced air pollution monitoring systems, environment-friendly buildings and urban planning, energy efficient smart street lighting, solid waste management, intelligent water management and drainage systems [41].

4 Contribution of Smart City Approach to Local Administration

Within the scope of the smart city approach, many organizations, projects and enterprises around the world commit themselves to providing a smart vision to the future of cities. In addition to global initiatives (Europe Smart Cities, Concerto cities, Europe Smart Cities Ranking, etc.), cities are developing their own smart city projects (such as Amsterdam, Barcelona, Songdo, Vienna, Toronto). Below is given information about one of the initiatives—"CONCERTO cities initiative"—and some smart city projects and their qualifications.

CONCERTO was established in 2005 with the aim of integrating energy efficiency measures and renewable energy systems into cities through a strong cooperation tool involving local authorities, technical consultants, academics and private companies throughout Europe. Fifty-eight cities, 23 countries, and 70 related cities benefit from the knowledge acquired. Within the scope of CONCERTO, 1.830.000 m^2 construction and rehabilitation has been done. About 530 tons of CO_2 emissions per year have been prevented. Changing the behavior of the users and providing the application of renewable energies into the city within the framework of the measures to be made for energy conservation form the basis of the applications [42].

The best example of smart city applications is in Amsterdam. Amsterdam's smart city definition is as follows: the smart city is the wise employment of a city's capital and infrastructure investments to ensure a high quality of life with sustainable economic growth and efficient use of natural resources. The smart applications of the city of Amsterdam date back to 1995 [43].

A pioneer union has been set up to define how energy can be protected between the inhabitants, private companies and the government now and in the future. Amsterdam Smart City focuses on innovative technologies to stimulate the change of behavior of co-residents within the framework of energy conservation [44]. It aims to regulate the energy consumption of Amsterdam inhabitants within the framework of climate targets. For this purpose, it is aimed to reduce CO_2 emissions by 20% in 2025 compared to 1990. Thus, the city will be entitled to the title of the most sustainable city in the world in 2025 [45].

Vienna, Toronto and Copenhagen are among the other cities that set an example for being a smart city in the world. In these cities, it is observed that the cities' administrations demonstrate their conformity to the smart city concept with the investments made especially for low carbon consumption. Reduction of carbon consumption in buildings (Vienna), use of natural gas operated garbage trucks and use of garbage from the city's dumpsite for the purpose (Toronto), 40% of the city's population using bicycles for transportation (Copenhagen) are among the examples that can be shown for the outstanding applications of smart city administrations [46]. In addition, via sensors placed in trash cans in Barcelona, garbage level is monitored remotely, and garbage collection services are optimized. It has also been decided to use smart water systems to manage irrigation in the green areas of the city. Sensors in street lamps adjust the light intensity according

to daylight [47]. About 9700 tons of CO_2 and 600,000 liters of water have been saved annually from smart traffic and park irrigation systems in Barcelona [48]. In the ongoing project in Songdo in South Korea, water and electricity usage is monitored through smart systems in buildings, sensors process vehicle signals to track traffic flow on the road and smart street lamps set for pedestrian traffic are employed [47].

There are many examples in the world that show that smart city applications lead to savings and efficiency. Some of these are highlighted below [48].

- In Los Angeles, a 35 reduction in stopovers, a 20% reduction in intersections, a 13% reduction in travel time and a 12.5% reduction in fuel consumption were achieved through smart transportation systems.
- With the smart street lighting system, the electricity consumption savings rate in Oslo is 70%.
- With a smart parking system, in a European city where 200,000 people live, the time to find a parking space can be reduced by 50%, contributing to the quality of life and fuel savings. While parking revenues increase, personnel expenses can be reduced by 5%–10 %.
- With the help of intelligent solid waste collection systems, the garbage collection system became more stable through sensors planted inside the garbage bins in Seoul's tourist attraction Buckhon, and a 43% reduction in garbage collection costs was achieved by saving from unnecessary travel times to empty bins.
- 15,000 smart meters installed in the United States, Georgia, Tennessee and North Carolina, led to 65 % reduction in operational costs and 50 % reduction in unpaid bill debts

The problems of local governments with increasing population and the smart solutions applied in these areas are given in Tab. 1.

5 Results

Smart city solutions are the systems that aim to increase the environmental sustainability, quality of life, welfare, participation, knowledge and intellectual capital of the city by using information and communication technologies. Technology is only a tool for solving problems in the city. The main issue is to analyze the more complex problems well, to create strong databases to be used in this analysis and to produce technology-oriented solutions.

Sustainable urban development has direct connections with the city's natural and human resource values, urban form, land use decisions and transportation relations. For this reason, sustainable urban planning should be at the center of the smart city strategies and in this context, the subjects that the city needs should be determined and every issue (e.g., energy, infrastructure, transportation, building, etc.) that needs solutions should be planned and organized in an integrated way.

The aim should not be to reduce operating costs and turn the city into a technological laboratory in smart cities; but it should aim to improve the quality of urban

Tab. 1: Urban Problems and ICT-Supported Smart Solutions [49]

	Problems	Smart Solutions
Transportation	Traffic congestion Time spent in traffic Harmful gas emissions	Adaptive traffic control Smart crossroads Park guidance systems Smart bicycles
Energy	High energy demand Expensive, inefficient energy consumption Illegal electric use	Smart meters Smart grids Building energy management
Water	High water demand Resources losing their sustainability capacity	Smart meters Water quality control Detecting leakages
Health	Low service quality Difficulties during service delivery	Home care service
Environment	Environmental pollution	Smart solid waste collection system
Safety	High crime rates	Smart safety systems (Surveillance cameras, camera sensors, software, laser, etc.)
Urban	Unplanned urbanization Time and place restriction removed in service delivery	City information system City guides Information desks Address and population systems Wireless Internet infrastructure

life, to save natural resources and energy, and to reduce negative environmental externality by backing up existing applications with technological assistance. At this point, there is a need for an innovative approach in terms of planning, operation and implementation in the process of transition to smart cities. This approach requires a deeper understanding of the concept of smart city and the implementation of practices in connection with this mind. The biggest task in this regard is on the shoulder of central and local governments. For, transforming a city into a smart state involves the use of high-end technological systems; and, this requires innovative steps. In addition, personnel with an associate degree and a bachelor's degree who can use this technologyto make the city smart and manage it by employing this technology should be trained.

When it is decided to transform the city into a smart city, the needs, innovative opportunities and objectives of the city should be clearly laid out. Since the management of the process of planning, operation and implementation involves multiple systems, coordination between units becomes important and the smart decision systems that form the basis of the process should be supported by a dynamic database. Components determined after various studies and implementations in the context of the strategic plans of cities should be laid out; cooperation should be established for public and private actors, and most importantly, urban participation should be ensured to design socially sustainable and livable smart cities. In addition, in the light of the information given in the central administrations, the following strategies which may contribute to the smart solutions of local governments should be transferred to urban plans and designs at different scales:

- Converting the natural and cultural landscape elements in city into numerical terms and preparing them for analysis within the scope of the necessary objectives,
- Developing urban green infrastructure plans and establishing relevant digital databases,
- Integrating conservation principles of natural resources and efficient water and waste management works into planning and design works,
- Overseeing climatization in the development of settlement texture (light, ventilation, water network, heat island effect),
- Ensuring the integration of renewable energy technologies into cities,
- Disseminating and promoting of zero energy structures or energy efficient structure applications,
- Encouraging compact urban development, referring to integrated living spaces, non-motorized transportation systems, comfortable and safe environment,
- Improving the quality of life and promoting participation in public life,
- Integrating of information communication and energy network technologies (smart grids),
- Developing information communication infrastructure and ensuring integration with urban space,
- Strengthening education, health, cultural activities, and individual security opportunities,
- Encouraging participation in decision-making processes.

References

1 United Nations (UN), *Report of the World Commission on Environment and Development: Our Common Future.* 1987 [cited 2018, 10 November]; Available from: http://www.un-documents.net/our-common-future.pdf.

2 Meadows, D., Randers, J. and Behrens, W. *The Limits to Growth, A Report For The Club Of Rome's Project on the Predicament of Mankind 1972.* Universe Books, New York. Library of Congress Catalog Card Number: 73-187907 ISBN 0-87663-165-0.

3 Suzuki H., Dastur A., Moffatt S., Yabuki N., Maruyama, H. *Eco² Cities: Ecological Cities as Economic Cities*, 2010. The International Bank for Reconstruction and Development/The World Bank, Washington DC..

4 United Nations (UN), *The World's Cities in 2016*. 2016. [cited 2018, 10 November]; Available from:http://www.un.org/en/development/desa/population/publications/pdf/urbanization/the_worlds_cities_in_2016_data_booklet.pdf.

5 Haughton, G. and Hunter, C. *Sustainable Cities*. 2004. Routledge Taylor and Francis Group, London and New York. ISBN 0-203-64556-1 Master e-book.

6 Platt, R. H. *Toward ecological cities: adapting to the 21st century metropolis*. Journal Environment: Science and Policy for Sustainable Development, 2004, 46(5), 10–27.

7 Kabisch, N., and Haase, D. *Green Spaces of European Cities Revisited for 1990–2006*. Landscape and Urban Planning, 2013, **110**, 113–122.

8 *Cittaslow International Charter* [cited 2018, 5 November]; Available from: http://www.bastamag.net/IMG/pdf/newcharter_1_-1.pdf.

9 Voytenko, Y., McCormick, K., Evans, J., and Schliwa, G. *Urban Living Labs for Sustainability and Low Carbon Cities in Europe: Towards a Research Agenda*. Journal of Cleaner Production, 2016, **123**, 45–54.

10 *Congress of NewUurbanism*. [cited 2018, 5 November]; Available from: https://www.cnu.org/sites/default/files/charter_english.pdf.

11 Komnikos, N. What Makes Cities Intelligent? *Smart Cities, Governing, Modelling and Analysing the Transition*, Deakin (Ed.), 2014. Routledge. Taylor and Francis Group, London and New York. ISBN: 978- 0- 415- 65819- 5

12 European Parliament (EP). *Mapping Smart Cities in the EU*, 2014, Prepared for European Parliament's Policy Department.

13 Cocchia, A. Smart and Digital City: A Systematic Literature Review, *Smart City How to Create Public and Economic Value with High Technology in Urban Space*, R. Paola, D. C. Rosenthal-Sabroux (Ed.), 2014. pp. 13–45, Springer International Publishing, Switzerland. DOI: 10.1007/978-3-319-06160-3_2.

14 Washburn, D., Sindhu, U., Balaouras, S., Dines, R. A., Hayes, N., and Nelson, L. E. *Helping CIOs Understand "smart city" Initiatives*. Growth, 2010, **17**(2), 1–17.

15 Caragliu, A., Del Bo, C., and Nijkamp, P. *Smart Cities in Europe*. In 3rd Central European Conference in Regional science – CERS, 2009, 7–9 October, Košice, Slovak Republic.

16 Chourabi, H., Nam, T., Walker, S., Gil-Garcia, J. R., Mellouli, S., Nahon, K., et al. *Understanding Smart City Initiatives: An Integrative Framework*. In 45th Hawaii International Conference on System Sciences, 2012, pp. 2289–2297, 4–7 January Maui, HI.

17 Cohen, B. *Smart City Wheel.* Retrieved from SMART & SAFE CITY. 2013 [cited 2018, 3 November]; Available from: http://www. smartcircle. org/ smartcity/blog/boyd-cohen-the-smart-city-wheel.

18 Dameri, R. P. Comparing Smart and Digital City: Initiatives and Strategies in Amsterdam and Genoa. Are They Digital and/or Smart? . Smart City . Paola, D. C. Rosenthal-Sabroux (Ed.), 2014. pp. 45–88, Springer International Publishing, Switzerland. ISBN 978-3-319-06159-7.

19 Anthopoulos, L. *Smart Utopia VS Smart Reality: Learning by Experience from 10 Smart City Cases.* 2017, Cities, **63**, 128–148.

20 Brunn, D., S., Ghose, R., Graham, M. *Cities of the Future, Digital Technology and the City,* Cities of the World: World Regional Urban Development, Brunn, D., S., Mitchell, H., M., Ziegler, D., J. (Ed.), 2012. pp. 577–590, Rowman & Littlefield Publishers, Plymouth.

21 United Cities Local Governments (UCLG). *Smart Cities Study: International Study on the Situation of ICT, Innovation and Knowledge in Cities. Bilbao.* 2012. [cited 2018, 1 November]; Available from: http://www.uclgdigitalcities.org/ app/uploads/2015/06/en_smartcitiesstudy.pdf.

22 Pezzutto, S., Mosannenzadeh, F., Grilli, G., and Sparber, W. *European Union Research and Development Funding on Smart Cities and Their Importance on Climate and Energy Goals.* In International Conference on Smart and Sustainable Planning for Cities and Regions, 2015, November, pp. 421–435,. Springer, Cham.

23 Mosannenzadeh, F., and Vettorato, D. *Defining Smart City. A Conceptual Framework Based on Keyword Analysis,* 2014, Journal of Land Use, Mobility and Environment.

24 Caragliu, A., Del Bo, C., and Nijkamp, P. S*mart Cities in Europe.* Journal of Urban Technology, 2011, **18**(2), 65–82.

25 Abella, A., Ortiz-de-Urbina-Criado, M., and De-Pablos-Heredero, C. *Information Reuse in Smart Cities' Ecosystems.* El profesional de la información, 2015, **24**(6), 838–844.

26 Batty, M., Axhausen, K. W., Giannotti, F., Pozdnoukhov, A., Bazzani, A., Wachowicz, M. *Smart Cities of the Future,* The European Physical Journal Special Topics, 2012, **214**, 481–518.

27 Lehr, T. *Smart Cities: Vision on-the-Ground.* Smart Cities. pp. 3–15. 2018, Springer, Cham.

28 Albino, V., Berardi, U. and Dangelico, R. M. *Smart Cities: Definitions, Dimensions, Performance, and Initiatives.* Journal of Urban Technology, 2015, **22**(1), 3–21.

29 Dameri, R. P. *Searching for Smart City Definition: A Comprehensive Proposal.* International Journal of Computers & Technology, 2013, **11**(5), 2544–2551.

30 Gil-Castineira, F., Costa-Montenegro, E., Gonzalez-Castano, F., López-Bravo, C., Ojala, T., and Bose, R. *Experiences inside the Ubiquitous Oulu Smart City.* Computer, 2011, **44**(6), 48–55.

31 Nijkamp P, Oirschot G, Oosterman A. *Regional Development and Engineering Creativity: An Instrumental Comparison of Science Parks in a Knowledge Society,* 1993, Free University, Research Memoranda, Amsterdam.

32 Giffinger, R., and Gudrun, H. *Smart Cities Ranking: An Effective Instrument for the Positioning of the Cities?* ACE: Architecture, City and Environment, 2010, **4**(12), 7–26.

33 Madakam ve Ramasmawy, Madakam, S., & Ramaswamy, R. *Smart Cities-Six Dimensions.* In International Conference on Advances in Computing and Information Technology-ACI, 2014, pp. 38–41.

34 Giffinger, R. *Smart Cities, Ranking of European Medium-Sized Cities,* 2007, Centre of Regional Science, Wien, 10–13.

35 Giffinger, R., Kramar, H., Milanović, N.P., Strohmayer, F. *Smart City Profiles,* 2014. PLEEC-Planning for Energy Efficient Cities, Deliverable 2.1 Part 1 P. 2-6.

36 Krassimira, P. *Smart Cities: A Nexus for Open,* Smart Cities, Governing, Modelling and Analysing the Transition, M. Deakin (Ed.), 2014. Routledge, Londra.

37 Kumar, T. V., and Dahiya, B. *Smart Economy in Smart Cities.* 2017 pp. 3–76. Springer, Singapore.

38 Vanolo, A. *Smartmentality: The Smart City as Disciplinary Strategy.* Urban Studies, 2014, **51**(5), 883–898.

39 Kaygısız, Ü., and Aydın, S. *Yönetişimde Yeni Bir Ufuk Olarak Akıllı Kentler-Smart Cıtıes As A New Dımensıon in Governance,* Mehmet Akif Ersoy Üniversitesi Sosyal Bilimler Enstitüsü Dergisi, 2017, **9**(18), pp. 56–81.

40 Giffenger, R., Kramer C., Kalasek and Natasa-Meijers, Smart Cities Ranking of Europen Medium-Sized Cities— Final Report 2007, 2008, Prepared by Vienna University of Technology.

41 Elvan, L. *Akıllı Şehirler: Lüks Değil İhtiyaç,* İTÜ Vakıf Dergisi,Temmuz-Eylül 2017, **77**, pp. 6–9.

42 Sınmaz, S. *Yeni Gelişen Planlama Yaklaşımları Çerçevesinde Akıllı Yerleşme Kavramı Ve Temel Ilkeleri.* 2013, Megaron, **8**(2), 76–86.

43 Dameri, R. P., and Rosenthal-Sabroux, C. Smart City and Value Creation, *Smart City,* Paola, D. C. Rosenthal-Sabroux (Ed.),2014. pp. 1–12, Springer International Publishing, Switzerland. ISBN 978-3-319-06159-7

44 Amsterdam Smart City [cited 2018, 5 November]; Available from: https://amsterdamsmartcity.com

45 Mora, L., and Bolici, R. The Development Process of Smart City Strategies: The Case of Barcelona, *Re-city: Future City - Combining Disciplines* J. Rajaniemi (Ed.), 2016. pp. 155–181, Juvenes Print, Tampere.

46 [cited 2018, 5 November]; Available from: http://digitalage.com.tr/dunyadan-akilli-sehir-ornekleri/

47 Köseoğlu, Ö., and Demirci, Y. *Akilli Şehirler ve Yerel Sorunlarin Çözümünde Yenilikçi Teknolojilerin Kullanimi.* Uluslararası Politik Araştırmalar Dergisi, 2018, **4**(2): 40–57.

48 *Akıllı şehir yol haritası* [cited 2018, 1 November]; Available from: https://www.sehirsizin.com/Documents/Deloitte-Vodafone-Akilli-Sehir-Yol-Haritasi.pdf.

49 *Bilgi Toplumu Stratejisinin Yenilenmesi Projesi Bilgi ve İletişim Teknolojileri Destekli Yenilikçi Çözümler Ekseni Mevcut Durum Raporu* [cited 2018, 1 November]; Available from: http://www.mobilsiad.org.tr/img/ebulten/aylik/Nisan_2013_bilgi_ve_iletisim_teknolojileri_destekli_yenilikci_cozumler_mevcut_durum.pdf.

Res. Ass. Demet Ülkü Gülpınar Sekban, Res. Ass. Duygu Akyol, and Assist. Prof. Doruk Görkem Özkan

14 Sustainability of the Urban Landscape: An Investigation of the Effects of Ecology on Urban Parks within the Scope of Integrated Landscape Approach

Abstract: The changing living conditions, conservation of natural resource assets, increasing population and deteriorating environmental conditions in order to adapt to landscape design/projects, come up with the concepts contained in it. In this context, sustainability and ecological planning/design policies play a very important role in 21st century landscape designs. In landscape projects, sustainability, and ecology philosophy are a driving force for many innovative approaches in the field of design and planning, so landscape architects take into account ecological design and planning criteria considering the principles of sustainability in their projects. The landscaping projects that try to reveal the solutions in a very wide area ranging from a small-scale garden to the city, from the city to the nature restoration, are based on the principles of sustainability and within the framework of the infrastructure of ecology, especially in urban scale. In this chapter, it has been studied that the sustainability of urban landscape has an integrated structure and the necessary studies for the analysis of this structure should be created and the ecological thought should be reflected in the projects. Within the scope of the study, the theoretical infrastructure was processed through urban park samples. Within the scope of ecological landscape design criteria, three urban park areas in the study were evaluated and suggestions were made on how to transfer ecological thought to landscape.

Keywords: Sustainability, Urban landscape, Ecology design, Integrated landscaping

1 Introduction

1.1 The Sustainability of the Urban Landscape

Developments in domains of technology and science instigated the industrial revolution, and industrial revolution encouraged the rural population to migrate to the emerging industrial areas. Hence, as a result of industrial revolution, a rapid urbanization process started due to the migration. Numerous problems, mainly environmental pollution, energy demand, need for infrastructure systems and solid waste recycling, emerged due to the uncontrolled and rapidly developing urbanization process. Parallel to the increase in the effects of the abovementioned problems and the areas influenced by these problems, the "ecological footprint" of the urban areas increased and the need for urban sustainability turned out to be a focus of discussions (Wackernagel & Rees, 1996). Consequently, in order to solve these global problems, it became inevitable to approach concept of sustainability at the urban scale.

Having the concept of sustainability examined at the urban scale, the urban areas are expected to be self-sufficient and in unison with the surrounding ecosystems. In order to meet these basic expectations, urban landscape becomes highly significant. Urban landscape facilitates quality, healthy and environmentally friendly settings in order to support spatial organizations that could create appropriate environments to meet both the physical and the social needs of the user (Demircan & Sezen, 2018). In this respect, the urban landscape approach is capable of making significant contributions to the sustainability of an urban area, through being inspired by nature, adopting natural processes and providing systematic solutions in accordance with the structural and ecological qualities of the urban areas (Korkut, Kiper, & Topal, 2017; Turner, Donato, & Romme, 2013). It is essential that the urban landscape, which contributes to the sustainability of the urban areas, is shaped within the framework of certain criteria. These criteria are considered critical in order to establish both the shaping of the urban landscape and the sustainability of the urban areas on strong foundations (Atıl, Gülgün, & Yörük, 2005). These criteria vary according to the morphology of the area; approaches based on the micro-climatic data; and the effective, active and appropriate use of approaches such as clean energy, vegetation landscape elements, natural resources and recycling. Hence, such criteria facilitate the evaluation of sustainability at an urban scale. On the other hand, the evaluation of the sustainability concept at urban scale uncovers several novel sustainable urban approaches; however, it is essential that these approaches are established with respect to diverse criteria such as ecology, ecological events and ecological integration (Haughton, 1997; Kandziora, Burkhard, & Müller, 2013; Müller, 2005).

1.2 Ecology and Urban Landscape

Urban areas are subjected to pressures and threats, such as air pollution, lack of natural resources, population density, traffic congestion, and noise intensity, which affect the quality of life negatively over extended periods of time. Such pressures and threats seriously disturb the ecosystem of the urban landscape. Due to the fact that urban areas are significant on a micro-scale to ensure sustainability in the macro-scale, the systematic construct and the texture of the urban settings should be based on ecological foundations in order to ensure a sustainable urban development (Korkut et al., 2017). The systems established on ecological basis are constructs that are capable of meeting current needs without preventing the future generations to meet their needs, that respect the nature and that aim being environmentally sensitive and achieving certain levels in terms of quality of life.

Approaches based on ecological processes are defined as the planning and design decisions that originate from considerations based on minimizing the possibility of damaging the nature and the environment through the balance of preserving and utilization. In other words, such approaches aim to protect biological diversities, water circulation lines, natural resources, the topography of the areas and plant and animal habitats (Dong & Cheng, 2014). Briefly, the approaches, which take ecological and urban landscapes together into consideration, consider the human and the nature as a whole. The design and planning decisions, therefore, are established upon nature and environment (Dong, Zhou, Xin, Tan, & Wang, 2018). It is highly essential that the use of nature remains within a balance of preserving and utilization. Therefore, concepts such as reuse, recycling, reduction of the use of natural resources and renewable energy are frequently used. These concepts focus on supporting the sustainability of urban areas together with the ecosystems that constitute the nature and the environment (Dong & Cheng, 2014). Consequently, it becomes essential that the urban landscape elements in urban systems are planned and designed in integration with the ecological systems and concepts.

1.3 The Integration of Ecology with the Landscape

The ecological approach in land use decisions is currently acknowledged as a common approach. The foundations of the ecological approach in planning was established due to the developments and changes that emerged in the 1960s, named as "environmental concerns" until the 1980s and as "quality of life and sustainability" after 1980s. The rapid depletion of resources, the effects of human activities reaching the polar zones on horizontal level and beyond the atmosphere on vertical level necessitated the concept of "returning back to the nature," and acting, protecting and developing in accordance with the demands and requirements of the nature.

Ecological approach requires a shift in methods and techniques of spatial planning and management. Accordingly, the nature and its elements should be taken into account with respect to their intrinsic values of existence, rather than being considered solely as a setting that is utilized for meeting the needs of humans (Cengiz & Gönüz, 2011). In other words, the actions of planning and design should be developed and implemented based on the capacity of placing oneself in the position of others and the capacity of integrating with the object in question.

Ecology and landscape are both multifaceted concepts. The multifaceted nature of landscape stems from three main points, namely planning, design and management. Ecology, on the other hand, is a concept that needs to be evaluated within the scope of these three points and should be integrated with them. In such an evaluation process, the right definitions of these two concepts and considering them as a whole would directly facilitate their integration. The concept, recently founded as "integral urbanism" in literature, defines the prerequisites in the design and planning of urban areas. A similar concept is mentioned for landscape designs and through the integrated landscape concept, the multifaceted nature of landscapes is once more emphasized, addressing the fact that landscape cannot be considered outside the limits of ecology (Schova et al., 2013). Such integration is only possible through the incorporation of two concepts during the planning and design phases. In order to integrate the ecological considerations to the landscaping of parks,

Tab. 1: The Criteria of the Integral Overview of Ecological Landscape Design

	Energy Efficient Landscape Design	Natural Resource Protection	Bio Comfort
Structural Applications	1. Bicycle path and park 2. Biological pond 3. Structural floor shading 4. Heat absorption low surface coating	1. Waste sorting 2. Use of ground water and wastewater 3. Use of ecological material	1. Accessibility 2. Wildlife protection 3. Structural floor shading 4. Heat absorption low surface coating
Vegetal Applications	1. Selection of local plant species 2. Wind corridor 3. Structural floor shading 4. Wind shield	1. Selection of local plant species 2. Increasing biological diversity 3. Biological pond formation 4. Mulching 5. Compost fertilizer production	1. Wind corridor 2. Wind shield 3. Building shading 4. Structural floor shading 5. Improving air quality
Irrigation	Automatic irrigation system use	1. Rainwater storage 2. Automatic irrigation system use	Micro climatic additive
Ligthing	1. Use of solar panel 2. Selection of energy saving lighting elements 3. Solar tunnels	Suitable lighting with flora and fauna	Suitable lighting with flora and fauna

especially designed in urban areas, similar criteria were defined in numerous studies, which focus on not disrupting the landscape-ecology relationship or to integrate ecology and landscape, and these criteria were categorized under main headings. Within the scope of the present study, a criterion table was obtained through the examination of the relevant studies (Tab. 1).

2 Material and Method

2.1 Materials

Three urban parks, which were designed and completed in recent years, were examined within the scope of the present study. The selection criteria of the urban parks were based on the frequency/preference of utilization in urban scale and on adopting environmentally sensitive design approaches. Accordingly, "Fareground," "South Park" and "Nature at mooring" are selected and studied in this chapter.

Tab. 2: Fareground Urban Park's Information (Landezine, 2018)

Project Location	**Congress Avenue, Austin, Texas, USA**
Design Year	2016
Year Built	2017
Event Areas	Grand Staircases, Cloudscape, Shade Canopy, Wood Decks, Jewel Box Restaurant, Amphitheater/Hill, Bike Racks, Green Wall, Event Plaza, Ada Ramp

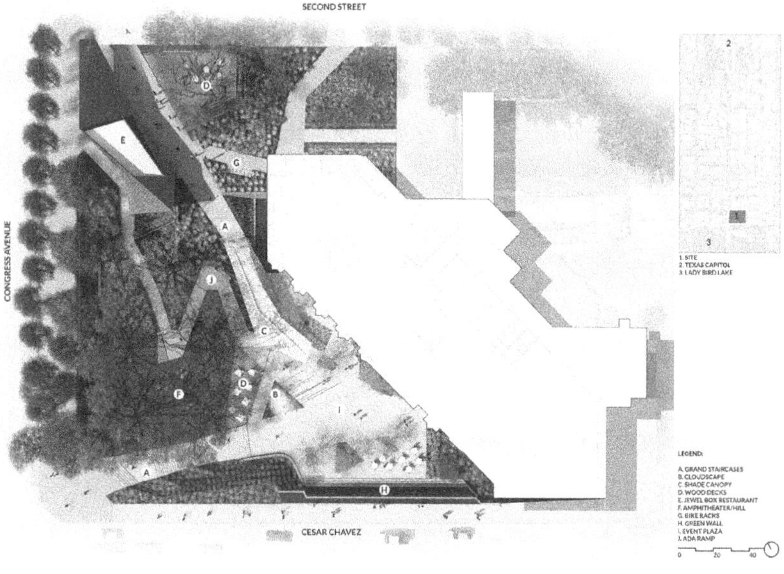

Fig. 1: Fareground Urban Park's plan (Landezine, 2018).

"Fareground" Urban Park

Fareground is an urban park designed to provide a breathing spot for the urban community in the urban center, especially where office functions were located (Tab. 2). The design decisions of Fareground urban park were structured around activities such as gathering, socializing, relaxation, recreation and resting (Fig. 1). In addition, the park was designed by taking environmentally sensitive approaches into consideration. With respect to such qualities, Fareground was honored with a 2018 Texas NEVER Merit Award, the 2018 ULI Impact Award for Project Innovation and the 2018 ABJ CRE Award for Office Development Rehab (Studiodwg, 2018).

Tab. 3: South Park Urban Park's Information (Fletcher Studio, 2018)

Project Location	**San Francisco, CA**
Design Year	2014–2016
Year Built	2017
Event Areas	Flexible Space, Circulation Patterns, Access Points, Pathway, Playground, Decks, Rainwater Collecting Area.

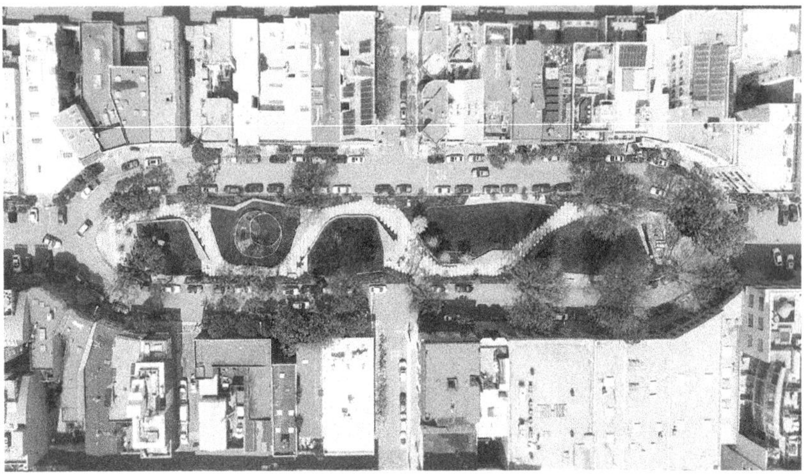

Fig. 2: South Park's plan (Fletcher Studio, 2018).

Fareground was designed and implemented through the efforts of a multidisciplinary team and was based on the concepts of dynamism, transformative potential and provision of diversity for ensuring a sustainable urban landscape (Landezine, 2018).

"South Park" Urban Park

South Park is one of the oldest public spaces available for public use since the late nineteenth century and was designed to provide picturesque walks for the urban community. The park served for a wide range of users (migrants, refugees, artists, designers, young and senior individuals, etc.) for almost 165 years. Currently, the park was redesigned as a common space of the user from various cultural backgrounds (Tab. 3) (Fletcher Studio, 2018). The redesign included several functional properties that were present in the old version of the park (Fig. 2).

Tab. 4: Nature at Mooring Urban Park's Information (Agence Ter, 2018)

Project Location	**Boulogne – Billancourt, France**
Design Year	2006
Year Built	First phase: 2011; Second phase: 2017
Event Areas	Playground, Sports Areas, Multifunctional Grass Fields, Gardener's Storage, Public Bathrooms, Gallery, Meeting Room

In the redesign of South Park, various innovative designs and functional decisions were made in order to provide an ecologically sustainable landscape design. The most significant are utilization of plants that endure drought and irrigation system that utilizes rainwater (Fletcher Studio, 2018).

"Nature at Mooring" Urban Park

Nature at Mooring Urban Park was designed within the scope of the urban transformation efforts at the urban center, where industrial areas and population were dense (Tab. 4). Functions based on the movement of the water landscape element were commonly used in the design of the park. The park was used as a water basin, therefore several boundaries within the park were designed to be flooded when the water level would rise (Fig. 3) (Agence Ter, 2018).

The design project of the park was particularly paid attention to during the phases of design and realization. There exists no artificial lighting in the park. The idea behind such a design decision was based on preventing the light pollution and utilizing a user-based lighting approach (Agence Ter, 2018).

2.2 Method

The three urban parks selected within the scope of the present study were evaluated through a literature review with respect to the criteria of integrating the ecological factors within the urban landscape. The evaluation criteria, presented in Tab. 1, were compiled from several studies, as an outcome of the literature review. Urban parks, designed with respect to these criteria and considered as sustainable areas, were evaluated to determine their level of integration with ecological considerations.

3 Findings

In this section, predetermined ecological design criteria are scrutinized in depth for the selected urban parks and an examination of whether or not ecological considerations were actually integrated into the landscape was conducted. The first example was

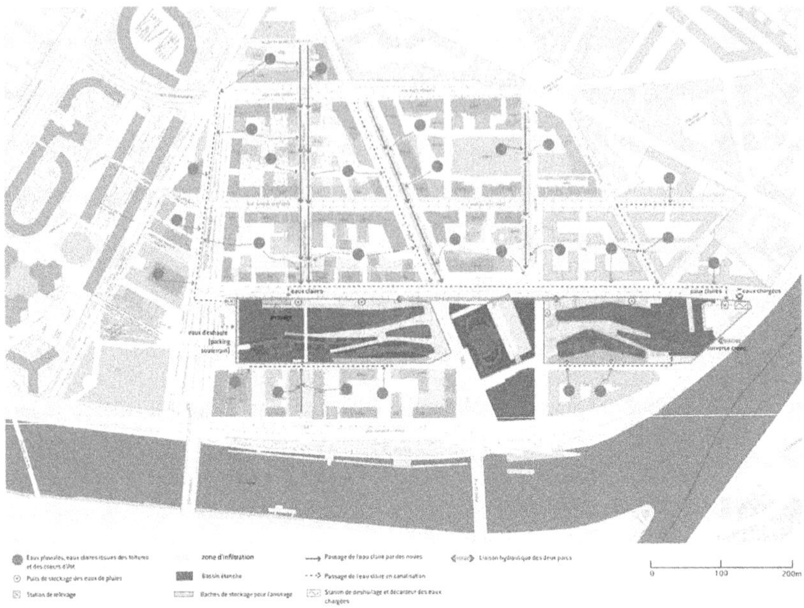

Fig. 3: Nature at Mooring Urban Park's plan (Agence Ter, 2018).

the "Fareground Urban Park," designed in Texas, USA. In the park, structural ground shading for efficient energy use was provided by using both shading elements and plant material. Local plant species and elements of the local flora were selected; thus a contribution towards the continuation of biodiversity in the region was achieved. However, there was no effort towards using a water collection system or a biological pond. On the other hand, the existence of an automated irrigated system pointed out the consideration of efficiency in water consumption (Tab. 5).

The second urban park selected within the scope of the present study, South Park, was designed in the city of San Francisco in California, USA. The first significant characteristic of this urban park was the making use of the existing ground water and the infiltration system installed for waste water management. In addition, the existence of a biological pond, prioritizing accessibility and providing accessibility with flexible play materials were the other remarkable features of the park. Moreover, the lighting system designed to strongly illuminate the main arteries of the park during late hours was a consideration that follows the reduced user potential and night circulation, and for the secondary circulation axes a less strong lighting system was provided. Another important consideration that strengthens the ecological considerations of the urban park project was the selection of appropriate plant material, given the fact that project area was located in a xerophytic climate (Tab. 6).

Tab. 5: Integration of the Fareground Park in Ecological Design

	Energy Efficient Landscape Design	Natural Resource Protection	Bio Comfort
FAREGROUND PARK	**Structural Applications** Structural floor shading		**Structural Applications** 1.Heat absorption low surface coating 2.Structural floor shading 3.Building shading
	Vegetal Applications 1.Building shading 2.Structural floor shading	**Vegetal Applications** Selection of local plant species	**Vegetal Applications** 1.Selection of local plant species
	Irrigation Automatic irrigation system	**Irrigation** Automatic irrigation system	Irrigation
	Lighting Utilizing solar energy	**Lighting** Utilizing solar energy	**Lighting**

In the "Nature at Mooring" Urban Park located by the River Seine, the most outstanding detail was the selection of finishing materials with low heat absorptance. In addition, provision of shading and structural ground shading with plant materials and shading elements and taking advantage of the river bank as an accessible axis within the urban park area are the other readily noticeable features of the Nature at Mooring Urban Park. Providing natural storm water collection pits for storage and the selection of local plant species are also the details that could be emphasized as ecological considerations that were taken into account during the design process of the urban park (Tab. 7).

4 Conclusion

Currently, it is evident that the ongoing destruction of the natural areas causes deterioration of ecological balances and the non-natural living conditions in urban areas cause significant environmental problems. Especially in large urban areas, where planning and design considerations long overlooked the ecological concerns, the living conditions are increasingly becoming difficult. The large-scale sustainable solutions to the problems emerging in urban areas depend on the capacity to realize sustainable approaches on a small scale. Such dependency could be observed both for small-scale urban parks and for large-scale recreation

Tab. 6: Integration of the South Park in Ecological Design

	Energy Efficient Landscape Design	Natural Resource Protection	Bio Comfort
SOUTH PARK	**Structural Applications** 1. Biological pond 2. Structural floor shading	**Structural Applications** Use of ground water and wastewater	**Structural Applications** 1. Heat absorption low surface coating 2. Structural floor shading 3. Building shading 4. Accessibility
	Vegetal Applications 1. Building shading 2. Structural floor shading	**Vegetal Applications** Selection of local plant species (species of plant)	**Vegetal Applications** 1. Improving air quality 2. Wind shield
	Irrigation Automatic irrigation system	**Irrigation** Automatic irrigation system	**Irrigation** -
	Lighting Utilizing solar energy	Lighting	**Lighting** Suitable lighting with flora and fauna

areas within an urban context. Formerly, in landscape design works, completed by public institutions, local administrations and mass housing initiatives, the aesthetic approaches were more prioritized than the ecological approaches; however, in current landscape architecture practice there is a more widespread point of view to improve deteriorations in natural ecosystems caused due to urbanization as much as possible and to create resilient landscape designs that have the capacity to adapt to climate change. In several landscape projects, design ideas are developed in order to adapt the design to nature, to mimic nature, to provide resilience towards natural disasters, to highlight the significance of biodiversity and to utilize ecological functions. However, observing the circumstances in Turkey, it is possible to assert that, although various projects adopt an ecologically responsive title, the advantages they offer in terms of ecologically integrated landscape approaches are limited and inadequate. Especially in the context of landscape design, the use of large grass surfaces in urban parks, inadequacy in automatic irrigation systems, inappropriate plant material selections without considering the location of the project, the use of impermeable floor finishing materials, failure to facilitate lighting systems that utilize solar energy and provide savings, failure to provide storm water storage systems in areas with high precipitation could be considered as the main negative characteristics of urban parks in Turkey. In order to avoid all abovementioned problems, the integration of ecology into urban parks or urban landscapes should be taken into consideration with respect to a multifaceted

Tab. 7: Integration of the Nature at Mooring Park in Ecological Design

	Energy Efficient Landscape Design	Natural Resource Protection	Bio Comfort
NATURE AT MOORING PARK	**Structural Applications** 1. Biological pond 2. Building shading 3. Structural floor shading	**Structural Applications** Use of ground water and wastewater	**Structural Applications** 1. Heat absorption low surface coating 2. Structural floor shading 3. Building shading 4. Accessibility
	Vegetal Applications 1.Building shading 2.Structural floor shading	**Vegetal Applications** 1. Establishment of biological pond 2. Selection of local plant species	**Vegetal Applications** Increasing air quality
	Irrigation Automatic Irrigation System	**Irrigation** Rainwater Storage	**Irrigation** -
	Lighting Selection of energy saving lighting elements	Lighting	**Lighting** Suitable lighting with flora and fauna

design approach and an integrated landscape approach should be carried out for future urban landscape practices.

References

1 Agence Ter. (2018). Nature at mooring. Retrieved from http://www.landezine.com/index.php/2017/06/nature-at-mooring-boulogne-park-by-agence-ter/

2 Atıl, A., Gülgün, B., & Yörük, İ. (2005). Sürdürülebilir Kentler ve Peyzaj Mimarlığı. *Ege Üniversitesi Ziraat Fakültesi Dergisi, 42*, 215–226.

3 Cengiz, E.,Gönüz, A.(2001). Ekolojik Açıdan Kentsel Alan Kullanımları: Çanakkale Kent Merkezi Örneği. *Atatürk Üniv. Ziraat Fak. Derg.*, *42* (1): 79–89, 2011.

4 Demircan, N., & Sezen, I. (2018). Use of Green Spaces for Liveable and Sustainable Cities; Urban Allotment Gardens. *Iğdır Universitesi Fen Bilimleri Enstitüsü Dergisi, 8*(1), 263–270.

5 Dong, J., & Cheng, W. (2014). Based on the Characteristics of Respondents and the Voice of the Urban Neighborhood Public Space Business Facilities Noise Environment Evaluation Research. *Journal of Harbin Institute of Technology, 4*, 103–109.

6 Dong, J., Zhou, T., Xin, L., Tan, Y., & Wang, Z. (2018). *The Sustainable Expression of Ecological Concept in the Urban Landscape Environment Design.* Paper presented at the IOP Conference Series: Earth and Environmental Science.

7 Fletcher Studio. (2018). South Park, San Francisco. Retrieved from http://www.landezine.com/index.php/2017/06/south-park-san-francisco-by-fletcher-studio/ Access Date: 15.10.2018

8 Haughton, G. (1997). Developing Sustainable Urban Development Models. *Cities, 14*(4), 189–195.

9 Kandziora, M., Burkhard, B., & Müller, F. (2013). Interactions of ecosystem properties, ecosystem integrity and ecosystem service indicators—A theoretical matrix exercise. *Ecological Indicators, 28*, 54–78. doi:https://doi.org/10.1016/j.ecolind.2012.09.006

10 Korkut, A., Kiper, T., & Topal, T. Ü. (2017). Kentsel Peyzaj Tasarımda Ekolojik Yaklaşımlar. *Artium, 5*(1), 14–26.

11 Landezine. (2018). Fareground. Retrieved from http://www.landezine.com/index.php/2018/08/fareground-by-dwg/

12 Müller, F. (2005). Indicating Ecosystem and Landscape Organisation. *Ecological Indicators, 5*(4), 280–294. doi:https://doi.org/10.1016/j.ecolind.2005.03.017

13 Schova,V. B., Krauklis, A. A. & Snytko, V. A. (2013). Toward a Unification of Concepts and Terms Used in Integral Landscape Investigatons. *Eurasian Geography and Economics*, 16(9), 616–622. doi.org/10.1080/00385417.1975.10640099.

14 Studiodwg. (2018). Retrieved from http://studiodwg.com/projects/fareground/

15 Turner, M. G., Donato, D. C., & Romme, W. H. (2013). Consequences of Spatial Heterogeneity for Ecosystem Services in Changing Forest Landscapes: Priorities for Future Research. *Landscape Ecology, 28*(6), 1081–1097. doi:10.1007/s10980-012-9741-4

16 Wackernagel, M., & Rees, W. (1996). *Our Ecological Footprint: Reducing Human Impact on the Earth.* Philadelphia: New Society Publishers.

17 URL-https://acikders.ankara.edu.tr/pluginfile.php/74768/mod_resource/content/1/B%C3%96L%C3%9CM%204.pdf Access Date: 15.10.2018

Ercan Gökyer

15 Ecological Landscape Planning and Management: Approaches and Innovations

Abstract: In recent years, natural resources have been negatively affected due to technological developments and human activities. As a result of this, serious environmental problems have emerged around the world. Because of these negative effects, deterioration has accelerated in landscape. Today, rational and sustainable planning, effective protection and management studies should be carried out in the landscape that is important with its unspoiled natural and cultural structure. In this study, first phases of conservation, planning and management activities in landscapes are disclosed. Importance of using computer technology and related tools in studies of landscape ecology are highlighted.

Keywords: Landscape, Landscape ecology, GIS, Landscape planning, Resource inventory

Introduction

While people live in landscapes of different types and sizes, they also benefit from landscapes for different purposes. Landscapes could be of different types and sizes, and performing various functions such as agricultural activity areas, production and conservation areas of natural resources, habitat and activity area for wildlife and biodiversity, recreational and touristic activity areas.

In recent years, natural resources have been damaged due to technological developments, human activities and natural disasters and the pressures on landscapes have increased. As a result of this climate change and serious environmental problems have emerged. As a result of this situation, change and deterioration are accelerated in landscapes. In this way, landscapes cannot perform their functions sufficiently. Today, the human population in the world is constantly increasing. People who live mostly in urban areas, are affected by the negative conditions. Hence, in recent years landscapes are preferred such as rural areas and mountainous areas, which have low human impact and where local cultural features are preserved.

It is inevitable that negativities will arise with human impact; hence, rational and sustainable planning, effective protection and management studies should be carried out in the landscapes, which are important with their unspoiled natural and cultural structure. The evaluations are made considering the negativities in landscapes in the past experiences; today the necessity of evaluating the approaches of landscape ecology in studies related to landscapes is understood.

In this study: 1) Firstly starting with landscape concept, landscape ecology and related definitions and topics are mentioned as a summary. 2) The importance of

Fig. 1: Landscapes include different types and sizes of land use and geographic characteristics.

data collection, inventory studies and the use of Geographical Information Systems (GIS) in landscape planning are emphasized. In recent years there have been significant developments in computer technologies. Consequently, the use of GIS and related technologies in landscape planning and design studies has also increased. In this context, this section emphasizes the use of computer technologies and GIS in landscape planning studies.

Landscape and Landscape Ecology

Before the concept and content of landscape ecology which is a sub-discipline of ecology, it is important to understand the concept of landscape that constitutes our environment. Various definitions related with concept of landscape have been made up to the present day. The landscapes, starting from a narrow and confined space surrounding our living areas in cities, could be varied to include different types and sizes of land use and geographic characteristics (Fig. 1). The landscape was firstly described by the German geographer Alexander Von Humbdolt as "all the features of a piece of land" [1]. Forman and Godron (1986) describe landscape as forming a land of ecosystems that interact with each other and replicate this feature in large areas [2]. Characteristics which form a landscape are among the subjects of the landscape ecological studies. Especially the land use pattern and the land use changes in time and their effects on the function of the whole landscape are evaluated with the help of GIS and remote sensing.

The landscape ecology has been used since the early 1980s and has been studied continuously. It is preferred and used by people of many disciplines such as geographers, landscape architects, regional planners, and foresters [3]. Landscape ecology examines the interactions between landscape elements and ecosystems,

Fig. 2: Habitats are fragmented and isolated by roads.

their functional properties and occurring changes in time [4]. Map and numeric data which are produced by using geographical information are given in Fig. 1.

Systems and remote sensing techniques and technologies associated with these are used to understand and evaluate the relationships between ecological processes of landscapes that are characterized by structure, function and change in the studies related to landscape ecology [3, 5, 6]. In addition, measurements in the field, observations, examinations and information obtained from interviews with individuals are used to evaluate and interpret of data that are obtained from computer technologies.

Studies in landscape ecology focus on land use patterns, landscape change and the relationships between them. In landscape, the changes and differentiations which are emerged due to natural causes and human interventions affect the landscape structure and the function of the landscape. This situation effects human life and landscape health. Especially, changes and differentiations due to human interventions cause fragmentation and isolation in landscapes. As a result of these effects, large areas with similar features within the landscapes are fragmented or get decreased. In the fragmented landscapes, the habitats of wildlife and their areas of movementbecome smaller. The most important factor that leads to fragmentation and isolation of landscape is the vehicle roads. Habitats and movement areas of species are fragmented by the roads which are constructed for transportation (Fig. 2). The data that are obtained from land use maps which are produced using GIS, related to landscapes, are mostly associated with human activities in the evaluation of structure, function and change, which are the basic features of landscapes. For example, in northwestern part of Turkey, it has been determined that the fragmentation and isolation has decreased as the human population has declined here over time in these rural and mountainous areas of Bartın Province and its surrounding areas. This situation is considered as positive in terms of landscape health and the species that live in the area [7].

Landscape Planning

With the most general expression, landscape planning means all the analyses and evaluations which are made in order to make macro decisions for the use of land and resources. Landscape planning according to the European Landscape Convention [8] is defined as future-oriented actions which are aimed to increase the quality and value of landscapes and reducing the negativities. The relationship between land and resources and with the environment, interaction among them and the processes in the field should be analyzed and evaluated in the best way while making landscape planning decisions. Therefore, in landscape planning, the scale of study, boundaries of study area and the inventory of the resources should be made in the best way to increase the success and applicability of the study [7]. Some outputs have been obtained with the results of landscape planning studies. These products are presented in report of plan with current status and future decisions, maps and visual data. Readability of maps and visual data from the result products is very important. The scale of the maps, explanations and demonstrations should be prepared in accordance with the subject of study and standards.

Landscape Management

Landscape management is described in the European Landscape Convention as a guide to express sustainable development and adaptation to changes with human impact and natural processes [8]. It is necessary to manage the areas which are mostly protected, as natural and semi-natural areas, including human uses and activities as a whole with their own flora and fauna diversity with a plan to perform sustainability. Thus, activities and practices which will reduce the effects of adverse conditions in the field and contribute to preserving the natural and cultural characteristics and values related to the area are made within scope of management plans of these areas [9].

In landscapes with different characteristics, different management approaches and objectives are applied. For example, there are three important key approaches in the management of natural and protected landscapes. These include: 1) reducing the human impact on landscapes, 2) maintaining the integrity of large corridors (rivers) (Fig. 3), 3) to maintain the continuity of natural processes in the landscape. At the same time, it is aimed to reduce the isolation for interior species and to prevent harmful effects from the environmental areas with the management activities in the natural areas. It is aimed to maintain a balance between agricultural crops, pastures and natural vegetation grown, in the management of agricultural landscapes. In management, it is necessary to determine the human impact on the landscape and the changes in the area [1, 7]. Thus, the most appropriate management model is performed for the field to solve the problems that may arise in the future and solutions could be developed to these problems. Because the studies

Fig. 3: Stream corridors connect landscape elements to each other.

related to landscapes are related each other, landscape management and landscape planning studies involve similar stages and approaches.

Boundaries and Scale

In studies related to landscapes, it is important to determine the boundaries of the research area according to the natural thresholds for environmental relations. Thus, the boundaries are analyzed not only in the study area but also in relation with its surrounding area. Determining the boundaries of the study area according to the natural and physical thresholds allows us to make evaluations and analyses considering the environmental characteristics and uses. In related ecological studies, environmental assessments should be made according to the boundaries created through the peaks of the mountains that are the physical formations surrounding the study area (Fig. 4).

Suitable working scales should be determined for the maps to be formed within the scope of planning, protection and management works done according to the study area boundaries determined in landscapes. In the studies, scales of the maps which are produced for the whole area have differences from maps which are produced for special usage and areas. However, certain standards should be followed in map scales.

In landscapes, the scale refers to the spatial or temporal dimension of objects or processes and characterized by grain and extent. Grain is the smallest unit in satellite imagery, finest spatial resolution (smallest unit) in maps produced as polygon. Extent refers to the size (whole area) of the study area [3, 10]. According to this,

Fig. 4: Physical boundaries of the research area are determined by peaks of the mountains.

we could make general classification and evaluations on broader scale including the whole area. For example, while, small-scale forest units (stand) could be used to inventory wood species, large-scale (basin size) forest areas could be used to determine the general forest cover structure as a leafy, mixed or coniferous forest area. In landscape ecology studies, landscape types are evaluated with the patch level (smallest unit), and the landscape level which expresses the total area.

Resource Inventory (Collecting Data and Analysis)

In conservation, planning and management studies, resource inventory and data collection constitute one of the most important stages of the studies related to landscape. Three basic data are obtained as abiotic (physical), biotic (biological) and cultural (socio-economic) in order to be used in landscape studies. The data on resources within the boundaries of the landscape should be collected systematically in a specific order [11, 12]. The data should be transferred to the digital platform by using GIS within the computer system. In this way, many analyses could be made and maps could be generated in GIS platform via numerical data.

Data on the study area and its surroundings during the resource inventory and data collection phase: 1) abiotic (physical) data: geology, topographic structure, soil and climate, 2) biotic data: flora and fauna, 3) cultural data, land use pattern (past and actual), population, human activities and economic situation. Depending on the characteristics of the workspace, the content of the data to be provided for the features in the field may include variety. In the beginning of the study, these data should be evaluated and identified [7, 11, 12]. It is very important that relevant experts identify, investigate and report the features in the area.

New Approaches and Techniques in Landscape Planning

Geographic Information Systems, remote sensing and related tools and materials (landscape metrics, geo statistics) are used in the studies related to landscapes to increase success and provide effective and reliable results. Nowadays, landscape ecology and topics of study are progressing depending on developments in computer technologies. Studies in the landscape ecology are mostly made by land cover maps which are produced using remote sensing and GIS. The numerical data which are obtained by analyzing the land use maps (Fig. 5) provide understanding and evaluation of the changes in the land use pattern and the temporal processes in the area.

The numerical values which are obtained by using landscape measurements are among the important tools used in the studies of landscape ecology, landscape planning and landscape management. Many landscape metrics are developed for this purpose.

Results and Evaluation

In the beginning, there are similarities in the studies for conservation, planning and management activities in landscape. These studies generally include: 1) to provide data on the resources (abiotic, biotic and cultural), 2) to understand the ecological processes in the field and determine the changes, 3) to determine effects of human activities.

In landscape, elements of landscape interact with each other in boundary, which are formed according to natural and physical thresholds. This interaction is higher from the mountainous regions to the sub regions. Flow corridors from the high points of the area (corridor) are the elements that connect the landscape characteristics to each other. The rainfall in the up zone within the landscape could have positive or negative effects around our cities in the sub region. For example, flows which have occurred with long-term and heavy rains could damage our cities and environment (Fig. 6). Thus, the importance and necessity of evaluating the landscapes within the boundaries, which are determined by natural and physical thresholds, is understood.

In the planning and management studies related to landscape, obtaining of data on resources in the field is the most important stage of the study. It is important

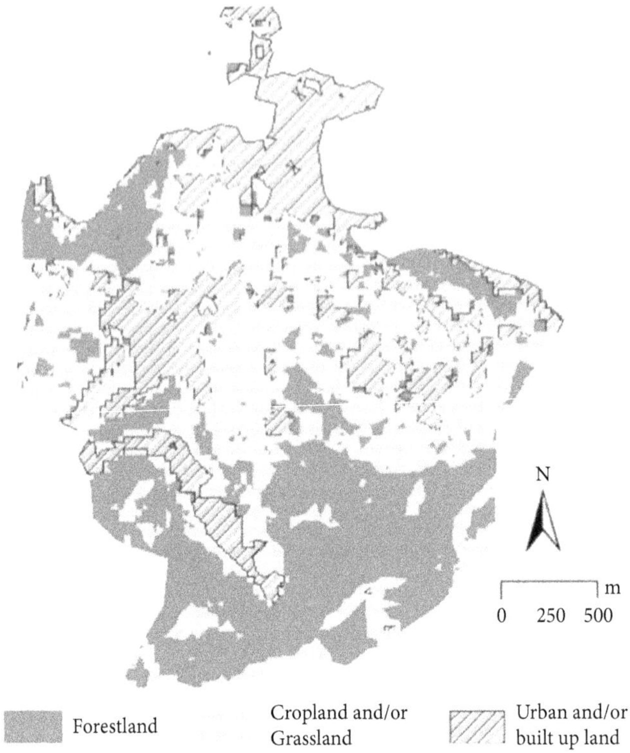

Fig. 5: Land use map (2015) (Adapted from [13]).

Fig. 6: Cities could be affected by surface flows from the upper zones of the basin.

to provide the data numerically or to transfer the non-numeric data to the digital platform at the beginning of the study. However, faster and more reliable analysis and evaluations could be made by using numerical data. With the development of computer technologies, landscape planning and landscape management studies have reached a different dimension. At the same time, the use of computer technologies has increased the visual quality of the results of products from the studies.

References

1 Forman, R.T., Godron, M. (1986). Landscape ecology. 619, John Wiley and Sons, New York.

2 Odum, P.E. and Barrett, W.G. (2008). Ekolojinin temel ilkeleri (Fundamentals of ecology). Çeviri editörü: Prof. Dr. Kani Işık. Palme Yayıncılık, Ankara.

3 Turner, M.G., and Gardner, R.H. (2015). Landscape ecology in theory and practice, pattern and process, Springer-Verlag, New York.

4 Dramstad, E., Olson, D.J. and Forman, T.T.R. (1996). Landscape ecology principles in landscape ecology and land use planning. Island Press, USA.

5 Gergel, E.S. and Turner, G.M. (2002). Learning landscape ecology a practical guide to concepts and techniques. Springer Verlag, New York, USA.

6 Farina, A. (2000). Landscape ecology in action. Kluwer Academic Publishers, Netherlands.

7 Gökyer, E. (2009). Bartın ili kent merkezi ve Arıt Havzası'nda peyzaj değerlendirme (Doktora Tezi, Basılmamış), (Landscape Assessment in Bartın City and Arıt Basin) (Doctoral thesis, Unpublished), Ankara Üniversitesi Fen Bilimleri Enstitüsü Peyzaj Mimarlığı Anabilim Dalı, Ankara.

8 Council of Europe (2000). The European Landscape Convention, European Treaty Series - No. 176, Florence, Italy.

9 Eurosite (1999). How to Use the Eurosite Management Planning Toolkit 99, Eurosite Programme & Development Office Postbus 1366 NL - 5004 BJ TILBURG Netherlands.

10 O'Neill, R.V., Smith, M.A. (2002). Scale and hierarchy theory (Edited by: Gergel, S. E., and Turner, M. G.,). Springer Science and Business Media, Switzerland.

11 Harris, W.C., Belknap R.K., Furtado, J.G., Forster, R.R., Blossom, H.D. (1967). Three approaches to environmental resource analysis, sponsored and published by the conservation foundation. Washington, USA. 59–74.

12 Şahin, Ş., Perçin, H., Kurum, E., Uzun, O. ve Bilgili, B.C. (2014). Bölge - alt bölge (il) ölçeğinde peyzaj karakter analizi ve değerlendirmesi ulusal teknik kılavuzu (National Technical Guide for Landscape Character Analysis and Evaluation in the Region – Subregion (Province) Scale). Müşteri Kurumların T.C. İçişleri Bakanlığı, T.C. Çevre ve Şehircilik Bakanlığı ve T.C. Orman ve Su

İşleri Bakanlığı olduğu, T.C. Ankara Üniversitesinin Yürütücü Kuruluş olduğu ve TÜBİTAK KAMAG 1007 Programı 109G074 No'lu PEYZAJ-44 Projesi Çıktısı, 148 Sayfa, Ankara.

13 Gökyer E, Öztürk M, Çorbacı Ö.L., Dönmez Y. (2018). Evaluating the effects of tourism on urban development: case study of Amasra (Bartın). International Journal of Recent Engineering Research and Development (IJRERD), 13(2), 3–8.

Asst. Prof. Dr. A. Esra Cengiz and Assoc. Prof. Dr. Demet Demiroğlu

16 Integrating Ecosystem Services into Spatial Planning

Abstract: Ecosystem services are an approach which was developed on the economic benefits of the natural capital. The use of an ecology-based term (ecosystem) together with economy may be acknowledged as a method to protect nature and natural resources in today's world because ecosystem services refer to the benefits of an ecosystem and its economic contributions in the short, medium and long terms. Drawing up spatial plans which allow utilizing natural and cultural resources on the basis of conservation and use depending on the planning approaches based on ecosystem services is a quite important approach for the continuity of ecosystem services and to create sustainable spaces. Firstly, the fundamental concepts about ecosystem and the definitions of ecosystem are described in the study. Secondly, the definition and brief history, classification and functions of ecosystem services (ESs) are provided. Thirdly and finally, evaluations on the concept of spatial planning and on how to integrate ecosystem services into spatial planning studies are made and the results are presented.

Keywords: Ecosystem services (ESs), Spatial planning, Ecological-based planning, Ecosystem, Sustainability

1 Introduction

Also expressed as the rate of people living in urban places, "urbanization" is growing rapidly, with values reaching about 80 % in most European countries today, and urban areas are expanding more rapidly than the urban population (1,2). The urbanization movements which developed simultaneously with the Industrial Revolution changed the cities physically, socially, culturally and economically. The lifestyles, government structure, capital power, population increase, migrations and economic conditions that changed over time also caused the living spaces in cities to differ (3).

More than half of the world population now lives in cities. In "the Sustainable Cities Development Report (2009)," prepared in 2009 by the United Nations (UN), it is stated that the urban and rural populations equaled each other in 2008 for the first time in the world history and that 70 % of the world population will live in urban areas in 2050. This has caused cities to be regarded as one of the starting points of environmental problems as well (4,5,6,7,3,8).

The basic problem caused by urbanization and its greatest impact on living things are the negative changes in habitats. The upsetting of the ecological balance and the loss or disintegration or isolation of natural areas for such reasons as the increase in building, the loss or improper use of natural areas, the proliferation of nonlocal/exotic species, and environmental pollution besides the sociocultural changes depending on the population increase are some of these impacts (1,9,10).

The dispersed development of the current cities and their spread over vast areas negatively affect natural habitats in particular. Urban expansion substantially affects natural resources and consumes agricultural lands. Hence, it negatively affects emissions, ecosystems and biodiversity. Additionally, urbanization negatively affects natural resources, also because it changes the land use by people and their ways of transport, industrial and agricultural production, consumption and social activities. Reducing the impacts of urban development on the environment is only possible through an ecological planning approach (11,2).

All the above-mentioned give rise to both the occurrence of large urban masses and the gradual increase in the pressure on ecosystems (12). Odum and Barrett (13) describe cities as parasites with low energy that feed on rural areas. Therefore, Kışlalıoğlu and Berkes (14) stated that the change in land use/land cover depending on the rapid urbanization worldwide altered the ecosystem services (ESs) offered by the ecosystems, reduced them by making them irregular and gave rise to new weaknesses in the socioeconomic structure accordingly.

Taking place as a natural result of the change in the land cover particularly caused by urbanization and the sectoral uses depending on urbanization, ecosystem degradation leads to the gradual decrease in biological diversity and ecosystem services (ESs) on the global scale.

Biological diversity is the name given to the diversity of ecosystems (natural capital), species and genes throughout the whole world or in a specific habitat. Biological diversity provides the services required for the continuity of the economic and social lives of human beings. Biological diversity is also essential for such ecosystem services such as pollination, climate regulation, flood protection, soil fertility and food, fuel, fiber and medicine production (15). It is seen that the competence in the sustainable management of unimpaired ecosystems and in the repairing of considerably damaged and impaired ones significantly improved in the last 30 years (16). Supporting the spatial planning approaches based on the ecosystem services defined as the benefits, processes, and products obtained either directly or indirectly by human beings as a result of the functions of the ecosystem and therefore having important roles in environmental protection and sustainable development is one of them (17). Determining the functioning of ecosystem services that are defined as the contributions of ecosystem functions to human welfare as well as the factors which lead to their change is among the factors which enhance the usability of the concept in plans and policies (18). Especially in the recent years, the scientists studying city planning have been trying to develop strategies about the integration of the approach to ecosystem services and the green infrastructure plans that secure these services into spatial planning (19).

As Çoban and Yücel (12) state, ecosystem services should first of all be successfully integrated into spatial (country/region/city) planning levels in order to carry out successful practices with the ecosystem services that focus on the connections between ecosystems and human welfare, as well as on the advantages they bring.

In conclusion, principles of ecology-based planning and design should be formulated from the country scale to the urban scale in the spatial planning process and the ecological, economic and sociocultural services of this approach should be presented at each level.

In summary, drawing up spatial plans which allow utilizing natural and cultural resources on the basis of conservation and use depending on the planning approaches based on ecosystem services is a quite important approach for the continuity of ecosystem services and to create sustainable spaces.

Firstly, the fundamental concepts about ecosystem and the definitions of ecosystem are described in the study. Secondly, the definition and brief history, classification and functions of ecosystem services (ESs) are provided. Thirdly and finally, evaluations on the concept of spatial planning and on how to integrate ecosystem services into spatial planning studies are made and the results are presented.

2 Ecosystem Services (ESs)

Their Brief History and Definition

Ecosystem services and natural capital have become prominent terminology in the ecological and wider literature, and yet convey different meanings to different communities (20).

Although the scientific past of the concept called "Ecosystem Services" (ESs) in the international literature but "ekosistem hizmetleri" (ecosystem services) in the literature in our country dates back to the 1970s, its use in the literature began to gain momentum in the 1990s (21,22,23,24). The concept was first used in the literature by Ehrlich and Ehrlich on the basis of Westman's (1977) term "Natural Services," and it was predominantly interpreted and defined together with the concepts of ecosystem functions and biodiversity (24). Later on, it was expressed as "the conditions and processes natural ecosystems and the species they make up them carry out to sustain human life" by *Daily*(21).

The concept became more widespread with the *Millennium Ecosystem Assessment (MEA) Report,* which was the product of a four-year study carried out by the United Nations (UN) and attended by 1360 experts from 95 different countries and which was acknowledged as the most comprehensive research on the world ecosystems. While ecosystem services were described as natural capital in the report, it was stressed that the capacity of the natural capital to support future generations was reduced by human activities (25,26). The second most important international study following the MEA report is the research entitled "The Economics of Ecosystem and Biodiversity (TEEB)," which occurred with the initiative of the UN Environment Program. One of the goals of the research, where biodiversity, economy and interdisciplinary study were emphasized, was "to lay emphasis on the increased costs of the loss of biological diversity together with ecosystem degradation" (27,28). In the recent years, significant developments have been experienced in the analysis and assessment of ecosystem services and the studies of integration into planning and management processes have gained momentum particularly with the new approaches brought to nature-human relationships by the Millennium Ecosystem Assessment (MEA) (29).

The numerous studies available in the literature on ecosystem services are studies which ecologically, economically and socially address, define and classify ecosystem services (30). However, many of the definitions follow the framework set out in the Millennium Ecosystem Assessment (MEA) 2005 (20).

Within the framework of the definitions made in various sources concerning ecosystem services, ecosystem services are also defined as the conditions and processes natural ecosystems and the species they make up them carry out to sustain human life (21,23); the benefits people are provided with from ecosystem functions or the direct and indirect contributions of ecosystems to the welfare of people (22,31,25,27); the benefits people obtain from ecosystems (26,32); and the ecological products directly consumed and utilized for human welfare (33). "Ecosystem services," a term which refers to the different ways through which nature supports human welfare, depend substantially on the proper functioning of ecosystems (12).

Considering these definitions, Ecosystem Services (ESs) were briefly evaluated as "all conditions, processes, functions, benefits, and products that ecosystems provide to sustain human life and to provide human welfare" by Albayrak (24).

The concept of urban ecosystem services, first used by Bolund and Hunhammar (32), is defined as "the value and benefits the inhabitants of a city are provided with by the internal ecosystems within a city" (19) and the scope of the system is evaluated under four main titles, namely ecological, geographical, socioeconomic and policies (Ruhl et al., 2007 cited in 24).

Their Classification

There are different views on the classification of ecosystem services. However, the most commonly used one among these classifications is the classification by the MEA Report 2005. The classification in the *Millennium Ecosystem Assessment Report* (25), which includes an ES-based assessment on the global scale, which allows steering the policies at the country level, and which enables the ESs to become an issue of priority on the political agenda besides the scientific agenda, significantly maintains its validity and forms the basis for the studies developed (34). In addition, the CICES (2013), the Common International Classification of Ecosystem Services, was prepared in order to quantify and assess ecosystem services and constitutes a section of the framework designed to support the 2020 Biodiversity Strategy of the EU (35). Furthermore, there are also classifications in various studies concerning the classification of ecosystem services and they were used as references in this study (21,22,25,33,23,27,36,28,37,38,39).

On the other hand, the classifications by the MEA Report (2005) and the TEEB (2010) were also taken as the basis within the scope of the study, since they also have a guiding quality in terms of making a spatial assessment and a comparison, as also stated by Tezer et al. (34).

Ecosystem services were assessed in four main ecological function groups as supporting, provisioning, regulating and cultural services in the MEA report but as habitat services, provisioning services, regulating services and cultural and comfort services in the TEEB report (Tab. 1).

Provisioning services are the products obtained from ecosystems; regulating services are the benefits obtained from the regulation of ecosystem processes; cultural services are the nonmaterial benefits people obtain from ecosystems; and supporting services are those services that are necessary for the production of all other ecosystem services (Fig. 1) (41).

Their Functions

The process-based description of ecosystems has led to improved understanding of the functions ecosystems perform in natural settings. The transformation of energy and materials into the vegetation structure, for example, provides habitat for other organisms. The decomposition of materials in the ecosystem builds soil structure. Each process underway in an ecosystem thus contributes to one or more of a set of functions associated with the ecosystem and with its relation to other ecosystems (42). The same basic biological and chemical processes occur in all ecosystems, but different conditions yield different functional representations (42).

Gõmez-Baggethum and Barton (43) analyzed and classified the important ecological services and the ecosystem functions and components that formed the basis for these services in 11 groups (Tab. 2).

Likely to be described as the contributions of nature to human beings or the blessings of nature as well, ecosystem services also provide the continuity of life on earth. The ecosystem resources, their benefits and the drivers of change are shown in the figure below (Fig. 2).

3 Sustainable Spatial Planning and Integration of Ecosystem Services into Spatial Planning

The concept of planning, with which the land use on earth today is significantly and substantially associated, is an index of systems which should also encompass extremely critical ecological perspectives from the country scale to the urban scale and the relevant spatial use decisions. Thus, as Gedikli (45) also states, the field of spatial planning has been encompassing new planning approaches and methods since the 1980s as the commencement of the rapid transformation of the socio-economic order in the world during these years gave rise to the origination of new phenomena and problems in urban areas and, as expressed by Akyol (46), the conventional planning methods began to be criticized on the grounds that they remained insufficient particularly for the new phenomena and problems seen in big cities.

Tab. 1: The Classification of Ecosystem Services (25,27,24,19,12)

Main Group of Ecosystem Services (MEA, 2005) Classification	Ecosystem Services (TEEB, 2010) Classification
Supporting • Food cycle • Water cycle • Primary production • Photosynthesis • Soil formation	**Habitat** • Lifecycle maintenance • Gene pool protection
Provisioning • Food • Biological raw material • Decorative resources • Genetic resources • Fresh water • Biochemicals and medical products	**Provisioning** • Food • Fresh water • Raw materials • Genetic resources • Medicinal resources • Ornamental resources
Regulating • Climate regulation • Air quality regulation • Water regulation • Erosion regulation • Water purification and waste treatment • Disease and pest regulation • Pollination • Moderation of extreme events/Natural risk reduction	**Regulating** • Air purification • Climate regulation • Disturbance prevention or moderation • Regulation of water flows • Waste treatment • Erosion prevention • Maintaining soil fertility • Pollination • Biological control
Cultural • Cultural diversity • Moral and ethical values • Information system • Educational value • Inspiration • Aesthetic values • Social relationships • Feeling of place and space • Cultural heritage asset • Recreation and ecotourism	**Cultural and Comfort** • Aesthetic information • Recreation & tourism • Inspiration for culture, art and design • Spiritual experience • Information for cognitive development

The running out of natural resources and of biological diversity draws attention as one of the most serious ecological problems today. The incomplete/improper practices in the content of spatial planning are also regarded as one of the most fundamental reasons for this problem, for the selection of locations for the land

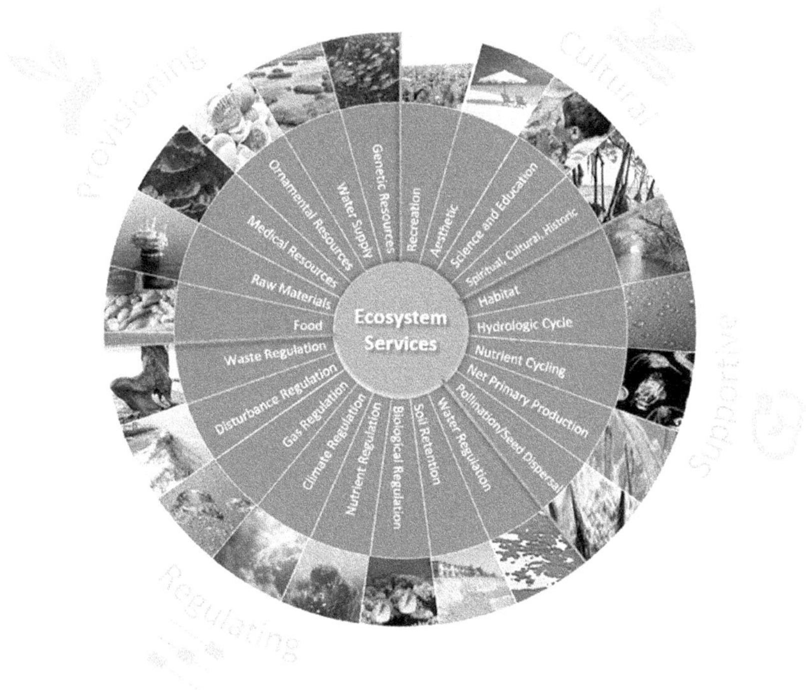

Fig. 1: The classification of ecosystem services (Source: 40).

uses in the current physical plans is performed mostly without taking the existence of natural resources into consideration. Regarded as one of the most important shortcomings of physical planning, this approach has now begun to make ecological planning approaches more valid and necessary upon the rapid running out of natural resources. Nevertheless, as Şahin (47) also states, natural areas continue to be damaged rapidly since "Ecology-Based Landscape Plans" have not been adequately included in the legal processes and upper-scale plans that are being implemented worldwide yet.

The most fundamental difference of ecological planning from physical planning is that it aims at balanced organization among land uses. This is achieved when all existing and potential uses in an area keep the negative impacts with the environment and within themselves at the minimum level (5). Besides, the sustainable use of the available natural and cultural resources within the conservation and use balance is another fundamental basis for the ecological planning approaches.

Some spatial awareness that particularly developed after the 1980s and that has quite intensified since the 2000s has formed in the arena of spatial planning.

Tab. 2: The Classification of Important Ecosystem Services in Urban Areas and Underlying Ecosystem Functions and Components (43,19)

Ecosystem service	**Functions and components**	**Examples**	**Examples of Indicators/ proxies**
Food supply	Energy conversion into edible plants through photosynthesis	Vegetables produced by urban allotments and peri-urban areas	Production of food (tons yr−1)
Water flow regulation and runoff mitigation	Percolation and regulation of runoff and river discharge	Soil and vegetation percolate water during heavy and/or prolonged precipitation events	Soil infiltration capacity; % sealed relative to permeable surface (ha)
Urban temperature regulation	Photosynthesis, shading and evapotranspiration	Trees and other urban vegetation provide shade, create humidity, and block wind	Leaf Area Index; Temperature decrease by tree cover × m^2 of plot trees cover (°C)
Noise reduction	Absorption of sound waves by vegetation and water	Absorption of sound waves by vegetation barriers, especially thick vegetation	Leaf area (m^2) and distance to roads (m); noise reduction dB(A)/ vegetation unit (m)
Air purification	Filtering and fixation of gases and particulate matter	Removal and fixation of pollutants by urban vegetation in leaves, stems and roots	O_3, SO_2, NO_2, CO and PM10 µm removal (tons yr^{-1}) multiplied by tree cover (m^2)
Moderation of environmental extremes	Physical barrier and absorption on kinetic energy	Storm, floods and wave buffering by vegetation barriers and heat absorption during severe heat waves	Cover density of vegetation barriers separating built areas from the sea

Tab. 2: Continued

Ecosystem service	Functions and components	Examples	Examples of Indicators/ proxies
Waste treatment	Removal or breakdown of xenic nutrients	Effluent filtering and nutrient fixation by urban wetlands	P, K, Mg and Ca in $mgkg^{-1}$ compared to given soil/water quality standards
Climate regulation	Carbon sequestration and fixation in photosynthesis	Carbon sequestration and storage by the biomass of urban shrubs and trees	CO_2 sequestration by trees
Pollination and seed dispersal	Movement of floral gametes by biota	Urban ecosystems provide habitat for birds, insects, and pollinators	Species diversity and abundance of birds and bumble bees
Recreation and cognitive development	Ecosystems with recreational and educational values	Urban parks provide multiple opportunities for recreation, meditation, and pedagogy	Surface of green public spaces (ha)/inhabitant
Animal sighting	Habitat provision for animal species	Urban green spaces provide habitat for birds and other animals people like watching	Abundance of birds, butterflies and other animals valued for their aesthetic attribute

The fact that the socioeconomic order in the world began to transform rapidly and that this process of change and transformation gave rise to the origination of new phenomena and problems in national, regional, urban and rural areas demonstrated that one could not challenge them with the conventional planning methods (48,49,50,51,52,53,54,46). This led to the occurrence of new approaches in planning. The sustainable spatial planning based on ecological approaches is one of them.

Sustainable spatial planning aims at better quality of life for the present and future generations, economic and social development, and environmental protection. It also pioneers the holistic conservation of the natural, artificial and cultural

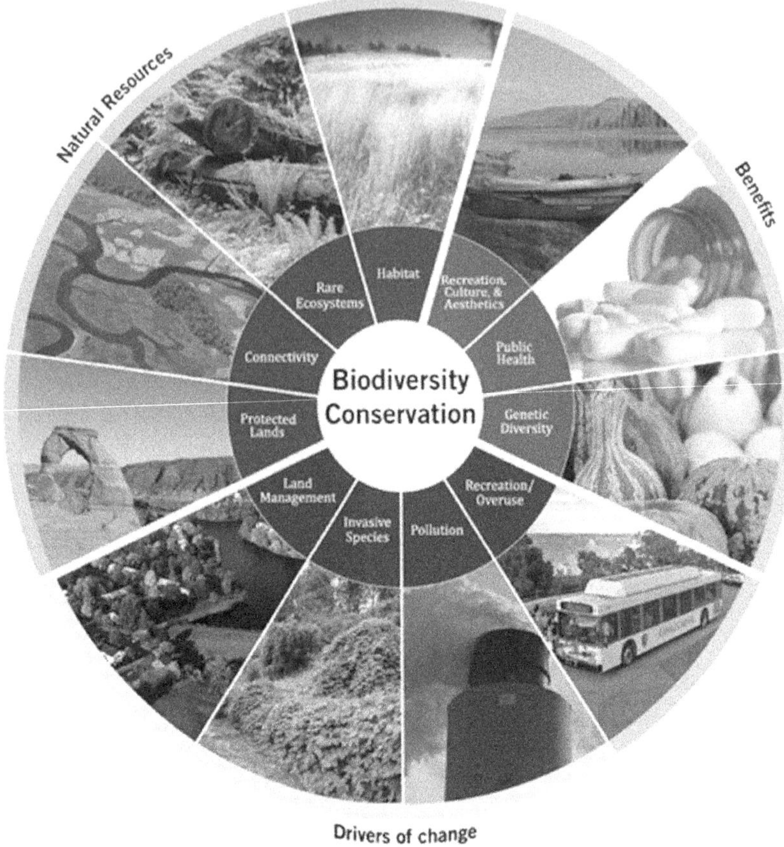

Fig. 2: Ecosystem resources, benefits and drivers of change (Source: 44).

environments (55,46). Undoubtedly, creating sustainable spaces also requires the more rationalistic and sustainable use of the available natural and cultural resources. This is possible through the accurate reading of the environmental assets of the space to be planned and when the human activities developing on it develop uses in conformity with the fundamental of environmental sensitivity (56,57). Sustainable spatial planning mediates in the sustainability of natural and cultural heritage assets and biological diversity; in ensuring the efficient and economical use of energy and resources; in making urban spaces sound and livable by enhancing the quality of urban life; in eliminating the negative impacts of the urban production/consumption/waste/recovery chain on environmental resources and assets; and in recycling or making the possibilities of reuse widespread. Within this scope, it is based on the understanding of a long-term vision and comprehensive strategic planning that includes environmental, spatial, sociocultural, economic, and administrative objectives (11,58).

In this context, sustainable spatial planning is beyond conventional land use planning. Sustainable spatial plans should be plans with sociocultural and ecological contents which not only have the physical planning understanding but also show the distribution and interrelationships of the elements of the built and natural environments and which provide an opportunity for spatial development suitable for this (59).

Ecosystem services are an approach which was developed with respect to the economic benefits in the protection of nature and natural resources. Ecosystem services express the benefits of an ecosystem as the economic contributions it provides in the short, medium, and long terms and therefore reveal the importance of the fact that spatial development should be ensured with an ecological planning approach (12). For example, natural capital, such as forests and rivers, provides flows of ecosystem services such as non-timber forest products, clean water and nature tourism, which benefit people. Natural capital also supports production systems such as agriculture and fisheries, thereby supporting food security and economic productivity (60) (Fig. 3). The benefits ecosystems offer within the supporting, provisioning, regulating and cultural scopes require managing these resources in line with the principle of sustainability (61).

For these reasons, in order to provide the ecological balance of nature today, basing the decision support processes at the stage of drawing up spatial plans likely to be regarded as the instrument of human beings to intervene in nature on the fundamentals of ecological planning is getting more and more important (34). At this point, the integration of ecosystem services from the country plan to urban plans in the spatial planning process is striking as an extremely important approach.

It is also emphasized in many studies that the ES-based spatial planning approach will yield more rational results in terms of environmental protection and development. Since the supporting services out of the four main categories determined by the MEA (2005) are not directly quantifiable, the first three categories are generally taken into consideration in the assessment of the ESs in the spatial planning process especially in the upper-scale (e.g., country and region) planning process. Provisioning, regulating and cultural ESs are spatialized by using land use and land cover data; change analyses can be made in the context of temporal data; and the land cover can therefore be used as a spatial decision support tool in the context of its ES potential (17,62,63,64,65,34). In this context, the "multifunctional" approach, an essential implementation strategy for landscape design and spatial planning, may also contribute to ES-based planning and management practices as a rational instrument. In the context of spatial planning, the ESs can be assessed considering the functional groups of ecosystems (25).

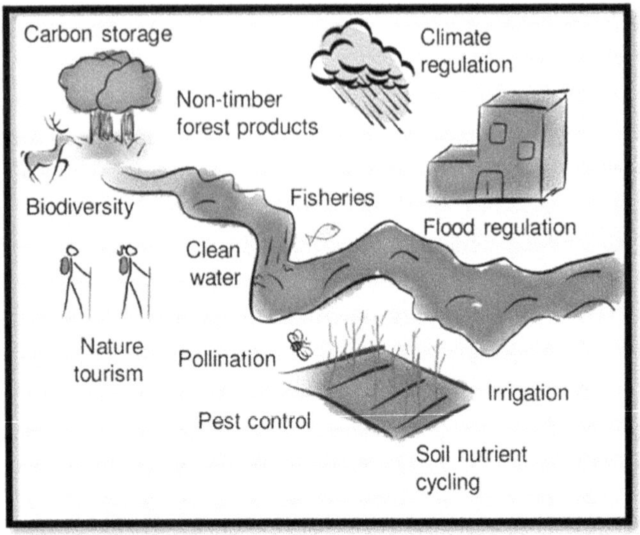

Fig. 3: The links between natural capital, ecosystem services and benefits to people (Source: 60).

The majority of the scientific studies on the calculation of ecosystem services are studies on the regional or national scale which are about the ecosystems in natural and rural landscapes and the domestic and foreign studies on the ecosystem services in urban areas are rather limited (66,43,67,68,69). However, the most fundamental reason why ecosystem services are impaired or lose their function is urban ecosystems.

Cities are dependent on nonurban ecosystems but benefit from urban ecosystems too (32). Urban ecosystem services are services which offer significant contributions to the quality of life of city dwellers (such as food supply, regulation of the flow of water, noise reduction, climate regulation, and recreational and social benefits) (19). Nevertheless, the provisioning, regulating and cultural services included in the classification of ecosystem services are damaged and the sustainability of cities becomes impossible as a result of the rapid increase in urbanization and the occurrence of building masses. At this point, first of all, green infrastructure plans should be prepared in order to guarantee the ecosystem services of cities by adhering to the upper-scale plan decisions (12).

The concept of green infrastructure is extremely associated with the disciplines about country planning, regional planning, urban planning-urban design, nature protection, and rural planning. Even though the scales of green infrastructure studies vary, this concept refers to the integration among the networks formed by the natural areas based on landscape ecology on the rural scale. On the urban scale, however, it defines a multifunctional network of green areas. By scale, the definition of green infrastructure was grouped under two titles as a. Green infrastructure on the urban scale and b. Green infrastructure on the landscape scale (regional, national and international scales)(70).

The concept of green infrastructure shares common objectives with sustainable land management planning on the spatial planning scale. Green infrastructure systems were developed in order to reduce the impacts of the contradictions between economy and ecology (71,72,73,74). In order for the green infrastructure system to succeed and provide the benefits expected from it, it should not be assessed separately from the existing land uses (for instance, settlements, agricultural lands and recreational areas) but with a holistic approach in the spatial planning process (75,76,77).

In this context, when the spatial planning process is evaluated today, it turns out that ecosystem services and green infrastructure plans accordingly should be included in all plan levels from the upper scale to the lower scale. As required by an ecology-based approach, land use decisions should be made with an approach which prevents habitat disintegration and at the same time considers ecosystems and their contributions. Furthermore, natural, rural and urban landscapes should be evaluated as a whole on the basis of landscape habitat corridors in this process. These evaluations should be turned into a project base and guide the next stage(43).

However, as urban areas display non-linear and dynamic interactions within growth patterns, how complicated systems can be integrated into urban planning practices concerning urban ecosystems turns out to be an important question(78).

Thus, regarding the urban planning scale, firstly, the urban ecosystem character should be presented by considering the natural and cultural landscape features of the city and the current situation should be evaluated (19). Secondly, the quantities of urban green areas should be determined in line with the contributions that the green areas in the city will make (68,69). Following this, green infrastructure plans should be prepared; the principles of ecological planning should be adopted in these plans; and the green area centers (reserves, society parks and natural areas, tilled lands, areas protected for agricultural purposes, regional parks and groves, and managed natural areas) and green links (green belts, green roads, conservation corridors and landscape links) in the city should be associated and projected with a holistic approach (43,67). Ecological and sustainable design solutions such as roof gardens, vertical gardens, rain gardens, drainage solutions and plant selection for the sustainable use of water should be offered on the basis of green infrastructure plans. So, the continuity of the natural ecosystem in the city should be ensured depending on the fundamental that the city is an ecosystem which is not separated from the natural and rural areas in city planning (12). In this way, one will simultaneously contribute to many ecological, social, economic and cultural areas of the city (e.g., microclimatic comfort, efficient use of surface water, development of urban biodiversity, improvement of the air quality, reduction of urban heat islands, enhancement of social interaction, social health, providing of social unity, economic benefit/an increase in real estate and rent values through green areas, and a low maintenance cost/less irrigation and less fertilization by creating ecologically sustainable landscapes) through the green areas in which biological diversity is supported, along with ecosystem services. Supporting biological diversity in urban green areas also means that city dwellers find nature immediately next to them as well as the development of the ecological quality of cities (10,19).

Nevertheless, on the condition that the available parks, green roads, natural areas, road trees, small-scale home gardens, apartment building and gated community gardens, pavements and many jointly and privately owned open and green areas created to support biological diversity and hence ecosystem services in urban areas are designed and managed with ecological principles, they might become potential areas. Moreover, design and maintenance approaches should be implemented to support ecosystem services also in façade greening, as well as in roof and balcony terrace arrangements (10).

To sum up, green infrastructure systems are one of the most efficient ways for ecosystem services to fulfill and carry on their function in the spatial planning process. As Çetinkaya (79) also states, the natural habitats, green corridors and open and green areas located in urban areas and their close vicinity are associated with each other with the green infrastructure system. The planning approach based on this relationship forms a framework for nature conservation and sustainable economic development. It contributes to the enhancement of the quality of life of the urban people and to the creation of sustainable cities. It provides the people in the country, in the region and in the city with various benefits at the national (preservation of biological diversity), regional (determination of nature conservation areas, making a connection among nature conservation areas and protection of natural resources), and local (climate control, clean water, food safety and recreation) levels. The benefits concerned contribute to the sustainability of the urban landscape. Thus, there is a need for the integration of the green infrastructure system into landscape planning and related planning policies. As Odum and Barrett (13) also state, the natural ecosystem and the technological ecosystem should absolutely be integrated in the understanding of sustainable city planning.

4 Conclusion

The land use that changes through human interventions owing to the increased need for food, clean water, clean air, houses and employment upon the growth of cities and the gradual increase in the population living in cities accordingly causes biodiversity to be damaged and leads to the upsetting of the ecological balance as ecosystems lose their qualities.

In this context, the 2020 Goals of the European Union, which aim to halt and recover the loss of biological diversity, are quite assertive. Attainment of these goals requires implementing efficient policies, intersectoral coordination, ecosystem management approaches and understanding the value of biological diversity better (15).

Therefore, for the economical use and sustainability of natural resources, it is now a must to eliminate the pressures on resources and to plan sustainable spaces in agreement with this (80,81).

Demiroğlu ve Karadağ (19) also states ecosystem services are an approach which was developed on the economic benefits of the natural capital. The use of an ecology-based term (ecosystem) together with economy may be acknowledged as a method to protect nature and natural resources in today's world because ecosystem services refer to the benefits of an ecosystem and its economic contributions in the short, medium and long terms.

Whilst the pressures experienced in urban areas lead to the understanding of an urban society segregated from ecosystems, the demands concerning the natural capital and ecosystem services are gradually increasing in urban areas. The functions and components of the ecosystem services in urban areas are a summary of the ecological evaluation of services (12). Even if only some of the urban ecosystem services are provided, the local, social and economic values of the services provided may be amazingly high. In this sense, it is crucial to reveal urban ecosystem service indicators and measurements. Making these evaluations on ecosystem services is necessary to implement the approach in the spatial planning process (43). In general terms, the integration of ecosystem services into urban planning can easily be accomplished due to the similar nature of ecosystem services and spatial planning which share common goals (78).

In conclusion, the ES-oriented spatial planning process, which considers the multiple functions and benefits offered by ecosystems depending on the land use/land cover, is of vital importance for sustainable development. Employing innovative tools in the planning, management and design studies based on ecosystem services will also assume a role that will support the sustainability of ecological services. The products and benefits people obtain from natural resources may be determined more realistically with the spatial planning approaches based on ecosystem services. Sustainable spatial management approaches that provide the conservation and use balance may be realized by scrutinizing the spatial distributions and interrelationships of ecosystem services (66).

As Bolund and Hunhammar (32) also state, it is extremely clear that the ecosystem services created nationally, regionally and locally have a significant impact on the quality of life in urban areas and should absolutely be addressed in land use planning.

The process of executing the ecosystem services will be started after it is put into the levels of spatial planning. When the planning process of today is evaluated, it is clearly seen that ecosystem services should be in all the planning levels, from upper scale to subscale. In this context, green infrastructure plans stand out as plans to increase ecosystem services in cities by preserving natural processes.

References

1 Antrop, M., *Landscape change and the urbanization process in Europe.* Landscape and Urban Planning, 2004. **67**: pp. 9–26.

2 Pouya, S. and S. Pouya, *Biyolojik çeşitliliğe ve ekosistem hizmetlerine katkı sağlayan kentsel projeler.* Journal of Urban Academy, 2017. **10** (3): pp. 306–322.

3 Altuntaş, A., *Sürdürülebilir toplumlar ve metropollerin baskılarından kurtulmak için alternatif bir yol: sürdürülebilir kentler.* Mustafa Kemal Üniversitesi Sosyal Bilimler Enstitüsü Dergisi, 2012. **9** (17): pp. 135–148.

4 Aydın, B., *Gelişme Alanlarında Ekolojik Kentsel Yerleşim Kriterlerinin Belirlenmesi ve İmar Planı Kapsamında Yorumlanması: Ömerli Havzası-Sancaktepe Örneği.* 2010, Yüksek Lisans Tezi, İTÜ Fen Bilimleri Enstitüsü, İstanbul.

5 Kocataş, A., Ekoloji-Çevre Biyolojisi. 2010, ISBN: 978-975-483-177-1, Ege Üniversitesi Basımevi: Bornova-İzmir. 597 s.

6 UN, *World Urbanization Prospects: The* 2009 *Revision, UN Department of Economic and Social Affairs, Population Division.* 2010 [cited 2015, 4 March]; Available from: www.un.org

7 Aksu, C., *Sürdürülebilir Kalkınma ve Çevre.* 2015, Güney Ege Kalkınma Ajansı Raporu, 33 s. [cited 2015 10 January]; Available from: http://geka.gov.tr/Dosyalar/o_19v5e00u1ru61bbncf2qmlcpv8.pdf

8 Anna, K., K. Jaan-Henrik, K. Jakub and H. Dagmar, *Ecosystem services in urban land use planning: integration challenges in complex urban settings-case of Stockholm.* Ecosystem Services, 2016. **22**: pp. 204–212.

9 Berry, B. J. L., *Urbanization,* in *Urban Ecology An International Perspective on the Interaction between Humans and Nature,* J. M. Marzluff, E. Shulenberger, W. Endlicher, M. Alberti, G. Bradley, C. Ryan, U. Simon, C. ZumBrunnen, Editor. 2008, Online ISBN: 978-0-387-73412-5, Springer Science Business Medya: NewYork. pp. 25–48.

10 Uslu, A. and N. Shakouri, *Kentsel peyzajda yeşil altyapı ve biyolojik çeşitliliği destekleyecek* olanaklar. Türk Bilimsel Derlemeler Dergisi, 2013. **6** (1): pp. 46–50.

11 Mansuroğlu, S., P. Kınıklı and B. Saatci, *Antalya'da kentsel gelişimin ekolojik açıdan değerlendirilmesi ve sürdürülebilirlik kapsamında önerilerin geliştirilmesi.* Ege Üniv. Ziraat Fakültesi Dergisi, 2012. **49** (3): pp. 255–264.

12 Çoban, A. and M. Yücel, *Kent planlamasında ekosistem hizmetlerinin rolü.* Düzce Üniversitesi Bilim ve Teknoloji Dergisi, 2018. **6** (2018): pp. 444–454.

13 Odum, E. P. and G. W. Barrett,. *Ekolojinin Temel İlkeleri.* Prof. Dr. Kani Işık, Editor. 2008, ISBN: 978-9944-341-74-5, Palme Yayınları: Ankara. pp. 374–412.

14 Kışlalıoğlu, M. and F. Berkes, Çevre ve Ekoloji. 2010, 12. Basım, ISBN: 978-975-14-1413-7, Remzi Kitabevi: İstanbul. 280 s.

15 AÇA, *Biyolojik Çeşitlilik-Ekosistemler.* 2018 [cited 2018, 20 August]; Available from: https://www.eea.europa.eu/tr/themes/biodiversity/intro

16 Alexander, S., J. Aronson, O. Whaley and D. Lamb, *The relationship between ecological restoration and the ecosystem services concept.* Ecology and Society, 2016. **21** (1): p. 34.

17 De Groot, R. S., *Function-analysis and valuation as a tool to assess land use conflicts in planning for sustainable, multi-functional landscapes.* Landscape and Urban Planning, 2006. **75** (3–4): pp. 175–186.

18 Tokgöz, G. and N. Say, *Kentsel ekosistem hizmetlerinin haritalanması için kullanılan göstergeler, yöntemler ve geliştirilen araçlar.* Adana Bilim ve Teknoloji Üniversitesi Fen Bilimleri Dergisi, 2018. **1** (1): pp. 1–8.

19 Demiroğlu, D. and A. A. Karadağ, 2015. *Ecosystem Services Approach to Spatial Planning in Turkey* in *I. Uluslararası Kent Araştırmaları Kongresi, 16–17 April 2015.* 2015: Eskişehir, Turkey. pp. 153–168.

20 Robinson et al., *What are ecosystem services and natural capital, and how does this apply to soil science?* in *1st International Conference and Exploratory Workshop on Soil Architecture and Physical-Chemical Functions "CESAR,"* 30 November - 2 December, 2010. 2010: Tjele: Denmark. pp. 32–330.

21 Daily, G. C., *Introduction: What are ecosystem services?* In *Nature's Services: Societal Dependence on Natural Ecosystems,* G. C. Daily, Editor. 1997, Island Press: Washington, pp. 1–10.

22 Costanza, R., et al., *The value of the world's ecosystem services and natural capital.* Nature, 1997. **387**: pp. 253–260.

23 Daily, G. C., et al., *Ecosystem services in decision making: time to deliver.* Journal of Frontiers in Ecology and the Environment, 2009. **7**: pp. 21–28

24 Albayrak, İ., *Ekosistem Servislerine Dayalı Havza Yönetim Modelinin İstanbul-Ömerli Havzası Örneğinde Uygulanabilirliği.* 2012, Doktora Tezi, İTÜ Fen Bilimleri Enstitüsü, İstanbul.

25 MEA, *Ecosystems and Human Well-Being: Synthesis.* 2005, ISBN 1-59726-040-1, Millennium Ecosystem Assessment, Island Press: Washington DC. 137 p. [cited 2018, 24 October]; Available from: https://www.millenniumassessment.org/documents/document.356.aspx.pdf

26 MEA, *A Toolkit for Understanding and Action-Protecting Nature's Services, Protecting Ourselves,* Millennium Ecosystem Assessment, Island Press: Washington DC. 28 p. [cited 2018, 24 October]; Available from: https://www.unpei.org/sites/default/files/PDF/ecosystems-economicanalysis/MEA-A-Toolkit.pdf

27 TEEB, *The Economics of Ecosystems and Biodiversity: Ecological and Economic Foundations. Biodiversity, Ecosystems and Ecosystem Services (Chapter 2),* 2010 [cited 2018, 25 October]; Available from: http://www.teebweb.org/our-publications/teeb-study-reports/ecological-and-economic-foundations/

28 TEEB, *TEEB Manual for Cities: Ecosystem Services in Urban Management.* 2011 [cited 2018, 25 October]; Available from: www.teeb.org

29 Willemen L., P. Verburg, L. Hein and E.F. Mensvoort, *Spatial characterization of landscape functions,* Landscape and Urban Planning, 2008. **88** (1): pp. 34–43.

30 Arslan Muhacir, E. S., İ. Tazebay, *Kırsal turizm türlerinin belirlenmesinde bir araç: ekosiztem hizmetleri yaklaşımı.* Türkiye Ormancılık Dergisi, 2017. **18** (1): pp. 74–81.

31 De Groot, R. S., M. A. Wilson and R. M. J. Boumans, *A typology for the classification, description and valuation of ecosystem functions, goods and services.* Ecological Economics, 2002. **41**: pp. 393–408.

32 Bolund, H. and S. Hunhammar, *Ecosystem services in urban areas.* Ecological Economics, 1999. **29**: pp. 293–301.

33 Boyd, J. and S. Banzhaf, *What are ecosystem services? The need for standardized environmental accounting units.* Ecological Economics, 2007. **63** (2–3): pp. 616–626.

34 Tezer et al., *Ekosistem Servislerine Dayalı "Havza Koruma Alanları" Tanımlamasının Önemi ve Kapsamı: Düzce – Melen Havzası.* 2018 [cited 2018, 20 October]; Available from: http://www.skb.gov.tr/ekosistem-servislerine-dayali-havza-koruma-alanlari-tanimlamasinin-onemi-kapsami-duzce-melen-havzasi-s28364k/

35 MAES, *Mapping and Assessment of Ecosystems and their Services: Indicators for Ecosystem Assessments Under Action 5 of the EU Biodiversity Strategy to* 2020. 2014, 2nd Report – Final, Publications Office of the European Union, Luxembourg. 80 p. [cited 2018, 25 October]; Available from: http://ec.europa.eu/environment/nature/knowledge/ecosystem_assessment/pdf/2ndMAESWorkingPaper.pdf

36 Hermann, A., S. Schleifer and T. Wrbka, *The concept of ecosystem services regarding landscape research: a review.* Living Reviews in Landscape Research, 2011. **5** (1): p. 5–37.

37 Chan, K. M. A., T. Satterfield and J. Goldstein, *Rethinking ecosystem services to better address and navigate cultural values.* Ecological Economics, 2012. **74**: p. 8–18.

38 McPhearson, T., Z. A. Hamstead, and P. Kremer, *Urban ecosystem services for resilience planning and management in New York City.* Springer, 2014. **43** (4): p. 502–515.

39 IPBES, *Intergovernmental science-policy platform on biodiversity and ecosystem servicesreport of the plenary of the ıntergovernmental science-policy platform on biodiversity and ecosystem services on the work of its fourth session.* 2016 [cited 2018 20 October]; Available from:http://www.ipbes.net/sites/default/files/downloads/pdf/ipbes_4_19_en.pdf

40 GeoAdaptive, *Ecosystem Services Categories.* 2015 [cited 2018 15 October]; Available from: http://geoadaptive.com/rlwi_southwest_florida/

41 WWF, *Living Planet Report/Risk and Resilience in a New Era.* 2016 [cited 2018 20 October]; Available from: http://awsassets.panda.org/downloads/lpr_living_planet_report_2016.pdf

42 Ruhl, J. B., S. E. Kraft and C. Lant, *The Law and Policy of Ecosystem Services.* ProQuest Information and Learning Company, UMI Number: 3229831. 2006, Island Press: Washington, USA. 311 p.

43 Gõmez-Baggethun E. and D. N. Barton, *Classifying and valuing ecosystem services for urban planning.* Ecological Economics, 2013. **86**: pp. 235–245.

44 EPA, *Ecosystem Services in EnviroAtlas/What are ecosystem services, and why are they important?* 2016 [cited 2018 30 October]; Available from: https://www.epa.gov/enviroatlas/ecosystem-services-enviroatlas

45 Gedikli, B., *Stratejik Mekansal Planlama: Planlamada Yeni Anlayışlar, Yöntemler ve Teknikler* in *Kentsel Planlama Kuramları,* M. Ersoy, Editor. 2. Baskı, ISBN: 978-975-533-521-6, 2012, İmge Kitabevi Yayınları: Ankara, pp. 237–290.

46 Akyol, D., *Stratejik mekânsal planlama yaklaşiminin temel özellikleri ve diğer planlama yaklaşimlarindan ayrildiği yönleri.* Uluslararası Sosyal Araştırmalar Dergisi, 2018. **11** (58): pp. 588–593.

47 Şahin, Ş., *Peyzaj Ekolojisi Kavramsal Temelleri ve Uygulama Alanları* in *Peyzaj Yönetimi.* A. Akay, M. Demirbaş Özen, Editor. 2009, No: 354, ISBN: 978-975-8918-36-2, Türkiye ve Ortadoğu Amme İdaresi Enstitüsü Yayını: Ankara. pp. 31–57.

48 Albrechts, L., P. Healey and K. R. *Kunzmann, Strategic spatial planning and regional governance in Europe.* Journal of the American Planning Association, 2003. **69** (2): pp. 113–129.

49 Harris, N. and A. Hooper, A., *Rediscovering the spatial in public policy and planning an examination of the spatial content of sectoral policy documents.* Planning Theory and Practice, 2004. **5** (2): pp. 147–169.

50 Sartorio, F. S., *Strategic spatial planning a historical review of approaches, its recent revival, and an overview of the state of the art in Italy.* DISP – Te Planning Review, 2005. **162** (3): pp. 26–40.

51 Schmidt, S., *Land use planning tools and institutional change in Germany: recent developments in local and regional planning.* European Planning Studies, 2009. **17** (12): pp. 1907–1921.

52 Faludi, A., *European spatial planning: past, present and future.* Town Planning Review, 2010. **81** (1): pp. 1–22.

53 Polat, E., *Planlamada yeni bir paradigma baskısı: stratejik mekânsal planlama (SMP).* Çağdaş Yerel Yönetimler, 2010. **19** (4): pp. 1–18.

54 Allin, S. and C. Walsh, *Avrupa Kent-Bölgelerinde Stratejik Mekânsal Planlama: Paralel Süreçler mi Farklılaşan Yörüngeler mi?* Planlama Dergisi, 2013. **23** (3): pp. 111–124.

55 Özcan, A., *Ekolojik Temele Dayalı Sürdürülebilir Kentsel Gelişme: Malatya Kent Örneği Üzerinden Bir Değerlendirme* in *38. ICANAS Uluslararası Asya ve Kuzey Afrika Çalışmaları Kongresi, 10–15 September 2007.* 2007: Ankara, TURKEY. pp. 689–710.

56 Sılaydın, M. B., *Ekolojik Dengenin Korunmasını Hedefleyen Yeni Bir Şehir Planlama Süreci Önerisi* in *Peyzaj Mimarlığı 3. Kongresi,* 22–25 Kasım 2007: Antalya, TURKEY. pp. 154–160.

57 Karadağ, A., *Kentsel ekoloji: kentsel çevre analizlerinde coğrafi yaklaşım.* Ege Coğrafya Dergisi, 2009. **18** (1–2): pp. 31–47.

58 Özcan, K., *Kent sürdürülebilirlik gündemi: bir kavramsallaştırma denemesi.* Avrasya Terim Dergisi, 2016. **4** (2): pp. 7–17.

59 Cengiz, A. E., A. A. Karadağ, and D. Demiroğlu, *Sürdürülebilir Mekanlar için Peyzaj Planlamanın Önemi* in *International Congress of the New Approaches and Technologies for Sustainable Development,* 21–24 September 2017. 2017: Isparta, TURKEY. pp. 4–15.

60 Conservation International, *Monitoring Natural Capital and Human Well-being in Madagascar: A National Demonstration of Indicators for Sustainable Development.* 2015, Technical Report, Moore Center for Science and Oceans and Madagascar Country Program, Conservation International, Arlington VA. 46 p. [cited 2018, 24 October]; Available from: https://www.conservation.org/global/madagascar/Documents/METRICS%20Policy%20report%20Madagascar%20(english).pdf

61 Eker, Ö., *Su Kaynakları Yönetiminin Ormancılık ve Ekosistem Hizmetleri Kapsamında Sosyo-Ekonomik Bakımdan Değerlendirilmesi* in 2023*'e Doğru 4. Doğa ve Ormancılık Sempozyumu,* 3–6 December 2017. 2017: Antalya, TURKEY. pp. 363–371.

62 De Groot, R. S., et al., *Challenges in integrating the concept of ecosystem services and values in landscape planning, management and decision making.* Ecological Complexity, 2010. **7** (3): pp. 260–272.

63 Burkhard, B., F. Kroll, S. Nedkov and F. Müller, *Mapping ecosystem service supply, demand and budgets.* Ecological Indicators, 2012. **21**: pp. 17–29.

64 Burkhard, B., M. Kandziora, Y. Hou and F. Müller, *Ecosystem service potentials, flows and demands –concepts for spatial localisation, indication and quantification.* Landscape Online, 2014. **34**: pp. 1–32.

65 Tezer, A., et al., *Ömerli Havzası'nda Ekosistem Servislerine Dayalı Bütünleşik Havza Yönetim Planı'nın Geliştirilmesi Projesi.* 2015, İSTKA TR10/14/DFD/0039 No'lu İSTKA Projesi Nihai Raporu, İstanbul, TURKEY.

66 Derkzen, M. L., A. J. A. Van Teeffelen and P. H. Verburg, *Quantifying urban ecosystem services based on highresolution data of urban green space: an assessment for Rotterdam, the Netherlands.* Journal of Applied Ecology, 2015. **52**: pp. 1020–1032.

67 Coşkun Hepcan, Ç. and Ş. Hepcan, *Ege Üniversitesi lojmanlar yerleşkesinin hava kalitesinin iyileştirilmesine yönelik düzenleyici ekosistem servislerinin hesaplanması.* Ege Univ. Ziraat Fak. Derg., 2017. **54** (1): pp. 113–120.

68 Coşkun Hepcan, Ç. and Ş. Hepcan, *Kentsel yeşil altyapı analizi: Bornova örneği.* Mediterranean Agricultural Sciences, 2018. **31** (1): pp. 37–43.

69 Demircan, N., *Kentsel tarımın kent ekosistem servisi olarak değerlendirilmesi ve iklim değişikliğindeki hafifletici etkileri.* Atlas International Referred Journal on Social Sciences. 2018. **4** (9): pp. 485–491.

70 Yazgı, D., K. T. Yılmaz, Yeşil Altyapı Kavramının İlgili Yasal Düzenlemeler İçerisindeki Yeri ve Uygulamaya Yönelik Öneriler in 6. Peyzaj Mimarlığı Kongresi "Söylem ve Eylem," 08–11 Aralık 2016. 2016: Antalya, TURKEY. (cited 2018, 10 November]; Available from: https://www.researchgate.net/publication/324796992

71 Weber, T., *Landscape ecological assessment of the Chesapeake Bay watershed.* Environ. Monit. Assess., 2004. **94**: pp. 39–53.

72 Ahern, J., Green Infrastructure for Cities: The Spatial Dimension, in Cities of the Future: Towards Integrated Sustainable Water and Landscape Management, V. Novotny, P. Brown, Editor. 2007, IWA Publishing: London. pp. 267–283.

73 Wickham, J. D., K. H. Riitters, T. G. Wade and P. Vogt, *A national assessment of green infrastructure and change for the conterminous United States using morphological image processing.* Landscape and Urban Planning, 2010. **94** (3): pp. 186–195.

74 Mell, I. C., *Green infrastructure planning: A contemporary approach for innovative interventions in urban landscape management.* Journal of Biourbanism, 2012. **1**: pp. 29–39.

75 Benedicts, M. A. and E. T. McMahon, *Green infrastructure: smart conservation for the 21st century.* Renewable Resources Journal, 2002. **20** (3): pp. 12–17.

76 Çetinkaya, G., *Kentsel Peyzaj Planlamada Yeni Yaklaşımlar: Yeşil Altyapı ve Yeşil Koridorlar* in *5. Peyzaj Mimarlığı Kongresi "Dönüşen Peyzaj", 14–17 November 2013.* 2013: Çukurova Üniversitesi (Adana), TURKEY. p. 265–277.

77 Çetinkaya, G. and O. Uzun, *Peyzaj Planlama.* 2014, ISBN: 978-975-511-608-2, Birsen Yayınevi: İstanbul. 183 s.

78 Tezer, A., et al., *Integrated planning need for the resilience of urban riverine ecosystems: Istanbul-Omerli watershed case.* Ecohydrology & Hydrobiology, 2012. **12** (2): pp. 153–163.

79 Çetinkaya, G., *Kentsel peyzaj ekolojisinin sürdürülebilirliği için yenilikçi bir yaklaşım: yeşil altyapı ve planlama politikası.* İdeal Kent Dergisi, 2014. **12**: pp. 218–245.

80 Marsh, W. M., *Landscape Planning Environmental Applications.* 1997, ISBN: 0-471-24207-1, John Wiley&Sons: New York. 434 p.

81 Makhzoumi, J., Pungetti, G., *Ecological Landscape Design and Planning.* 1999, ISBN: 0-419-23250-8, Taylor & Francis: New York. 330 p.

Assoc. Prof. Dr. Ferhat Hacialibeyoğlu and Assoc. Prof. Dr. Deniz Dokgöz

17 New Approaches in Design within the Context of Architectural Production Practice of Local Governments: The Case of Luleburgaz Yildizlari Women's Academy

Abstract: From the point of view of local governance in the context of production of public space, Lüleburgaz Municipality is remarkable with its high quality approaches in the social and spatial intentions. Of special interest are their projects that take care of and target the development of mainly childen, youth and women in a social and spatial context and focus on the concept of Academy such as; *Lüleburgaz Yıldızları (LY) (Stars of Lüleburgaz in English) Football Academy, Lüleburgaz Yıldızları Motorcycle and Bicycle Academy, Lüleburgaz Yıldızları Women's Academy (LYWA), Lüleburgaz Yıldızları Swimming Academy, Lüleburgaz Yıldızları Art Academy.* In addition to their contributions to social and social development goals, these projects stand out with the spatial qualities and architectural insights they intend to or have. This work examines the Lüleburgaz Municipality's method of producing public space in the last decade from the perspective of the relationship of "architecture and space," while investigating the concept of Academy and the approach that brings this way of production into being through "space and social connection." In this regard, this chapter investigates the concept of Academy that Lüleburgaz Municipality proposed as a new education model to the city and its citizens, the concept of competition that local governments prefer/choose as a process of procuring projects and the contexts of architecture/space/social connection/relationship via the concept of competiton; from the lens of the design that won the first prize in the Lüleburgaz Yıldızları Kadın Akdemisi (Lüleburgaz Yıldızları Women's Academy LYWA) architectural project competition that took place in 2015.

Keywords: Space production, Competition, Local government, Academia, Public space

1 Introduction

The main issues discussed in the formation and development process of the city are absence of holistic methods that deal with design and planning approaches, and the inability of achieving a whole with high quality from individual high quality products. Besides, a comprehensive and holistic perspective can create high quality living areas by taking the space-space producing practices, social living conditions and the social relations network of the city into consideration; in addition, it positively affects the whole city in many areas. This approach can more likely to be realized and sustained especially in local scales, in which communication and interaction potentials are strong. At this point, local government and decision-making mechanisms can reach refined results in an either spatial or social sense by moving with a pluralistic etiquette. From the point of view of local governance in the context of production of public space, Lüleburgaz Municipality is remarkable with its high quality approaches in the social and spatial intentions. Of special interest are their projects that take care of and target the development of mainly childen, youth and women in a social and spatial context and focus on the concept of Academy such as; *Lüleburgaz Yıldızları (LY) (Stars of Lüleburgaz in English) Football Academy, Lüleburgaz Yıldızları Motorcycle and Bicycle Academy, Lüleburgaz Yıldızları Women's Academy (LYWA), Lüleburgaz Yıldızları Swimming Academy, Lüleburgaz Yıldızları Art Academy.* In addition to their contributions to social and social development goals, these projects stand out with the spatial qualities and architectural insights they intend to or have. This work examines the Lüleburgaz Municipality's method of producing public space in the last decade from the perspective of the relationship of "architecture and space," while investigating the concept of Academy and the approach that brings this way of production into being through "space and social connection." In this regard, this work investigates the concept of Academy that Lüleburgaz Municipality proposed as a new education model to the city and its citizens, the concept of competition that local governments prefer/choose as a process of procuring projects, and the contexts of architecture/space/social connection/relationship via the concept of competiton; from the lens of the design that won the first prize in the Lüleburgaz Yıldızları Kadın Akdemisi (Lüleburgaz Yıldızları Women's Academy LYWA) architectural project competition that took place in 2015 [1].

2 Academies in the Context of the Local Governance Vision of Lüleburgaz Municipality

While "Academy" is a concept that prioritises the field of education, it became prominent in the context of local governance for Lüleburgaz Municipality. What the Municipality defines as "good governance" in its 2017–2021 strategic plan is a process, in which Academies play a significant role [2]. This process deals with describing the route of the city via analyzing its workings, and spatialising social functionality in a way particular to the city. Lüleburgaz Municipality nearly offers a new "education" model to the city and its citizens with the idea of Academy in accordance with the Municipality's main principle of increasing the life quality for local citizens. This model incorporates new facilities that deal with increasing the citizen consciousness for the age groups and users it involves. The Academies, which are set up under the main caption of "Lüleburgaz Yıldızları (Stars of Lüleburgaz)" target individuals of all ages; however, the youth have a special place under this caption. The process started with the Lüleburgaz Yıldızları Football Academy (LYFA) that deals with the physical activities of individuals of all ages, the Lüleburgaz Yıldızları Motorcycle and Bicycle Academy (LYMBA) that looks for solutions to the traffic and transportation problems, and the Lüleburgaz Yıldızları Flavor Academy (LYFA). It is continuing to diversify with Information Academy (LYIA), Women's Academy (LYWA), Art Academy (LYAA) and Swimming Academy

(LYSA). Lüleburgaz, with its setup to create "impact," aims to create ties between the academies, and the city and the citizens through the activities that will be realized in the Life Academies. It portrays new common living spaces through academies with an original, analytical, systematic, participatory, holistic and multilateral approach. At this point, the academies that will be opening in a short while such as Happiness Academy (LYHA), Tennis Academy (LYTA), Saloon Sports Academy (LYSSA) and Creativity Academy (LYCA) have to be assessed in this context

Within the Academy format of Lüleburgaz Municipality in the last decade, its way and method of producing "public space" also consolidate the Academies' sayings about the city. This setup can be defined through the relationship between "architecture and space," and it exhibits the spatial transformation the Academy format generates in an urban understanding and the link between "spatial and social relationship" that this way of production creates. This link integrates the perceptions of the concept of Academy and high quality spaces and production of the built environment, which takes the public to the foreground. Therefore, this approach communicates with the city and the citizens in the fields of art, sports and culture, and aims to bring the citizen into existence as an actor equipped with academic thought. It comes into being not only through content/function but through the idea of creating this content in high-quality spaces. This situation supports the perception in the relationship between space and social perception through quality.

3 Competitions in the Context of Local Government's Space Production Practices

An architectural project competition is a pluralistic activity arena and production practice that plays an active role in enhancing the architecture environment in a country; it also provides an arena for putting out and discussing new expansions in the field of architecture [3]. The history of this method in Turkey leads to the 1930s; the production of many high-quality spaces, the production of public buildings in particular, in either the building or urban scale was observed via competitions. More than one actor exists in this process; the actors, who demand, design and evaluate. The common goal of each actor can be defined as reaching a design/space of high quality. Besides its actors, the competition method can also be expressed as a pluralistic environment in terms of products, so that multiple alternatives can be compared, discussed and evaluated in an environment created by diversified and productive approaches. This is significant in space production/acquirement. In this sense, the method of competition contributes to the architectural culture and built environment through the production of buildings/spaces, at the same time it positively affects the perception of architecture and space culture in the social area.

Looking at the ample number of public buildings that were constructed each year, the few number of public buildings that were constructed via the pluralistic environment of competitions are conspicuous. Contrastingly, even though the number of buildings acquired by the competition method is small, their quality levels are quite high. Many local governments obtain either service buildings or public buildings of various functions for citizens' use via competition method, yet most of these projects are not applied. Therefore, besides the idea of acquiring projects through competition, the continuity, sustainability and application of this concept into the built environment is quite significant.

Lüleburgaz Municipality has been acquiring and applying projects of buildings that became prominent with their public quality and the concept of Academy through competitions since 2013. Within this approach, idea and project competitions were organized for Lüleburgaz Intercity Bus Terminal in 2013, Lüleburgaz Yıldızları Women's Academy and Lüleburgaz Yıldızları Art Academy in 2015, Lüleburgaz Yıldızları Swimming Academy in 2016 and Tosbağa Stream Recreation Area in 2017. As a result of these competitions, the Bus Terminal (in 2016) and the Womens' Academy (in 2017) were built and opened for use; meanwhile the constructions for Art Academy and Swimming Academy are continuing. Moreover, different ideas obtained through the competition process for the Tosbağa Stream Recreation Area provided a stimulating ground for future studies. Looking at these competitions as a pluralistic production environment in the context of the presence of different approaches and ideas, they were selected and applied amongst a number of projects; 94, 48, 27 and 33 for the Bus Terminal, Womens' Academy, Art Academy and Swimming Academy, respectively. In addition, five promising project proposals were identified for the Tosbağa Stream Recreation Area idea competition within 59 different ideas and approaches [4].

In approximately a decade, the Municipality has shown that it gives importance to the idea of a sustainable development that aims at the development of both the society and the individual, as well as high-quality spaces in the relationship between user-space relationships; in addition, it adopted competitions as the space production medium to realize this thought. The reasons that Lüleburgaz Municipality chose competitions as a method for public space production emphasize the necessity for the user to feel happy and at peace in the place, and the significance of the relationship between space, users and the activity. At the same time, the idea that public production issues should be discussed and evaluated by experts gain prominence; in addition, the idea that participating projects and their evaluations during the competition process have created a learning medium for the Municipality is expressed [5]. In recent years, Lüleburgaz Municipality came up as an exemplary of local governance both with the discourse it has set regarding the city and citizen through the idea of Academy, and with the built environment comprising of high-quality spaces it has created through the competition method. The aforementioned two production approaches set forth can be described as a value that can be read through public space and social development.

4 Architecture in the Context of Social Benefits: The Case of Lüleburgaz Yıldızları Women's Academy

Lüleburgaz Municipality has set the year 2016 as "women's year" in order to create awareness with sensitivity to gender equality in its development of

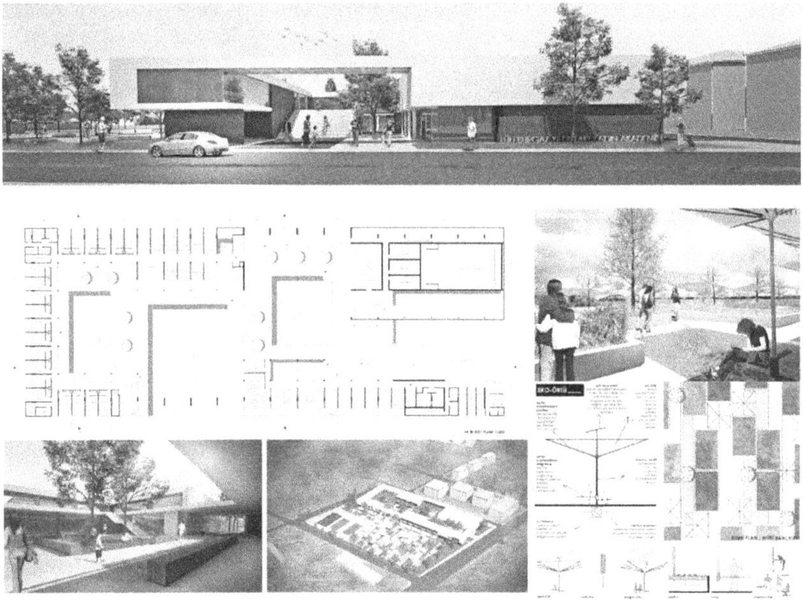

Fig. 1: The project proposal that was presented to the competition and won the first prize.

policies and year-long activities that socially and politically empower women; in addition, it decided to establish a multifaceted Academy for women [6]. This Academy will embody a sustainable development model by providing educational and social services, offering counseling and guidance, organizing professional and artistic courses, and helping with active participation in social life to the women, children and families, who live in the district but need support. With this model, the Municipality multilaterally reexamined the "shelter for abused women" concept that deals with the spatial issues created by an environment of violence against women; moreover, it wanted to obtain the spatial consequence of its Academy model via competing and opened a competition on the subject. The requirements in the Lüleburgaz Yıldızları Women's Academy (LYWA) competition consisted of enclosed spaces to shelter battered women, kindergarden and day care center, common social spaces, cafeteria and recreation areas, and open spaces containing parks, hobby gardens, sport fields and open car parks. The site has an area of 20,846 m^2 and the total construction area of the building would be 5796 m^2 (Fig. 1).

LYWA was designed by Orhan Ersan, Ferhat Hacıalibeyoğlu and Deniz Dokgöz. The project has won the first prize in the national architectural project competition

Fig. 2: Views from construction process of LYWA.

in 2015 and was built till 2017. LYWA is structured to encourage women's participation and production in social life, provide solutions and support to other familial problems. It is a center where children and women can participate in activities related to development and production by communicating with experts, and can produce solutions with the participation of the public through various congresses and seminars. The center that harbors shelter for battered women, kindergarden and day care center ensures its users to have a strong relationship with the building and use it all 24 hours of the day. Therefore, the center creates a life pattern different from what the main city has. This life pattern will establish a new kind of human-nature-life relationship with the effect of sports fields, ecological farming areas and production workshops; thus in a manner of speaking the Academy proposes a communal form of living (Fig 2).

LYWA was designed outside the city center of Lüleburgaz, in an empty building lot, within a newly developing residential zone. Approximately 50% of the building island can be utilized in accordance with the designated boundaries, while the remaining areas can be developed pursuant to open space suggestions. Istasyon Street, which passes to the east of the lot is the most important transportation artery that links the building with the city. While the Academy's main entrance is from this street, the entrance to the shelter for abused women and children-related units are from the street to the north.

The design incorporates three main courtyards and their relationships between platforms and the ground to enhance the spaces scaled according to life and lifestyle

Fig. 3: Perception of the building from the entrance façade and expression of general building tectonic.

requirements; these relate to a number of design sensitivities. The building is low-rise, has courtyards in which it offers activities and a chance to relate to nature, gives its users happiness, and creates surprising spaces with inner streets (Fig 3).

The open space design of the building was obtained by the reinterpretation of the traditional courtyard typologies, it makes itself evident based on the relations between the ground and upper ground floor of the courtyard-space organization. The elements that put fragmented courtyard-space organization together are the solution of function groups in a single mass plasticity. Both the internal relations of the program components and the relations of these components with open space are integrated with this element and the canopy-building constitution that combine this element and open spaces together. The spaces that the building gives

Fig. 4: Views from the kindergarden and entrance courtyards.

to Istasyon Street and hobby gardens form public openings for the city. The design provides openings to the direction of the Istasyon Street with a large amphitheater and to the direction of the park grounds with an event courtyard. These openings stand out as strong public openings created by the mass tectonics of the canopy. Thus, the public can easily access the entrance taken from the Istasyon Street and social spaces located closeby. The amphitheater climbs to the open life platform, which was formed at the upper level and can be described as a new ground floor, and strengthens the public character of the structure with its perception from the main entrance and invitation to the upper ground (Fig 4).

The planimetric design of the project is based on increasing the number of ground floor planes. The plane created at the +4.00 elevation describes another second floor with both its proportions and the distribution of the function groups. While the women's shelter, nursery and day care center provide a controlled access to the courtyards from the (+ -0.00) elevation that is the ground floor of the building, common social spaces, cafeteria and recreation area stand out as the main functions that support the courtyards. While the distribution of the function groups in the building allows access to the upper ground level (+4.00), the women's shelter in this level is distinctive with its own inner courtyard.

Even though the children's spaces and the courtyards of the women's shelters can be opened to the outside, the presence of a sheltered courtyard in particular to the children's spaces and the women's shelter is thought to have a positive impact on the users' psychology. The sequence of spaces that transport women and children users to open terraces, courtyards and then to open spaces forms the general spatial character of the building. Both the public and point access facilities that link the courtyards in the vertical make it possible for the upper and lower floors to work together. This situation nearly describes a second ground floor with its potential of connecting the operation of the programs under the canopy to the upper floor (Fig 5).

Fig. 5: Building tectonics and material organization.

The selection of material organization in the building was simplified to a level that would bring out the building tectonics and not obscure it; this simple mentality supports and reveals the simple mass composition of the building. Wooden surfaces and white washed surfaces are designed to show this simple tectonic form and reinforce its effects (Fig 6).

The spatial arrangement of the building is provided through the continuity of closed, semi-closed and open spaces. In addition to the semi-open spaces such as the courtyards and platforms that were defined as a part of the building, open areas including sport fields, children playgrounds and hobby gardens also exist in the totalitarian setup of the building. While the changes in altitude for sport areas animate the use of outdoors, energy-efficient equipment proposed in open spaces create ecological solutions with functionalities such as opening and closing in order to collect rain water according to climatic changes.

5 Conclusion

Public spaces, which shape living environments, can be built with high quality. While the creation of public spaces with high quality is the responsibility of professional experts, the governing bodies and decision mechanisms in a city have a

Fig. 6: Open space uses, views form hobby gardens, playgrounds and sport fields.

large-scale impact and also responsibility. Yet the subject of a space is human, the individual. The socio-cultural mold of an individual, the individuals' communication with each other and the dynamics of relationships in the urban scale form the social structure. In this context, cities can be discussed and evaluated on their social and spatial productions and the values they exhibit. Lüleburgaz is the topic of this work with the values it exhibits in the spatial and social arena in the context of local governance. The quality of public space was scrutinized in the context of the place, worth and gains of women in the social area in particular to Women's Academy (LYWA).

LYWA can produce value for the city and citizens mainly through elements for women, such as gaining visibility in the social and urban life, participating in all areas of daily life, discovering life-long learning opportunities, encouraging participation in cultural and artistic events. Besides, LYWA's contribution to social life can be seen through its properties like contributing to culture of entrepreneurship, unearthing abilities through society's productivity targets and encouraging production opportunities, discovering values and establishing confidence in individuals.

The main concept that brings these values into existence is space. The quality of space also gives quality to the activity taking place inside the space. To establish architectural culture in the social field and to produce value in the spatial context is possible through the quality of physical spaces that generate the built environment and the utilization opportunities it serves to society. Thus, while space creates the mode of relationship between architecture and social life, it also provides a communication medium between the stakeholders.

References & Notes

1 Lüleburgaz Municipality Women's Academy was designed by Orhan Ersan, Ferhat Hacıalibeyoğlu and Deniz Dokgöz. The project has won the first prize in the national architectural project competition in 2015 and was built. The applied design also won the *Raci Bademli Good Practices Award* in 2018 and a merit award as part of *Turgut Cansever National Architecture Awards Program* in 2018.

2 URL: https://www.luleburgaz.bel.tr/yonetmelik/LYWA.pdf

3 Alpaslan, H.İ., Hacıalibeyoğlu, F., *Kamunun proje elde etme yöntemlerinin yapı stoğunun niteliğine etkisi yarışmalar örneği*, in *TMMOB İzmir Kent Sempozyumu*. 2009: p. 509–514. İzmir, Turkey.

4 Yarışmayla yap, 2015 yarışma raporu kitapçığı

5 URL: http://www.arkitera.com/soylesi/855/yapilari-bir-mimarla-anlasip-yaptirabilirdik-ancak-tarifimiz-dogru-olur-muydu_?

6 Lüleburgaz Belediyesi Kadın Akademisi mimari proje yarışması şartnamesi.

Assist. Prof. Dr. Fulya Üstün Demirkaya

18 Reconstructing the Cultural Identity: Transformation of Architecture-Transformation of City during the Tanzimat Era

Abstract: Eighteenth and 19th centuries are the period in which rapid transformations in a wide range of areas, namely, economic, social, military and architecture emerged mainly in the European states and a number of regions as well as the closest neighbor of Europe, the Ottoman State, in a relatively short span of time. With Tanzimat era that is entered with the edict that was heralded on 3 November 1839, novel and unfamiliar concepts such as Western, which was contemporary, new, modern and reform that never existed in traditional Ottoman life before gradually resonated in the Ottoman social and daily life and particularly in the central administration. In this context, cities as the centers of architectural production and application sites became the most salient tool to directly unite "Westernization" message in its most striking way with the society and to make it more visible in order to achieve the mission. Formed through architecture and coordinated retreatment of public domain, cities met an unprecedented process of transformation in which it was viable to instantly monitor the transitions taking place in all parts of the state. The urban reforms of Tanzimat era cities maintained their transformation within the context of modernity and modernization on economic, military and judicial levels. Structures such as government palaces, post offices, business centers, piers, apartments, schools and towers that were new concepts in the literature of Ottoman architecture integrated to the structural inventory of the specific period. In this study, Tanzimat era has been analyzed as a period in which a decisive split was observed in the structure of traditional Ottoman society due to the rise of novel concepts, habits, demands and emerging institutions. Instead of discussing the political meanings of this period, our study expands on the reflections of political and social developments/transformations leading to Tanzimat Firman and its consequential results in the domain of architecture and urban texture.

Keywords: 19th Century, Tanzimat era, Westernization, Transformation of city, Architecture

Introduction

Tanzimat Firman (Imperial Edict of Reorganization—Transformation of the State) is one of the most critical documents of modernization history of the Ottoman Empire. Heralded by Reşid Pasha on the 3rd of November 1839 in an official ceremony in Gülhane, one of the last statements of the firman "*ve keyfiyyât-ı meşrûha, usûl-i atîkayı bütün bütün tağyir u tecdîd dimek olacağından*(Sağlam, 2007) proclaimed with the word transformation (*tağyir*: to transform) and modernization (*tecdid etme:* to modernize) that a list of old procedures in state administration and in social life would then be forsaken after adopting new approach (Akyürek, 2011).

İnalcık (2018) defined Tanzimat such; "the conclusive phase of the reorganization attempt in Ottoman Empire of which main institutions were damaged in the face of the dominating power of Europe that rose up as a brand new civilization all-ready to attack". Within the time span that started with the passing of Sultan II. Mahmud; ensued by the coronation of Sultan II. Abdülhamid in 1876 and continued till the proclamation of I. Meşrutiyet (First Constitutional Era) a wide range of reforms in the domains of law, tax system, education and various other social practices as well the improved relations with European states rendered numerous impacts on the Ottoman-social life including architecture practices (Akyürek, 2011). Hence during this period that is defined as Westernization in late-Turkish history, novel and unfamiliar concepts such as Western, contemporary, new, modern and reform that never existed in traditional Ottoman life before gradually resonated in the Ottoman social and daily life and particularly in the central administration. An absolute requirement "catch the modern age" these reforms mainly aimed to help the Ottomans keep up with the advanced Western world. Cities, as the centers of architectural production and application sites, became the most salient tool to directly unite "Westernization" message in its most striking way with the society and to make it more visible in order to achieve the mission. Formed through architecture and coordinated retreatment of public domain, cities met an unprecedented process of transformation in which it was viable to instantly monitor the transitions taking place in all parts of the state.

Tanzimat practices initially took place in port districts such as İzmir, Mersin and Trabzon which constituted the key point of improved relations with foreign world and also in the city of İstanbul that featured as the capital city rather than a coastal city.[1] Also in a number of settlements connected to these ports via railways those practices were eminent (Acar, 2000).

In this study, Tanzimat era has been analyzed as a period in which a decisive split was observed in the structure of traditional Ottoman society due to the rise of novel concepts, habits, demands and emerging institutions. Instead of discussing the political meanings of this period our study expands on the reflections of political and social developments/transformations leading to Tanzimat Firman and its consequential results in the domain of architecture and urban texture.

1 Regarding the change of the port cities from the 16th to the 19th century; the study of Eldem, H., Goffman D. And Masters, B., *Doğu ile Batı arasında Osmanlı Kenti Halep, İzmir ve İstanbul*, çev. Sermet Yalçın, 2012. İstanbul, İş Bankası Publications" focused on Halep, İzmir and İstanbul which are three impotant port cities of the period, is a remarkable work.

Triggering Factors Tanzimat Firman

It is inevitable that, as an effect of newly staged economic relations and administration styles emerging from the largest to the smallest unit in which human beings lead their lives, there would be a rising demand for a new substructure system and new institutions. Eighteenth and 19th centuries are the period in which rapid transformations in a wide range of areas, namely, economic, social, military and architecture emerged mainly in the European states and a number of regions as well as the closest neighbor of Europe, the Ottoman State, in a relatively short span of time. The triggering factors for the Ottoman State to experiment such a major transformation are as listed below:

Military: During the period between the start of 17th century and the 19th century the "Western" world increasingly outpaced Ottomans in military, political, economic, cultural, social and spatial dimensions. After the 1699 dated Treaty of Karlofça (Karlowitz) the balance of powers changed against the Ottoman State in the face of Western world. Due to such changes it was mandated to conduct reforms particularly in military domain.

Administrative and political: Until the 19th century Ottoman Empire held an advantage over foreign states. Yet in parallel with the advancements in military domain the state lost its political dominance and freedom of will against the foreign states during this era. As a consequence of the weakness in the face of other states, foreign citizens possessed an increased power within the state; thereby, it was inevitable to meet a political transformation process that also embraced democracy concept.

Economic: Industrialization in trade and industry that the Ottoman State adopted due to close economic ties with European states then evolved both the quality and quantity of trade and industrial commerce by means of mass production. At the end of 19th century commercial agreements signed with the industrialized Western countries turned Anatolia into an open market; hence, artisanship and small industry collapsed (Aktüre, 1981). Besides, as a result of the increased population of foreigners in cities due to improved commercial ties with Europe and the outbreak of Crimean War gradually populated the cities when on-duty soldiers and materials were dispatched to Black Sea coasts, primarily to İstanbul. It then became mandatory to form new organizations, since there was not a favorable transportation-transmission system to solve this commercial density and the hospitals were inadequate to treat contagious diseases.

Education: As is the case in all institutions of traditional Ottoman society, Islamic culture played a vital role also in education which is the most significant tool in attaining knowledge. To ensure that targeted transformations were realized in all institutions and to secure the coordination and continuity, it was essential to train knowledgeable personnel to adapt the new system. To educate these personnel, it was required to replace the old educational system with modern educational institutions.

Other factors: Frequent fires in cities resulted in potent damages. Defined by Moltke as (1995) "thousands of houses like small matches are tightly and unorderly lined and it is easy to fathom the devastating consequences of a fire that outbreaks in even one square meter." Those fires, as stated above, agonized administrators of the age since they also spread to European lands" (Aydın, 2006). Indeed the note shared by Mustafa Reşid Paşa (Baysun, 1960) while he was in a European Embassy "there are frequent fires in Istanbul and other places. European newspapers criticizes us and ask 'why do you build wooden structures by wasting so much money; they all end up in a fire; are there no stones and bricks in your country?" heralded the growing frustration and the changes expected to take place soon after in these cities.

In short the military defeats, population increase, the need to take precautions against damaging fires in cities, introducing new transportation vehicles and necessity to transform roads to suit the vehicles were the underlying factors in starting construction activities (Acar, 2000).

First Steps to Urban Construction

Until Tanzimat era, Ottoman cities were not governed by their residents. Instead civil servants appointed by the Sultan ruled the cities in which administrative and civic functions were separated. Kadı (Muslim judge) as a scion of central bureaucracy served as the authority of jurisdiction, law enforcement, financial tasks and city administration in the district where he was assigned to work for a definite period (Ortaylı, 2011). In a different saying, considering that particularly in monumental–public construction works there was a strong influence of wakif–imaret connection (Özcan, 2006). Construction process that shaped urban spaces was realized under the authority of Muslim judges either through the state or a foundation or an institution (Cerasi, 1999). In that sense, the greatest novelty being introduced by Tanzimat to Ottoman cities where religious law and traditional sense of administration dominated was solving problems by law and order.

Thus in parallel with the change in the economic and administrative structure of the country a number of legal regulations were introduced in urban governance and municipality services, construction materials, terms and conditions for the density and height of buildings, ownership regulations such as land division or rights of nationalization, using new transportation vehicles such as trams and road improvement works to fight against frequent fire outbreaks so as to design urban structuring process were introduced. (Çadırcı, 2011; Ortaylı, 2011) A few of these works can be elaborated as such; partial or local-scaled plot plans named as fireplace plans for the sites exposed to fires. Road widths of a number of cities, İstanbul in particular, were reorganized and a new design of road network was projected based on grid plans and right-angled road schemes (Özcan, 2006).[2] Additionally the

2 The streets were arranged in accordance with regulation no. 1848, in the form of big streets with minimum 7.6 m width, streets with 6 m width and streets with 4.5

new road system that was organized in line with pedestrian scale also changed as an effect of the need for wider streets especially in cities due to the rise of new model cars copied from Europe. The new transportation network and vehicle system let each place be accessible and new suburban neighborhoods' expansion outside of the city center (Acar, 2000).

Besides, urban transportation is no longer pedestrian or boat transportation in 16th century. For this reason, there has been a change in urban texture with dead end and narrow streets and high density. Steamboats and tramways got started to be operated by foreigner companies as well; the inclusion of horse cart in transportation system is probably the most important progress that affects all urban texture. These developments necessitated for streets to be wider. This urban expansion was no longer to be through the division of existing parcels. Private property within agricultural lands around the city was legitimatized by laws. The outer walls of castles in some cities were pulled down and an expansion to plains happened (Tekeli, 2011).

Another reason that enforced transformation of road system was the problem of settling immigrants who moved to the existing cities and district surroundings as an outcome of internal and external migrations starting at the end of 18th century and continuing until the 20th century. The formation of geometric-shaped immigrant neighborhoods beside traditional Ottoman neighborhoods with an organic composition was ordered by specific rules. It was enforced to design uniform, single-lined streets of equal widths and rectangular to each other (Aktüre, 1981).

In the face of commercial density outbreak after the Tanzimat it was again essential to revise transportation-road system. Even more importantly, hospitals failed to effectively work against the outbreak of contagious diseases that surfaced due to high intensity of contact. Transformation of trade in this direction and resultant problems led to the emergence of appropriate and supportive building types that could adapt into the new system of cities. Also the new commercial structures that entered during this age to Ottoman city life (shops, stores, office blocks, etc.) forged the new trade center by lining opposite directions on the main transportation axle of the extension of old city (Aktüre, 1981). In addition to the solution proposals for the raised concerns, intellectual and administrative changes witnessed in Ottoman State transformed the perspective of society towards the city. City squares, parks and gardens were also integrated to the urban space. Urban green spaces were no longer agricultural production areas that contributed to citizens' nutrition. While some of these spaces were turned to new housing areas, municipal parks used by all citizens also begun to appear. Municipality parks that were created by transformation of graveyards remained within the borders of the city through the expansion of the city (Tekeli, 2011). Besides, as the symbol of cities and Westernization in general, clock towers were built in different quarters of Anatolia (Çelik, 2015).

---

m width. In regulation no. 1882, the width of big streets was enlarged to 11.5 m, the streets with 4.5 m of minimum width were enlarged to 7.6 m. (Çelik, 2015).

In addition to the above listed transformations aimed in urban scale, this large-scale and comprehensive modernization also introduced the need for a new formation in a corporate structure related to architecture and demanded novel approaches in the education of architects (Acar, 2000). Imperial Ottoman Guild of Architects, based on the relation of master-apprentice and functioned as a practical performance, failed to meet the construction work of new building types emerging after the Tanzimat. Hence all construction under the authority of public works and construction body were united in one establishment named as Ebniye-yi Hassa Müdüriyeti (Guild of Building Direction). These works were executed by public officers assigned to cities by the central government and foundations or similar institutions; on the other hand in small districts there was not a structural transformation that essentially necessitated a rigid control of construction works (Ortaylı, 2011) Nevertheless the lack of qualified personnel demanded in the construction of much needed buildings with new performance functions was envisaged to be compensated by architects who received academic training in Europe or non-Muslim architects familiar with Western architectural style and European architects.[3]

Architectural Representation during Tanzimat Era: New Building Types

Having kicked off in the military domain and gradually spread to all domains of social life the attempt of Westernization introduced with itself the need for new concepts and new building types that would adopt new functions in response. Although a vast majority of buildings performing traditional functions maintained their continuity, they still lost their former dominance in construction activities; some have lost their prominence and even faced the threat of extinction. In addition to the works in city construction, the new building types that emerged out of necessity and entered to the repertory of Ottoman architecture as well as a list of unique structures saved their rank in Ottoman architecture. As a result of new structure type based on the changes in the order of state and society, the new structures that maintained function, form and plan designs that were not needed by the society before would then rise. These structures that should be analyzed as a reflection of the needs and political initiatives of the specific age on the physical domain are military structures in particular followed by official buildings, health establishments, industrial facilities, entertainment places and trade and accommodation buildings.

3 The need in this field was stated thus: "To make sure that we could make use of Western educated architects in the future too, it would be reasonable to send about ten or fifteen talented Turkish youngsters to Europe for now to learn the art and science of architecture" (Akpolat, 2002).

Fig. 1: Trabzon Arsenal building and the patrol on the western side of the building (Bölükbaşı, 2006).

Military Structures

As an effect of the first steps of modernization on the road to Westernization in the military domain, there was a pressure for new spaces to unite the changes with institutional structure. Thus, arsenals, patrols, arm factories and barracks were designed.

Barracks: The occurrence of a change in the administrative organization of army also led a change in the spatial structure of the city, and new barracks were built as a continuity of main transportation axes (Aktüre, 1985). Formerly constructed in wooden material barracks was inspired from the 19th-century European style barracks architecture. There was a wide patio in the center and it was designed in rectangular and square plan reflecting the approach of structuring with masonry material with the gates on two or three story symmetrical directions.

Patrols: Patrols were installed in the major districts of cities in order to transmit urban security service to all districts. In Ebniye Nizamname (Building Regulation) it was decreed to reserve a land to build patrols in newly established districts (Fig. 1).

Prison: In the Ottoman State detection and punishment of a crime was based on the principle of Islamic law and law was then executed in religious courts. Upon the enactment of the first law on penal procedure after Tanzimat in 1879, the institute for detaining violators of law was also subordinated to a legal system. Until the time of introduction of modern prisons, fortified sites such as castles were utilized to detain violators of law (Kalebendlik). After the mid-19th century new buildings named as prisons were built to serve detainment purposes or existing

Fig. 2: The road from Sinop Government Palace to the port; municipality, clock tower (not completed yet), Gendarmerie Command built in 1920 (Özcan Canım Archive).

buildings were transformed to perform this function (Yıldız, 2012). A partial division of İbrahim Pasha Palace was transformed into a Public Prison. In 1870 the first prison was structured in Sultanahmet. Next new prisons were erected in almost all quarters of the city. To choose the best construction site for prisons the peak central point of the city was selected and specific attention was paid to select a site nearby the Government Palace and post office-telegraph office.

Official Structures

Government palace: After the Tanzimat, transformation spread to the administrative structure and rising demand for buildings also known as state offices corresponded to the structuring of Government Palace in the domain of architecture. Government Palace symbolized central administration concept and attempt to manifest state's administrative power in all domains of the country. As the first structures erected after the Tanzimat, it was aimed to select the site for Government Palace outside the old city structure; a site that would designate the advancement of city and even more importantly which would point to new second center.[4] Next this structure was jointed by buildings such as courthouse, telegraph office and patrol. It means vital public structures were organized in this new center (Yazıcı, 2008) (Fig. 2).

Municipality Building: Aside from the new administrative style that also accounted for the emergence of Government Palaces it was also required to

4 As a different practice, a few models chose the old city center like Ortahisar district in which the new Trabzon Government Palace was erected.

Fig. 3: The texture formed by the period structures at Çorum (old Çansaat Square) (Ömer İskender Tuluk Arşivi).

organize the heavy crowd and mobility being emerged as a result of Crimean War. Hence in capital city İstanbul where the density was most heavy Şehremâneti (Municipality) was established to pioneer the first step of Ottoman municipality approach. The first municipality that could be termed as modern was structured on Paris model; it had independent income sources, entertained its own municipality building and municipality court in İstanbul city. Offering service to the residents of Galata and Beyoğlu, its name was Altıncı Dâire-i Belediyye (Sixth Office Municipality) (Çelik, 2015). It was witnessed that in due course municipality buildings were structured in a number of Anatolian cities outside İstanbul. In parallel with the administration system of Tanzimat era, these structures were mostly located within a holistic texture that encompassed government palace, prison and public structures such as telegraph/post office. Specifically in commercial cities, the locations with frequent commercial activity were selected as construction sites in order to check artisans and manufacturing sites and perform the tasks of tax collection (Kolay, 2018).

Post office and Telegraph office: The need to establish a new communication system as a result of expansion to the outside world corresponded to the need to plan post office and telegraph office buildings in the 19th century-port cities in particular (Fig. 3). Mostly jointed with the government palaces telegraph and post office buildings formed a central administrative core together with the public structures already present in the texture they joined.

Education Structures

Schools: It is undoubted that any modern-age state organization is based on educational reforms. Training the bureaucratic staff that is highly demanded in response to the overall transformation and the necessity to provide professionalization-institutionalization, it was essential to also conduct reforms in the domain of education. These reforms not only took place in the education system itself, the new system also mandated construction of new buildings. Medresahs that represented traditional Islamic education system were replaced with new school buildings in Western style. Although there was initially a cautious approach to such a transformation, old-style Koran schools continued to exist, but civilian schools such as iptidai (primary), rüştiye (secondary) and idadi (high school), sultani (advanced high school) darülfünun (university) also saved their spots in the new educational system. In addition, through advocating an Ottoman nation with equal rights for all minority schools that were founded by non-Muslim communities, it was a requirement in this system to grant equal rights of education. 1869-dated Directory on Education stipulated the places where these schools would be opened and the types of these schools. In this directory it was specified to found in any neighborhood and village (in case of need in more than one or two neighborhoods or villages) minimum one primary school, in villages and neighborhoods where the population is mixed with Muslims and non-Muslims separate schools for the communities, in communities larger than five hundred residents if all residents are Muslim rüşdiye (secondary) schools to teach Islam and if the entire population is Christian an all-Christian school would be opened for children. In places with mixed religious groups separate secondary schools would serve for Muslim and Christian children and in townships larger than one thousand houses and wherever deemed essential, there were high schools in which secondary school graduate Muslim and non-Muslim kids would study together. It was also envisaged to found Mekteb-i Sultanî (advanced high schools) in the city center or central district of each city where all high school graduates could collectively attend (Çadırcı, 2011; Koyuncu, 2013) (Fig. 4)

Commercial Structures

Shops-Stores: As an effect of the improved commercial relations with Europe goods were freely circulated in every corner of Ottoman cities which then turned into an open market, so it was required to open a number of small shops and stores which lined in a street to sell these products. Aktüre (1981) reports that during that era there was one shop to serve every —three or four houses. In conjunction with the population and economic condition of the age it is striking to see that there were large quantities of shops and stores in the cities.[5] Almost all of the structures that emerged to meet the new functional needs after Tanzimat were primarily organized outside the old center of trade in such a favorable way to form a new center.

Office block: "Westernization" that started with Tanzimat reflected its effect on architecture in administrative structures and most specifically in commercial structures. This texture formed by covered bazaars and inns that formed the trade center of a Traditional Ottoman city was now replaced with the new business center. These business centers contained the buildings exhibiting new functions

5 Although during that time there was not a measurable population increase in regions, there was still a climb in the number of shops; refer to Aktüre, S. for a detailed discussion on the reasons.

Fig. 4: Mektebi-Sultani (advanced high schools), Sultan Abdulaziz Period, today's Galatasaray High School (Yıldız Album).

that were deemed essential in the new business relations. Within that context firstly office blocks were built to warrant that foreign trade dealers' direct connection with the product could be secured (Acar, 2000)

Passages: Like the office blocks they symbolized the new building types having emerged in response to the changing business relations in that age. Particularly inspired from Paris, these passages served the consumption branch of trade activities.

Banks: Currency reforms to stabilize the national economy and meeting of the empire with the foreign capital brought up the concept of bank in urban life. Currency transactions that were previously performed by foundations, covered bazaars and guilds were now taken over by banks founded next to the office block. Within the context of Westernization program the very first bank was opened in 1863 with the name "Bank-ı Şahane-i Osmani" (Ottoman Bank) to the end of designing the modern substructure essential to view the gradual transformation of the Empire (Eldem, 1999) (Fig. 5). In 1888 the first state bank Ziraat Bankası (Bank of Agriculture) was established to regulate agricultural crediting under the state control (Yetiz, 2016).

Transportation Structures

Invention of steam machines and using steam technology in ships and railway vehicles enabled to fuel revolutionary advancements. New structures that would

Fig. 5: Bank-ı Osmani-i Şahane, Facade of Haliç (Ottoman Bank Archive).

serve to the new transportation vehicles were also built. Port structures were among these types of structures. Especially in port cities where international trade was frequent, in addition to port and custom buildings enabling the transactions of large-scale ships, it was also needed to build passport office, quarantine, passenger hall, patrol and similar other structures.

Railways that connected cities are also major initiatives of transportation system that created a big transformation in physical texture. By constructing the railways functioning as the key tool in access, development and centralization, it was projected to empower military control of the state, prevent any revolts and vandalism by dispatching troops on the spot and by transmitting agricultural goods to the market it was also aimed to strengthen production capacity of the locals (Şen, 2003). The biggest contribution of all the stations erected in large cities to the urban texture was, in parallel with the main objectives, it could become viable by the hotels and other commercial buildings that operated in the close neighborhood (Ertuğrul, 2009).

Accommodation Structures

Hotels: A range of innovations were observed in the domains of communication and transportation as an outcome of the potential introduced by trade life of the 19th century. Hence there was a pressing need for space in Ottoman cities of which new commercial and cultural relations with the region and the world to accommodate foreign visitors. Previously accommodation to European visitors was offered in the caravansaries on the intercity highways and city center inns, but now they were replaced by a new structure type added to the Ottoman architecture culture in Tanzimat era; they were positioned in the city in connection with the new commercial center that was developing right next to the traditional trade center.

Civil Structures

Embassies: Civil structures were the domain in which the traces of changing life culture of the 19th century were most visible. In this domain the first transformation was seen in embassy buildings of Ottoman lands that adopted a Western standard after the new urban practices of Tanzimat. Embassy buildings reflected the power and prestige of the builder state. In particular, European architects who visited İstanbul to start the construction work of European state embassies met with Sultan and other high echelon bureaucrats. In those meetings where they were free to discuss their projects, the architects also took over other critical tasks from the Ottoman state (Akyürek, 2009).

Residences: Built under the influence of Europe, those residences differed from the traditional Ottoman residence typology with respect to size, plan, position, etc., as seen in the messages on the identity of their dweller. Transformation could be felt starting from the top level of state. Aside from the main palace structures of the classical Ottoman style, they were again mostly built by European architects using masonry material in an identical form with the European models that reflected the power of king and state. They were mostly small palaces that exemplified the architectural style of the era (Beylerbeyi Palace, Çırağan Palace, Ihlamur Mansion etc.).

In the newly developed quarters of the city, row-houses and apartments that were especially popular among non-Muslims, foreigners and administrative class Ottomans started to be constructed, although they did not exist in the Ottoman architecture practices before. In Ebniye Nizamname (Building Regulation) too, in the newly developed quarters of the city, apartments were built by using the means of masonry construction technique that was meant to be a tool in fire prevention. These row-house lines which spread nearby the business centers that emerged in the industrial cities of the 19th century were formed of units that seemed repetitively ordered (equidimensional, same-planned and faceted; built in the same time period) (Sağdıç, 1999). Taking part as a new architecture typology in the urban texture row-houses also played a remarkably vital role in the history of planning since they were the very first planning experiences in which road arrangements and structuring pattern were designed in conjunction with each other (Batur et al., 1979).

Health Structures

Military Hospitals: Performing urban services and in addition to solving transportation, water and residential problems, offering healthcare solutions were the top wishes of the administrators of this age. Recognized as a social charity institution in Ottoman society Ottoman health establishments that were also named as Bimarhane, Dâr-üş Şifâ, Timarhâne functioned as charity organizations donated to the society (Yavuz, 1988). There was an attempt to build health organizations in parallel with the modern health approach in the West and erect new hospitals. These reforms first took place in the military system that was one of the pioneer practice domains of Tanzimat and military hospitals were then constructed.

Quarantine station: Due to outbreak of epidemics that surged in Crimean War and escalated global trade as of the 19th century, the Ottoman state started to take certain measures in the domain of health. Particularly in port cities quarantine[6] stations were constructed to inspect the health condition of incoming visitors and to supervise for a certain length of time to the end of detecting sick ones. Quarantine stations were then erected on the shores nearby the large ports.

Quarantine stations have been one of the most significant health establishments built on the lands of this state, since the date Ottoman Health Council was founded to treat all kinds of epidemics, plague and cholera in particular. While some of them were temporary facilities that were used only when the diseases turned into epidemics, some others provided service for long periods of time thus they somewhat functioned as private hospitals to treat epidemics[7](Ayar and Kılıç, 2017).

Industrial structures

Defined as small enterprises employing limited numbers of staff manufacturing the goods in ateliers, the traditional model of industry was heavily reliant on human labor. However due to the radical change and transformation in the entire globe after the Industrial Revolution in the first quarter of 19th century this model failed to be competitive in the new international order. In this so-called industrial system of pre-Tanzimat era, companies in private sector predominantly served in the field of small handicrafts while industry sector in state control mainly produced goods for militaristic and defensive purposes. With the onset of 19th century a strong shift was observed from small-scale industry that relied on small handicrafts to factory-made mass production. Within that context which can be defined as industrialization policies of Ottoman state during Tanzimat era factory complexes that can also be described as organized industrial sites produced a wide

6 Disinfecting the goods and spaces, inspecting the dead bodies and burials were bound to specific rules in quarantine practices which were not much welcomed among Muslim community. It was practiced by the Ottomans for the first time in 1831 during the cholera epidemic spread in İstanbul and ships that came from the Black Sea were quarantined (Yıldırım, 2006).

7 The earliest examples that offered service in Ottoman State during 19th century were quarantine stations located in Kamaran, Beyrut, Basra, Trablusgarp, Klazomen, Sinop and Kavak.

range of goods from minerals to textiles. In addition to Hereke Factory a number of factories to meet the needs of army were founded by the state. It is possible to see that aside from state-led industrialization attempts, private sector also started to emerge during Tanzimat era and private sector also attempted to found a number of industrial factories (Seyitdanlıoğlu, 2009). Production activities often performed around city center in 16th century tended to be performed outside city in the form of new organization. Although factories,which are a new element of urban surroundings, are few in number, they are located in easily accessible places around the city (Ejder, 1975; Tekeli, 2011).

Miscellaneous Structures

Museum: During Tanzimat era, in parallel with the changes witnessed in all domains of society, there was a corresponding change and development in the perspective to ancient works. Mostly as an effect of the critiques about the works of art smuggled to foreign countries and lack of a museum in the capital city to exhibit the pieces of art brought up the issue of classic works and museology concept started to gain attention. Next museology started to be recognized as an institution (Türkseven, 2010).

Theater: Theaters lacked any continuity with the traditional culture but introduced a brand new function as a complete innovation that was introduced to the urban texture by Westernization process. Although conceptually a theater resembles the traditional improvised play shown during the Ottoman age in indoor places such as coffee houses or public squares, there was not a special place catered for only traditional improvized plays (orta oyunları). So, after becoming more popular in Tanzimat the theater then became a new type of entertainment for the Ottomans.

Fire and Clock Towers: With respect to typology, fire towers were not a novel concept but after Tanzimat different construction techniques started to be employed. Beyazıt Fire Tower is the most popular sample model. Its original material is reportedly wooden and recently the tower has been renovated in masonry structure. It was aimed to allow fast intervention on any fire outbreaks in İstanbul (Ertuğrul, 2009).

Next to the government palaces that mirrored transformation in bureaucratic system, another building type that gathered attention was clock towers. Indicating the change in time perception after the improved commercial relations with Europe, these towers offered an alternative to the traditional time concept that was based on Azan (prayer) times in a day. In addition to functioning as the focus of city center and showing the time these structures were envisaged and structured to be the urban symbols of political authority (Ertuğrul, 2009). The words of Fani Efendi from period poets saying that "it is such a work that it is unique; seemingly clock strikes, in fact government calls on"[1] (Anonymous, 1981) are very explanatory of the fact that clock towers represent the power and authority of the state.

[1] The original text is as follows: *"Bir muazzam eserdir ki, misli yok, naziri yok; Zahiren saat çalar, manen hükümet seslenir."*

Conclusion

In the Ottoman age urbanism and urbanization works such as urban planning, development, improvement, populating, establishing and maintaining peace and security were formed within the framework of adapting traditional system and also keeping continuity with the owned properties of the Ottomans while blending with the new offers in the cities Ottomans conquered.

Hence, particularly as of 15–16th centuries direct reflection of transformations experienced in socio-economic aspects on the formation of city life could be actualized in a remarkably long span of time. As a symbol Tanzimat era is a worthy period in which this slow transformation gained speed and a radical term of change and transformation was started in the Ottoman Empire. In the 19th century the transformation that was emergent in all state institutions, military in the first place, also took place in architecture domain and throughout this century all architecture works were executed in the control of different establishments and associations. All of the new regulations engulfed in the term Tanzimat were also reflected in the physical structures of the Ottoman cities aside from institutional transformations. All these arrangements and changes covered by the term of Tanzimat were also reflected on urban texture, with the involvement of structure types and typologies in the physical texture, which weren't included in the traditional Ottaman urban structure.

Between 1839 and 1882, known as the urban reforms of Tanzimat era, cities maintained their transformation within the context of modernity and modernization on economic, military and judicial levels. Structures such as government palaces, post offices, business centers, piers, apartments, schools and towers that were new concepts in the literature of Ottoman architecture integrated to the structural inventory of the specific period. These emergent structure types in Ottoman urban texture constituted early trials of Western origin city-architecture organization by virtue of the new attraction centers they forged in urban space.

From that viewpoint in the Tanzimat era cultural and social life transformation could be seen in these structures that implemented newly adopted functions to mirror the architectural style of the era. While the power of state was exemplified in the military and official structures such as barracks, patrols and government palaces, similarly a number of structures such as post office, municipality and schools in which functions were more valued than aesthetics were also constructed. These were new types of buildings not popular in the Ottoman architecture program before, and it is viable to claim that as an effect of the transformation process realized in city administration and organizational structure Ottoman architecture began to forgo its religious identity.

The communication, which had been established with world through steam ships and railways until that time, wasn't in military class power any more, and begun to be established with open public delivery system. What intended by new relation channels were station buildings, new docks and post office buildings. This change which caused change in the relations with environment caused also appearance of structures with different functions. New transportation system and increasing international commercial relationships necessitated the establishment of warehouses, and to build hotels for accommodation of travelers. The change in military class and central government was that state affairs was begun to be performed not from palace/mansion, but from public buildings such as government office, municipal to be newly constructed. Starting from the west, all changes took place in military and political areas, and the reflection of all these chances on physical texture also caused changes in social admittances, and theaters, museums and luxury consumption shops, which had never been encountered in Ottoman culture before they appeared in urban centers.

Both the position of the construction area of these new buildings involved in the period literature (especially public buildings) in urban texture, and their relationships with existing buildings, are noteworthy for period conditions. The message of the state saying catch the era, "of which fundamental institutions disrupted and which was rising with a new civilization" was given at buildings built with stones in the regions of the city which weren't zoned for construction yet or very little zoned for construction, and at the new urban center constructed in the place of the former urban center. In cities not topographically suitable for such an expansion, this message was fulfilled in a management island created with integration of public buildings to former urban center in current texture. When seaports are a matter of discussion, the tendency to form a management site representing the local foot of central management separates in other cities, with positioning of municipal buildings in regions where business is intensively done.The buildings like clock towers, pointing out the necessity of a change in traditional society norms, were also begun to be involved in this new texture organized within borders of urban areas.

It is seen that all mentioned planning activities of buildings, constructed as an adaptation of Western settlement culture, and that bring solutions in themselves, such as prevention of fires, arrangement of urban road network so as to allow the passage of new means of transport and the integration of migrants into urban spaces were directed by Turkish engineers and administrators having observed planning activities in the West. Those who were engaged in architecture services were predominantly non-Muslims, foreigners and Levantine architects of foreign origin.

Jail, municipality building, post office, telegraph office, museum, hotel, restaurant, theater, office block, passage and bank entered to the Ottoman world not only as building types but also with their functions. Hence, by exhibiting a new approach different from the habits of traditional society and on top of that by adopting a transformative function of the society, these structures maintained their functional validity up until present day through constant development.

References

1 Acar, G., *Tanzimat Dönemi Fikir ve Düşünce Hayatının Mimari Alana Yansıması*, Doctoral dissertation, Mimar Sinan Fine Arts University, 2000. İstanbul.

2 Akpolat, S. M., *Tanzimat Sonrası Osmanlı Mimarlığı*, Türkler, 2002. 15, pp. 350–359

3 Aktüre, S., *19. Yüzyıl Sonunda Anadolu Kenti Mekansal Yapı Çözümlemesi*, 1981. Ankara: METU Faculty of Architecture Printing Studio.

4 Aktüre, S., *Osmanlı Devleti'nde Taşra Kentlerindeki Değişimler*, Tanzimat'tan Cumhuriyet'e Türkiye Ansiklopedisi, 1985. vol. 4, pp. 896–899.

5 Akyürek, G., *Bilgiyi Yeniden İnşa Etmek: Tanzimat Döneminde Mimarlık*, Bilgi ve İktidar, 2011. İstanbul: Tarih Vakfı Yurt Publications.

6 Akyürek, G., *Tanzimat Döneminde Mimarlığın Değişen Bilgisi: Fenn-i Mimari*, Gazeteler ve Diğerleri, Türkiye Araştırmaları Literatür Dergisi, 2009. 7(13), pp. 93–120.

7 Anonymous, *Adana*, Yurt Ansiklopedisi, 1981. 1, p. 56.

8 Ayar, M. and Kılıç, Y., *Osmanlı'da Vebanın Sona Erişine Dair Bir Değerlendirme*, Türk Dünyası İncelemeleri Dergisi [Journal of Turkish World Studies], 2017. 17(2), pp. 163–181

9 Aydın, M., *Osmanlı'da Ev Kültürü*, Osmanlı'da Kültürel Hayat, 2006. İzmir: Işık Publication.

10 Batur, A., Yücel, A., and Fersan, N., *İstanbul'da Ondokuzuncu Yüzyıl Sıra Evleri Koruma ve Yeniden Kullanım İçin Bir Monografik Araştırma*, METU Journal of the Faculty of Architecture, 1979. 5(2), pp.185–205.

11 Baysun, M. C., M. *Reşid Paşa'nın Siyasi Yazıları*, İÜEF Tarih Dergisi [Turkish Journal of History], 1960. 111(15), pp.174–190.

12 Bölükbaşı, A., *Anılarda Trabzon*, 2006. Trabzon: Serander Yayınları.

13 Çadırcı, M. *Tanzimat Döneminde Anadolu Kentlerinin Sosyal ve Ekonomik Yapıları*, 2011. Ankara: TTK Press.

14 Cerasi, M. M., *Osmanlı Kenti Osmanlı İmparatorluğu'nda 18. ve 19. Yüzyıllarda Kent Uygarlığı ve Mimarisi*, 1999. İstanbul: Yapı Kredi Publications.

15 Çelik, Z., *19. Yüzyılda Osmanlı Başkenti Değişen İstanbul*, trans. Selim Deringil, 2015. İstanbul: İş Bankası Publication.

16 Ejder, L., *Factory Districts in Bursa During the 1860's*, METU Journal of the Faculty of Architecture, 1975. 1(1), pp. 85–94.

17 Eldem, E., *Osmanlı Bankası Tarihi*, 1999. İstanbul: Tarih Vakfı Yurt Publications.

18 Eldem, H., Goffman D. and Masters, B., *Doğu ile Batı arasında Osmanlı Kenti Halep, İzmir ve İstanbul*, çev. Sermet Yalçın, 2012. İstanbul, İş Bankası Publications.

19 Ertuğrul, A., *XIX. Yüzyılda Osmanlı'da Ortaya Çıkan Farklı Yapı Tipleri*, Türkiye Araştırmaları Literatür Dergisi, 2009. 7(13), pp. 293–312.

20 İnalcık, H., *Devlet-i Aliyye : Osmanlı İmparatorluğu Üzerine Araştırmalar IV Ayanlar, Tanzimat, Meşrutiyet*, 2018. İstanbul: İş Bankası Publication.

21 Kolay, E. *Osmanlı Yerel Yönetim Sisteminin Mimari Alana Yansıması: Tanzimat'tan Cumhuriyet'e Belediye Binaları*, Doctoral dissertation, Ankara University, 2018. İstanbul.

22 Koyuncu, A., *Tanzimat Dönemi Eğitim Reformları Işığında Çanakkale'de Gayrimüslim ve Ecnebi Mektepleri*, Çanakkale Araştırmaları Türk Yıllığı, 2013. 15, pp.77–127.

23 Moltke, H. V., *Moltke'nin Türkiye Mektupları*, trans. Hayrullah Örs, 1995. İstanbul: Remzi Kitabevi

24 Ortaylı, İ., *Tanzimat Devrinde Osmanlı Mahalli İdareleri (1840–1880)*, 2011. Ankara: TTK Press.

25 *Ömer İskender Tuluk Arşivi*

26 *Özcan Canım Archive*

27 Özcan, K., *Tanzimat'ın Kent Reformları: Türk İmar Sisteminin Kuruluş Sürecinde Erken Plânlama Deneyimleri (1839–1908)*, Osmanlı Bilimi Araştırmaları [Studies in Ottoman Science], 2006. 7(2), pp. 149–180.

28 Sağdıç, Z., *Osmanlı Mimarlığı'nda Akaretler Sıraevlerinin Yeri ve Önemi*, Master Thesis, İstanbul Teknik University, 1999. Çanakkale.

29 Sağlam, M. H., *I.Tertip düstûr kılavuzu: (Osmanlı devlet mevzuatı) 1839–1908*, 2007. İstanbul: Tarih Vakfı Yurt Publications.

30 Seyitdanlıoğlu, M., *Tanzimat Dönemi Osmanlı Sanayii (1839–1876)*, Tarih Araştırmaları Dergisi, 2009. 28(46), pp.53–69.

31 Şen, L., *Türkiye'de Demiryolları ve Karayollarının Gelişimi*, 2003. Ankara: TESAV Publication.

32 Tekeli, İ., *Anadolu'da Yerleşme Sistemi ve Yerleşme Tarihleri*, 2011. İstanbul: Tarih Vakfı Yurt Publications.

33 Türkseven, H., *Osmanlı Devleti'nde Eski Eser Politikası ve Müze-i Hümayun'un Kuruluşu*, Master Thesis, Çanakkale Onsekiz Mart University, 2010. Çanakkale.

34 Yavuz, Y., *Batılılaşma Döneminde Osmanlı Sağlık Kuruluşları*, METU Journal of the Faculty of Architecture, 1988. 8(2), pp. 123–142.

35 Yazıcı, N., *Trabzon Örneğinde Tanzimat'tan Cumhuriyet'e Hükümet Konağı Binaları*, Uluslararası Sosyal Araştırmalar Dergisi [The Journal of International Social Research], 2008. 1(5), pp. 943–959.

36 Yetiz, F., *Bankacılığın Doğuşu Ve Türk Bankacılık Sistemi, Niğde Üniversitesi İktisadi ve İdari Bilimler Fakültesi Dergisi*, 2016. 9(2), pp. 107–117.

37 Yıldırım, N., *Osmanlı Coğrafyasında Karantina Uygulamalarına İsyanlar: Karantina İstemezük*, Toplumsal Tarih, 2006. 150, pp.18–27.

38 Yıldız, G., *Mapusâne Osmanlı Hapishanelerinin Kuruluş Serüveni (1839–1908)*, 2012. İstanbul: Kitabevi Press.

Assoc. Prof. Dr. Funda K. Açıcı

19 Contemporary Approaches in Museum Space Design

Abstract: The inclusion of the public/visitors in museum functions has led to the replacement of the concept of museum by that of contemporary museum, that is, it has resulted in the emergence of new museology, which has transformed museums into places that promote broad public access to real-life experiences and emotions. What distinguishes contemporary museology from traditional museology is that museums today engage in interactive communication with the public. Therefore, visitors of museums, which have been designed based on modern functions and new understanding, are more open to different experiences and enjoy them more. Contemporary exhibitions of artworks attract more visitors, and activity areas in museums turn into meeting points where visitors can sip their coffee and enjoy a chat. It is, therefore, clear that contemporary museums are and will be important landmarks that bring together history, art and culture.

Keywords: Museum, Contemporary museum, Contemporary approaches, New museology

1 Introduction

The term "museum" has the connotations of "protection," "safekeeping" and "tradition." Everything in museums constitutes the world of protection, knowledge, culture and aesthetics. Museums, which are socio-cultural laboratories in a sense, serve as repositories of historical and social developments (1).

Museums are the most important institutions bringing the history of cities to today. Museums also play a key role in keeping cultural heritage alive by enabling us to preserve the past in our memories and handing it down to future generations. As İslamoğlu (2) states: "The safeguarding of cultural heritage allows for cultural continuity and creates a sense of history in the minds of people, who, then, can better understand the present."

Museums not only serve the general public, but also respond to the needs of various researchers and experts. Museums have different stakeholders with different perspectives such as visitors, scientists, historians, students, educators, and local/central public or private museum managers and staff, and therefore, can be regarded as institutions that serve different purposes. For some, the mission of museums is to contribute to the education of society and to increase its intellectual capacity by advancing and expanding knowledge while some see them as a venue for intellectual entertainment and leisure, and some others consider them to be research centers.

Globalization has also facilitated the change in people's perception of museums. Museums are one of the most important venues in reflecting the cultural heritage of cities, which are also the focal point of globalization. Bringing new meanings and roles to cities, this position also results in dramatic transformations in a wide spectrum ranging from urban life to urban spaces (3). Globalization has accelerated communication and liberated people by virtually removing borders among nations. This has found its reflection in people's lives and in spaces as well (4). This reflection has given rise to a contemporary museum concept.

2 Contemporary Museum Concepts

The second half of the 20th century witnessed the developments that lay the foundations of contemporary museology. Although museology was an object-oriented discipline until the 1960s, it made use of knowledge and methods from other disciplines as well. The 1960s was a revolutionary period in which new museology models and practices were introduced. The division of museology into sub-fields such as museum management, restoration/conservation, exhibition design, education, visitor relations, research, archiving and communication is considered the second revolution in the field of museology (5). The third revolution has been the emphasis on the social functions of museums. The changes in museum management approach and the redefinition of museum professionalism in this context are contemporary innovations in the field of museology. As for the social function of museums, it is more important than ever to organize and manage the relationship between museum and audience, the latter of which has now been rendered the "agent" (6).

2.1 Development of Contemporary Museum Concept

According to Hall (7), what museums are is nowadays defined by what they are not. Their unique identities are determined by descriptive elements that they do not seem to possess. With the new understanding of museology, museums have gone beyond being just a temple where muses are locked in (8). As Fleming (9) puts it, museums are "a place" but "not just any place." In line with this, Huyssen (8) states that "the need for auratic objects, for permanent embodiments, for the experience of the out-of the ordinary, seems indisputably a key factor of our museumphilia."

In contemporary museology, museums should publicize their activities and attract people's attention in order to achieve integration with society. Museums should reach out to the public through different media and undertake strategic planning to expand their sphere of influence and appeal to a wider audience than just the narrow circle of artists, scientists, intellectuals, art lovers and students (1). Contemporary museums serving as non-formal education institutions today have turned to public relations to reach a wider audience for the transmission of culture and knowledge, and for the education of society.

Reeve and Woollard (10) emphasize that, along with the diversification of functions of museums in the 21st century, many terms describing museum visitors have been proposed. Some of these terms are "visitors, spectators, audience, consumers, mass, shareholders, participants, users and customers." According to Hudson and Ritchie (11), the main factors accelerating the change in the functions of museums are as follows:

1. People's increased expectations of social life
2. Increase in GDP per capita in Western countries
3. Changes in museum experts' job descriptions
4. Increased number of independent museums

According to Hatton (12), the changes in the functions of museums in favor of society began to take shape in the classical age of Europe. Museums used to be regarded as the temples or treasure chambers of the classical world. The changes in their social functions have come out in favor of the audience thanks to postmodernism. The new understanding of museology has been integrated with a contemporary educational approach which views visitors of all ages as "active participants" and focuses on creating inclusive museum environments and facilities to respond to visitors' diverse needs and interests. Therefore, the experience of the audience (the agent), in other words, the "museum process" becomes crucial in contemporary museology. The museum process is a visitor-oriented experience that focuses on visitors' knowledge and experience to enable them to be active participants in museum exhibitions (6).

The inclusion of the public/visitors in museum functions has led to the replacement of the concept of museum by that of contemporary museum, that is, it has resulted in the emergence of new museology, which has transformed museums into places that promote broad public access to real-life experiences and emotions.

3 Contemporary Approaches to Museums

Nowadays, museums have stopped thinking of their collections as fixed entities and come to the realization that exhibition and presentation are often more important than the collections themselves. The shift from the strict canons of the traditional museology to the innovative presentation and exhibition methods of the new museology has manifested itself in contemporary approaches to recent museum designs. In this context, museums aiming to share technological developments and different views with a broader audience incorporate visual educational materials (computer screens, holograms, virtual tours, interactive projection systems, three-dimensional scanning, simulators, three-dimensional modeling and digital applications) into their exhibitions (6).

Developments and contemporary approaches in the field of museology are an inevitable consequence of the era in which we live because we see new developments in every sphere of life every day. It is, therefore, clear that we should go beyond the common perception of museums as sites whose sole objective is to exhibit the artifacts of ancient history.

According to Gurian (13), museums are "necessary" and "hybrid" organizations that interact with their immediate environment and remove borders between other institutions. Viewers no longer want to visit exhibitions in which they are supervised by someone else. Contemporary museums should give the audience a purpose, help them find answers to their questions and be organized in a way that promotes participatory design processes, which are crucial for meaning-making. Participation from within the museum refers to cultural heritage creators (through museum objects or by direct participation), museum experts and other museum stakeholders contributing to the process of learning, entertainment and inspiration. Participation from outside the museum refers not only to visitors participating in exhibitions, screenings, educational events, live presentations and guided tours, keeping up with museum publications and using museums' information services but also to non-museum institutions and organizations (public institutions and organizations, non-governmental organizations and universities) cooperating with museums (6).

Contemporary museums should include all the aforementioned concepts and activities and be open to new ideas and creative approaches in order to respond to the needs of the time and take it further.

3.1 Contemporary Museum Approaches

The relationship between architecture and museum design has attracted much attention nowadays, and to a lesser extent in the past. In the history of architecture, the design of museum buildings has always sought more than functionality alone. Since the Enlightenment, thinkers have approached the design of a utopian city of universal knowledge as an architect and placed the museum at the center of the city. With the sacralization of art and heightened focus on purely aesthetic pleasures, ostentatious architectural designs that overshadow the works of art displayed in museums are generally replaced by plain and simple designs. One of the most remarkable examples is New York's Museum of Modern Art (1931) designed by Philip S. Goodwin and Edward D. Stone in 1939. Both the building design and the white walls in the interior mark the beginning of the era of the "white cube" (14), (Fig. 1).

Mies van der Rohe designed Berlin's Neue Nationalgalerie (1968) as an envelope for artworks, and created an ideal setting for their viewing by minimizing the visual impact of the building (Fig. 2). Although Frank Lloyd Wright reacts to this process with his distinctive design of New York's Guggenheim Museum, the design of the building still raises questions about whether it has the ideal conditions for the viewing of artworks (14) (Fig. 3).

The transition to a new architectural style—nurtured by the debates of new museology, legitimized by the rise of contemporary art and perceived by many as a rupture—is marked by Paris' Centre Georges Pompidou designed by Renzo Piano and Richard Rogers in 1972 (Fig. 4). Distinguished not only by its design, but also by its style of exhibition, the Centre Pompidou ushers in a new era where museums are part of the culture industry. Operating within the discourses of emancipation from bourgeois constraints, democratization and going public, the

Fig. 1: Museum of Modern Art, New York (1931), (15).

Fig. 2: Neue Nationalgalerie, Berlin (1968), (16).

Fig. 3: Guggenheim, New York (17).

Fig. 4: Centre Georges Pompidou—France's Museum of Modern Art (19).

Centre Pompidou blurs the boundaries between entertainment and education. The impressive, subjective and sculptural design of the museum building is presented as the most effective formula to emphasize this blurring. However, the museum building could not have been conceived, if it had not been for the transnational capital pouring into global cities. Competitive cities in the global economy turn

Fig. 5: Guggenheim Museum Bilboa, Spain (20).

to prestige-oriented urban landscape designs rather than to urban planning (18). Therefore, museum buildings become the most popular symbols of this process with their distinctive designs, unusual styles and artistic expressions (14).

Thanks to the decanonizing features of the fine arts expanding the realm of freedom available to architects, they can translate their artistic visions into unique works of art, independently of their function. What might have been a disadvantage in the past has placed architects in great demand by bureaucrats who strive to make their cities more fashionable. Frank Gehry's Guggenheim Museum in Bilbao, Spain, is considered such a successful implementation of this formula that the term "Bilbao Effect" was coined to emphasize the key role architectural and cultural projects have in revitalizing cities (14), (Fig. 5).

So far, we have discussed how contemporary museum approaches have progressed to today. Doing this, we have demonstrated the latest approaches in line with the 21st-century contemporary art

Located in the Flaminio quarter of Rome, built on the grounds of the former Montello military barracks and designed by Zaha Hadid, MAXXI is considered one of the most important structures of the contemporary period. In fact, the project was pitched in 1999 when Gehry was to build the Guggenheim Museum, but it could be completed in 11 years due to budgetary and some other problems (14)

Speaking of contemporary museum approaches, MAXXI is worthy of further discussion in this section as it won two awards and proved its success.

3.2 A Contemporary Museum: MAXXI National Museum of the 21st Century Arts

The National Museum of 21st Century Arts, alias MAXXI, was designed by Zaha Hadid, and was opened in May 2010. Being the first national contemporary art museum in Italy, it won two consecutive awards. Having been previously

Fig. 6: MAXXI, National Museum of 21st Century Arts (21).

shortlisted three times, Zaha Hadid won the 2010 RIBA Stirling Prize, which is one of the most prestigious architectural awards, against a strong candidate, David Chipperfield, who restored the Neues Museum. The 2010 RIBA Stirling Prize Jury described MAXXI as "a building for the staging of art, and whilst provocative at many levels, this project shows a calmness that belies the complexities of its form and organization." In November of the same year, Zaha Hadid won the World Architecture Festival (WAF) World Building of the Year (14).

MAXXI was designed by Zaha Hadid and completed in 2009. Since then, it has been one of the most important museums of Rome, known for its cultural and artistic richness. MAXXI is the first national museum of contemporary art in Italy (Fig. 6).

In designing the museum, Zaha Hadid set out to establish an "art campus" with an interior design consisting of passageways that would overlap and connect to create interactive spaces. She has created a layout based on flexibility. The continuity of the exhibition areas allows these spaces to be used for temporary and mobile exhibitions. She avoided designing walls or any obstacles that would impede the continuity of the exhibition areas. The fundamental elements of the museum building are in the courtyard at the entrance. The sinuously curving concrete walls and suspended black staircases connecting the floors form the main architectural identity of the building (Fig. 7).

Fig. 7: MAXXI, Suspended Black Staircases (21).

Some exhibition areas in the courtyard and museum are naturally illuminated by sunlight. Zaha Hadid has designed a geometric layout with different perspective points and a fragmentary space to provide inter-site flow. This system is not regular but chaotic, as in life itself. This chaotic system and the reconstructed flow model are in contrast with the classical, historical and static structure of Rome. However, she used classical touches in her design and formed a bridge between the two disparate worlds.

The smooth, curved walls and the general configuration of the building are consistent with the classical symmetrical facade design. In addition, the structure of the building was characterized by the condition of the surrounding buildings. MAXXI's geometry was designed according to the axles of the surrounding buildings to make it compatible with the city's urban fabric. Special attention was given to make sure that MAXXI is illuminated by natural light from the glass openings between the concrete ceiling beams. The ceiling beams are also used to hang artworks. The beams, staircases and floor lighting provide a route for visitors (22).

A large atrium with full height leads to the reception areas, cafeteria, bookstore, and auditoriums and galleries hosting permanent collections and exhibitions (Fig. 8).

Fig. 8: MAXXI, Temporary Exhibition Halls (21).

While glass, steel and concrete give the exhibition spaces a neutral appearance, the movable panels ensure flexibility (Fig. 9).

MAXXI contains flexible spaces for sculptures and installations, which are probably the most common form of expression in contemporary art. MAXXI was defined as a collection-free project when opened in 2010. Exhibitions and collections of contemporary art and architecture were presented after it was opened. In this sense, it is clear that MAXXI is a work of art that already exhibits itself.

4 Epilogue

As Schubert states, the lesson to be learned from the past of museum is that it is not at all true that the relationship between the present and the future is a constant rule. What this only means is the replacement of the present with the precise rules of the past. No matter what the future holds, museums will be capable of extraordinary adaptation and continue to be both healthy cultural structures and open to external influences. With their openness, flexibility, versatility, transparency, imperviousness to corruption and appeal to a wide audience, museums are fundamental institutions of democratic societies (24).

Countless examples of museums from the past and present show that the main function of museums has begun to change with modern approaches. Since the general objective of museums is to fund artworks, their designs are generally plain and simple with empty spaces. The main criterion for the design of museums is that they do not overshadow the artworks themselves.

Gradual transformations in museum designs have brought about different approaches that seek the "new" and capture the "contemporary" in the 21st century. This approach, in a sense, has given birth to art itself as a defiance. Museums designed based on contemporary approaches have created an aesthetic perception as a work of art and come into prominence in this era. These approaches are steps taken in the right direction for making sure that architectural structures that will be passed down to the next generation reflect the characteristics of the period they pertain to.

Fig. 9: MAXXI Exhibition Areas (23).

MAXXI is a new manifestation of the formula of star architect+brand+contemporary art. Even Italy, one of the few countries that do not compromise on museology and protectionism, writes the story of its defeat by the culture industry and globalization. Designed as a work of art, MAXXI promises to perform an art show on stage and achieves it. The fact that it was Zaha Hadid's MAXXI, not David Chipperfield's Neues Museum that won the RIBA Stirling Prize indicates that this trend is also appreciated by architects (14).

In conclusion, what distinguishes contemporary museology from traditional museology is that museums today engage in interactive communication with the public. Therefore, visitors of museums, which have been designed based on modern functions and new understanding, are more open to different experiences and enjoy them more. Contemporary exhibitions of artworks attract more visitors, and activity areas in museums turn into meeting points where visitors can sip their coffee and enjoy a chat. It is, therefore, clear that contemporary museums are and will be important landmarks that bring together history, art and culture.

References

1 Karabıyık, A., *Çağdaş Sanat Müzeciliği Kapsamında Türkite'deki Müzecilik Hareketlerine Bir Bakış*, Atatürk Üniversitesi Sosyal Bilimler Enstitüsü Yüksek Lisans Tezi, 2007, Erzurum

2 İslamoğlu, Ö., The Importance of Cultural Heritage Education in Early Ages, International Journal of Educational Science, 2018, Vol. 22, pp. 19–25.

3 Yalçınkaya Erol, Ş., Küreselleşme ve Mekanda Dönüşüm: Dışa Kapalı Siteler. *Dönüşüm Konusunda Uygulamalı Bir Araştırma: Galata Çalıştayı 2011-Beykent, 2012*, s. 134–143, İstanbul: Beykent Üniversitesi Yayınları.

4 Aras, A., Analysis of the Effect of Globalization Users' Color Preferences Over Living Space Furniture. The Turkish Online Journal of Design, Art and Communication - TOJDAC ISSN: 2146-5193, July 2018 Volume 8 Issue 3, pp. 447–456

5 Mensch, P., Museology and Management: Enemies or Friends? Current Tendencies in Theoretical Museology and Museum Management. In *Euope in Museum Management in the 21st Century. 2004* (pp. 3–19). Tokyo: Museum Management Academy.

6 Karadeniz, C., Okvuran, A., Artar, M., Çakır İlhan, A., *Yeni Müzebilim Bağlamında Müze Eğitimine Çağdaş Yaklaşımlar ve Müze Eğitimcisi*, Ankara Üniversitesi Eğitim Bilimleri Fakültesi Dergisi, 2015. 48(2) 203–226.

7 Hall, S., Modern Sanat Müzeleri ve Tarihin Sonu. *Müze ve Eleştirel Düşünce. Tarih Sahneleri-Sanat Müzeleri II* (Ed: A. Artun) (Çev: R. Akman). 2006. İstanbul: İletişim Yayınları, ss.300–312.

8 Huyssen, A., Bellek Yitiminden Kaçış: Kitle İletişim Aracı Olarak Müze. *Müze ve Eleştirel Düşünce. Tarih Sahneleri-Sanat Müzeleri* II (Ed: A. Artun) (Çev: R. Akman). 2006. İstanbul: İletişim Yayınları, ss.259–290.

9 Fleming, D., Creative Space. Reshaping Museum Space. Architecture, Design, Exhibitions. (Ed: S. Macleod). 2005, London: Published by Routledge. pp. 53–61.

10 Reeve, J.,& Woollard, V., Influences on Museum Practice. In C. Lang ve J. Reeve, (Eds.), *The Responsive Museum* (p:5–17). 2006: Hampshire: Asgate Pub. Ltd.

11 Hudson, S., & Ritchie, R., J., Promotoing Destinations via Film Tourism: An Empirical Identification of Supporting Marketing Initiatives. *Journal of Travel Issues*, 2006. 44(4), 387–396.

12 Hatton, A., *The Development of Museums Social-Purpose*. Paper presented at Baltic Museology School: the museum's social purpose, 2008 June, Palmse, Estonia.

13 Gurian, H. E., *Civilizing the Museum: The Collected Writings of Elaine Hheumann Gurian*. 2006: New York: Routledge.

14 Köksal, A.H., Mimarın Fendi Müzeyi Yendi: MAXXI, Bir Güncel Sanat Müzesi'nin Vaatleri. Yapı Aylık Mimarlık Kültür Sanat Dergisi Sayı: 350, s. 48–51. Ocak 2011, İstanbul.

15 The Museum of Modern Art (MOMA) in NYC. [cited 2018, 2 November]; Available from: https://www.timeout.com/newyork/museums/museum-of-modern-art

16 Yeni Ulusal Sanat Galerisi, Neue Nationalgalerie in Berlin, [cited 2018, 2 Nowember]; Available from: https://www.museumsportal-berlin.de/tr/muzeler/neue-nationalgalerie/slideshow/#2

17 Guggenheim Museum in New York, [cited 2018, 1 November]; Available from: https://www.timeout.com/newyork/museums/guggenheim-new-york

18 Harvey, D., From Managerialism to Enterpreneuralism: The Transformation in Urban Governance in Late Capitalism, Geografiska Annaler, Series B, Human Geography, 1989. Vol.71, No.1, s: 3–17.

19 Centre Pompidou, [cited 2018, 3 November]; Available from: https://www.parisinsidersguide.com/centre-pompidou-paris.html

20 Guggenheim Museum in Bilboa, [cited 2018, 1 November]; Available from:https://www.guggenheim.org/about-us

21 Baan, I. Maxxi Museum, 2017. [cited 2018, 4 November]; Available from: https://www.arkitektuel.com/maxxi-muzesi/

22 Arkitektuel, Maxxi Muzesi, [cited 2018, 4 November]; Available from: https://www.arkitektuel.com/maxxi-muzesi/

23 Zaha Hadid, The Maxxi Museum Rome. [cited 2018, 4 November]; Available from:https://www.inexhibit.com/case-studies/zaha-hadid-the-maxxi-museum-rome-part-1/

24 Schubert, K., Küratörün Yumurtası. Müze Kavramının Fransız İhtilalinden Günümüze Kadar Olan Evrimi. (Çev: R. Simith). 2004. İstanbul: İstanbul Sanat Müzesi Vakfı Yayınları.

Assist. Prof. Dr. Hülya Öztürk Tel

20 An Evaluation of Urban Texture Components in the Example of Şanliurfa Hasimiye Square with Regards to Urban Identity and Urban Aesthetics

Abstract: Urban equipment elements are indispensable elements of urban open spaces. They are the objects that facilitate the life of the individuals within the urban texture, provide communication between the individuals and add value and identity to the space functionally and aesthetically. Sanliurfa is a very ancient city its history dating back to 10,000 BC. In this study, urban furniture of Hasimiye Square, located in the old city of Sanliurfa, was evaluated in terms of urban identity and aesthetics. The purpose of this chapter is to evaluate the existing urban furniture of Hasimiye Square, and make new suggestions suitable to the traditional city texture. Within this scope, Sanliurfa Hasimiye Square and Sarayönü area were evaluated in this study. The urban furnitures in Hasimiye Square and Sarayonu area in the traditional city center of Sanliurfa were classified and problematic areas related to design and location selection decisions were determined. It has been determined that the Urban furniture elements of Hasimiye Square do not reflect the traditional square charecteristics of Sanliurfa; it has been observed that serial production based urban furniture are used which do not give importance to design. The meaning of urban furniture for the city and the relation they have established with the city have been evaluated in the context of Sanliurfa city identity and new suggestions suitable with traditional city texture have been made.

Keywords: Urban components, Urban identity, Sanliurfa, Hasimiye Square

1 Introduction

Urban texture components are objects facilitating personal and social life of people within urban texture while ensuring interpersonal communication, adding functional and aesthetic meaning and defining and completing the location. Urban equipment components have an uplifting influence in urban landscape in addition to their functional purposes [1].

Urban texture components are indispensable elements of urban open spaces. It is the object that facilitates the life of the individuals within the urban texture, provides communication between the individuals, adds value to the mecha functionally and aesthetically. It is the object that facilitates the life of the individuals within the urban texture, provides communication between the individuals and adds value and identity functionally and aesthetically.

Instead of being abstracted from environment, urban fixtures need to be designed as a component of the city in such a way as to reflect the identity of the city. Especially in the traditional spaces that reflect the identity of a city, the design of urban furniture requires a more attentive approach.

Urban equipment components, turning a city into a livable and perceivable place, need to have specializing characteristics that define and determine the environment and make it special. Equipment components are the most critical tools, contributing to formation of urban identity [2].

The images, generated by the city we live in and objects that make up this city, are determinants of urban identity. On the other hand, the most vital determinants of urban identity are urban equipment components.

The purpose of this chapter is to evaluate the existing urban furniture of Hasimiye Square, and make new suggestions that are in sync with the traditional city texture as mentioned before.

Within this scope, Sanliurfa Hasimiye Square and Sarayönü promenade were evaluated in this study. The urban infrastructure elements in Hasimiye Square and Sarayönü promenade in the traditional city center of Şanlıurfa were classified, and problematic areas related to design and location selection decisions were determined.

It has been determined that the Urban Furniture elements of Hasimiye Square do not reflect the traditional urban location of Şanlıurfa; it has been observed that serial production-based reinforcement elements are used which do not give importance to design.

The meaning of urban elements for the city and the relation they have established with the city have been evaluated in the context of Şanlıurfa city identity and suggestions have been made.

2 Material and Method

The study material consisted of urban furniture, used at Haşimiye Square in Şanlıurfa traditional urban texture and in external locations situated on Sarayönü promenade. The research field is one of the oldest settlement centers in Şanlıurfa, Mesopotamia. It has had a strategic significance throughout history, thanks to its location next to water resources and trading routes. The city has a 12,000-year historical background, and it experienced Phrygian, Lydian, Hellenistic, Iranian, Roman-Byzantium and Turkish periods.

There are a great number of real cultural assets in Şanlıurfa traditional urban center (Fig. 1,2).

There are 1, 2, 3 grade archeological site areas, natural site areas, urban site areas and 1104 registered structures located at Şanlıurfa city development plan boundaries of protection purposes. Haşimiye Square and Sarayönü Avenue Şanlıurfa urban center, selected as the research field, are positioned at a cross-section of a historical khan inn and worship spaces. There are numerous historical inns and a shopping center located at the south of the field, in addition to primarily Gümrük Inn, Hacı Kamil Inn and Mencek Inn. Moreover, there is Özdiker Passage, with no historical characteristic. Mevlevihane Mosque is situated at the west of the

Fig. 1: The research field, Haşimiye Square.

square and the north of the square continues with Sarayönü Avenue. Many stores are located at Sarayönü Avenue part. The square has lost its identity because of increasing number of motorcycles in the city in recent years, and it has turned into a motorcycle parking lot. There are no equipment components included in the field to reflect the urban identity either.

Survey, data collection, analysis and synthesis-based landscaping research methods were used in the study. In the first stage of the research, the materials to be evaluated were designated and their photographs were taken. In the next stage, the "equipment components" used at Şanlıurfa Haşimiye Square were grouped among themselves by benefiting from the previous categorizations made in this subject, and the furniture were examined for aesthetic and functional purposes based on the analyses and observations performed at the site (Fig. 3).

3 Effects of Urban Equipment Components on Urban Identity and Urban Aesthetics

Characteristics and designs of urban equipment have a critical role for formation of urban identity and aesthetics. Urban identity is shaped in a considerably long time frame. A mixture of urban geography content, cultural level and architecture, local customs and living style shape a city [4].

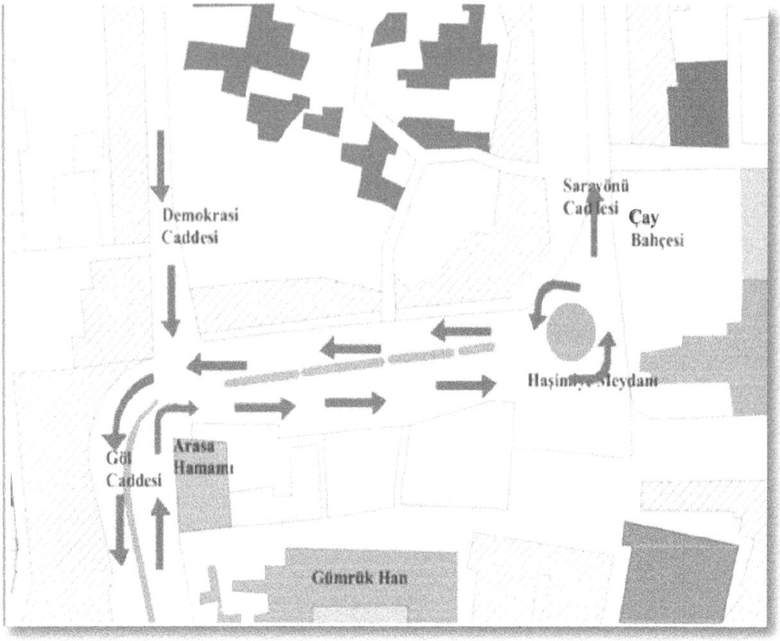

Fig. 2: The research field traffic scheme.

Fig. 3: Photos taken at Haşimiye Square during the years 1940–1980 [3], (left), its view in 2018 (original 2018) (right).

It is necessary to accurately perceive and define the environment primarily in order to set the components making up city identity. For designating the urban identity, which develops as a result of mutual relations of natural environment, social structure, socio-cultural structure and construction, these components must be considered as a whole [4].

Urban identity concept is concerned with the type of city structure, the way this structure is perceived by people and the way people attached meaning to it [5].

City identity can be defined as a content integrity extending from the past to the future, affecting the urban image and is generated by a great process with a different scale and interpretation in any city and shaped by physical, cultural, socio-economical, historical and formal factors and formed by the city people and their life style, making the constantly developing and sustained urban concept live [5].

Cities, in addition to their topographic area and climatic and other physical features, differ culturally and in terms of their needs. They caused the opinion that unique and special solutions should be developed for any city as far as the urban furniture is concerned. It is necessary that furniture must be used in any city, equalizing with the city and reflecting the identity of that city, in place of using urban furniture with serial manufacturing.

Urban furniture must be considered as a wider integrity with its other urban components within the environment, and must bear functional and aesthetical value in an environment and along with other city components.

Urban equipment components (sitting components, shading components, plastic components, parterre, bus stops, pools, fountains, sculptors, telephone booths, snack bars, lighting equipment, separators, garbage cans, routing and signs, information and communication boards, etc.) are extremely highly influential on urban identity and aesthetics with their physical and visual features.

It is a must that equipment components, as one of the distinguishing components and as a part of the system supporting it, must have continuity in the city both technically and visually. Equipment components make up the cognitive and symbolic dimensions of a city with their relation with the city identity, in addition to their functional assignments [7].

Wrong selection and use of these objects, which add meaning to urban identity, lead to confusion and create chaos and visual confusion. It is necessary that urban equipment components that are to be used in historical cities especially must be in harmony with the historical urban identity in terms of form, color and texture, and in terms of materials. They must be reinforced with unique designs, not causing visual confusion and disharmony for urban identity and aesthetics.

It is seen that great attention has been paid to urban furniture, which is considered as an inseparable part of urban identity, in cities that come forward with their historical identity particularly. Designs that reflect urban identity are remarkable in urban furniture design. Hence, maintenance of identity, which is formed during historical process, is also achieved [6].

It is necessary that urban furniture integrate with the environment and become an element of the city, reflect and reinforce the existing urban identity.

Urban equipment components must be evaluated with the history of urban place and along with its environmental and cultural textures and must be considered as a component adding aesthetics to the urban space and enriching the urban texture.

Results

Haşimiye Square was selected since it is a central location giving direction to urban identity and aesthetics at the historical city center, as far as the effect of urban equipment components on urban identity and aesthetics are concerned.

In our study, we grouped urban furniture that are situated at Haşimiye Square in the scope of urban furniture; components related with the infrastructure (lighting components, telephone booths, bus stops and fountains), components that did not relate with the infrastructure (sitting components, plant cases, garbage cans, separators, bill boards, trade panels, street panels, sculptures and plastic components) [8]. It was analyzed whether urban equipment components situated in the study field conformed to the urban identity and aesthetics.

- **Equipment Components related with the Infrastructure**

Lighting Components and Telephone Cabins: Lighting components are standard components, lighting roads, squares, piers, bridges, streets, parks and similar locations that are open to public use and are assigned for public use and placed by electric distribution companies and municipalities [9].

Considering the lighting components at the research field, it was seen that lighting with double-sided and tall halogen bulbs was inadequate, but intense store lights and vehicle lights had made up for this deficiency. It was determined that the used lighting components were built in modern style, not adapting to the traditional texture. The lighting components did not reflect the historical and cultural background of the square.

Despite the diminished use and frequencyof telephone booths because of cellular phone popularization, telephone units are still among the most important urban equipment components appealing with their aesthetic designs in many Western cities.

Telephone booths in the research field were of standard window cabins used in Turkey. It was seen that the telephone cabins were extremely poorly maintained and they did not display an image in harmony with the location's historical texture (Fig. 4).

Bus Stops: bus stops can be defined as passage components where buses take and leave passengers and where passengers wait for bus [8]. During bus stop design stage, factors must be taken into account, including conformity to local climate, purposes and aesthetic values, environment friendliness, ease in their construction, their ability to provide safety to people and protection against environmental conditions, user intensity, use frequency, and users' waiting time. Bus stops located at the study field have a modern and casual look with the applications

Fig. 4: Lighting components and telephone booths at the research field.

Fig. 5: Examples of bus stops and fountains.

made in recent years. The use of heat-absorbing materials will not be appropriate because the region is considerably hot. The historical city will gain more identity in case bus stops are constructed with Urfa stones, the local material in the area where the historical texture is situated.

Fountains: it was seen that the fountain culture, having a different and special place in Turkish culture, was still attempted to be sustained at Hasimiye Square and on Sarayönü Avenue, contained in Şanlıurfa traditional urban texture. Water tank fountain of Hüseyin Pasha Mosque on Sarayönü Avenue and Fevziye Nusret Büyükhatipoğlu Fountain are unique designs, constructed with traditional Urfa stone material and consisting of Urfa stone decoration designs. We can say that these designs affect urban identity and aesthetics positively thanks to both their design and local material usage (Fig. 5).

- **Equipment Components Not Connected to the Infrastructure**

Sitting components: Sitting components are used to meet an action need, including resting, chatting, waiting or just for leisure purposes. They are also used to create a social space, gather people at a desired location and create communities [8].

Sitting components must contain ergonomic materials with few parts as resistive to intense use, in addition to their aesthetic beauty. It was determined that the sitting units situated in the research field were not ergonomic and aesthetic and sitting components were needed at Haşimiye Square particularly. It was seen that the existing sitting groups were not very comfortable since they lacked back-supports. The utilized sitting units were constructed of concrete and casually, and not adhering to the design. Furniture would be produced to improve urban identity by using Urfa stones, located in the region naturally, in place of concrete.

Plant cases: plant cases of various shapes are used to soften the cold appearance of concrete somewhat and to add aesthetics to that space in cities where green fields are decreasing and concrete use is climbing rapidly, especially at pedestrian spaces, on narrow avenues and when plants cannot be grown in soil directly [9].

It was seen that the plant cases located at the research field were left empty and no plants were planted and some of them were used for sitting purpose, with modification of their function by people.

Garbage cans: their functions are collection of human-generated garbage and protection of it until transportation with the necessary vehicles. The internal parts of garbage cans at the study field were emptied and lost their function. Furthermore, among the areas with intense use, there was an underground lift container and it turned out that the underground lift garbage container at Haşimiye Square with usage density was a reasonable selection for a place used considerably.

There were lift garbage containers at the center and it turned out that underground lift garbage container was a sound preference for the intensely used place. Since garbage cans were not emptied as required, they caused visual pollution and hygienic problems caused by pollution. (Fig. 6).

Separators: Separators; these are equipment components used at the place exterior to determine the boundaries and to ensure privacy in the location. Today, they are used commonly to separate vehicle and pedestrian traffics and on streets, pavements and fields where vehicle entry is not preferred. These equipment components used on Sarayönü Avenue were made of metal- and plastic-based materials and were not aesthetic and natural looking.

Billboards: these consist of billboards, electronic panels, advertisement towers, pisa panels, racket panels, panel-changing billboards, light billboard, LED screens and similar [10].

These are communication components that could affect people's feelings, opinions and habits and are effective on them by means of teaching and directing to a specific direction.

Fig. 6: Sitting components, plant cases and garbage cans in the research field.

Fig. 7: Separators at the research field.

Billboards and street signs located in the study field were randomly used ordinary components. It can be said that some of the billboards were placed on the routes as to hinder people's passage. It is important that panels are positioned at proper places as not to prevent vehicle and pedestrian traffic-flows and to be seen very clearly. Panels must be positioned at the eye level of a standing person. It is also required that heights of different panels standing side by side must be in proportion with each other [11]. It can be said that commercial billboards at Haşimiye Square were made of different materials and colors and were placed irregularly and caused visual pollution. Street signs were used randomly (Fig. 7).

Sculpture and Plastic Components: The components increase visual value of urban landscaping and making the design more effective and perceivable owing to their color, form, shape and measurement characteristics [12].

The plastic element situated at Haşimiye Square, reminding the Mırra culture, contributes to partially explain the bitter coffee culture in Şanlıurfa.

Fig. 8: Statues and plastics in the research field.

Water flows through plastic components made of marble material, descending from the top towards the bottom in Haşimiye Square; however, they do not operate because of neglect. Marble material destructs the urban identity and aesthetics since it is not the local material of Şanlıurfa and because it does not conform to the traditional urban texture. It would be more appropriate if it was designed by using Urfa stones, the local resource of Şanlıurfa, as adapting to the traditional architecture (Fig. 8).

Conclusion and Recommendations

The major problems at the beginning of the use of urban equipment components within the research field are design errors, unqualified and unaesthetic material use, destruction due to unconscious use and poor maintenance. The most critical factor found among the equipment components at Şanlıurfa Haşimiye Square and on Sarayönü Avenue was the intense usage of equipment and rapid deformation due to inadequate number of equipment and failure of users to pay the necessary attention. Thus, it is necessary that local administrations carry out their maintenance in certain intervals.

It was seen that various unrelated equipment components were brought and a chaos was created in the designs carried out on different dates in the research field. One of the most important causes of this was that local administrations used ordinary urban furniture without caring for urban identity and aesthetics.

It was determined that randomly selected equipment in the research field was not in harmony with the historical place. Urban furniture to be designed in a historical city, for example, Şanlıurfa, should adapt to the urban identity. Considering Şanlıurfa's traditional urban texture, Şanlıurfa urban identity is damaged due to the inclusion of equipment components that do not conform to this traditional texture and lack the maintenance, functionality, aesthetic and ergonomic aspects.

Correction of these problems can be achieved with the coordinated study of the concerned institutions in the city and with the design studies considering the historical and natural textures of the city. Products must be designed by carrying out multidisciplinary studies for equipment component design.

It is important that designs are created by considering the local climatic conditions and by using local materials.

Unique urban furniture designs beautify the place where they are positioned by improving visual quality and in addition, they would affect the psychological states of user by improving visual quality. Though it can cause problems such as not to be perceivable at the beginning, they will generate a user community who can question the design quality, environmental and urban furniture design quality as open to innovations [13].

It is necessary that unique urban furniture is designed to make the urban identity visible in the traditional urban texture of Şanlıurfa. The use of Urfa stones, a local material, to design urban furniture will improve the harmony with the traditional texture.

Changing of local administration constantly and economic restrictions are critical hindrances for the support of urban equipment components [14]. It can be said that urban identity and aesthetics were not affected positively due to bringing different type of equipment components to Haşimiye Square and by causing chaos. A collaboration and coordination is needed between the designer, producer and municipality to design urban equipment components of great significance.

References

1 Güney, A., Erdem, Ü., Zafer, B. ve Hepcan, Ş., Peyzaj Konstrüksiyonu, Ege Üniversitesi Ziraat Fakültesi Yayınları No:514, İzmir, 1996: p.147

2 Güremen, L., Kent Kimliği ve Estetiği Yönüyle Kentsel Donatı Elemanlarının Amasya Kenti Özelliğinde Araştırılması, E-Journal of New World Sciences Academy, Cilt: 6, Sayı: 2, 2011.

3 Kürkçüoğlu, A. C., ''Urfa ''Fotoğraflarla Evvel Zaman İçinde'' Ankara, 2011.

4 Önem, A.B. and Kılınçarslan, İ., Haliç Bölgesinde Çevre Algılama ve Kentsel Kimlik, İTÜ Dergisi/A, Mimarlık, Planlama ve Tasarım, Cilt 4, Sayı 1, 2005: pp.115–125.

5 Türkoğlu, D., Kentsel İmge: İstanbul'dan Bulgular, İTÜ Dergisi/a cilt 1, Sayı 1, İstanbul, 2002.

6 Güner, E., Kent Kimliği İlişkisi Bağlamında Kent Mobilyaları Sultanahmet Meydanı Örneği, Mimarlık Ana Bilim Dalı Yüksek Lisans Tezi, İstanbul Arel Üniversitesi, 2015.

7 Bayrakçı, O., "Kent Mobilyaları Tasarımında Kimlik Sorunu ve Kent Kimliği İçindeki Yeri" Kamu Mekanları Tasarımı ve Kent Mobilyaları Sempozyumu, Mimar Sinan Üniversitesi, İstanbul, 775–779, 1991).

8 Akyol, E., Kent Mobilyaları Tasarım ve Kullanım Süreci. Yüksek Lisans Tezi, İstanbul Teknik Üniversitesi, 2006.

9 Aslanboğa, İ., Kentlerde Yol Ağaçlandırması, Ege Üniversitesi Ziraat Fakültesi Yayınları, İzmir, 1990.

10 Durmuş. Ç., İstanbul Kent Mobilyaları Üzerine Bir Aaştırma, Yüksek Lisans Tezi, İstanbul: Yıldız Teknik Üniversitesi FBE, 2008.

11 Bilen, S., Ö., "Kentsel Dış Mekanların Tasarımında Antropometrik Verilere Bağlı Olarak Peyzaj Elemanlarının Ankara Örneğinde araştırılması," Ankara Ün. Fen Bil. Ens. Peyzaj Mimarlığı Ana Bilim Dalı, Doktora Tezi, 2004.

12 Uzun, G., Peyzaj Konstrüksiyonunda Yapı Materyalleri, Ç. Ü Ziraat Fakültesi Ders Kitabı, No: 40, Adana, 1990.

13 Aksu, Ö.V., Kent Mobilyaları Tasarımında Özgün Yaklaşımlar İnönü Üniversitesi Sanat Ve Tasarım Dergisi Inonu University Journal Of Art And Design Cilt/Vol. 2 Sayı/No. 6 2012: pp. 373–386

14 Güneş, S., Kent Mobilyası Tasarımında Disiplinler Planlama/TMMOB Şehir Plancıları Odası Yayını, Sayı: 33, 2005.

Assoc. Prof. Dr. Kemal Demir

21 An Analysis of the Spatial Needs and Flexibility in Postgraduate Centers of Theology in the Framework of Kayseri Case

Abstract: Religious education in Turkey is within the scope of public services. In order to increase the quality of this service, the Directorate of Religious Affairs conducts training at the Postgraduate Centers of Theology (PCsT). The PCsT that operate in different regions of Turkey are today progressing and the new ones are opening. In comparison to existing centers, PCT of Kayseri offered a unique approach in terms of its architectural design. In this study, the spatial development of PCT of Kayseri was examined; the novel perspective of this center was introduced.

Keywords: Education, Specialization, Religion, Flexibility in architecture, Postgraduate centers of theology, Education building

1 Introduction

It is an acknowledged fact that to achieve social development, in addition to education [1], production and sharing, a belief system also plays critical role. Accordingly, transferring religious belief system of the past to modern age as well as to the future on the basis of written resources and documents stands before us as one of the most salient aims of public service. In Turkey this particular public service is executed under the authority of Directorate of Religious Affairs. In-service trainings of the employed personnel in the Directorate (muftis and preachers primarily; inc. all graduates of theology school) spread over a period of 30–33 months in a Postgraduate Center of Theology (PCT).

Today in-service trainings are provided in PCTs stationed in Ankara, İstanbul, Konya, Trabzon, Erzurum, Kayseri, Rize, İzmir, Diyarbakır and Kocaeli. Directorate of Religious Affairs aimed to open a new PCT. Hence in this paper spatial data from current centers have been analyzed and the new PCT in Kayseri of which architectural design was drawn by the author of this paper has been scrutinized. In comparison to existing centers, the said PCT of Kayseri offered a unique approach thanks to its design, objective, scope, method and application and also introduced a novel perspective.

2 Spatial Development of Postgraduate Centers of Theology

The diagram below exhibits data on the spatial positioning and capacity to analyze with respect to PCTs, the first of which was opened in 1974 as PCT of İstanbul Haseki.

Next the centers spread to nine Turkish cities. Although the centers were built at varying dates, still their land size, land volume, positioning, spatial capacity and quantitative and qualitative features of the inner spaces of PCTs showed similarities (Fig. 1).

Data in Fig. 2 indicates that in the early dates these centers served as PCTs but during the development stage they operated in the buildings that replicated public construction method. Among these PCTs it stands out that Kayseri PCT offered service on the basis of a unique design and application method.

An analysis of spatial data reveals that, as expected, there were variances on the basis of the functional diversity of Client's Requirement List; data on units, space dimensions and capacity; thereby it is feasible to claim that design decisions and construction processes were implemented with no functional competency or any principal resolutions (Fig. 2).

Regarding the spatial-development processes of educational buildings that were established within the framework of public institutes' investment principles and targets, Karabey lists the factors as below [2]:

1- Planning,
2- Programming,
3- Designing,
4- Constructing (application),
5- Usage-Maintaining-Repairing-Operating,
6- Assessment.

In the design process of Kayseri PCT, a different path than the design and construction processes of traditional public buildings was abided by. In the programming stage of design process, active participation of Provincial Directorate of Religious Affairs, a non-governmental organization and donators was secured. Within that context the formation and application process of executed Kayseri PCT was treated with respect to items below;

a) Programming
b) Design and application
c) Current usage
d) Demand for Development/Transformation.

3 Formation Process of Kayseri Postgraduate Center of Theology

A) Programming

To the end of detecting the needs of projected building/buildings, relevant analyses were conducted in programming process. In order to transfer the analyses into usable data for design, the analyses specified below; employer demands; functional, dimensional and equipment-based features of the buildings were identified as regards the quantity and quality.

SPATIAL DATA OF POST GRADUATE CENTER OF THEOLOGY												
NAME OF THE BUILDING	PROVINCE/ DISTRICT	YEAR	LAYOUT PLAN	TYPOLOGY	LAND AREA m2	BUILDING FLOOR AREA	BUILDING TOTAL AREA	EDUCATION	DORM	TRAINESS BOARDING	INSTRUCTORS BUILDING	MOSQUE
1 İSTANBUL HASEKİ P.C.T.	FATİH/ İSTANBUL	1976			1780	600 / 0.45	3000 / 1.70	3000	X	X	X	X
2 KONYA SELÇUK P.C.T.	SELÇUKLU/ KONYA	1989	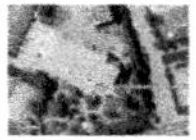		53000	1200 / 0.02	4800 / 0.09	4800	X	X	X	X
3 TRABZON P.C.T.	AKÇAABAT/ TRABZON	1994			5612	1200 / 0.22	3595 / 0.64	1346	1659	X	X	X
4 ERZURUM ÖMER NASUHİ BİLMEN P.C.T.	ERZURUM	1996		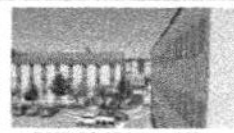	8382	1800 / 0.22	5400 / 0.64	3900	1500	X	X	X
5 KAYSERİ P.C.T.	TALAS/ KAYSERİ	2002	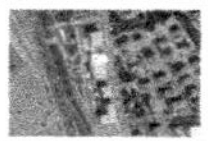		15936	3471 / 0.22	13305 / 0.84	4758	4820	8217	2616	380
6 RİZE MÜFTÜ YUSUF KARALİ P.C.T.	RİZE	2009	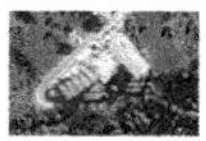		7000	5600 / 0.80	10400 /1.49	400	250	X	X	X
7 İZMİR P.C.T.	KARABAĞLAR /İZMİR	ongoing	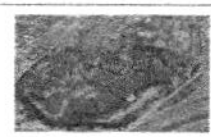	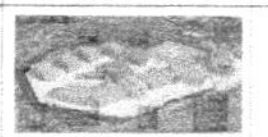	6000	4680 / 0.78	20300 / 3.38	10800	5200	2800	X	1500
8 DİYARBAKIR P.C.T.	YENİŞEHİR/ DİYARBAKIR	ongoing			7000	2000 /0.29	6000 / 0.85	850	780	X	X	X
9 KOCAELİ P.C.T.	BAŞİSKELE/ KOCAELİ	ongoing			40000	2752 / 0.1	5000 / 0.12	2000	3000	X	X	X

Fig. 1: Spatial data from the postgraduate centers of theology (Google Earth).

NAME OF CENTER / YEAR / TOTAL BUILDING AREA	CONSTR. PROCESS		BUILDING TYPES AND THE UNITS OF THE BUILDINGS OF POSTGRADUATE CENTER OF THEOLOGY																								
			EDUCATION										HOUSING														
													DORM						TRAINESS BOARDING				INSTRUCTORS BUILDING				
	PUBLIC	COOPERATION	GOVERNANCE	MEETING ROOM	STAFF ROOM	LECTURERS ROOM	RESTING ROOM	MASJID	LIBRARY	SEMINAR HALL	CONFERENCE HALL	CLASSROOMS	LOBBY	ADMINIST.	LIBRARY	DINING HALL	STUDY ROOM	APART ROOM	LIVING KITCHEN	KIDS BEDROOM	MASTER BEDROOM	SHOWER TOILET	LIVING KITCHEN	KIDS BEDROOM	MASTER BEDROOM	SHOWER TOILET	MOSQUE
1 İSTANBUL HASEKİ P.C.T.. 1976 3000 m2	●			●	●	●	●	●	●	●	●	●	○	○	○	○	○	○	○	○	○	○	○	○	○	○	○
2 KONYA SELÇUK P.C.T. 1989 4800 m2	●		●	●	●	●	●	●	●	●	●	●	○	○	○	○	○	○	○	○	○	○	○	○	○	○	○
3 TRABZON P.C.T. 1994 3595 m2	●		●	●	●	●	●	●	●	●	●	●	●	●	●	●	●	●	○	○	○	○	○	○	○	○	○
4 ERZURUM ÖMER NASUHİ BİLMEN P.C.T. 1996 5400 m2	●		●	●	●	●	●	●	●	●	●	●	●	●	●	●	●	●	○	○	○	○	○	○	○	○	○
5 KAYSERİ P.C.T. 2002 13305 m2		●	●	●	●	●	●	●	●	●	●	●	●	●	●	●	●	●	●	●	●	●	●	●	●	●	●
6 RİZE MÜFTÜ YUSUF KARALİ P.C.T. 2009 10400 m2	●		●	●	●	●	●	●	●	●	●	●	●	●	●	●	●	●	○	○	○	○	○	○	○	○	○
7 İZMİR P.C.T. ongoing 20300 m2	●		●	●	●	●	●	●	●	●	●	●	●	●	●	●	●	●	●	●	●	●	○	○	○	○	●
8 DİYARBAKIR P.C.T. ongoing 1630 m2	●		●	●	●	●	●	●	●	●	●	●	●	●	●	●	●	●	○	○	○	○	○	○	○	○	○
9 KOCAELİ P.C.T. ongoing 5000 m2	●		●	●	●	●	●	●	●	●	●	●	●	●	●	●	●	●	○	○	○	○	○	○	○	○	○

● YES ○ NO

Fig. 2: Building types and the units of the buildings ofPCTs.

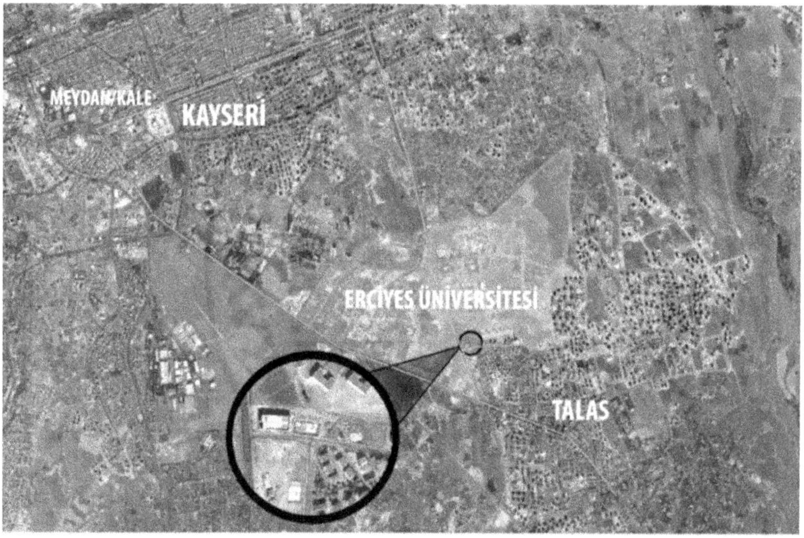

Fig. 3: Aerial photo.

1 Data Related to Area

15.936 m² of land allocated by Talas Municipality is situated inside the city park zone in the development plan. The land has a unique position between Aşık Veysel Boulevard and Şaban Yılmaz Street functioning as the main arterial road drawing the border between Talas and Erciyes University. Lying in the direction of north-south with a gradient of 4% in the form of horizontal rectangle, the land moves towards the park on the west, to Ali Mountain on the south and to the district of Talas on its northern and eastern directions. In the development plan data, its function is defined as "Area for Vocational Training and Social Facility" and the land exhibits a building density of E=1 (15.902 m²) see (Figs 3–4).

2 Employer Demands

With respect to design and application process, below-listed items were concomitantly resolved among Religious Affairs Kayseri City Council, Association for Serving Science and concerned donators;

2.1 Client's Requirement List

a) Education Building

Entrance: reception, lobby, 100-person capacity TV and resting hall, 100 m² library, materials and attendants' room, tea house.

Education: 10 classrooms (40–50 m²), restroom, Internet room

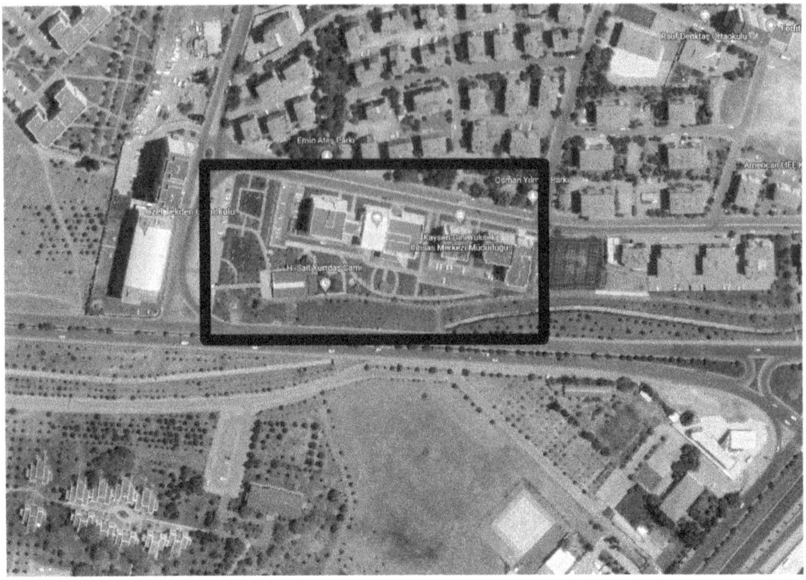

Fig. 4: Aerial photo.

50-person capacity prayer room, ablution, restroom, trainers' room consisting of 2-staff 10 units.

b) Dormitory: (70 people)

Administration: Management Office (reception, meeting room, restroom, resting room), supervisor room, 3-unit admin staff room), 30-person meeting hall, restroom.

Entrance: Information desk, lobby, TV-resting room, sports hall

150-person capacity cafeteria, kitchen, admin office, on-duty teacher, attendants' room, infirmary, warehouse.

Sleeping/Living: 2-person capacity 30 units, suite rooms (bathroom-WC integrated), 2-person 5-room unit visitor-trainers' room, one lobby on each floor, prayer room, Internet room.

c) Instructors' Boarding: (150 m^2) 10 residences

d) Trainees' Boarding: (80 m^2) 80 residences

e) **Conference Hall:** (300 people) Affiliated with the Education Building

f) **Mosque:** (380 people) indoor (900 people) Outdoor Praying Center

2.2 *Design and Application Demands*

- Education center components listed in the Client's Requirement List; Religious Affairs Kayseri City Council, under the coordination of association & donator or to be financed independently and to be undertaken by builders.
- Positioning of the education building independently from the dormitory and boarding residences.
- Via a closed or semi-open passage accessing from the dormitory to education building.
- Designing and building the Trainees' Boarding and Instructors' Boarding as separate residential buildings.
- In the design and application of the Mosque, contributing to the end of implementing and positioning the traditional type of 350-person mosque that was designed and executed by the Directorate of Religious Affairs.
- To help the mosque also offer an outdoor public service opening the ablution fountain and outdoor praying center as positioned in the park zone; usage of the park and center zone as one integrated space.
- To make the cost of the building economical and the operation of the building economical and long lasting.

B) Design and Application

a) About the Layout Plan

The Requirement List of the building has been developed in accordance with the demands of the employer. It has been developed by ensuring common participation into design and application process (Fig. 5).

Accordingly it was projected to fixate designed buildings by creating a wall effect on the eastern border of the land. On its western direction, the buildings would unite with the city park and outdoor public zone would thus be extended. The mosque, in addition with an outdoor praying center, was positioned as a public zone on the land division of the city park expanding on western border (Fig. 6).

b) About the Design Decisions

In line with the demands, integral buildings group of PCT were independently programmed to achieve the design-application process.

To that end in the programming process, identical features on functional-positioning and structural-model, users' quantity and quality regarding the space capacity, frequency and density of space usage, projected number of personnel and trainees to be employed in the near and distant future, training/meeting/culture/resting hall spaces and the quantity and capacity of the equipment forming these spaces, future development and transformation of the training center that was previously specified in its spatial organization have collectively been taken into account. In that respect the design would be implemented as below:

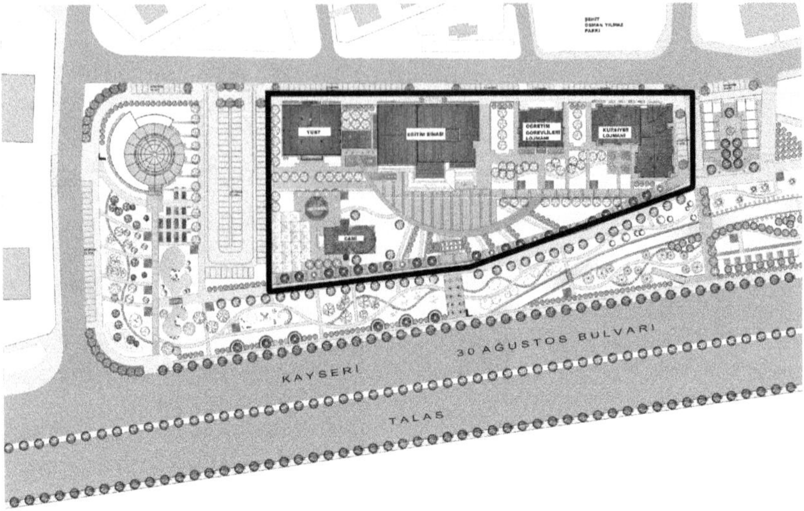

Fig. 5: Layout plan (Demir, 2002).

a) Education Building: A 3-floor Education Building and 300-person capacity Conference Hall were placed within the same structure unity in accordance with the training objectives. Basement (ground floor) of Education Building that contained administration/training/personnel/culture and resting hall spaces were projected to be used independently from the outer space.

b) Dormitory: It was connected with the Education Building in closed and semi-closed passage. In ground-entrance floor control/resting room/culture spaces; in the 1st floor, cafeteria and service spaces; in the 2, 3, 4, 5th floors; 73-person apart sleeping units with their living, cooking and workshop compartments and on the basement floor technical components were positioned.

c) Instructors' Boarding: 10 residences to meet the needs of married couples with 2/3 kids were designed to emphasize neighborhood relations. On ground floor independent units from the entrance, public training spaces were positioned while on the basement technical components and social spaces were integrated.

d) Trainees' Boarding: Addressed to the needs of childless married couples/trainees with kids, there are 48 residential units on the 6th floor. On the ground floor, boarding entrance and lobby units were positioned while on the basement technical components and social spaces were projected.

e) Mosque and outdoor praying center: In parallel with the employer demands the mosque space, fountain and outdoor praying center were positioned as per their function.

Fig. 6: Front silhouette (Demir, 2010).

c) Design Resolutions about Flexibility

It is viable to argue that the emergent transformations and transformations witnessed in employers' demands in parallel with the past-dated and ongoing socio-cultural, economic and technological developments, as of the second half of the 20th century, flexibility started to heavily influence architecture domain. As regards the definitions on the concept of flexibility,W. Gropius states that architects are required not to perceive the buildings they design as a monument; rather they should be flexible in a way to adapt to shifts in life itself and keep pace with the vibrant dynamism of modern life [3].

Schneider and Till assert that flexibility refers to the existing and potential needs. Flexibility, in its first stage, entails the diversity of architectural units. Next, the concept refers to the transformation ability and adaptability of such units due to the newly rising demands in due course. Lastly it means that current buildings should be open to novel functions; accordingly, to ensure that these buildings exhibit flexibility feature it is essential to take into account near-future needs of its potential users [4].

Flexibility is the capacity of buildings serving to a specific function to satisfy the emerging and transforming needs and conditions during and after their usage process.

In the design processes of educational buildings, although it is essential to design in accordance with function, it is also required to design them flexible enough to meet the potential development and transformation in the quantity and quality of program methods and systems. In educational buildings, spaces should be organized in parallel with the training systems and training methods. It is suggested that large classrooms could be turned into smaller meeting rooms or halls where students can engage in scientific discussions; in other words they need to be flexible since learning process would significantly improve provided that education buildings were flexible [5].

Currently users' participation in design-application and usage processes seems to be essentials for the capacity of flexibility approach to adapt to the developing/transforming needs.

Ersoy states that user-centered design is a design process in which building and user's behaviors are defined with scientific data; behavioral objectives of users are decisive factors for design; user behaviors guide the designer and are formed through a constant exchange of feedback between the designer and user[6].

Suggested flexibility concepts on design process are transformability, changeability, extendibility, contractibility, divisibility, additionality, and removability. Within that context flexibility strategies can be identified as design, application, usage and sustainability.

Design flexibility is to meet the increase in the quantity of user needs, demand for transformation and transformability, equipment needs and increasing the spatial capacity to cope with lacks, transformation of spaces and building a new space it is required to design divider and separator building elements,

Application flexibility is extendibility or contractibility of space size and structural pattern via building envelope modular divider-building elements.

In usage flexibility is the diverse functions ability of main spaces and circulation spaces—to be permanently or temporarily used—and to achieve that providing modular divider panels, curtain walls and building elements to enable alternative training, meeting, exhibition, activity and several other purposes.

Sustainability is within the context of flexibility by long-term and effective usage of the building to prevent functional wear outs; environmental sensitivity by consuming fewer energy and water supplies; overall conservation of the environment thus positively contributing to the welfare of ecological environment.

Derivative quality refers to the data on development and transformation; emergent information that can allow to take precautions for being sensitive to the potential development and transformation connected with the usage tools and objectives of buildings and spaces in the future. It refers to designing flexible spaces for potential extendibility and contractibility changes. They must be equipped with features that can be used for different objectives when required and such information can be analyzed within the scope of data on derivative quality [7]

Accordingly, as regards Kayseri PCT, resolutions taken on the design flexibility have been listed as such:

- By considering transformation and development in the function of education building, integrating a number of potential reserved spaces in its design; hence on the basement of this building; (ground floor) a multipurpose reserved space to function as an exhibition hall to meet potential needs, on the basement the reserved space is the area projected to be meeting/seminar break-time hall across the classrooms,
- About the dormitory; projecting that basement, having connected with the education building and 1st floor bridge passages are transformable to a closed space; by the same token basement passage is transformed into a resting hall of dormitory and 1st floor passage is transformed into a cafeteria,

- Based on the expected spatial and functional requirements that could emerge in Client's Requirement List as a result of the increased number of trainees in the future, it was adopted in Kayseri PCT as a design principle to integrate additions and joints to all the precautions related to the entire structure departments; accordingly;

Future-relevant extra needs of potential space for all buildings complementing Kayseri PCT were taken into account; roof floors, a high parapet wall leveled with the height of floor and also a bearing system structuring that considers potential floor additions and mechanic & energy systems have been designed and applied (Fig. 7)

C) Current Usage

Since its first opening in 2016 to that date, below-listed usages have been observed in the Complex:

Education Building; In the Basement; exhibition hall transformable to a resting hall; library to a study/chatting hall; attendants' room to a tea-house niche; on the 1st floor meeting hall transformable into a library unit,

On the Basement, 1st and 2nd floor; classrooms are, as required by training objectives, formed in (U)-shaped seating and equipment pattern.

Dormitory; In the basement; library space turned into exclusive-VIP classrooms and Internet room turned into wet area and staff rooms; lobby was widened and by blocking the passage to the education building, it was turned into a resting hall.

On the 1st floor; the cafeteria that was divided to serve separately the male trainees, instructors and female trainees was now unified as one total space (Figs. 7,8,9).

D) Development/Transformation

a) About the Education Building

During the interviews with users, in parallel with space needs, projected development and transformation objectives on varied supplementary education units are as defined below.

- In the basement; contracting the exhibition/multipurpose hall and forming independent spatial units as study/chatting and resting halls,
- In the Basement; contracting the lobby space and by blocking gallery space turning it into a library; transforming break-time hall across the classrooms into smaller seminar halls,
- In the 1st floor; it was demanded to turn the library into a large seminar hall and initiating the design and application process to meet this demand (Fig. 10).

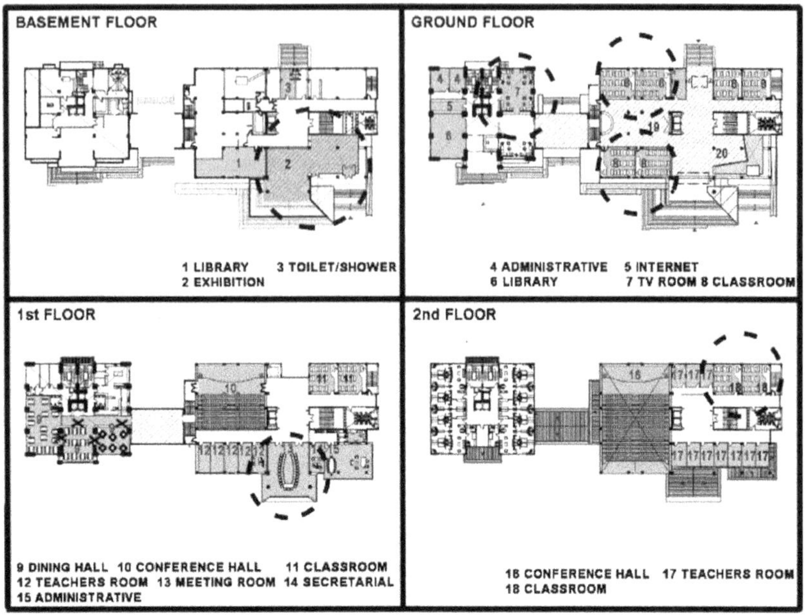

Fig. 7: Original design documents (Demir, 2002).

b) About Increasing the Capacity of Current Buildings

In buildings, the potential rise of the quantity of users in the future was pre-considered and within the context of additional floors onto the final roof floor, a parapet in the height-level of the floor was designed and applied (Figs. 11–12).

Conclusion

The Complex was first opened as the Advanced Training Center affiliated with the Directorate of Religious Affairs. Nonetheless in its developmental process due to the emergent need to initiate Religious High Education that were common in academic institutes it became essential to continue education in spaces that specifically served training purposes. Accordingly, as spatial usage means of PCT are analyzed it is detected that data on Client's Requirement List, units and capacities were formed in accordance with the obtained data on land and relevant legislation; therefore, it surfaced that training could not be provided in the required quantity and quality to meet its educational objectives but rather it exhibited dissimilar capacities and functions. In that sense, unlike

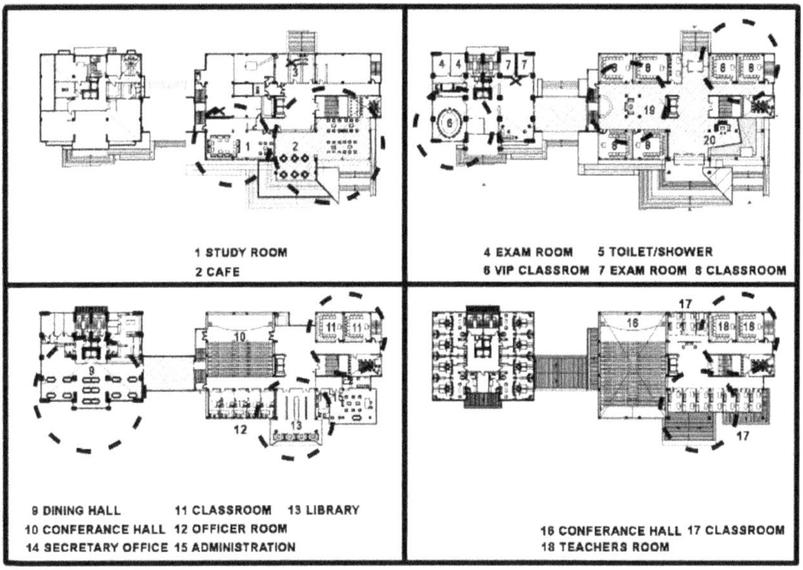

Fig. 8: Current usage (Demir, 2002).

other centers spatial development process of Kayseri PCT differed with respect to the factors below:

- As for positioning it is suggested that data on the land and improvement responded the developed plan and Client's Requirement List; and inside the whole city park in which training, housing and culture structures were built in the maximal flexibility to be adaptive for potential spatial developments in the future, via creating a wall effect on the park borderline the center elevated open public space usage to maximum level and matched with the urban texture of its immediate vicinity.
- Positioning the mosque inside the park zone as required by its semi-public and semi-private usage framework and being open to visual perception from the outer space. It is suggested to add landscape components between PCT and city park in a way not to prevent public usage in between the ownership border which would eventually expand current park zones and PCT borders.

Acknowledgement:

The author would like to acknowledge Miss Beyza Büyüknalbant who assisted in drawing the figures.

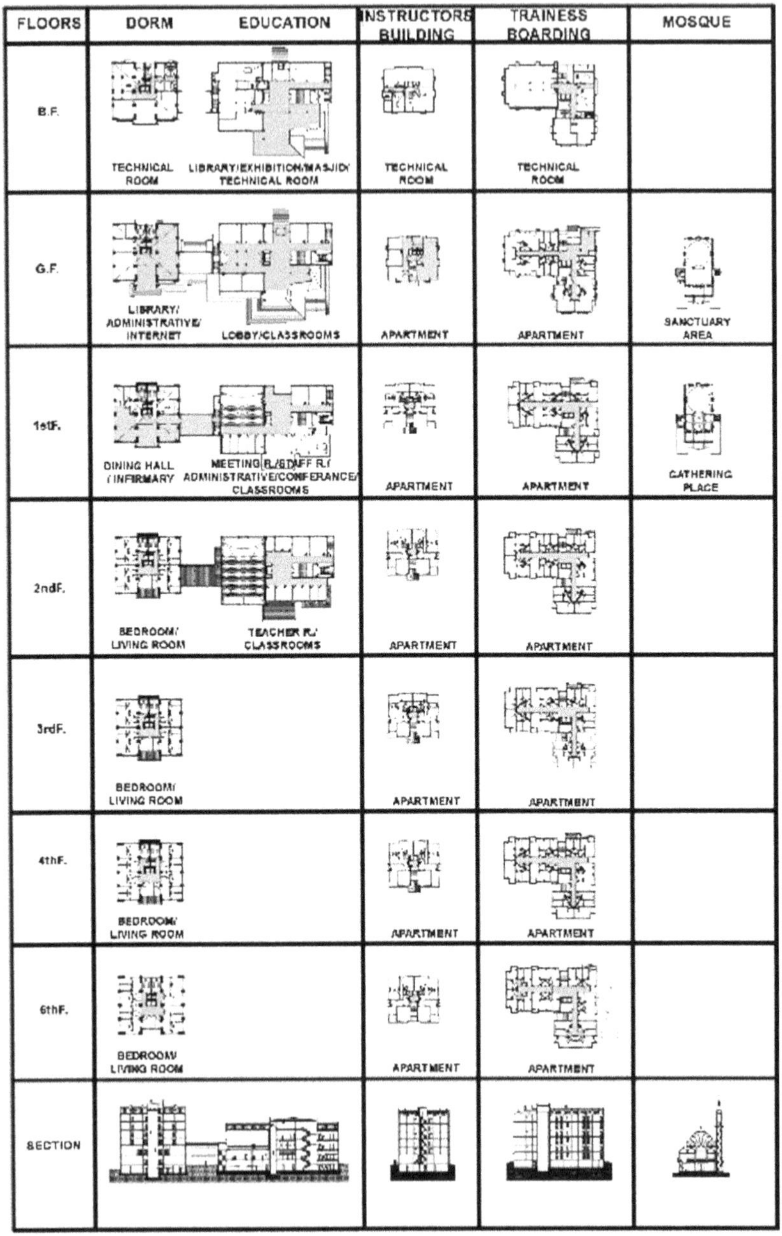

Fig. 9: Layout plans of Kayseri PCT (Demir, 2002.)

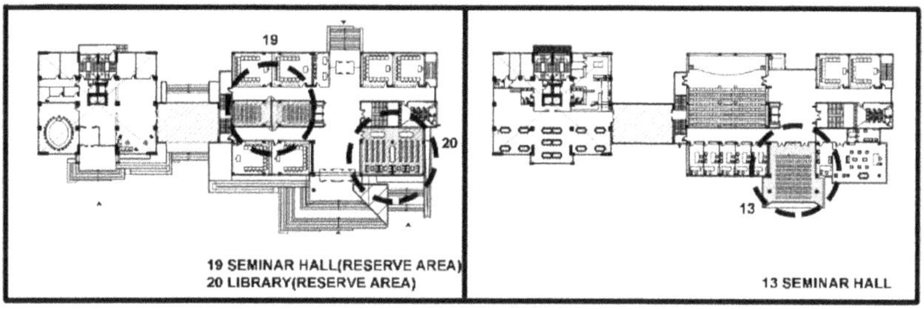

Fig. 10: Demand for development and transformation (dormitory-education building) (Demir, 2018).

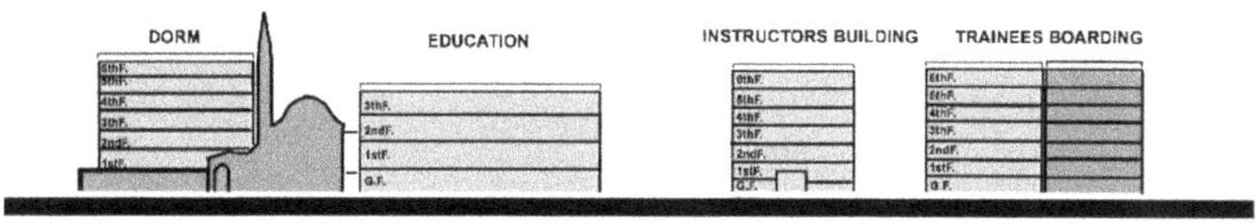

Fig. 11: Design/application decisions for possible additions (Demir, 2018).

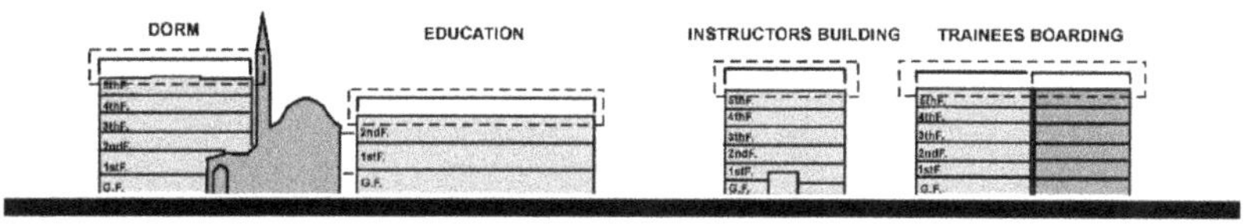

Fig. 12: Possible additions to the buildings (Demir, 2018).

References

1 Dönmez, Y., Türkmen, F., Çabuk, S., *Üniversite Öğrencilerinin Planlarına İlişkin Öğrenci Görüşleri*, Uluslararası Hakemli tasarım ve Mimarlık Dergisi, 2015, **6**: 1–14

2 Karabey, H., *Eğitim Yapıları Geleceğin Okullarını Planlamak ve Tasarlamak Çağdaş Yaklaşımlar, İlkeler*, 2014, Literatür Yayınları: İstanbul

3 Forty, A. *Words and Buildings A Vocabulary of Modern Architecture*, 2000, Thames Hudson İnc.: New York

4 Schneider, T., Till L., J. *Flexble Housing*, Oppurtunities and Limits, 2005

5 Demirkan, H., *Eğitim Kalitesine Uygun Öğrenme Mekanları Tasarımı*, 1995, MPM Yayınları No:570: İstanbul

6 Ersoy, Z. *Mimari Tasarımda Kullanıcı Odaklı Süreçler*, Mimarlık, 2010, **351**, 19–23

7 Türeyen, M. N., *Üniversite Yapıları*, 1999, Dokuz Eylül Yayıncılık: İzmir

Demir, K., Drawings and visuals from the Architect's archive, 2010

Demir, K., Drawings and visuals from the Architect's archive, 2018

http://egitimhizmetleri.diyanet.gov.tr/sayfa/458, 2018

Assist. Prof. Dr. Mahire Özçalık

22 Cemeteries in Uyghur Culture and the Importance of Cemeteries in Open-Green Area System

Abstract: Cemeteries; It is a form of land use which must be absolute within or near residential areas such as housing, industry and trade. However, it should be considered as open green areas along with the cemetery function. As a result of rapid and distorted urbanization, the exhaustion of existing green areas or the increase of rent in urban areas and the difficulty of creating new green spaces, cemetery greenery has also gained new functions. The inadequacy of open-green areas or the transformation of existing open-green areas into different usage forms is the result of unplanned and rapid urbanization. In this sense, planning of the cemeteries as green areas is an important issue of urban planning. In this direction, in order to emphasize the importance to be given in planning and protection of cemeteries in Uighur culture, some cemeteries and importance in the Uighurs were emphasized and place in the light-green system of cemeteries was emphasized and some suggestions were made.

Keywords: Cemetery, Grave, Uyghur cemeteries, Uyghur graves, Tomb and cupola

1 Introduction

Death is an inevitable end that comes after birth and is accepted as a transitional period; in many societies it is believed that with death everything is not over; death is just a space change. It was accepted that after the death of the human being, it was not destroyed and it migrated from this world to the other world. However, as adopted in many cultures, death is not really an end, but the beginning of the end; ending of the material existence of the human being and the beginning of a new life of spiritual existence. The tombs and cemeteries, which are the most important signs left by the dead, also symbolize the social structure (Tunçel, 1999). The grave culture has developed differently in different civilizations and cultural frameworks around the world. The oldest remains of the history of humanity, the cemeteries where some people look scared, some people think deeply, have been the research subjects of historians, art historians, architects since ancient times and the research subject of todays' landscape architects with the architectural features of burial customs and tombstones. In most of the studies, the architectural features of the cemeteries have been evaluated and their cultural identities have been revealed, but there have been very few subjects in the scientific studies as the open-green areas which constitute an important part of the urban texture.

In Uyghur language, the meaning of the grave means "kabir." In Central Asia, the tombs show the tombs of historical figures and legendary individuals, and the places where the Uyghur Turks have prayers for desires, such as having children, fertility and rain. In this study, some grave-cemetery samples, grave decorations, cemetery development and structural, plant characteristics and changes of cemeteries were mentioned. In this context, some arrangements and maintenance proposals have been presented for the graveyards to be embodied to the city as an open-green area.

Cemeteries: It is a form of land use which must be absolute within or near residential areas such as housing, industry and trade. However, it should be considered as open-green areas along with the cemetery function. As a result of rapid and distorted urbanization, the exhaustion of existing green areas or the increase of rent in urban areas and the difficulty of creating new green spaces; cemetery greenery has also gained new functions. The inadequacy of open-green areas or the transformation of existing open-green areas into different usage forms is the result of unplanned and rapid urbanization. In this sense, planning of the cemeteries as green areas is an important issue of urban planning.

In addition to the religious and traditional dimensions of the cemeteries planned according to certain criteria, the ecological, functional and aesthetic contributions of the graveyard green texture should be taken into consideration. In addition to the graveyard green, it is another issue that should be emphasized in the determination of vegetative arrangements at the level of individual graves, determination of plants suitable for this purpose and planning. The rigorous work in this direction has a great effect on the conversion of cemeteries into a park and the increase of visitor interest. In addition to sufficient abundance of green, it is very important to have the annual maintenance operations in such areas and to provide the necessary maintenance personnel and to make the infrastructure facilities such as irrigation and lighting.

2 Cemeteries in Uyghurs

Because of the influence of different beliefs at different times in Uyghurs, the cemetery culture showed different characters as in other Turkish communities. The cemetery cultures of the Uyghurs were influenced by beliefs such as Shaman, Buddhist and Hindu which they believed before Islam. After the acceptance of Islam, the ancestors of modern Uyghurs, the Karahans as the first Turks to accept Islam, Islam's grave structures were adopted.

The tomb structure, tombstones and burial tradition of Uyghurs throughout history have been set to religious principles with the adoption of Islam. It is known that the tomb stones of the Uyghurs have different flower motifs and sometimes animal figures for decorative purposes.

In today's Mongolia, the Uyghurs living on the eastern shores of the Selenga River, under the sovereignty of Gokturks, replaced the Gokturks and established the Uyghur state in 745 BC. The Uyghurs accepted Buddhism, but they buried their dead in hill and like places, like the Gokturks and the Kyrgyz. The dead were buried in the pits below the tumulus, heads placed to east or north direction (Saracoglu, 1950). These graves of which roofs are dome shaped were called stupa. Stupas were

Fig. 1: The view of Sultan Saltuk Buğrahan's Tomb (Anonymous, 2018).

used both as a burial site and as an altar place. It is thought that the architectural structure of these Buddhist graves was influenced by the tent graves of Hun and Gokturks. It is also thought that the stupas are the inspiration for the tombs and cupolas built in Anatolia (Çoruhlu, 1999). On the other hand, the Uyghur Turks, who were Manichaean, did not maintain the old grave tradition (Başkan, 1996).

Uyghurs have erected graves of famous people and have helped sustain these people's names in history. The oldest examples of this can be seen on the Orkhon Valleys' gravestones, which demonstrate heroism. In transferring the information of many historical people to the future, the tombs play a role of monument. For this purpose, tombs and cemetery researches should be a part of the Uyghur history and civilization research and should be protected with appropriate methods.

2.1 Tomb of Saltuk Buğrahan Cemetery and Tomb

The tomb of Sultan Saltuk Buğrahan, the founder of the Karahanlı State, is located in the Meşhet quarter of the Suntağ village, 3 km away from the south west of the city of Artuş in the north of Kaşgar. After the death of Sultan Saltuk Buğrahan, It was built by his eldest son Musa Binni Abdulkerim for the remembrance of his father. It has been restored and rebuilt many times since the time of the Karahans (Fig. 1). The first construction of the tomb was made of wood and fired block brick. This structure is the first commemorative structure and tomb of the Islamic religion in East Turkistan (Defined as Xinjiang Uyghur Autonomous Region by Chine). It is clear that the following statement is written on a locked high-timbered arched gate to the east of the tomb: "restored in 1938 by Zohuriddin who is the Sultan of Kashgar." The Sultan Saltuk Buğrahan tomb has been restored many times in

Fig. 2: Hebbi Haccem cemetery (Original, 2012).

the past 1000 years with the support of rich people and individuals. The size of the cemetery is 10,000 square meters including courtyard, graveyard and orchards (Rahile, 2001).

2.2 Hebbi Hacem Cemetery

The Hebbi Hacem Cemetery is called as "Paşayım" among the people. This cemetery is located in the northwest of the Üstün Artuş Village (Upper Artush Village) of Artuş (Artush) City. It belongs to the 19th century and is one of the legacies of Islamic history because of its architectural features. There are several poplar trees and willows in the immediate vicinity of the cemetery and cemetery has the feature of open-green space to meet the needs of our age (Fig. 2).

Occupants have seen as a threat the Uyghur's cemetery culture since history, so they have not made possible more than a few cemeteries; also these situations have been continuing same as today.

3 The Place and Importance of the Cemeteries in the Culture

The rule of the burial of the human corpses under the land is being continued without any change from prehistoric age to today. Methods that differ from community to community for the burial of human corpses have been recorded in history. Mass burial of human corpses in trenches have been made, human corpses have been left to thepredatory animals for being eaten by them and human corpses have been buried under the land directly or in a container. The increase of human corpses with increase of human population and the unpleasant smell and spreading of contagious diseases from decaying bodies have led to new methods of burial. So, cemeteries were created for burial of human corpse as a result of religious prohibitions, beliefs of communities and rule of the health (Uslu 1997). Different communities have followed different burial customs. Natural caves were used as cemetery at prehistoric ages. Then, rock graves were seen. Gardens of churches were used in medieval age in Europe; then people needed more area because these gardens were not enough in the 19th century. Graves were a big structured monument with dome in Turkish States before the Islam religion came to existence. Cemetery style was changed partially with Islam. Cemeteries existed out of the city walls in Roman and Byzantine cities. However, cemeteries were the as city park in the settlement area with its characteristics at Ottoman. It cannot be seen the artistic and literal structure of the cemeteries used in the history in today's cemeteries. Cemeteries are perceived as very rarely visited by city people and as an area of general burial fields (A. Odabaş, S. Açıksöz ve S. Atturay, 1994).

In general meaning, cemeteries are defined as the burial area of the cities. In addition to this, areas in defined proportion should be allocated for pedestrian ways for visitors, recreation, ceremonies, parking, fountains, buildings for administrative and service clerks, florists, marble-masons and basic needs for visitors. Further to that, furbishing items like warning, guiding and introducing plates, sitting benches, sitting items should be placed.

Uyghur continued the Hun and Gokturk's cultural heritage and was the first Turkish tribe that chose permanent settlement. They had continued to live their culture very long time during the rule of the State as well as after the collapse of their State with the aspect of Middle Asia Turkish history. Richness in the beliefs and lives of today's Uyghur can be seen in their cemeteries, which have a very long cultural past.

4 The Place and Importance of the Cemeteries in Open-Green Areas System

Cemeteries have a very important place in every civilization from prehistoric ages to today. Humans have left the cemeteries as a sign of their life to future generations. Cemeteries are the places that show the sociocultural situations, economic circumstances, city spread situations of civilizations and communities. Some communities used splendid materials because of their religious requirements or philosophical considerations, left beside special objects and sculptures to escort the dead person to afterlife; some communities defined the cemetery areas and constructed the graves with items defining as plain perspective. In the socioeconomically developed countries, cemeteries serve as green areas for utilizing its green texture, quiet and tranquil environment, providing a resting and meditating place for the alive more than the dead (Gönen, 1992).

Our requirements for green areas are increasing very largely day by day because of rapid population increase and distorted and unplanned urbanization. Cemeteries are physically important green city areas; they are different from their surrounding from the aspect of their dense plant cover, and their functional green areas are not increasing nor decreasing. From this perspective, cemeteries should be converted to areas providing functional and passive recreation opportunities.

Open-green area systems have been effective for cities from the past ages to today. "Green Way Movement" emerged in America as a new movement for the planning of open area systems at the end of 19th century. Then, this movement spread rapidly to Europe. In this period, city planners and administrators considered landscape corridors to connect with each other for submitting more than the park and recreation area's functions. Today, most of countries of America and Europe are giving great importance to open area planning for not losing the open area in cities, saving the ecological balance and creating the livable environment. In this respect, differential using of cemeteries which different green areas are made.

5 Results

As said by Aksoy and Özkardeş 2015), cemeteries are very important areas which reflect our history, religion, beliefs, cultural identities, artistic considerations, customs, traditions, conventions, sadness, anger, hope, ambition, economy and daily life. These are the very important written record left by these lands to us because we have very less written history.

This study is prepared for emphasizing the importance of planning of cemeteries as a part of the open green area texture in today's Uyghur culture. In this study, open green area principles are evaluated, current situation is revealed and suggestions are developed for preventing the future and existing historic Uyghur cemeteries. Death is perceived differently in different cultures at different ages as a part of natural life. Religious beliefs and philosophic considerations have played a great role in the shaping of the fear and beliefs related to death and cemetery culture; the perception method has created the cemetery culture of community from history to today.

Graves are the souvenir which reflect in very good manner the artistic pleasure and social feeling of communities at specific historical period and also shows its time accurately. General perspective of the city is completed by cemeteries. They can undertake the functional of open green areas, and can provide big contributions to city ecosystem and green texture. Unfortunately, undertaking these functions cannot be achieved with current planning, application and communication works. When we can make the cemeteries well-ordered, well-cared, green open areas, well-preserved with its trees, flowers, structures and converted to the situation being very useful for city health, it is not logical to make these into unused and reluctantly visited stone bulked areas. During creation, shaping and using of the cemeteries, religious beliefs and cultural structure of communities have the big influence as well as maintenance, repair and operation.

Planning of cemeteries will may make direct great contributions to open-green area of city as well as will preserve and transfer to future generations the graves and cemeteries which are the cultural and historical heritages of the city. Today Uyghur cemeteries should be constructed with the taking into account that population increase rate, death rate, selecting enough area, selecting the area having required features, dominant religious beliefs of community, changing processes and requirements of the city, contemporary requirements, contemporary considerations.

References

1 Aksoy, Y. ve Özkardaş, V. (2015). Study of Karacahmet Mezarlığı with the Aspect of Landscape Design and Handling. Journal of İstanbul Aydın University Year 3, Issue 12, pp. 83–102.

2 A. Odabaş, S. Açıksöz ve S. Atuturay, 1994, "Cemeteries Under the Scope of City Planning," 4th National Regional Science/Regional Planning Congress, 231–234, Trabzon

3 Başkan, S., 1996. "Grave Stone of Karamanoğulları Period," (Journal of Turkish Culture Ministry ISBN 975-17-1422-2, Prime Ministry Press, Ankara.

4 Çoruhlu, Y., 1999. The Influence of the "Transition from Cairn and Tent to Dome and Tomb" to Cemetery Culture and Human Life from History to Today18–20 December 1998, Cemetery Foundation Journal, İstanbul.

5 Gönen, G. (1992). "Cemeteries," Landscape Architecture Journal, (III), Ankara.

6 Rahile,D. (2001). "Uyghur Cemeteries," Xinjiang Public Journal. Urumçi.

7 Saraçoğlu, N., 1950. "Study about the Turkish Cemeteries," İTÜ Architecture Faculty, Proficiency Study, İstanbul Press, İstanbul.

8 Tunçel, G., 1989. "Mosque Description Grave Stones at West Anatolia Region," Turkish Culture Ministry Press: 1056, Demonstration Work Series No: 20 Mas Press, Ankara.

9 Uslu, A. (1997). Anatolia Cemeteries in History and a Study on the A Sample Cemetery Planning for Ankara City with a Contemporary Approach. Ankara University Science Institute, Ph. D. Dissertation, Ankara.

Melih Öztürk and Nadim Kamel Copty

23 Landscape Planning for Sustainable Groundwater Resources

Abstract: Landscape planning is associated with the ecological land use management in order to sustain these lands together with their natural and cultural settings. These natural resources basically involve the components of vegetation, soil, water and the underground. Groundwater is one of the vital hydrological resources of the lands. Large fraction of the precipitation reaching the earth surface is enriched with minerals as it infiltrates into the soil and rocks, ultimately recharging groundwater resources. Accordingly, both the quantity and quality of the groundwater is influenced by the physical and chemical characteristics of the lands and its usages. Groundwater resources are often the primary source of water, especially in arid and semi-arid regions. Groundwater is used for drinking purposes and other domestic usages, agriculture and as a source of water for livestock. Hence, groundwater resources are one of the major ecosystem services within both the urban and rural landscapes. However, groundwater resources are under threat due to human-induced pollution and the sharp land use and climate changes of the recent decades. Consequently, landscape planning must be conducted to simultaneously achieve sustainable land use management objectives and to conserve and protect this vital resource.

Keywords: Landscape ecology, Groundwater resources, Sustainable management, Land use, Ecosystem services

1 Principles of Landscape Planning

Earth broadly consists of complex geological and soil formations, water and living organisms that are thoroughly interconnected and interwoven within a surrounding atmosphere [1]. The original ecological interactions and unique intrinsic dynamics generate systematically regulating distinctive environmental units: the ecosystems. Landscape is basically defined as the entire land that is within the field of vision through the naked eye, visual instruments and satellite images [2]. Hence, the landscape can be made up of either a sole ecosystem, or a mosaic of multiple ecosystems [3]. These ecosystems, in the form of land uses in particular, are utilized by humans, who can make decisions on these ecologically and functionally differentiating land resources.

Planning is conceptually termed as the act and process of decision-making and projection towards the future, on a particular subject or issue. Plans that rely on a sound basis are likely to produce their intended objectives. Around this concept, landscape planning covers all the prospective decisions at a multitude of scales from the basins, regions and even to the continents, which comprise the macro environment of the urban and rural land uses, considering their ecological processes—characteristics, socio-economical systems and historical values [adapted from 4].

An integrative approach to landscape planning includes the conservation and development of the land and its resources, dependent upon their spatial characteristics and temporal dynamics. According to the land use/land cover (LULC) classification system of the (United States Geological Survey) USGS , the major LULCs are urban/built up areas, agricultural areas, rangelands, forests, water, wetlands, barren lands, tundras and perennial snow/ice globally [5]. Additionally, there are subcategories under these major LULCs.

Taking all the necessary precautions before and during the usage, together with the required reclamation practices after the usage, is imperative for the sustainable landscape planning and application objectives. Consequently, the concept of sustainable landscape planning principally intends to propose ecological land uses compatible with their nature, to enable resource utilization for the sake of both humans and the environment [6]. Around this conservative utilization concept, it is essential to develop restoration projects in order to ensure resource replenishment and land rehabilitation together with the encouragement of land management approaches not only considering the current and long-term socio-economic measures but also taking the necessary environmental precautions.

2 Hydrological Cycle and Groundwater

One of the major ecosystem processes is the hydrological cycle, which not only defines the water interchange between the land and atmosphere but also plays a crucial role on other ecosystem processes. Hence, some of these basic ecosystem processes are either driven by the water or treated within the body of the hydrological processes [7]. In fact, physicochemical processes such as the fracture of rocks, formation of soils, and sedimentation of substances are explicitly shaped through water and its movement. Material and nutrient-based ecological processes—in particular the oxygen, carbon, nitrogen, phosphorus cycles—are directly associated with the hydrological cycle. Therefore, the regular global circulation of water, in a sense, indicates the steady and healthy operation of an ecosystem.

The hydrological cycle primarily begins with precipitation that reaches the earth surface particularly in the form of rainfall and snow. At the earth surface, precipitation may be intercepted by the individual leaves and/or by the plant crown, dependent upon the vegetation cover present over that surface. The intercepted precipitation then either evaporates or drips gradually onto the lower leaves before experiencing the throughfall. Besides evaporation, transpiration is the release into the atmosphere of water that has been uptaken by plant roots [8] together with the dissolved nutrients. The precipitation that reaches the earth surface either directly dropping on, or subsequently by throughfall, flows principally along two directions: down slope as overland flow or percolates into the soil and is referred to as infiltration. Precipitation may also be temporarily stored at the ground surface when liquid water finds its way and accumulates in any depression, or in the form of snow accumulation until it melts into liquid form which can then evaporate into the atmosphere, infiltrate through the soil or appear as surface runoff. Overall flowing waters that do not infiltrate into the soil join to form streams which ultimately feed surface water bodies such as lakes, reservoirs and seas.

Water that infiltrates into the soil, termed soil moisture, is enriched with nutrients and minerals from the soil. Soil moisture is partly uptaken by plant roots or evaporates from the soil. Soil water flows mostly downwards or to a lesser extent in a lateral direction as interflow that can feed streams or reappear at the surface. The water infiltration that does not evaporate or is not uptaken by plant roots reaches the saturated zone below the water table, recharging groundwater resources [9]. It is estimated that more than 30% of the global fresh water resources are in the form of groundwater, with polar ice forming about 69% while surface waters do not exceed 1% [10].

Aquifers are geologic units that readily allow the transmission of groundwater. Aquifers are commonly classified as unconfined and confined. Unconfined aquifers have basically undulating water table, which varies with the permeability, pumping, discharge and recharge conditions whereas confined aquifers are restricted by an overlying relatively impermeable layer [11]. There is constant exchange between stream water and groundwater; from the groundwater to the stream flow in the form of baseflow in the dry season, and from the stream flow to the groundwater during the wet seasons in particular [12]. The chemical constituents and characteristics of groundwater are largely affected by the chemistry of the geological formations through which water flow occurs [13].

3 Groundwater in the Ecology of Landscape

Each landscape intrinsically forms a basis for the various ecological cycles and processes dependent upon its natural characteristics. These cycles and processes generate mainly among the land, ocean and atmosphere. Although the rocks and soils constitute the major portion of the land, both surface and subsurface waters play a crucial role within the material and mineral cycles amongst those solid parts. As mentioned in the previous section, the part of the precipitation that reaches the soil surface, infiltrates into the soil, carrying nutrients and minerals through the pores it percolates. Dependent on the permeability of the deeper layers, soil water may percolate deep into the underlying formations. This groundwater further incorporates materials and minerals from the rocks it contacts, into its physicochemical constitution [14,15]. The physicochemical constituents of groundwater defines its quality and, thereby, its suitability for the intended usages. Potable groundwater should meet associated drinking quality standards, whereas agricultural groundwater should meet appropriate irrigation standards without causing any adverse health effects. Any groundwater supply for the agricultural livestock should also satisfy the sanitary conditions that will prevent the occurrence and spread of the animal pests and diseases.

Whereas the quantity of the groundwater is primarily influenced by the physical characteristics of the landscape (such as soil and rock permeability), the quality of the groundwater is principally influenced by the chemical characteristics of the soil and rocks within which it flows. The impact of the biological characteristics of the landscape on both the quantity and quality of the groundwater is also notable. Numerous scientific studies have evaluated the role of biosystems on groundwater quantity [e.g., 16]. Forest landscapes usually play relatively conservative role on both the quantity and quality of the groundwater. They relatively favor the infiltration of the precipitation into the soil rather than flowing overland, therefore promoting evapotranspiration [17] and subsurface flow compared to the surface runoff. Furthermore, forest landscapes are to some extent protected from human interventions and relatively away from pollutant sources. These comparatively natural characteristics of the forest landscapes help the preservation of the underlying groundwater quality [18].

The situation is more challenging for agricultural biosystems particularly under the influence of pesticides and fertilizers [19]. In this case, the agricultural soil together with the vegetation is more prone to chemical deterioration, threatening the quality and sustainability of the groundwater resource. Agricultural land uses pose the risk of elevated nitrate concentrations in the groundwater particularly due to the application of fertilizers [20]. The situation is also valid for the land use changes that intend to convert the forests and grasslands to urban and agricultural lands [21]. Similarly, the quantity of the groundwater is affected by the agricultural activity since there is a constant exchange between the soil and groundwater [22]; from the soil to the groundwater through leaching and from the groundwater to the soil through irrigation. Excessive irrigation using saline groundwater has been shown to lead to soil salinization, which is the accumulation of salts in the soil as a result of over-irrigation and the evaporation of excess water, particularly in arid and semi-arid regions, leaving behind high levels of salts. These elevated salt levels can lead to sharp decrease in crop yields, threatening food security and the sustainability of agricultural lands [23]. Another example of the impact of excessive groundwater extraction is the problem of arsenic poisoning in India and other countries. As the result of water table lowering due to groundwater extraction, naturally occurring arsenic in the soil is exposed to oxygen, which converts arsenopyrite, a harmless form of arsenic, to soluble ionic form of arsenic. The arsenic gradually accumulates in the body over many years leading to arsenic poisoning which can cause cancer, respiratory and vascular diseases, and liver fibrosis.

4 Ecosystem Services and Drawbacks

Ecosystem services are generally defined as the optimum benefits and uses that are supplied by the ecosystem for the material, social, recreational needs of human, without harming other living organisms and/or the environment [developed from 24]. The utilization of the ecosystem resources is significant for the sustainability of other ecological processes and basic life activities of some creatures.

However, these resources are limited and principally, their limitedness and potential scarcity should always be considered. Therefore, overconsumption of these resources will affect the usage of these resources by other organisms.

The situation is crucial particularly for shallow groundwater areas and for regions that suffer from water scarcity. Regions that receive relatively low precipitation, rely predominantly on groundwater resources rather than the surface water. Such regions suffer salinity and drought, which eventually result in poverty and food scarcity. Associated with the drought and overconsumption, groundwater depletion results in serious diseases dependent upon the lack of clean potable and irrigation water. Hence, agricultural cultivation should not rely on irrigated farming there. On the contrary, dry farming would be more suitable for the sustainable usage of the threatened groundwater resources.

Coastal aquifers are also under threat of salinization due to salt seawater intrusion into the groundwater resource [25]. The situation consequently influences human consumption of the resource both for drinking and irrigation purposes for agriculture. The excess saline water adversely affects human health. On the other hand, anthropogenic groundwater contamination can influence the quality of the seawater particularly in coastal areas.

Groundwater resources within urban and suburban landscapes are substantially valuable from the point of ecosystem services. Municipal groundwater serves not only for domestic purposes but also for the multiple needs of the urban open spaces ranging from the irrigation of parks, gardens, city farms, horticultural nurseries, green houses to the supply of water to decorative pools, ponds, and other waterworks in particular. However, these urban groundwater resources are often under the threat of contamination risks due to seepage from sewage systems and of the sealing risks due to construction activities. Consequently, depending upon the degree of human intervention, once the urban groundwater resource is exposed to any pollution or leakage, relatively long term and large efforts and costs are needed to recover or replenish the resource [26]. The rural landscapes, where settlements are initiated, also suffer from groundwater contamination risks due to the use of improperly maintained septic systems [27].

5 Sustainable Management of Groundwater

Groundwater, because of its significance for all living organisms of the landscape, must be accurately identified and characterized, particularly with regards to the spring source, route and spatial extent. Detailed spatial and temporal extents of this resource including flow paths and rate measurements must be identified, maintained and monitored continuously. Spatio-temporal inventories need to be acquired and mapped in order to conceptualize and help protect this vital resource. Indeed, the delineation of watersheds and catchments and underlying groundwater resources will assist in further decision-making processes for the sustainable usage of these resources within that landscape. Catchment scale mapping and inventory analysis are significant components for the sustainable management and planning of groundwater resources [28]. Furthermore, changes to mapped spatio-temporal groundwater data can be modelled in relation to long-term maps and satellite images of the land uses and meteorological data in order to evaluate and mitigate potential future climate and land use change scenarios and projections. Such models must consider landscape planning and sustainable management of both local and regional groundwater resources. Examples of recent model applications consider the efficiency of conservation measures and their impacts on vegetation and climate change factors [e.g., 29].

Groundwater resources within resilient landscapes carry the inherent potential to overcome depletion and contamination problems to a large extent. Sustainability of the resources comprising the groundwater can often be achieved under the influences of the land use and climate change within those resilient landscapes. On the other hand, drought-sensitive landscapes and vulnerable ecosystems suffer from adverse impacts of any land use conversions and climate alterations on groundwater resources [30]. For arid or semi-arid agricultural landscapes, overconsumption of groundwater for irrigation may lead to water scarcity, salinization, contamination and even land subsidence. Urban sprawl and industrial growth, which actually have the potential to produce heat islands, pose significant further threats on already depleting groundwater resources. These threats may result in the sealing impacts on the groundwater resource together with the mentioned negative effects on its quality and quantity [31]. In particular, deforestation to expand agricultural activities, urbanization and industrialization can drastically interfere with the sustainability of groundwater resources.

On the other hand, sustainable management of the groundwater resources in principle involves the judicious allocation of these resources among the associated stakeholders. The rightful sharing and fair allocation of the groundwater among those public users are essential in order to avoid conflicts and forfeiture of the neighboring field and areas within that landscape. The common use of the groundwater resource must always remain paramount for the sake of the sustainability of the social benefits and ecosystem health.

References

1 E. J. Tarbuck, F. K. Lutgens, D. Tasa: *Earth Science. Twelfth Edition. Pearson International Edition. Pearson Education Inc. New Jersey*, USA. (2009).

2 M. Öztürk: *Driver and Passenger Perception along the Roadside Landscapes: Safety and Recreation. In: Sustainable Planning and Design (Ed. Murat Özyavuz).* 255–268. Peter Lang Publication. Frankfurt, Germany. (2017).

3 V. Ingegnoli: *Landscape Bionomics, Biological-Integrated Landscape Ecology. First Edition.* Springer-Verlag, Milan, Italy. (2015).

4 W. M. Marsh: *Landscape Planning, Environmental Applications. Fifth Edition.* John Wiley and Sons, Inc. USA. (2010).

5 J. Randolph: *Environmental Land Use Planning and Management. First Edition.* Island Press, Washington, USA. (2004).

6 F. Steiner: *The Living Landscape: An Ecological Approach to Landscape Planning. Second Edition.* McGraw-Hill Professional Architecture. New York, USA. (1999).

7 R. C. Ward, M. Robinson: *Principles of Hydrology. First Edition.* McGraw-Hill Publications, London, U.K. (2000).

8 G. Bonan: *Ecological Climatology, Concepts and Applications. Second Edition.* Cambridge University Press, Cambridge, U.K. (2008).

9 G. F. Pinder, M. A. Celia: *Subsurface Hydrology. First Edition.* John Wiley & Sons, Inc. Publication. New Jersey, USA. (2006).

10 R. W. Healy, T. C. Winter, J. W. Labaugh, O. L. Franke: *Water Budgets: Foundations for Effective Water-Resources and Environmental Management. U.S. Geological Survey Circular 1308,* 90 *p.* Virginia, USA. (2007).

11 D. K. Todd, L. W. Mays: *Groundwater Hydrology. Third Edition.* John Wiley & Sons, Inc. Massachusetts, USA. (2005).

12 O. Dumlu, H. T. Yalçin, E. Bozkurtoğlu: *Groundwater Geology and Hydraulics (Yeraltısuyu Jeolojisi ve Hidroliği; in Turkish). First Edition.* Literatür Publishing, İstanbul, Turkey. (2006).

13 C. W. Fetter: *Applied Hydrogeology. Fourth Edition.* Prentice-Hall Inc., New Jersey, USA. (2001).

14 F. W. Schwartz, H. Zhang: *Fundamentals of Groundwater.* John Wiley & Sons, Inc. New York, USA. (2003).

15 N. Kresic: *Hydrogeology and Groundwater Modeling. Second Edition.* CRC Press, Taylor & Francis Group. Florida, USA. (2007).

16 Z. Zomlot, B. Verbeiren, M. Huysmans, O. Batelaan: *Trajectory analysis of land use and land cover maps to improve spatial-temporal patterns, and impact assessment on groundwater recharge. Journal of Hydrology. 554*: 558–569. (2017).

17 M. Öztürk, N. K. Copty, A. K. Saysel: *Modeling the impact of land use change on the hydrology of a rural watershed. Journal of Hydrology. 497*: 97–109. (2013).

18 R. T. T. Forman: *Land Mosaics: The Ecology of Landscapes and Regions. First Edition.* Cambridge University Press, Cambridge, U.K. (1995).

19 M. G. Turner, R. H. Gardner: *Landscape Ecology in Theory and Practice: Pattern and Process, Second Edition.* Springer-Verlag, New York, USA. (2015).

20 R. F. Valle Junior, S. G. P. Varandas, L. F. Sanches Fernandes, F. A. L. Pacheco: *Groundwater quality in rural watersheds with environmental land use conflicts. Science of the Total Environment. 493*: 812–827. (2014).

21 T. S. Narany, A. Z. Aris, A. Sefie, S. Keesstra: *Detecting and predicting the impact of land use changes on groundwater quality, a case study in Northern Kelantan, Malaysia. Science of the Total Environment. 599–600*: 844–853. (2017).

22 M. Zhang, S. Geng, S. L. Ustin: *Quantifying the agricultural landscape and assessing spatio-temporal patterns of precipitation and groundwater use. Landscape Ecology. 13*: 37–53. (1998).

23 H. Lambers: *Dryland salinity: A key environmental issue in southern Australia. Plant and Soil. 257(2)*: 5–7. (2003).

24 J. Tuinstra, J. Van Wensem: *Ecosystem services in sustainable groundwater management. Science of the Total Environment. 485–486*: 798–803. (2014).

25 S. P. Ranjan, S. Kazama, M. Sawamoto: *Effects of climate and land use changes on groundwater resources in coastal aquifers. Journal of Environmental Management. 80*: 25–35. (2006).

26 J. Qiu, C. B. Wardropper, A. R. Rissman, M. G. Turner: *Spatial fit between water quality policies and hydrologic ecosystem services in an urbanizing agricultural landscape. Landscape Ecology. 32*: 59–75. (2017).

27 J. A. Lagro JR, B. T. VowelS, B. Vondra: *Exurban housing development, onsite wastewater disposal, and groundwater vulnerability within a changing policy context. Landscape and Urban Planning. 167*: 60–71. (2017).

28 K. Glanville, T. Ryan, M. Tomlinson, G. Muriuki, M. Ronan, A. Pollett: *A method for catchment scale mapping of groundwater-dependent ecosystems to support natural resource management (Queensland, Australia). Environmental Management. 57*: 432–449. (2016).

29 Y. A. M. Van Der Knaap, M. De Graaf, R. Van Ek, J-P. M. Witte, R. Aerts, M. F. P. Bierkens, P. M. Van Bodegom: *Potential impacts of groundwater conservation measures on catchment-wide vegetation patterns in a future climate. Landscape Ecology. 30*: 855–869. (2015).

30 M. Pulido-Velazquez, S. Peña-Haro, A. García-Prats, A. F. Mocholi-Almudever, L. Henriquez-Dole, H. Macian-Sorribes, A. Lopez-Nicolas: *Integrated assessment of the impact of climate and land use changes on groundwater quantity and quality in the Mancha Oriental system (Spain). Hydrology and Earth System Sciences. 19*: 1677–1693. (2015).

31 D. N. Lerner, B. Harris: *The relationship between land use and groundwater resources and quality. Land Use Policy. 26S: S265–S273*. (2009).

Dr. Meltem Güneş

24 Transformation to Water-Sensitive Cities in the 21st Century: Sustainable Stormwater Management

Abstract: Stormwater management has become one of most crucial issues of the 21st century for increasing permeable surfaces in urban area, improving water resources and controlling flood due to severe weather events caused by global warming and climate change. In this respect, it is a well-known fact that the preservation and management of water is of utmost importance as a result of the rapid urbanization. Previous studies indicated that the traditional methods of stormwater management did not effectively improve water quantity and quality. Concordantly, the best practices for sustainable water development are sustainable stormwater management approaches. This chapter presents an overview of the concepts that emerged in the 21st century on the basis of NBS for sustainable stormwater management in urban planning and design processes. Initially, the concepts are defined and subsequently a brief overview of sample applications and practical solutions to urban problems in the context of sustainable stormwater management are presented. In conclusion, as a nature-based solution, it has been emphasized that the widespread adoption of sustainable stormwater management is vital to water demand, one of the most important issues of the future. It is discussed that the necessity of an integrated approach from sub-scale to upper scale plan should be used throughout the planning and designing process. Accordingly water should be primarily taken into consideration as a key element in both today and future's cities.

Keywords: Nature-based solutions, Stormwater management, Water-sensitive urban design, Nature-based solutions

1 Introduction

Fast-paced population growth and thus rapid and unplanned urbanization in the 21st century, dramatic reduction of permeable surfaces in urban areas as a result of land conversion and expansion of grey infrastructure (the human-engineered infrastructure) alongside adverse effects of climate change have been among the major underlying causes of reduced water supply [1].

The higher level of greenhouse gases in the atmosphere resulted in increased global temperatures and from 1880 to 2012, average global temperature increased by 0.85°C [2]. Given that increase in global average temperature of about 0.5°C were observed after 1950, this is a crucial matter that needs to be carefully considered [2]. Since 1977, the six warmest years in the global record have occurred after 2010. The year 2017 has been confirmed as the third warmest on record after 2015 and 2016 since 1880 by World Meteorological Organization (WMO). National Centers for Environmental Information (NCEI) (2018) highlighted that 2017, the warmest year on record, caused extreme events in many countries around the world such as tropical cyclones, floods and droughts [3]. Consequently, there is proof the earth has been getting warmer.

An increasing global temperature has led to global warming and accordingly polar ice-melt and the quality and amount of groundwaters have decreased due to high consumption [1]. According to the 2018 edition of the UN World Water Development Report, more than two billion people still do not have access to safe drinking water and more than double that number lack access to safe sanitation [4]. With the rapid growth of the world's population, water demand is projected to increase one-third of the global population [4].

The water cycle that should be under natural conditions is deteriorating in urban areas [5]. Water in urban areas is polluted by increased runoff and cannot infiltrate the ground due to impervious surfaces and is discharged to the public draining systems without evaporation. This clearly means that effective stormwater management is required [5].

In addition to addressing problems such as housing, transportation, socialization, different infrastructure developments and diminishing natural resources with increasing population density, urban design in the 21st century should offer solutions to water issues as well. Water is having a considerable impact on every aspect of urban construction environment, ranging from city scale to neighborhood scale, that is to say, the water should be taken into account to guide urban design [6]. In recent years, significant attention has been paid to sustainable stormwater management in order to increase pervious surfaces in cities, to enhance water resources as well as control of natural disasters such as flood and overflow.

The United Nations World Water Development Report, Nature-based Solutions (NBS) for Water, launched on 19 March 2018, illustrates how nature-based solutions play a vital role in achieving sustainable development [4]. Regarding the water resources management and stormwater management, NBS [4] use or mimic natural processes, enhance water availability, improve water quality and reduce risks associated with water-related disasters and climate change [4].

Currently, traditional techniques in stormwater management are not efficient [7]. Nature-based solutions include green infrastructure in parallel with grey infrastructure in a cost-effective manner [8]. The main objective is to decrease runoff utilizing natural drainage and to manage stormwater near its origin to enhance infiltration, thereby maximizing system efficiency and reducing costs [7]. Therefore, NBS use two infrastructures effectively [4].

This section gives the overview of the concepts based on NBS for sustainable stormwater management in urban planning and design processes. Firstly, the concepts are defined and then, sample applications and solutions that can be applied in the urban area within the scope of sustainable stormwater management are mentioned. As a result, as a nature-based solution, the widespread adoption of sustainable stormwater management should be ensured to meet growing water demand.

2 Sustainable Stormwater Management

The goal of the sustainable stormwater management is to respond to water on-site as much as possible and reduce stormwater runoff. This concept aims to rebuild the natural water cycle. In this respect, green infrastructure, low impact development and water-sensitive urban design, which can be applied to any scale (district, city and neighborhood) [8–10], has come to forefront in recent years as sustainable stormwater management practices. The terms defined below are used worldwide to describe sustainable stormwater management approach (Tab. 1).

Tab. 1: The Prominent Concepts within the Concept of Sustainable Stormwater Management [5, 11]

Low Impact Development - LID (USA and Asia)	Addresses systems and practices that use or mimic natural processes that lead to infiltration, evapotranspiration or use of stormwater to meet water quality goals and associated aquatic habitat [9]. They are the applications that use the basic principles created by inspiration from the natural environment (12). They manage the flow of surface water and the use of urban water at the source, using the principles created by inspiring the natural environment.
Best Management Practices-BMP (Europe and USA)	BMPs are structural, vegetative or managerial practices used to prevent or reduce water pollution [13]. Yet, the term BMP is not only limited to natural processes, but also to technical solutions [14].
Water Sensitive Urban Design-WSUD (particularly Australia)	WSUD is an approach to planning and designing urban areas employed to reduce adverse effects of urban development on the natural water cycle [15].
Sustainable Urban Drainage Systems- SUDS (UK)	SUDS are specifically designed to manage the drainage of surface water through storage, slow runoff down before it enters watercourse and to improve the quality of the run-off [16].
Green Infrastructure- GI (USA)	GI applications mimic natural habitats. At the same time, they both absorb the excess water run-off, reduce the amount of pollution in the waterways and conserve water while maintaining water quality [17].
Decentralized Rainwater -DRWM (Germany)	Decentralized Rainwater/Semi-Natural Stormwater Management describes aims and techniques as to rainfall in urban areas [14].
Integrated Urban Resource Water Management- IURWM (Global)	Provides insights to identify the quantity and quality of water resources, to forecast the present and future evolution of water demand and the effects of climate change [18].

As mentioned in the literature, nature-based solutions offer integrative strategies for ecological, economic, social and cultural sustainability. NBS in stormwater management integrates the needs of sustainable stormwater management with urban design and rebuild natural water cycle [5].

3 Sustainable Stormwater Management Practices and Solutions in Urban Areas

Sustainable Stormwater Management sample applications used to decrease polluted stormwater runoff are stated in Fig. 1 below:

Fig. 1: Golden Gate Park, San Francisco, sustainable stormwater management practices [19].

Some of the proposed green infrastructure practices recommended on a local scale for the urban area are described and illustrated as follows [8, 20].

Fig. 2: Vegetated swales or biofilters, Minnesata, USA [22].

They are vegetated channels that contribute to water management by improving water quality and decreasing overflow [21] (Fig. 2).

Fig. 3: Bioretention areas, parking lots in Gadsden, Alabama [24].

These areas are shallow, landscaped depressions that direct runoff into the bioretention area. Collected stormwater is filtered through the soil within a few days [23] (Fig. 3).

Fig. 4: Constructed wetlands, Yuantian village, China [26].

They are shallow basis that includes water tolerant vegetation filled with soil or pebble along with low construction and operation costs. They are a clean-up system that uses natural processes to improve water quality [25] (Fig. 4).

Fig. 5: Sustainable transportation features, Cycling Embassy of Great Britain [28].

It is possible to improve drainage on streets ensuring the permeable paving for bike paths as well as green space planning [27] (Fig. 5).

They are best for pavement and parking lots. By adding curb cuts, runoff is directed into the vegetated areas [29] (Fig. 6).

Fig. 6: Curb breaks, NE Glisan St, Oregon, USA[29].

Tree wells that are often constructed in the sidewalk and near pavement and protected by tree grates or creating soil conditions can contribute significantly to stormwater management [30] (Fig. 7).

Fig. 7: Tree wells, Ottawa, USA [31].

It is an ideal method to reduce runoff volume through storage and to treat water quality. Permeable paving techniques include porous asphalt and interlocking pavers [32] (Fig. 8).

Fig. 8: Permeable pavers, Morton Arboretum, Lisle, IL, USA [33].

Fig. 9: Above ground cistern US Virgin Islands [35].

It harvests rainwater from rooftops and collects it using underground storage tanks [34] (Fig. 9).

Fig. 10: Rain gardens, Broken Hill Subdivision, Pewaukee, WI, USA [33].

They are designed to divert gutter water into rain garden and include water-tolerant plants [22] (Fig. 10).

Fig. 11: Roof gardens Salt Lake City, Utah, USA [36].

Roof gardens are specifically designed to create buildings in harmony with environment and have a big impact on biodiversity combining aesthetics with ecology. Green roof does not require drainage systems and collected stormwater could be used [30] (Fig. 11).

Fig. 12: Infiltration trench, Clark County, Washington [37].

Trenches could be applied to urban settings. They distribute the stormwater in a stone filled trench. Stormwater runoff is infiltrated into the ground after a couple of days [34] (Fig. 12).

Fig. 13: Underground stormwater chambers below parking lot [38].

They provide solutions for stormwater management. They store stormwater and divert it into public stormwater system or ensure underground infiltration [37] (Fig. 13).

Fig. 14: Rain barrel, Courtesy of the city of Oakland, California [39].

It stores rainwater from rooftop and collect it in tanks through downspout and pipes [34] (Fig. 14).

The advantages of sustainable stormwater management to cities were detailed as follows [5, 10]:

1. To protect and enhance water quality of surface and ground waters or improve water quality and natural water system
2. To protect natural resources and ecological processes (e.g., provision of low cost drainage—sewer system—and reduced infrastructure cost) and economic contribution to urban development on a long-term basis (protection-usage balance)
3. To prevent or minimize wastewater discharges to the natural environment
4. To prevent natural disasters such as flooding and thus spread of pollutants
5. Provision of ecosystem services ensuring local climate through the opportunity of evaporation and natural drainage
6. To integrate water into the landscape to enhance visual, social, cultural and ecologic values
7. To decrease carbon emission and build energy efficiency through stormwater management
8. To build a more green, well-kept and visual environment and thus create more attractive spaces for urbanized people and tourists
9. To add value to local character and society providing solutions particular to urban area.

4 Conclusion

Stormwater management is one of the global issues of the 21st century in terms of increasing pervious surfaces in urban areas, improving water resources and controlling floods caused by extreme weather and climate events linked to global warming. For this reason, management and conservation of urban water have become significant in the face of rapid urbanization. Past studies revealed that the traditional approach to stormwater management had yielded ineffective results for water quality enhancement. In this sense, sustainable stormwater management approaches are best practices for sustainable water development. Through the thoughtful implementation of sustainable stormwater management, it is possible to enhance water cycle, preserve water resources and ecological processes in cities, create more resilient cities against natural disasters (earthquake, overflow, erosion, etc.) and climate change with cost-effective solutions, promote visual and attractive spaces, develop local structure and maintain more resilient and livable cities in terms of visual, ecological and economic aspects [40,41].

To conclude, in today's world and future, stormwater management as a significant component of urban planning and design should be handled primarily. It is therefore of utmost importance to develop innovative practices in the related site to preserve and improve water resources in line with urban form and infrastructure [42]. In addition to that, it is suggested to adopt a participatory and multidisciplinary approach in cooperation with public authorities and institutions, private sector, civil society organizations and the community. As a nature-based solution, widespread adoption of sustainable stormwater management is crucially vital for the water demand, one of the most important issues of the future. Further, it is necessary to adopt an integrated approach, from subscale to upper scale plan, to stormwater planning [41] and close attention should be paid to water challenges in cities of today and tomorrow.

References

1 UNESCO-United Nations Educational, Scientific and Cultural Organization. *The Impact of Global Change on Water Resources: The Response of UNESCO's International Hydrological Programme.* 2011. Printed by Stipa. France [cited 11.10.2018]; Available from: http://unesdoc.unesco.org/images/0019/001922/192216e.pdf

2 IPCC-Intergovernmental Panel on Climate Change. *Summary for Policymakers.* 2018. Printed by the IPCC, Switzerland. ISBN 978-92-9169-151-7.

3 WMO-World Meteorological Organization. *Climate Breaks Multiple Records in 2016, with Global Impact.* 2017 [cited 02.09.2018]; Avaible from: https://public.wmo.int/en/media/press-release/climate-breaks-multiple-records-2016-global-impacts

4 WWAP-United Nations World Water Assessment Programme/UN-Water. *The United Nations World Water Development Report. Nature-Based Solutions for Water.* 2018. ISBN 978-92-3-100264-9. Published by the United Nations Educational, Scientific and Cultural Organization (UNESCO), Paris, France.

5 Hoyer, J., Dickhaut, W., Kronawitter, L. and Weber B. *Water Sensitive Urban Design Principles and Inspiration for Sustainable Stormwater Management in the City of the Future. 2011. ISBN 978-3-86859-106-4.* Published by jovis jovis Verlag GmbH, Kurfürstenstraße 15/16, D-10785 Berlin.

6 Yu, S. *Shaping the Sustainable City through Water Oriented Urban Design.* 2012. The Graduate School of the University of Florida in Partial Fulfillment of the Requirements, the Degree of Master of Science in Architectural Studies University of Florida. Gainesville, FL 32611 USA.

7 Tsegaye, S. et al., *Transitioning from Gray to Green (G2G)—A Green Infrastructure Planning Tool for the Urban Forest.* 2018. Urban Forestry & Urban Greening, https://doi.org/10.1016/j.ufug.2018.09.005

8 Güneş, M. *Yeşil Altyapı Kapsamında Yeşil Ağ Planı ve Kent Kimliği Etkileşiminin İrdelenmesi: Ankara Cumhuriyet Dönemi Sınırı Örneği.* 2017. Ankara Üniversitesi Fen Bilimleri Enstitüsü Peyzaj Mimarlığı Anabilim Dalı. Ankara.

9 EPA-United States Environmental Protection Agency. *Urban Runoff: Low Impact Development* [cited 25.10.2018]; Available from: https://www.epa.gov/nps/urban-runoff-low-impact-development

10 The European Climate Adaptation Platform (Climate-ADAPT). *Water Sensitive Urban and Building Design (2016)* [cited 04.09.2018]; Available from: https://

climate-adapt.eea.europa.eu/metadata/adaptation-options/water-sensitive-urban-and-building-design

11 Barbosa, A. E., Fernandes, J. N., David, L. M. *Key Issues for Sustainable Urban Stormwater Management.* 2012. Pages 6787–6798.

12 Pima County Regional Flood Control District. *Design Standards for Stormwater Detention and Retention.* 2014. Runoff Detention Systems Floodplain and Erosion Hazard Management Ordinance. Supplement to Title 16, Chapter 16.48.

13 Stormwater Management: Best Management Practices (BMPs) [cited 10.09.2018]; Available from: https://www.michigan.gov/documents/deq/ess-nps-savvy-bmp_209386_7.pdf

14 The Stormwater Experts. *Terms of Rain Water Management.* Available from: https://www.sieker.de/en/fachinformationen/dealing-with-rainwater/article/begriffe-der-rw-bewirtschaftung-77.html

15 Kuller, M., Bach, P. M., Ramirez- Lovering, D., Deletic, A. Framing *Water Sensitive Urban Design as Part of the Urban Form: A Critical Review of Tools for Best Planning Practice.* 2017. Environmental Modelling & Software. Volume 96, Pages 265–282.

16 British Geological Survey [cited 04.07.2018]; Available from: https://www.bgs.ac.uk/research/engineeringGeology/urbanGeoscience/suds/what.html

17 EPA-United States Environmental Protection Agency. *Stormwater Management and Green Infrastructure Research* [cited 10.09.2018]; Available from: https://www.epa.gov/water-research/stormwater-management-and-green-infrastructure-research.

18 Bahri, A. *Integrated Urban Water Management.* 2012. Global Water Partnership Technical Committee (TEC) Background Papers, No. 16, ISBN: 978-91-85321-87-2. Printed by Elanders

19 The San Francisco Bicycle Coalition. *Connecting the City* [cited 07.09.2018; Available from: http://www.connectingthecity.org/routes/##SFBC##

20 EPA-United States Environmental Protection Agency. *Stormwater Management Practices at EPA Facilities* [cited 25.10.2018]; Available from: https://www.epa.gov/greeningepa/stormwater-management-practices-epa-facilities

21 Tennessee Stormwater Management. *Management and Design Guidance Manuel.* 2015 [cited 12.10.2018]; Available from: http://tnpermanentstormwater.org/manual.asp

22 Minnosata Stormwater Manual. *Vegetated Filter Strip* [cited 10.09.2018]; Available from: https://stormwater.pca.state.mn.us/index.php?title=File:Vegetated_filter_strip_4.png

23 Müftüoğlu, V. 2016. *Kentsel Tasarım Rehberlerinin Peyzaj Mimarlığı Açısından Ekolojik Çerçevede İrdelenmesi Üzerine Bir Yöntem Araştırması.* Doktora Tezi. Ankara Üniversitesi Fen Bilimleri Enstitüsü Peyzaj Mimarlığı Anabilim Dalı. Ankara.

24 Michigan State University Extension. *Either Detention or Retention Equals Stormwater Runoff Reductions.* [cited 12.10.2018]; Available from:http://www.canr.msu.edu/news/either_detention_or_retention_equals_storm_water_runoff_reductions.

25 EPA-United States Environmental Protection Agency. *A Handbook of Constructed Wetlands.* 2015 [cited 25.10.2018]; Available from: https://www.epa.gov/sites/production/files/2015-10/documents/constructed-wetlands-handbook.pdf

26 WWF-World Wide Fund for Nature. *Constructed Wetland Does More Than Spruce up the Garden* [cited 10.09.2018]; Available from: http://wwf.panda.org/?195532/constructed-wetland-does-more-than-spruce-up-the-garden

27 The Blog by Copenhagenize Design CO. *Bicycle Superhighways in Copenhagen Capital Region* [cited 11.10.2018]; Available from: http://www.copenhagenize.com/2017/06/

28 Cycling Embassy of Great Britain. *My Dream Cycleway.* Available from: https://www.cycling-embassy.org.uk/blog/2014/08/06/my-dream-cycleway

29 Wilkes East Neighborhood. *Stormwater Planters, Improving Our Environment and Our Neighborhoods*[cited 04.08.2018]; Available from: http://www.wilkeseastna.org/node/628.

30 Müftüoğlu, V. 2012. *Peyzaj Mimarlığında Mühendislik Uygulamaları 1 Ders Notları (Drenaj-Yüzey Akış Hesapları ve Düşük Etkili Gelişim).* Ankara Üniversitesi Ziraat Fakültesi Peyzaj Mimarlığı Bölümü.Ankara.

31 Landscape Online. *Protecting Tree Growth and Fertility in the Streetscape: Chicago's Navy Pier, Delta City Centre, Ottawa and Hatcher Road, Phoenix.* cited 09.09.2018]; Available from: http://www.landscapeonline.com/research/article-a.php?number=27514

32 Washington, D.C. Department of Energy & Environment. *Permeable Pavement Systems* [cited 12.10.2018]; Available from: https://ddoe.dc.gov/sites/default/files/dc/sites/ddoe/publication/attachments/Ch3.4PermeablePavement_0.pdf

33 Town of Delafield. *A Perfect Environment. Examples of Various Stormwater Management Practices* [cited 10.10.2018]; Available from:http://www.townofdelafield.org/pdfs/stormwater/Examples%20of%20Various%20Storm%20Water%20Management%20Practices%20SM.pdf

34 National Service Center for Environmental Publications (NSCEP). 1999. *Low-Impact Development Design Strategies An Integrated Design Approach.* Prince George's County Development of Environmental Resources Programs and Planning Division, 140 p., Maryland, USA.

35 Christopher J. Webb, P. E. 2007. *Cisterns for Stormwater Detention, Rainwater Cisterns for Stormwater Detention, Rainwater Storage, and Re Storage, and Re–Use, Seattle, WA.* Chris Webb and Associates, Inc., PS Bellingham, Washington.

36 Green Roofs. *Church of Jesus Christ of Latter-Day Saints (Lds) Conference Center* [cited 04.08.2018]; Available from: http://www.greenroofs.com/projects/church-of-jesus-christ-of-latter-day-saints-lds-conference-center/
37 Minnesota Stormwater Manual. *Best Management Practices (BMPs) for Stormwater Infiltration* [cited 25.09.2018]; Available from: https://stormwater.pca.state.mn.us/index.php?title=BMPs_for_stormwater_infiltration
38 Water World Magazine. *Underground Solutions for Stormwater Management* [cited 10.09.2018]; Available from: https://www.waterworld.com/articles/uwm/articles/print/volume-2/issue-3/features/underground-solutions-for-stormwater-management.html
39 Santa Clara Valley Urban Runoff Pollution Prevention Program. *Rain Barrels and Cisterns Stormwater Control for Small Projects* [cited 25.10.2018]; Available from: http://www.mywatershedwatch.org/wp-content/uploads/BASMAA_Rain_Barrel_Fact_Sheet_082312_APPROVED_online_ver.pdf
40 Natural England. *Green Infrastructure Guidance.* 2009 [cited 10.10.2018]; Available from: http://publications.naturalengland.org.uk/publication/35033.
41 The Community for Sustainable Drainage-Susdrain. *Water Sensitive Urban Design in the UK –Ideas for Built Environment Practitioners.* ISBN 978-0-86017-726-5. Published by CIRIA, Classic House, 174–180 Old Street, London EC1V 9BP (CIRIA project RP976).
42 The Government of Western Australia, The Department of Water and Environmental Regulation. *Urban Water Design* [cited 12.08.2018]; Available from: http://www.water.wa.gov.au/urban-water/urban-development/urban-water-design

Murat Akten, Ayhan Akyol, and Sibel Akten

25 Theoretical Approaches on Aesthetics and Urban Aesthetics Concept

Abstract: Urban aesthetics has been the concern of many academic researches, and there have been now more than hundred definitions of urban aesthetics. It is crucial to understand that aesthetics is more than just about the form and physical qualities of a place. Likewise, urban aesthetics is not only the concerns of academics, but also the concern of urbanites of cities and daily users of urban space.This research is based on the assumption that it is important to understand how the city people sense the aesthetics concept and how public spaces they prefer visually and aesthetically besides academic studies and urban places theories upon aesthetics.

Keywords: Aesthetics, Urban aesthetics, Urban public space, Urban identity

1 Introduction

What distinguishes human from other living things is the cultural structure that it owns. That cultural structure, which privileges human from other creatures, reserves many components including aesthetics. Hence ones such as the way of people living, value judgments, belief systems, etc., provide information about their aesthetics judgment structures, as well as their cultural structures. While societies constitute a set of aesthetic values according to their cultural structures, on the other hand, these aesthetic values becomes determinative factors for their "aesthetic judgments."

The universal dimension of the aesthetic judiciary depends on the conditions that will fundamentally enable the environment to be aesthetic. As a second step, some characteristics related to the history of that culture can be exhibited. These can be sculptures, architectural elements, or anonymous and historical forms within the city. But before such details, implementations which promise that people will be able to move freely, breathe, see, walk, stop, have a safe relationship with traffic, find their way easily, and make signs easier to read will be more prominent aesthetic conditions than all ornaments and artificial designs (1). Although aesthetics in the city varies according to the characteristics of the user population, it also gets formed as the beauty and its reflections on human emotions (2).

The concept of environmental aesthetics, which was previously related to nature and nature arrangements, has included also the urban spaces which are the scene of daily life, in the field of research and discussion since the second half of the 20th century. As urban space becomes one of the main research subjects of environmental aesthetics, it is seen that the relationship between environmental quality and public interest is one of the initial points of environmental aesthetics.

Today, environmental aesthetics, which focuses on environmental quality and the concepts of public interest, is observed to be divided as to research objects and developed in different directions (3).

In the modern cities of today, the feeling of incarceration in place and time leads to semantic shifts, snatches urban spaces from human values, and gradually makes human feel strange towards its environment. The estrangement of the people who cannot communicate in the cities built by the modern technology, the feeling of being out of place, losing their sense of belonging, makes it necessary to be more sensitive to facts which needs today's environmental aesthetics and ethical responsibility.

For that purpose, there must be basic assumptions in physical planning and design dimension of urban aesthetics. The subjects such as the occupancy ratio which will be in harmony with the traditional texture, and the specialties such as the color, texture, form, etc., to be used in new building facades and urban reinforcements can be mentioned among these assumptions. As urban planning and design affects human physical health, urban aesthetics affects human mental health (2).

This research is based on the assumption that it is important to understand how the city people sense the aesthetics concept and how public spaces they prefer visually and aesthetically besides academic studies and urban places theories upon aesthetics.

2 About Aesthetics Concept and Its Historical Process

2.1 Aesthetics Concept and Aesthetics Science

It is necessary to investigate the meanings of that word, its concepts, ideas suggested about aesthetics and extant "aesthetics concept" through the historical process to understand the concept of aesthetics science indeed. Considering aesthetics as a science, it is too hard to reach at a definition which everybody can agree with. However it is not possible to make a certain definition consisting of its all meanings; it is thought that there may set down a concrete view of its whole meaning in mind as scrutinizing from word to definition and considering it together with intellectual history.

The word "aesthetics" comes from the ancient Greek word of "aisthesis" or "aisthanesthai." As "Aisthesis" means "sensation" and "sensory perception," "aisthanesthai" means "being perceived by sense." Aesthetics, as a sensible perception, is thought of as a science of information provided by sentimentality (4).

Before that definition by Alexander Baumgarten (1714–1762) who gives the name to that word, the concepts such as beauty and aesthetics become a source of inspiration for lots of artists and had a chance to be argued by many works of art. Before and after it reveals as a science, its interest area sometimes was narrowed and sometimes widened and changed by the assigned meanings during that time.

Conceptual definition of *aesthetic* is expressed in different ways (5):

- *Aesthetic* is a form of reflexive thought. In other words, it is the human mind contemplating about its own action which allows it to create all the temples, cathedrals, palaces, statues, paintings, melodies, symphonies, and poems.
- The perspective of the artist, who forms the beauty or the beauty of the work of art, is the way he articulates what he says. This means the transformation of reality existing with man into artistic reality and the creation of aesthetic beauty except the beautiful or ugly person. Thus, *aesthetics* is the investigation of the art event and the evaluation of the artistic works in the works of art.

Every definition about aesthetics consists of "beauty" and its own values. As understood from its definitions, "beauty" is an inseparable part of aesthetics and its issue at the same time. But the concepts of "beauty" and "aesthetics" should not be jumbled together. Thus, as defining aesthetics, it is necessary to firstly determine the meaning of "beauty" and the point which it takes part in aesthetics. Hence the sense and sensation expression which is actually expressed by aesthetics sometimes loose their meaning and leaves its own place to word of "beauty." In many of sources, aesthetics concept is defined as the science of "beauty" at art and nature. But aesthetics appreciation matches up with not only the concept of "beauty."

As mentioned above, while the aesthetic appreciation is based on a universal basis, the concept of beauty is limited to the taste of a period, a nation, or a group. Therefore, "the Science of Beauty" approach leaves some absent points in the definition of aesthetics.

2.2 Aesthetics as a Science

"Aesthetics" is based on a 2500-year history as a concept. The first attempts were made by the Greek philosophers of the Pythagorean school to understand the essence of "Beauty." The three famous philosophers of Ancient Greece, Socrates, Plato, and Aristotle produced various ideas about the essence, beauty, and perception of being. For the first time, the question of "what is beautiful?" was asked by Plato (427–348). Plato searched for the relationship of beauty with other concepts such as goodness, benefit, and harmony in his various writings. Aristotle (384–322), Plato's student, seeks the rules of beauty in material structure, in this world. According to Aristotle, the beauty is the symmetry and limitation. This approach is quite mathematical and so far from today's aesthetics concept. It is a recent story that the retroject aesthetics was conceptually named as science.

2.2.1 Classical–Traditional Aesthetics

It is Alexander G. Baumgarten (1714–1762), one of the students of Wolff, who created the science we say as aesthetics and named it that way. Alexander G. Baumgarten, for the first time, grounded such a science, determined its topic and drew the limits of that science by his *The Book of Aesthetics* which was published between the years of 1750–1758. Even in 1735, in his doctoral thesis named as “Some Philosophical Thoughts on Poetry,” he mentioned the possibility of such a science. The word of *Aesthetics* was mentioned as science name in that book for the first time. In other words, the identification of such a science takes place with the publication of *Aesthetica* (4).

Baumgarten explains the aesthetics in *Aesthetica* such as: Aesthetica is theory of the liberal arts, gnoseology of the lower faculties, the art of beautiful thinking, and the science of sensate cognition. In such a description, for the first time, we come across with a definition of science that we call aesthetics. All these different elements tending to determine aesthetics can be motivated by a basic determinant; this basic determinative motivation is cognition sensitiva, that is, sensory cognition.

The sensitive cognition and sensitivity as basis of the aesthetics, occupies a large wide area. Whenever aesthetics is considered as only “beauty,” “the art of thinking about beauty,” it will be placed in a narrow meaning as being so far from sensitive cognition.

The sensitive cognition and sensation which form the ground of aesthetics spread a wide large area. When the aesthetics is considered as only “beauty” and “the art of thinking about beauty,” it will be taken far away from its other sensitive concepts and be squeezed into a narrow meaning. As mentioned before, sensation consists of values such as funny, sublime, elegant, tragic, interesting and ugly in the meaning of aesthetics defined with sensitive concepts as well as it includes beauty. All of these concepts contribute to value development of aesthetics as much as beauty concept and they are to be considered together with beauty.

Today Ludwig Wittgenstein opposes understanding aesthetics as a “beauty science.” Wittgenstein says that aesthetics is a large wide concept but the word of “beauty” easily leads to misunderstandings. As to him, the reason for leading these misunderstandings is that a science such an aesthetics having a wide area is restricted as being tied up to only “beauty.” Then aesthetics should not be grounded as just a beauty science for our today’s mentality.

All opinions suggested about aesthetics and narrowed or widened definition of “aesthetics” become definite by Kant and Fr. Schiller’s final word, and it gets its concept as “aesthetics,” “aesthetics science or things related to aesthetics science.”

2.2.2 Contemporary Aesthetics

The aesthetics suggested by A.G. Baumgarten as a science in the 18th century were discussed by Kant, Fr. Schiller, Karl Rosenkranz, Wittgenstein, Hegel, and many other thinkers, and as a trace of classical-traditional aesthetics, not just as a science of beauty, but it is accepted as a trace left in the minds of sensible perceptions. However, by the 20th century, the Italian philosopher Benedetto Croce (1866–1952) expressed new opinions about aesthetics.

As a science, aesthetic history and aesthetic problems do not jibe with each other. These heterogeneous problems also distinguish between classical-traditional aesthetics and contemporary aesthetics at one point. Traditional aesthetics finds aesthetic research in either beauty or art. However, at the beginning of our century, an aesthetic view opposed to such an aesthetic approach was proposed. This view is B. Croce's aesthetic basis. According to Croce, the subject of aesthetics is sentiment which has a plenary entity. Sentiment is the most simple cognition form that comes before conceptual cognition and provides its basis. Perceptions provide us the real; however, sensations present imaginary one as well as the real ones (4).

As to Croce, the judgments on art works are not the attributions of joint judgments, because the art works are physical appearances of an expression (beauty) arising for a moment in artist's soul. Every expression in the way of art contains an individual specialty. Thus evaluations of each person about that way of expression may be different. Then there cannot be an agreed aesthetics judgment. As there may be different perceptions of aesthetic judgments as to differences of taste and culture owned by persons, there also may be different perceptions since persons are different physiologically. Art work is spiritual acquisition, that is, a product of a subjective experience. Similarly, the person looking at that work will also look at it as a result of his own subjective experience (6).

However aesthetics concept being tried to set in determinative boundaries could not have a chance to express itself actually and traditional-classic aesthetics concept has attracted more than contemporary aesthetics. As considering the modern aesthetics approach, the meaning Croce assigns to aesthetics differs from Baumgarten and Plato at essential points. Plato grounds aesthetics on only sensation and accepts it as far from reality. Also in Baumgarten's approach, aesthetics is considered as a tool which tries to reach competent cognition with logic in gnoseology of the lower faculties. As to Croce's aesthetics, the subject of aesthetics is perceived only as a spiritual activity in intuition and expression, and agreed aesthetic judgments are rejected (5).

2.2.3 Tomorrow's Aesthetics

Tunalı (1996) describes the tomorrow's aesthetics as: Tomorrow's aesthetics will not be a luxury science for persons and society anymore, not like the traditional aesthetics; contrary to that, it will be a science that examines the human being in every form and will be "plenary universal" as depending on human, society, and technology but at the same time will be "popularis." However it is hard to plan our contemporary aesthetics in these norms, maybe there may lead to forming a new different way of aesthetics values as planning future's aesthetics.

As based on the definition of Tunalı, it is possible to address the tomorrow's aesthetics in terms of philosophical aesthetics. Philosophical aesthetics defines the "structural elements" constituting this quality in order to understand the aesthetic reality and to comprehend the integrity of this reality. Each aesthetic fact is related to a subject, that is, agent, and it is there for the taste of the agent (5).

According to Wittgenstein, the problem that aesthetics needs to investigate is the question of aesthetic judgment. Often, aesthetics is said to be a branch of psychology. Aesthetic problems usually have nothing to do with psychological experiments. Of course, there are emotions experienced by man in taking an aesthetic attitude, but examining them psychologically does not mean to examine them aesthetically. The explanation that psychological research wants to reach is a causal explanation, because psychology is a positive natural science. On the other hand, "the aesthetic explanation is not an explanation based on causality," because it is not an experimental, positive natural science; it is a philosophy science and the subject of that philosophy is directly the aesthetic judgments. Aesthetic science, which investigates aesthetic judgments, can never be explained as relating with the positive natural science, psychology (4).

In Wittgenstein's approach, the common opinion is that the object contributes to the formation of aesthetic value rather than the subject. The research subject of aesthetics is the "aesthetic judgment" and the aesthetic explanation does not include "causality."

As to that approach, "beauty" becomes an inseparable part of aesthetics value. The common point of taking a subjective and objective attitude also is value of "beauty." The concept of "beauty" as a value that participates in the aesthetic fact is an idea, and makes the aesthetic phenomenon absolutized and limited. The association of subjects and objects is provided not only in the "beauty" concept but also in the "judgment concept." Because "being beautiful" or "not being beautiful" which assigns to aesthetics value is a judgment, and it is the crossing point of both approaches too.

Tunalı (1996) summarizes the aesthetics concept explained in detail above as follows: Thus four basic structure elements were defined as to ontic (logic) integrity of aesthetics concept. These are:

- Aesthetics subject,
- Aesthetics object,
- Aesthetics value or beauty, and
- Aesthetics judgment

Aesthetics concept or aesthetics existence forms by the ontic integrity of these four elements. Here, that integrity forms the subject of philosophical aesthetics. Philosophical aesthetics is not only a psychology of aesthetic emotions, nor an art philosophy, nor beauty philosophy, nor aesthetic judgements. The concept of philosophical aesthetics is above all in terms of its content. Philosophical aesthetics surrounds and embraces the whole sphere of existence. The task of philosophical aesthetics is to investigate this aesthetic existence in terms of ontic elements.

3 What Is Urban or Environmental Aesthetics?

The aesthetic, defined by the process of sensing the formal specialties of the existence, when it considers not only appreciating the beautiful one or subjective appreciation, and also, in a general meaning, the way and process of human perception in the relationship with nature, the basic field of inquiry becomes an act of

Tab. 1: Evaluation of Yesterday, Today, and Tomorrow's Aesthetic (4,5)

	Traditional Aesthetics	Contemporary Aesthetics	Tomorrow's Aesthetics
Philosophers	• Baumgarten • Hegel • Kant	• Croce	• Tunalı
Subject	Subject contains the philosophy of beauty and the philosophy of art	As a subject of aesthetics, it focuses on sentiment concept that is a plenary existence.	The subject will be to search for the aesthetic reality and structural integrity and to try to base this integrity on a philosophical and ontological basis with a philosophical aesthetic.
Research Area	In parallel with logic investigates the competence of the intellectual cognition and facts, aesthetics investigates the accuracy of the sensitive cognition, i.e., the beauty. As to that, as beauty is the verification of sensitive cognition, aesthetics is considered as the logic of the sensitive cognition.	Sensation and expression are spread over a wide range from daily sensations and expressions to art sensation and art expressions. It is the aesthetics science that explores this broad field of sensation and expression.	It makes the philosophical aesthetics a research subject as investigating all human forming ways, and using a plenary universal way depending on human, society and technology in the light of being human-specific science.
Problem	The traditional aesthetic finds the aesthetic research field either in beauty or in art. This definition leaves missing points in reflecting the perception of aesthetic value.	Grounding of such a plenary expression science has not been effective much and the aesthetics have continued to be understood in the traditional understanding way.	For today, it is impossible to plan such an aesthetic because it has not yet acquired such an aesthetic existence.

Solution: As forming a new aesthetics and a new system of values:
- Transforming aesthetics into a reality as keeping it away from being a luxury science
- Examining all forms which human shapes
- Depending on human, society, and technology
- Leading to plenary and universal science

environmental and environmental interpretation beyond the theory of appreciation for art and an art work. In this case, aesthetics becomes a field of thought and action that is formed by social processes rather than being a kind of individual practice in the classical sense. The factor transformed to an act by aesthetic thought is the very environment concept. As a matter of fact, the environment concept directly refers to the urban environment as an artificial environment as well as the natural environment. As the urban environment is the result of the consciousness of shaping the environment differently from the natural environment, the concept gains an operational content. In this sense, environmental aesthetics offers an alternative perspective in the framework of aesthetic conceptualization for cities and urban spaces which are the largest works of art produced by human settlements and common civilization, rather than offering a judgment of appreciation for the natural environment. That perspective has to be an indispensable action area of urban planning and design, which is the highest stage of humanity since the enlightenment in urban space production process (technical and intellectual)(1).

It is an indisputable fact that awareness, cognition, and sensitivity level for that stage are not sufficient in terms of theoretical and practical aspects in our country's planning and architecture. In fact, in almost all urban and architectural plans produced, environmental aesthetic concepts are not among the basic principles and criteria that form the plan. We should emphasize that critical discourse against those is not fictionalized by an aesthetic point of view. This is closely related to the fact that the conceptual perspective level of our planning practice needs to be improved. Urban aesthetics has been the subject of many academic researches and hundreds of different definitions have been made. In the study of urban aesthetics, it is necessary to understand that aesthetics have more meanings than form. At the same time, urban aesthetics does not only concern academic environment, but it also concerns the citizens and daily users of urban spaces. This research is based on the assumption that it is important to understand how the citizens sense the aesthetics concept and what kind of public spaces they prefer visually and aesthetically, as well as academic studies and urban space theories related to aesthetics (7).

The analysis of the cities in the historical process provides some clues about urban aesthetics. These can be summarized as follows (8);

- Urban aesthetics, as considering macro scale, depends on the cultural accumulation of the soil structure. It is generally classified as "pre-industrial city" and "industrial city."
- The most effective factor forming the urban aesthetics has been natural environment. Vegetation, geomorphology, and climate are the elements that generate the natural environment. These elements play an important role both in the selection of the places where the cities are established and in forming the cities. It was emphasized that these elements should be compatible with each other in terms of urban aesthetics.
- Urban environment comprises of structures and spaces that these structures define. The harmony and relations between them determine the specialty of

urban aesthetics. The roads, which are considered as positive elements, come together and form negative elements which are outdoor spaces. Negative elements are defined as open spaces along with circulation systems. Roads and squares form parts of the circulation system, and breathing spaces of buildings such as courtyard, front yard and side garden, and green areas constitute parts of open spaces.

The strokes forming the surfaces of negative and positive subjects; texture, material, color, that is, the facades and coverings at outside spaces are the important components of urban aesthetics. Outside strokes forming facades are the ratios of doors and windows, vertical and horizontal strokes, structural material, color and elements of decoration. On the other hand, the outside aesthetic subjects are floor covers, urban and natural landscape materials, and urban furniture. All of these factors come together and form the macroform and silhouette related to whole city. The beauty of the whole city depends on the beauty of the city parts separately, as well as the beauty of their relations with each other.

4 Relationship between Urban Aesthetics and Urban Identity

After the industrial revolution, the scale and form diversity in urbanization were observed and cities grew; as the new functions appear, small-scale cities being able to be perceived at a glance disappear and instead industrial cities lodging large populations show up. Urban centers have grown and a period of aesthetic confusion has emerged as a result of the use and function-oriented applications. As a reaction to those new applications deprived of aesthetics, new urban approaches such as "city beautiful," "garden city," "industrial city,"and "city efficient" have come to the stage, and there passed a period of dissolution and contradictions in the context of both urban structuring and environmental and urban aesthetics (9).

When the historical development of urban aesthetics is evaluated, the most important result obtained is that the most important factor constituting the urban aesthetic is natural environment through all history. It is seen that the subjects being the natural environment, components such as geomorphology, topography, vegetation, climate, water presence, etc., play an important act in both choosing the place which cities were built and forming the cities. The aesthetic values carried by the cities that are in harmony with these elements are at a higher level. The cities which are well integrated with its environment in terms of materials, structuring, and natural elements have more specialized environments. Urban aesthetics is a phenomenon that is directly related to the social, cultural, technological, and economic structure of the period in which it was produced as it is in urban structuring. In addition, it is observed that the periodic characteristics of the buildings and their facade features are effective in urban identity. It is seen that the use of natural building materials is the most important factor in environmental harmony and the use of outdoor spaces defined in urban usage, as a social requirement, has been found in space installation in different periods and cultures as including subjects such as squares, forums, and courtyards throughout the historical process.

Today, the idea that human beings and the environment are an indivisible whole emphasizes that man-made environment should not only be a result of personal, social, and cultural differences, but also a product of the ecosystem. This approach aims to ensure the biodiversity and quality of life on a local and global scale by reducing the effects of activities in the cities on natural resources and the environment (10). Thus it aims to leave healthy, clean, and sustainable cities to next generations as relating human, nature, and city with each other (11). However urbanization policies, which are followed in the direction of sustainable development and development, often confront the economy with ecology and lead to the destruction of the natural environment. Although ecology is one of the most important subjects that should be taken into consideration in the creation of urban environment, urban aesthetics and urban identity are related to that. For this reason, priority should be given to policies aiming at sustainability in urban areas, protecting and improving the natural environment and aiming to increase the quality of life by providing an ecological identity to the city (12).

Many people in the world want to see cities with a unique aesthetic appeal with their natural, cultural, and built environment or visit the settlements having urban image more. As determining its own cultural and aesthetics specialties for each city, featuring those specialties and paying attention to preserve them are also important for the cultural sustainability of that city (13). At this point, the questions like "what does the existence of a city's identity mean?" or "what kind of concepts they contribute to city?" come into mind. The identification of a city's identity allows it to be recognized by distinguishing it from other cities. The inhabitants of the city have the consciousness of living with their environmental values that reflect their culture and surround them in their preferred living areas. The governors control all investment and infrastructure projects to be made in the city in order not to destroy the elements that give identity to the city. In addition, governors create a city-specific growth model to ensure the sustainability of the city identity in the future. As a result, it may be mentioned that the social awareness and urban consciousness of the people living in the city developed (14), and there is an increasingly peaceful and healthy urban living environment.

Such an approach having those specialties should have a basic design principle that accepts the human scale as basic data and aligns the human movement in the environment with the environment in a certain design setting. In parallel, the production of symbolic value that will refer to collective memory and enable the individual to produce new concepts should be another area of action of urban aesthetics. Thus, while the physical relationship with urban space creates new awareness in spiritual sense, getting strange which criticism focuses on is also prevented.

At this point, today's criticism made about the aestheticization experience of urban life roots in the dominant aesthetic point of view that perceives the city as a stage decor which encourages deep socio-economic distinctions as the main reason for the alienation of the citizen to the city and urban space.

5 Conclusion

Considering the environment as an aesthetic whole and to perceive the human as an integral whole with its environment constitute the main philosophy of environmental aesthetics (1). If the aesthetic is related to the whole perceived by the human senses, then the environmental aesthetics is based on the idea that the human being is dependent on him as a person who perceives his surroundings as his senses and his body. In addition to being a part of nature, human integrates with his perceptions; it is a subject who experiences more than a passive audience and actively participates in it. In this respect, environmental aesthetics is closely related to perception phenomenology and the theories of perception and cognition of psychology. The use of these theories in the formation of the theoretical framework of urban design should be the main reason for the conceptual proximity between environmental aesthetics and urban design theories(15).

As a result of the aesthetic evaluation of the environment, the individual determines the aesthetic attitude towards the environment and spatial behavior is shaped in the direction of this attitude. Based on the aesthetic evaluations of individuals about the environment, it is possible to obtain clues to which emotions and moods of individuals and to which spatial behaviors they lead. As a result of reflecting the obtained data to environmental designs, it can be ensured that the satisfaction of the environment and the harmony of the environment-behavior can be achieved.

Throughout history of humanity, societies have shown their cultural background in cities. The cities which people present to history by using their regulatory, constructive powers have gained importance not only with their formal characteristics but also with their social and cultural values.

In recent years, it is possible to say that there are big projects for which marketing is the priority and are far away from the integrity, depth, planning strategy, and horizon width in our cities. While large cities do not have upper scale plans and the approach and content of upper scale planning studies are not discussed, especially in Istanbul, and in our other big cities, big projects which are brought into the agenda by forgetting their place in the urban system with their production, employment, and social qualities should be examined especially. The fact that our natural, cultural assets and values which are to be protected are sacrificed to short-term economic returns on the basis of political authority and investor determination without any social negotiation and reconciliation processes stands before us as the most important priority and problem area.

References

1 Erzen, J., 2006. Çevre Estetiği. ODTÜ Geliştirme Vakfı Yayıncılık ve İletişim A.Ş. Yayınları, ISBN: 975- 7064-96-3, Ankara, Türkiye.

2 2. Çelik, D., Açıksöz, S., 2008. Tarihi Kentlerde Kent Estetiği Kaygısı: Bartın Örneği. Bartın Orman Fakültesi Dergisi Cilt:10 Sayı:14.

3 Koramaz, E.K., 2014. Kentsel Çevrede Estetik Değerlendirme ve Mekansal Davranış. İstanbul Ticaret Üniversitesi Fen Bilimleri Dergisi. Sayı: 26, s. 141–160.

4 Tunalı, İ., 1996. Estetik. İstanbul: Remzi Kitabevi.

5 Coşkun, T.D., 2008. Kent Estetiği Oluşumunda Yer Oluşturma İlkelerinin Koblenz/İzmir Kentleri Bağlamında Karşılaştırmalı Olarak İrdelenmesi. Dokuz Eylül Üniversitesi Fen Bilimleri Enstitüsü Yüksek Lisans Tezi, İzmir.

6 Ceylan, T.M., 1988. Fotoğraf, Estetik ve Görüntü Üzerine Denemeler. İstanbul: İfsak yay.

7 Pehlivanoğlu, Y., 2011. Understanding Perceptions Regarding the Aesthetics of Urban Public Space: Tunalı Hilmi Street, Ankara. A Thesis Submitted to the Graduate School of Natural And Applied Scıences Of Middle East Technical University.

8 Günay, B. ve Selman, M., 1994. Kentsel Görüntü ve Kentsel Estetik Örnek Kent:Ankara. Kent, Planlama, Politika, Sanat. ODTÜ Mimarlık Fakültesi Yayını.

9 Erdoğan, E., 2006. Çevre ve kent estetiği. ZKÜ Bartın Orman Fakültesi Dergisi, 8 (9), 68–77.

10 Görgülü, Z., 2009. Sürdürülebilirlik, Planlama ve Politikalar. Uluslararası Ekolojik Mimarlık ve Planlama Sempozyumu, 22–25 Ekim 2009, Antalya, Türkiye.

11 Whitehead, M., 2003. (Re) Analysing the Sustainable City: Nature, Urbanisation and Regulation of Socio-Environmental Relations in the UK. Urban Studies. 40 (7): 1183–1206.

12 Aytis, S., Altuncu, D., Polatkan, I., 2010. Kentlerde Ekolojik İlkeler Çerçevesinde Sürdürülebilir Planlama. GreenAge Symposium, Mimar Sinan Fine Arts University, Faculty of Architecture 6–8 December 2010, İstanbul, Türkiye.

13 Göğebakan, Y., 2012. Estetik Ve Kentsel Yapılanma Üzerindeki Etkisi. İnönü Üniversitesi Sanat ve Tasarım Dergisi. Cilt/Vol. 2 Sayı/No.4 (2012): 73–84.

14 Demir, C., 2006. Kent Kimliği Geliştirme Sürecinde Mekansal Model Tasarımı ve Kent Plancılarının Rolü. PLANLAMA 2006/3, s.11.-121, İzmir.

15 Bilsel, C., 2010. Urban Design and Environmental Aesthetics (Kent Tasarımı ve Çevre Estetiği). Union of Chambers of Turkish Architects and Engineers (TMMOB) Union of Architects Ankara Branch Dosya Magazine, 23: 6-

Assist. Prof. Dr. Nurgül Arısoy

26 Child-Friendly City Approaches

Abstract: Urban and public open spaces are the places where the individuals come together and enter into social interaction. Accurate and sustainable design of urban space is a priority for creating a healthy environment for the individual. Increase in traffic in cities, the insecurity of outdoor spaces cause children to spend more time indoors. Studies have observed that the games played outdoors support children's physical, spiritual and social development and creativity more than playing indoors. In this case, it is expected that the urban space meets the socialization environment for play which is the most basic need of the child and his self-recognition. In this sense, a number of improvements are being made in order to reuse the open spaces that are changing in terms of children's use. In this chapter, the concepts and initiatives of a child-friendly city, which are created in order to make the city more livable for children, are mentioned.

Keywords: Child, Play, Child-friendly city, Playing areas, Child-friendly environment approaches

1 Introduction

Children need spaces where they can experience and reflect their original experiences. Urban spaces are an indispensable necessity for the physical and mental development of children, and hence it is very necessary to create healthy and lively urban spaces.

The physical environment in which a child interacts has great importance for socialization and development. Children interact with their environment differently than adults do, so the environment has a significant role on every child's mental, physical and social development. Play is necessary for healthy development of the child. Play promotes cognitive, physical, social and emotional well-being, thus offering the necessary conditions for children to thrive and learn. Through play, the child can experiment, solve problems, think creatively, cooperate with others, etc., gaining a deeper knowledge about him/herself and the world [1]

In the 1977 Declaration on the Rights of the Child, which was announced to commemorate the World Children's Year, it was mentioned that the play has vital importance for health, nutrition, sheltering and education, as well as the development of the child [2].

The city is a place for children where they can enjoy being there, feel safe and be a part of society like others. Nowadays, it is seen that the time the children spend playing outdoors has decreased to a great extent. Outdoor play in public open spaces, which is important to children's development, has lost its place in the child's daily life. This is replaced by playgrounds created in indoor places.

Children tend to be kept inside and disconnected from the natural world because of crowded city life, traffic, air pollution and lack of open space for play. Thus, children have started to grow with a new play culture, by using technology.

Article 31 of the Declaration on the Rights of the Child, renewed by the United Nations Congress on 20 November 1989 emphasizes that children have the right to spend their time, rest, play in accordance with their age and participate in cultural and artistic activities.

The priority criteria should be considered as user needs and expectations in urban space design for children, and built environments should be designed on this basis.

A child-friendly city is defined as a city committed to fulfilling children's rights, including their right to influence decisions about their city; express their opinion on the city they want to live in; participate in family, community and social life; receive basic services; drink safe water; have access to proposer sanitation; be protected from violence; walk safely in the streets on their own; meet friends and play; have green spaces; live in an unpolluted environment; participate in cultural and social events and be an equal citizen of their city with access to every service, regardless of ethnic origin, religion, income, gender or disability [3].

It was declared in 1996 during the International United Nations Habitat II conference that: "The well–being of a child is the ultimate indicator of a healthy habitat, a democratic society and good governance" [3].

The child-friendly city (CFC) is the city that can support the child's physical and social development, so that the child can protect the urban life as an adult person in the future.

As a result of the studies conducted, it has been proved that the child's behavior is determined by the physical environment of the child rather than the personality and personality characteristics [4]

2 Child and the Importance of Outdoor Play

Physical environment in child development; excitation, information and influence spaces [5]. The child interacts with the physical environment and uses the physical environment as an active laboratory for making observations, trying and testing.

Play is defined as a natural and compelling activity that can be freely and spontaneously made; it promotes cognitive, physical, social and emotional well-being and provides pleasure; it is a source of happiness, and offers the necessary conditions for children to thrive and learn [6,1].

The child develops the ability to communicate through play. Play is a life learning tool for kids The child develops intellectually and physically in terms of play and gets an opportunity to discover and develop.

Playing with natural materials, exploring the outside world, having the opportunity to be included in daily life are extremely attractive play opportunities for children.

The time spent for activities that used to take place outside now increasingly are done indoors [7,8].

3 Children and the Playing Areas

The physical environment, as much as social environment in which the child is in contact, plays an important role in socialization and development of the child. Housing, housing environment, playgrounds and schools are the created child's physical environment.

3.1 Housing and Its Surroundings

The most proximate environmental influence on a young child's development being is the family home where they spend a large amount of time [9].

The first step of the physical environment of the child is the house where the relationship with the world is established [10].

The suitability of the dimensions and dimensions of the reinforcements in the creation of the created spaces and the size of the room is ignored and the child is ignored when choosing the location of the reinforcement and objects [11].

Gardens are convenient and accessible spaces for children's needs in which they spend quality time [12].

3.2 Street

The street is a place where the child is waiting to be discovered, the socialization of the child and the relationship with the real world. The streets serve active, passive, cognitive, creative and visual games to be exhibited, and can be used without the supervision of the parents, but children can be in the visual and auditory perception of the parents and other relatives [13].

However, today, especially in big cities, the streets are far from being safe places which can be used by children due to transportation traffic.

3.3 Open Spaces

Urban greenspace which contributes to the improved health of citizens helps preserve the natural environment, and improves the long-term sustainability of cities. It can take any number of forms, from parks and greenbelts, to gardens or even features like green roofs, and offers many benefits [14].

Open spaces offer freedom of movement for the child and helps the development of physical skills that cannot be used indoors. The play area in the outdoors is usually spacious, which allows the child to explore more of his surroundings which is fun for the child.

The shape of the terrain is another determinant; natural or artificial slopes, hills, pits for children have unique value during the play time. Interaction with natural materials within a naturalistic environment, children gain an awareness of the processes, structures and forms within that landscape, developing a greater appreciation for the natural environment and an ecological awareness [15].

3.4 Playground

Playgrounds can provide different opportunities to stimulate physical development for children of all ages. Play equipment can encourage children to be more active. Play is an important part of child's cognitive, physical and emotional development, characteristics affecting children's physical abilities and health [16].

The playgrounds provide children with opportunities to solve problems, explore, develop imagination and develop new games by playing with tools.

3.5 School Gardens

The school is considered to be the child's first basic socialization institution after the family. The importance of school and school gardens where children spend most of their time cannot be denied. The school gardens are generally separated from other places by offering the safest places for children in urban space and the open space potential in terms of play throughout the day.

The schoolyard is the child's city square and is one of the few places where many children can communicate with their peers in an open space [17].

4 Child-Friendly Environment Approaches

4.1 Historical Framework

The Save the Children Fund was established in the United Kingdom in 1919 in order to improve the lives of children through better education, health care and economic opportunities, as well as providing emergency aid in natural disasters, war and other conflicts [18].

The Save the Children Fund was founded in London, England, on 15 April 1919 by Eglantyne Jebb and her sister Dorothy Buxton as an effort to alleviate starvation of children in Germany and Austria-Hungary during the Allied blockade of Germany of World War I which continued after the Armistice [19].

Jebb created an initial draft for what would become the UN Declaration of the Rights of the Child in 1923. It contained the following five criteria:

1. The child must be given the means requisite for its normal development, both materially and spiritually.
2. The child that is hungry must be fed, the child that is sick must be nursed, the child that is backward must be helped, the delinquent child must be reclaimed, and the orphan and the waif must be sheltered and succored.
3. The child must be the first to receive relief in times of distress.

4. The child must not be put in a position to earn a livelihood, and must be protected against every form of exploitation.
5. The child must be brought up in the consciousness that its talents must be devoted to the service of its fellow men [19].

These five points were adopted by the League of Nations in 1924 and was thus known as the Declaration of Geneva. This was the first important assertion of the rights of children as separate from adults, and began the process that would lead to the UN Convention on the Rights of the Child, adopted by the United Nations in 1989 [19].

The 1989 UN Convention on the Rights of Children, CRC, is the background to many recent child-environment studies.

Child-Friendly Cities (CFC) project of UNICEF was initiated after two United Nations environmental conferences (The Rio Declaration on Environment and Development, United Nations Conference on Human Settlements (HABITAT II)). It is a UNICEF-led initiative that suggested a framework of nine building blocks. According to this:

1. Children's participation
2. A child-friendly legal framework
3. A city-wide Children's Rights Strategy
4. A children's rights unit or coordinating mechanism
5. Child impact assessment and evaluation
6. A children's budget
7. A regular state of the city's children report
8. Making children's rights known
9. Independent advocacy for children are the priority conditions.

A secretariat was established by UNICEF in Florence, called the United Nations Child-friendly City International Secretariat [20].

4.2 The Concept of Child-Friendliness

Four major issues that are critical to the creation and maintenance of a child-friendly community are safety, greenspace, access and integration [21]. McAlister indicated that for a community to be truly child friendly, it must both feel and actually be safe and healthy, it must promote accessibility, it must provide adequate natural green spaces and it must integrate all members of society.

Child-friendly environments refer to developing better conditions for children in the urban environment by focusing on child-friendly green spaces in the urban area [22].

Child-friendly spaces are defined as green spaces which are specifically designed in an urban area for children to interact with natural environment, have a positive impact on the development stages of children and add quality to the life of children [22–25].

Some urban design tools such as traffic calming help to design and develop child-friendly urban spaces.

4.2.1 Traffic Calming/Living Streets

Traffic calming/living street is a traffic management approach to reduce both the speed and dominance of motorized transport, as they move through residential neighbourhoods [26].

People have recognized that many residential streets are unsafe for children to play in because they are designed for motor vehicle access and parking. The advantage of street improvement methods is that it reduces the speed and noise of the vehicle. This goal can be achieved in two different ways. The first method is to design the road as a physical element that will reduce the speed of vehicular movement with constrictions, textured road pavements and speed bumps or redesign. The second method reduces speed through speed limits. Traffic calming/living street is a social space where people can meet and where children may also be able to play legally and safely. This is often achieved using the shared space approach, with greatly reduced demarcations between vehicle traffic and pedestrians. It became popular during the 1970s in the Netherlands, which is why the Dutch word for a living street (woonerf) is often used as a synonym. Country-specific living street implementations include: home zone (United Kingdom), residential zone (ru:Жилая зона, Russia), shared zone(Australia/New Zealand), woonerf (Netherlands, Flanders, South Africa, Namibia) and *zone résidentielle* (France) [27]. These streets are often built at the same grade as sidewalks, without kerbs. Cars are limited to a speed that defined to be pedestrian speed. These studies were first recognized in the Netherlands by the name *Woonerf* and later in the UK as home zone. In order to understand the general principles of the studies, it is necessary to explain the exit point of "woonerf" and "Home Zone" philosophy.

Woonerf

The *Woonerf* movement originated in the Netherlands in the 1970s as a way of re-balancing the relationship between people and the movement of vehicles.

Woonerf is a system for arranging the streets as a children's playground. Although the origin of the idea is originally English, the Netherlands makes this philosophy a reality. These streets are designed as places that take the street residents and pedestrians into account, forcing the drivers to go slowly, and are named *Woonerf* (Dutch word for a living street) by De Boer [28].

The Woornef system allows the development of playgrounds for children, recreation, seating spaces by reducing street traffic or traffic speed; thus, it is a system that aims to make the street livable for the child. The integration of vehicle traffic with pedestrian activity is the main principle in the planning and design of *Woonerf*.

Home Zone

A home zone is a residential street in which the living environment clearly predominates over any provision for traffic. It is an environment where the design of the spaces between homes provides space for motor vehicles, but where the wider needs of residents are also fully accommodated. Home zones were first suggested in the UK in the 1980s as a low-cost measure to reduce casualties amongst young children in residential areas and to allow them to play outside in safety [29]. The Home Zone 'initiative was launched in 1998 by the so-called Children Play Council (CPC) in the UK to raise awareness of the importance of the game in child development and the need to ensure adequate access to each child's better play opportunities and services [30]. The various elements in the street cause the vehicles to slow down and naturally the drivers drive safely in terms of pedestrians.

4.2.2 School Gardens

As a result of studies on school gardens, it is indicated that various elements can play an important role in the creation of school gardens such as water elements, to choose their own play actions and opportunities to create their own playgrounds, access to nature (trees, ponds, bushes, flowers, tall grasses, insects and animals), lands on which they can play, places and elements to sit on, places to lean or hide, and ripped material to play and the environments they can design [31,32,17].

4.2.3 Playgrounds

In the design and planning of children's playgrounds, great importance was given to safety, age-appropriate design, location selection, separation and positioning of game activities and elements, circulation, material selection and planting. The child has participated in both the design and implementation process.

4.2.4 Pedestrian Circulation for Children

It is important to provide clear, direct, barrier-free and safe access to communal outdoor space, parking areas, schools and public streets. Safety, amenity and durability should be used as key considerations for children.

Implementation and installation such as installing appropriate signage, or upgrading pedestrian crossings to include dropped kerbs, audible warnings and tactile paving are made for making circulation safer.

5 Conclusion

Play is a pleasure process in which children are willingly involved, with a particular subject, purpose, rule or not, but supporting all mental, emotional, social, physical development of child. Children's playgrounds and urban green spaces have an important role by contributing to children's physical and mental development.

For this purpose, the physical environment in which the child lives and interacts must be safer. The place to be used by children should be easily and dangerously accessible from all living units. Incorrect location choices include many risks such as unsafeness and accidents. Children have rights to use places as they want. Therefore, the participation of children in planning and design process is important. So, the planning must be done for children and with children. Planning and design disciplines consider children as users and partners in physical planning. In children's play how and by whom the space is used is important. Children need flexible areas and materials that they can change and do. It is clear that the child should not be ignored in the design of the physical environment in accordance with the importance of the child development. In this regard, it should be kept in mind that the child is a user in the design of all the spaces from housing scale up to the urban scale, the spatial perception, needs and preferences of the child should be taken into consideration and the design of the space should be considered in the light of this information.

References

1 Bento. G,. Dias. G. "The Importance of Outdoor Play for Young Children's Healthy Development." Porto. Biomed. J. 2017. http://doi.org/10.1016/j.pbj.2017.03.003.

2 Tekkaya, E. *Tasarlanmış çocuk hakları: Ankara çocuk oyun alanları*, Millî Eğitim Dergisi, 2001. c.151, s.25–37.

3 UNICEF Building Child Friendly Cities A Framework for Action, 2004. www.childfrienlycities.org

4 Horelli, L. "Creating Child-Friendly Environments, Case Studies on Children's Participation in Three European Countries," 1998. Childhood, 5(2), s. 225–239.

5 Olds A.R. (1987) "Designing Settings for Infants and Toddlers." In: Weinstein C.S., David T.G. (eds) Spaces for Children. Springer, Boston, MA

6 Aral, N., Gürsoy, F., Köksal, A. *Play in Preschool Education.* 2001.İstanbul: Ya-Pa Publishing.

7 Sibley, D. "Families and Domestic Routines: Constructing the Boundaries of Childhood." 1995. 123–137. London: Routledge.

8 Karsten, L. "It All Used to **Be** Better? Different Generations on Continuity and Change in Urban Children's Daily Use of Space." 2005 Children's Geographies no. 3 (3):275–290.

9 Christian H., Zubrick S. R., Foster S. Giles-Corti B., Bull F., Wood L., Knuiman M., Brinkman S., Houghton S., Boruff B., "The Influence of the Neighborhood Physical Environment on Early Child Health and Development: A Review and Call for Research. "Health & Place Volume 33, May 2015, Pages 25–36 https://doi.org/10.1016/j.healthplace.2015.01.005

10 Ertürk, S. (1992). Çocuk ve Mekân, Mimarlık ve Sehircilikte Mekân, Yıldız Üniversitesi, İstanbul.

11 Tandoğan Ş. İstanbul'da "Çocuk Dostu Kent" İçin Açık Alanların Planlama, Tasarım Ve Yönetim İlkelerinin Olusturulması. 2011. Doktora tezi, İstanbul.

12 http://www.dotrythisathome.com/the-importance-of-outdoor-play-and-therefore-gardens/

13 Gür, S. Ö., Lang, J., Özbilen, A., Ertürk, S., Ibis, T. Çocuk ve Çevresi: Çocuk Oyun Alanları Olarak Sokaklarımız, 1980. KTÜ Mimarlik Bölüm Bülteni, 5, s. 70–76.

14 Van den Berg. M. City with a future: How to build a Child-friendly City. 2013. Youth, Education and Society, Rotterdam.

15 Augusta Jackson E., Naturalistic Play Environments: Activating Children's Ecological Awareness, Development and Senses through Natural Materials, 2009, Master thesis, Washington State University.

16 Zhang H., and Li M. "Environmental Characteristics for Children's Activities in the Neighbourhood," Procedia - Social and Behavioral Sciences 38 (2012) 23 – 30 doi: 10.1016/j.sbspro.2012.03.320

17 Malone, K. and Tranter, P. "Children's Environmental Learning and the Use, Design and Management of Schoolgrounds," 2003. Children, Youth and Environments, Vol.13, no.2.

18 http://www.wikizeroo.net/index.php?q=aHR0cHM6Ly9lbi53aWtpcGVkaWEub3JnL3dpa2kvU2F2ZV90aGVfQ2hpbGRyZW4

19 Yates, Mick. "Eglantyne Jebb". LeaderValues. Retrieved 2018–10–09. https://en.wikipedia.org/wiki/Save_the_Children

20 Riggio, E. "Child Friendly Cities: Good Governance in the Best Interests of the Child," *Environment &Urbanization*, 2002.Vol **14**, No 2. Retrieved June 04, 2007, from http://www.bvsde.paho.org/bvsacd/cd26/enurb/v14n2/45.pdf

21 McAllister, C. "*Child Friendly Cities and Land Use Planning: Implications for Children's Health.*" Environments Journal, 2008. 35(3):45–56.

https://www.google.co.za/?gws_rd=ssl#q=McAllister%2C+C.++2008.++Child+friendly+cities+and+land+use+planning%3A+http://citeseerx.ist.psu.edu/viewdoc/download?doi=10.1.1.477.2834&rep=rep1&type=pdf

22 Goosen, Z. The planning and development of child friendly green spaces in urban South Africa, 2014, Master thesis, North-West University.

23 Howard, A. 2006. "What Constitutes Child Friendly Communities and How Are They Build?" Australian Research Alliance for Children & Youth. P1–57.

24 McDonald, L. Belfast healthy cities: child friendly spaces. 2012. https://www.gcph.co.uk/assets/0000/4136/Laura_McDonald_-_Child_Friendly_Cities_Glasgow.pdf

25 Wapperom, R. "Rotterdam, City with a Future: How to Build a Child-Friendly City." 2010. Paper presented at the Child in the City Conference, Florence, 28 October.

26 https://safety.fhwa.dot.gov/ped_bike/univcourse/pdf/swless11.pdf http://www.childfriendlycities.org/pdf/childstreet_leaflet.pdf

27 Living Streets, Wikipedia, the free encyclopedia https://en.wikipedia.org/wiki/Living_street

28 Hand, C. Woonerf: A Dutch Residential Streetscape. 2007

29 Butcher, L. "Roads: Home Zones," SN/BT/1137, Business and Transport file:///C:/Users/User/Downloads/SN01137 %20(1).pdf

30 Bristol City Council (2003) New Build Home Zone Design Guidelines, First Edition November 2003, Bristol City Council.

31 Fjortoft, I. and Sageie, J. (2000). "The Natural Environment as a Playground for Children: Landscape Description and Analyses of a Natural Landscape," Landscape and Urban Planning, 48(1–2), 83–97.

32 Moore, R. C. (1986). Childhood's Domain: Play and Place in Child Development. London: Croom Helm.

Dr. Oğuz Ateş and Azat Gül

27 Use of Digital Technologies in Landscape Architecture: 3D Printers

Abstract: 3D printing may seem like a relatively new technology with its recent popularity and accessibility, but it has actually been used around since the 1980s. When fused deposition modelling (FDM) patent expired in 2009, the technology became available to both small and large companies in every industry to take advantage of and the landscaping architecture is no different. Models are an important form of visual representation. Traditional 2D graphics are a vital part of the design process; however, a tangible 3D model does not only clarify the details of the space but also ensures the client fully understands how the end product will look. This chapter is about the role of 3D models in landscape research and design. These landscape models, next to drawings and maps, are very important in the process of design research in landscape architecture. These models offer to landscape researchers and designers different perspectives for visual thinking and visual communication.

Keywords: 3D visualization, Landscape architecture, 3D models

1 Introduction

Technological developments affect the production and economy, along with the life-styles of people, while leading certain professions to be erased from our lives, emergence of new professions, and adoption of new practices and approaches in the sector. One of these new technologies is the 3D printers. Notwithstanding that it is a quite new technology and that there are still many steps to be taken in terms of its advancement, it has already started to affect many areas. It is being used in many areas like industry, architecture, medicine, engineering, science, art and design, while expanding its sphere of influence as the days pass (1).

The term "3-D Printing" (3DP) is the name commonly used for a collection of additive manufacturing methods, methods which produce components via adding material layer by layer, in opposition to subtractive manufacturing methods which produce components by removing material from a block of material larger than the final component. Since first being introduced in 1981 by Hideo Kodama at the Nagoya Municipal Industrial Research Institute, Japan, the variety and quantity of methods and their speed, cost and accuracy have increased exponentially (6).

The 3DP market is rapidly expanding and is expected to have a value of $3.4 billion by 2015 (5) and estimated to be generating an economic impact of $230–550 billion a year by 2025. While the technology has been mostly used to produce prototypes to date, it is expected that by 2020, 50% or more of 3D printed production will be final products (12).

The architectural presentation techniques have been transformed into something new with today's developing 3D printing technology. Two-dimensional presentation techniques are one of the indispensable parts of the design process, but a 3D model, which is prepared with a decent technique, does not only describe the details of the design, it also ensures the final design to be clearly understood on how it will look in the end.

Designing process is based on knowing how to place the pencil on the paper. Architectural designers may explain their design concepts to customers by drawing their ideas. However, creating realistic architectural design models is much more challenging. "Fly-throughs" and 3D renders will relay nothing but the beauty and functionality of the design. A physical model lifts the lid on a design, thus ensuring its appearance to be clearly seen. This matter reveals the point 3D printing has reached today (7).

3D printers are the devices that turn models, which are designs on computer environment or scanned in 3D, into tangible objects within a quite short period of time, by using different materials, without needing a mold.

3D printers, using ABS and PLA thermoplastic polymer materials, operate based on FDM (FUSED DEPOSITION MODELING) or the technology of stacking via consolidation. It is built as layers by being melted with a nodule having a high-temperature filament material (4).

For production with 3D printer, 3D modeling programs are required. Some of these programs can be summarized as AutoCAD, Sketchup, 3Ds Max, Rhino3D, SolidWorks. Three-dimensional models designed with these programs are exported as STL files, thus ensuring the production with 3D printers.

2 Development History of 3D Printers

The first 3D printer was developed by Charles Hull in 1984, by the use of stereolithography (SLA) technology. In 1986, Charles Hull established the company "3D System" and developed the first commercialized 3D machine. The first printer was manufactured under the model name "SLA-250" by 3D System in 1988; then in the same year, Scott Crump invented the Scott Crump Fused Deposition Modeling, and established the company "Stratasys" in 1989 (9).

In 1993, Massachusetts Institute of Technology (MIT) developed the Three-Dimensional Printing (3DP) technology and started color printing with this technology. In 1995, Z Corporation purchased the MIT license and started to sell 3D printers using 3DP technology (3).

In 2006, Reprap projects developed a technology, which can copy its own parts, as well. The objective of this project was to ensure that everyone has a 3D printer. The first version of Reprap was completed in 2008 and it achieved to copy 50% of its own parts. In 2009, large amounts of home-type 3D models started to be sold like Cubify developed by 3D Systems and Maketbot in 2009 (9).

3 3D Printer Types

3D printer types vary depending on the area of use. There are seven types of 3D printers. These printers comprise of the following technologies: Stereolithography

(SLA), Digital Light Processing (DLP), Fused Deposition Modeling (FDM), Selective Laser Sintering (SLS), Selective Laser Melting, Electronic Beam Melting (EBM), Laminated Object Manufacturing (LOM) (11).

There are three types of 3D printer technologies, which are most frequently used in today's world. These are named as Selective Laser Sintering (SLS), Fused Deposition Modeling (FDM) and Stereolithography (SLA) technologies. Detailed information for these three printer modeling can be seen as follows:

3.1 Selective Laser Sintering (SLS)

Laser is a technique used as a power source. It is laid over the production tray with a sinter powdered cylinder mechanism. The laser runs over the granular material, melts the dust and conglutinates them, thus making the first layer. No supporting structure is required for the next layer in this method. With the movement of cylinder mechanism, the powder is re-laid for the second layer. The same process is repeated for the other layers. Once the production process is over, the tray is left to cool down. Lastly, the powder, melted on the tray, are cleaned for recycling, and the model production is completed (8) (Fig. 1,2).

3.2 Fused Deposition Modeling (FDM)

Wire-shaped solid material, which is named as filament, is transmitted to the heater with the spool and a motor. The filament in the heading becomes semi-fluid and the modeling, scanned by the software, follows a pre-designated specific track. Cooling and hardening right after getting out of the heading, the filament model constitutes the geometrical aspects, thus completing the first layer. The same process is followed for other layers and the product is left for cooling down. Cooled-down product is cleaned out of supporting parts, which makes it prepared (8) (Fig. 3,4).

3.3 Stereolithography (SLA)

This technique is based on curing up of certain parts over a liquid-form photopolymer resin layer by point-ultraviolet laser beams. After plastering the resin layer on the first one, the curing is maintained in the same manner, and once the layers are completed, the parts are removed out of the resin pool. The supporting units are mechanically removed from the part (Fig 5,6).

4 Materials Used in 3D Printers

Nylon filament, metal-like filament, ceramic filament, wooden filament, luminous filament, PET, PLA, carbon fiber PLA, ABS, polycarbonate ABS, UV-sensitive ABS materials are used in Fused Deposition Modeling (FDM) technology. Polyamide, glass reinforced polyamide and metal materials are used in Selective Laser Sintering (SLS) technology. Photopolymer resin and flexible resin materials are used in Stereolithography (SLA) technology (2).

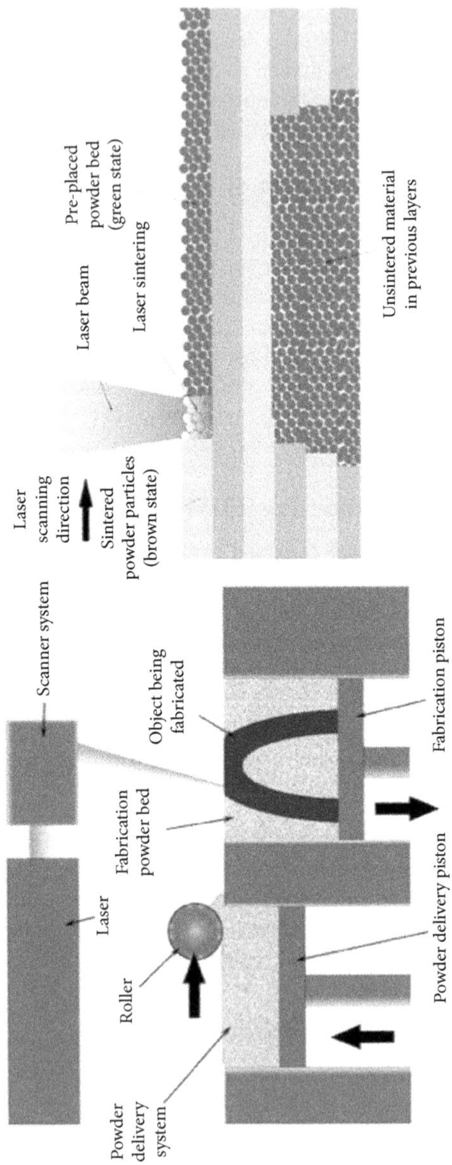

Fig. 1: SLS technique (8).

Fig. 2: A sample model produced via SLS technique (8).

5 Use of 3D Printers in Landscape Architecture

Emerging in the beginning of 1980s, 3D printer technologies are recognized to be in infancy despite its rapid development within the last ten years (10). Studies on 3D printers are in progress. It is estimated that 3D printers will supersede such traditional production methods like molding, forging and machining up to 2030 (13). Notwithstanding that the 3D printers have not completed its evolution in today's world, certain estimations can be made in the upcoming years.

Where the landscape architectures state to be competent for direct 3D modelling and 3D printing, they will have the potential to be included in the production and manufacturing process of any objects like banks and flagstones. This may affect the classical product development process, which may lead them to carry out production as they wish. Landscape architectures may increase their design powers both by breaking the routines of former methods and creating new forms.

All these technologies may ensure that the landscape architectures share their ideas with the customers and shareholders in a more appealing manner. Three-dimensional printed models bear the potential value of designing process, as well.

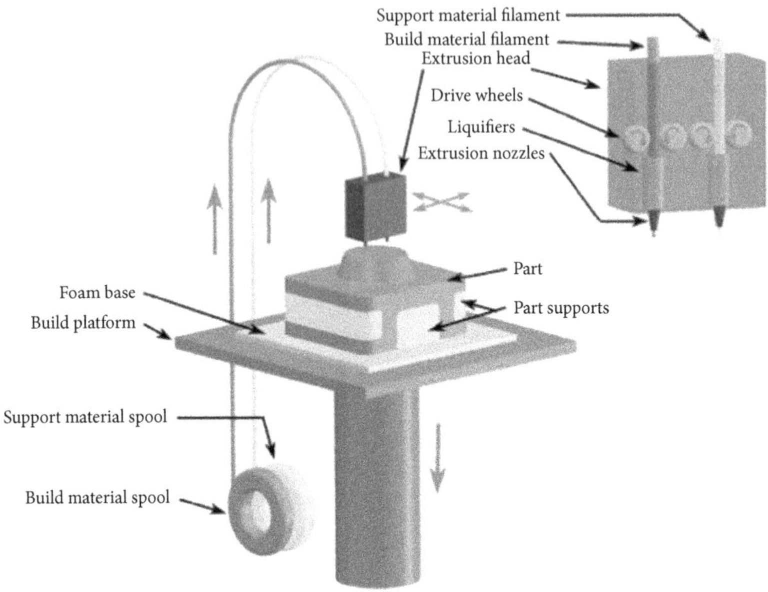

Fig. 3: FDM technique (8).

For example, in a study carried out in Harvard University, innovative, ergonomic and custom-designed banks were taken into production process following eight prototypes printed via 3D printers (Fig. 7).

In another research project, named "Mosspebble," carried out in the same university, "Cella" was produced with 3D printers. "Cella" is an object reminding of such organic objects like sea shells and porous rocks, to be used in rooftop gardens, along with regions with ecological problems (Fig. 8) (14).

6 Conclusion and Discussion

Three-dimensional models offer designers useful modes of operation for visual thinking and visual communication. The physical model provides for a "situated setting" (cognitive, perceptual and physical context) for reflective conversation with individuals and groups of people.

Three-dimensional models eliminate ambiguity when holding a real-life model of the concept. Test market potential discusses ideas with potential buyers or raises capital from investors. No need to wait for a supplier or make models by hand. Print it on your desk in hours. Easily recreate textures, finishes and colors to create a new dimension. You can print complex shapes and geometries.

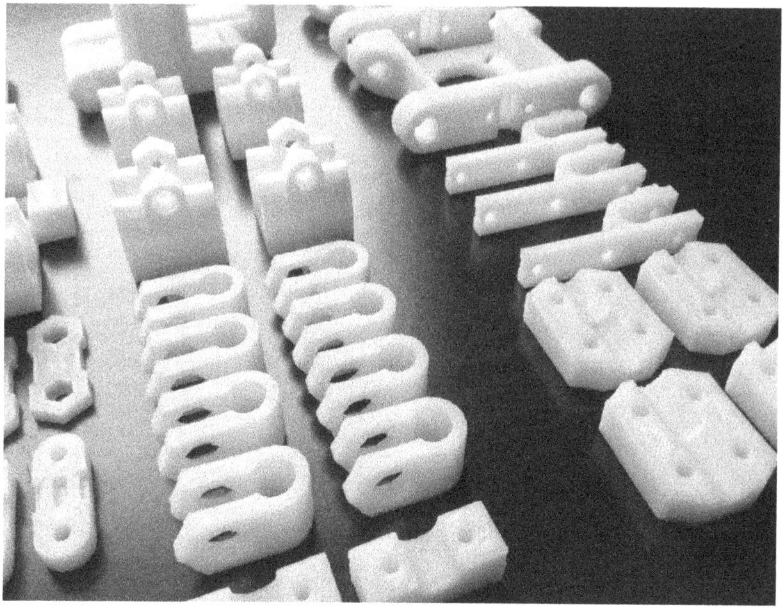

Fig. 4: A sample model produced with FDM technique (8).

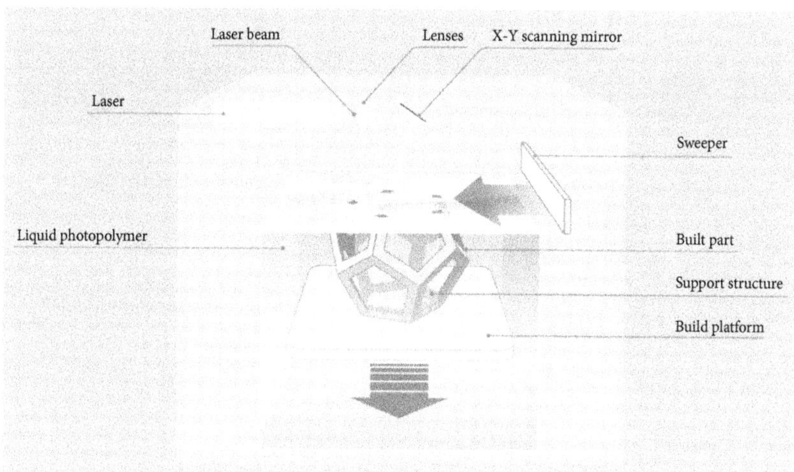

Fig. 5: SLA technique (8).

Fig. 6: A sample model produced with SLA technique (8).

Fig. 7: Custom-designed banks (15).

Fig. 8: "Cella" produced with 3D printers (14).

With technological developments, 3D printers have started to be used more than ever in our daily lives. It is quite obvious that 3D printers will be frequently used in the designing and planning phases of landscape architecture.

References

1 Balcıoğlu, Y.S., (2014). 3 Boyutlu Yazıcı ve Sinemada Kullanımı (Yayımlanmamış Yüksek Lisans/Doktora/Sanatta Yeterlik Tezi/Raporu). Yaşar Üniversitesi, İzmir.

2 Çetinkaya, K., "3D Yazıcı Sunumu," Karabük Üniversitesi, Mühendislik Fakültesi, Karabük. Sayfa: 4–13 (2015).

3 Gökçearslan, A. 2017, "Üç Boyutlu Yazıcının Grafik Tasarım Alanına Yansımaları," Fine Arts, Vol. 12, No. 2. Page 135–148.

4 İlgül, R. C. and Akgünoğlu, B., 3D Yazıcı, Karabük Üniversitesi Mühendislik Fakültesi Mekatronik Mühendisliği Lisans Bitirme Tezi, Karabük, 2016.

5 King, L. 2013, The race is on for 3D printing. VegasInc. [cited 2018, 18 September]; Available from: http://www.vegasinc.com/business/2013/sep/23/race-3-d-printing/.

6 Kodama, H. 1981, "Automatic method for fabricating a three-dimensional plastic model with photo-hardening polymer." Review of Scientific Instruments, Vol. 52, No. 11. Page 1770–1773.

7 Lozova, L., 3D printing in landscape architecture and pool design. [cited 2018, 18 September]; Available from: https://ultimaker.com/en/stories/37401-3d-printing-in-landscape-architecture-and-pool-design
8 3D Yazıcı Teknolojileri Çeşitleri, [cited 2018, 17 September]; Available from: https://www.poligonmuhendislik.com/bilgi-bankasi/3b-baski-teknolojileri
9 3D Yazıcı Tarihi ve Gelişimi. [cited 2018, 18 September]; Available from: http://priyoid.com/3d-yazici-teknolojisi/3d-yazici-tarihi/
10 Özgüven, S. (2015). Seramik Sanatında Üç Boyutlu Yazıcıların Yeni Bir İfade Biçimi Olarak Kullanılması. idil, 4 (18), s.167–183.
11 Şentürk, M. A., 3D Yazıcı Çeşitleri. [cited 2018, 16 September]; Available from: http://roboturka.com/3d-yazicilar/3d-printer-cesitleri/
12 Russell, P., 3D Printed Earthen Architecture A Sustainable Housing Solution for Displaced Populations. Master of Science in Product Design Innovation at Aston University. 2015: Birmingham, England.
13 Yılmaz, F., Arar, M. E., Koç, E. (2013). 3D Baskı ile Hızlı Prototip ve Son Ürün Üretimi. Dünya Gazetesi.
14 3DPrint.com, The Voice of 3D Printing/Additive Manufacturing. [cited 2018, 18 September]; Available from: https://3dprint.com/26042/cella-3d-printed-planters/
15 Architectural Design. [cited 2018, 12 September]; Available from: https://www.architectureanddesign.com.au/news/is-there-potential-for-3d-printers-in-landscape-ar

Ömer Lütfü Çorbacı and Türker Oğuztürk

28 Evaluation of Amasra's Visual Landscape Quality in Terms of Natural, Historical, and Cultural Values

Abstract: In this chapter; the visual landscape assessment of Bartın province, Amasra district is made, especially with the idea that the area is significant for its visual landscape value, and identifying areas of importance is also aimed. For this purpose; the scoring systems required for the evaluation of the visual landscape of Amasra are identified with the help of academicians and experts, and visually high-quality landscapes are defined. The landscape structure of an area is closely related to the perceived visual landscape quality. Visual landscape character has a significant effect on people's recreational activity preferences. The more natural, different, diverse, harmonious, intriguing, mysterious, perspective, reassuring, and regular the landscape structure of an area is, the higher visual value it provides. In this context, the role of landscape esthetical quality in environmental management should be taken into consideration and the visual esthetical structure of the landscape should be included in planning decisions.

Keywords: Amasra, Visual value, Landscape

Introduction

Before defining the value of the visual landscape, it is necessary to define the concept of landscape. According to the European Landscape Convention also signed by Turkey, which is a framework agreement for protection, management, and planning of natural and cultural landscape as a whole, landscape is, as perceived by people, the area formed as a result of the interaction and activities of area characteristics and natural and/or human factors (1). The interaction of a series of biotic, abiotic, visual, and cultural components creates the visual landscape (2). Landscape is a whole of ecological, socio-economic, and aesthetic values of natural and cultural assets that are in a field of view (3). "Visual quality" plays a key role in planning and designing to define the elements that make up the landscape or the structure they have. Therefore, determination of visual quality in an area is important in terms of determining the means of existing landscape structure and the potentials of the elements and components of the landscape in that area (4).

The visual landscape value is defined as a common product of certain (visible) landscape features that interact with the relevant psychological (perceptual, cognitive, and emotional) processes of the human observer. The concept of visual landscape value, which was brought forward and developed in the second half of the 20th century, has become an important research topic in environmental management and policies (5, 6). Visual quality assessment is an important component of environmental planning and management as well as being an active field of environmental perception studies, and is therefore one of the rings of the chain to be evaluated (7). Visual landscape evaluation is a method used to determine the changes in the landscape character. It aims to obtain data and information about the present visual characteristic of the area and to determine its sensitivity to possible changes. In the visual landscape evaluation, inventory studies and evaluation of the visual characteristics of the landscape are carried out for the purposes of planning, design, and resource management (8).

The purpose of visual landscape analysis is to determine the degree of sensitivity to possible changes by providing information on the current visual characteristics and status. This information will provide guidance for decision-making and strategy development stages of field use planning and resource management studies (9). The visual superiority of an area will increase the field's frequency of being preferred and together with field use planning, it will be a guide in decision-making, strategy development, and management stages (10). In addition, it has been mentioned that determining the visual features of the area during planning and analyzing it, is the most accurate method for the protection of visual resources (11).

Expert and user evaluations can be used in methods to determine the visual quality of landscapes. In the expert-based studies, visibility likes are stylistic, and they are carried out independent from the perception factor. In this evaluation, the contribution and opinion of many professions (landscape architecture, forestry, geography, and environmental engineering etc.) can be obtained. Experts generally use the scoring method to evaluate the visual landscape according to its biophysical characteristics (12). In such studies, experts or expert groups try to explain the current situation of landscapes. Visual quality is the evaluation of wide range of monitoring made by vehicle, by foot, or by aerial survey conducted in the area. Visual quality assessment is based on seven key factors. These are: 1. Morphological structure (land form), 2. Flora (vegetation), 3. Water, 4. Color, 5. Neighboring Landscape, 6. Rareness, 7. Cultural changes (13).

As landscape quality is generally defined in a wide range of environmental/ecological, sociocultural, and psychological factors for the environment, it has a highly perceptual and objective nature, and from this aspect, perhaps it is one of the hardest facts of environment to analyze and measure. Visual quality is equivalent to beauty; however, it has an objective characteristic too (14).

Today, the use of photographs in esthetical perception and preferences related studies is very common and it is a valid method for most experts. The reason for this is that it is faster, economical, and can be controlled more easily compared to real-world conditions. Studies show that there is a strong relationship between the ideas of the real landscape in the area and the thoughts about the photographic images of the beauty of the landscape (15, 16, 9).

The natural formations, water surfaces, important architectural, and historical objects etc., which stand out among the important visual landscape elements, should be determined during the field studies in the research area. In survey studies, different views and characteristics related to seasonal changes should also be considered. Photographs selected for the representation of sample areas during the evaluation examination, can be used for expert group evaluation. Besides, the observer develops solutions and strategies for visual landscape by paying attention to variables such as visibility range and angle of view for each unit (17–19).

The visual quality assessment of the landscape is used as a guiding tool, in the context of the visual data generated by the changes in the physical environment, in the planning and design of rural and urban environments and in the formation of some administrative policies. Assessment of the visual quality of a landscape unit is important for the determination of existing landscape structure, and for the potentials of the elements and components that form the landscape of that area (12).

Through the interpretation of visual quality form, the following comments can be made regarding the visual landscape value (20):

- Very Low and Low Class Visual Landscape Value (VLV) Areas: In areas where the visual quality of the landscape is low, plan decisions can be made which require major changes in the existing landscape character. For example, if such areas are ecologically compatible, and if the carrying capacity of the landscape is high, then they can be suitable for settlement and industrialization.
- Medium Class VLV Areas: In areas where the visual quality of the landscape is moderate, the existing landscape character should be partially preserved. When making changes in the landscape character, sudden changes and changes that are in contrast to the capacity of the landscape character should be avoided. The traces of the plan and management decisions to be taken in the field can be observed in the field; however, these should not play a dominant role on the overall landscape character.
- High Class VLV Areas: In areas where the visual quality of the landscape is high, the existing landscape character should be preserved. Changes in landscape character should not attract attention. The traces of the concrete results of planning decisions should be visible in the field but must be harmonized within the landscape character.
- Very High Class VLV Areas: In areas where the visual quality of the landscape is very high, it is important to maintain the existing landscape character within the conserved area system. As these areas have a very sensitive ecological structure, the nature of the plan decisions should be for conservation purposes. In these areas, it may be necessary to make decisions about landscape characters enabling intergenerational transfers. The traces of these plan decisions should not be apparent in the area.

In this study, the visual landscape assessment of Bartın province, Amasra district is made, especially with the idea that the area is significant for its visual landscape value, and identifying areas of importance is also aimed. For this purpose, the scoring systems required for the evaluation of the visual landscape of Amasra are identified with the help of academicians and experts, and visually high-quality landscapes are defined.

Material and Methods

The following resources, maps, and reports are used as materials:

- Topographic map of E28b4, E28b3, E29a4, E28c1, E28c2, E29d1 research areas with 1:25,000 scale (Ministry of National Defense, General Command of Mapping),
- 1/25,000 scale topographical map for the sites of E28b4, E28b3, E29a4, E28c1, E28c2, and E29d1 (Ministry of Defense, General Command of Mapping),
- 1:25,000 and 1:50,000 scale soil maps of the research area including the soil structure, erosion status, and land ability classes (former Ministry of Agriculture and Rural Affairs, General Directorate of Rural Services),
- The data regarding surface water and underground water presence in the research area, obtained from the Directorate General of State Hydraulic Works (DSI),
- CORINE land use/cover classification,
- Landsat satellite pictures,
- Visual landscape field study form,
- Esthetical-perceptual assessment field study form,
- Observations and photographs obtained during field study.

While determining the research siteborders of this study, we primarily used the administrative borders but also took the natural resources into account. Given that this study is primarily concerned with the district of Amasra, it is inevitable that the borders of the research site are determined by the administrative borders. Administrative boundaries are composed of the central district and its 30 village settlements. Amasra district of Bartın province is located at the western Black sea region of Turkey. Amasra is bordered by Kurucaşile to the north, Bartın to the west, Black sea to the north, and Arıt to the south. During the preliminary stage of the field study, we classified the resources that are used to evaluate the landscape of the research area into eight groups: water resources, wetland resources, topographic resources, historical and cultural resources, archeological resources, wildlife, service areas, and general landscape pattern.

In the study, the method developed based on expert evaluations is applied. The Landscape Character Analysis and the assessment in terms of Tourism/Recreation (PEYZAJ-44) is made in provincial scale, by group of experts formed by TUBITAK Project team (12 people) and Ankara University Department of Landscape Architecture faculty members (8 people). Among these source points, 69 visual quality points have been determined by taking into account the factors such as the landscape characters they have, the natural and cultural characteristics of the region, their usage density, and their level of authenticity. These points are also evaluated as esthetic and perceptual points. In this context, the land form and the esthetical-perceptual assessment land form are prepared. A general meeting is held on how to complete the forms before going to the field. Afterwards, the field is visited with a team of experts and the printed questionnaire forms are filled out at the determined points. The point coordinates of the areas are taken, their map sections are determined and the areas are photographed. Each of the questionnaire forms filled by experts is entered into a database and the parameter scored highest by a vast number of experts is considered valid for that area.

Visual assessment field form: The modified visual landscape field study form, which was originally prepared by (20) and the esthetic-perceptual assessment field study form are used for the determination of Visual Landscape Quality. Tab. 1 shows a visual landscape field survey form sample prepared for Bozköy Bay and Beach, one of the visual quality points in the research area.

Esthetical-perceptual assessment field form: The social perception of landscape plays an important role as it is a course of Landscape Character Analysis. In this context, literature reviews have been made and the most suitable criteria according to field characteristics of the research area have been selected in line with expert opinions. Tab. 2 shows an esthetical-perceptual assessment field work form sample prepared for Bozköy Bay and Beach, one of the visual quality points in the research area. There are two types of visual landscape assessments: objective and subjective. The objective method is used to determine the value of the visual landscape, and the subjective method is used to determine the esthetical and perceptual landscape values.

Objective method: The visual landscape value is a criterion that reveals the values of the existence of the source regardless of landscape elements, neighboring elements, or the characteristics of the observer. Visual diversity refers to naturalness and characteristic features (20). This study is based on the Visual Resource Management System (VRM) models used by (18) and (20). VRM is a method to maintain the visual value of the landscape and to reduce the effects of different field uses on this value. The landscapes are classified according to their quality by using the method. Finally, maps are created for the visual quality of the landscape.

Subjective method: It includes esthetical and perceptual landscapes created during field observations. Sixty-nine esthetical and perceptual points are determined in the area. The distributions of these points according to the characteristics are presented in the esthetical and perceptual criteria table. In order to determine the perceived landscaping types of the areas, point and vicinity-based assessments are also made. The ArcGIS 10 software is preferred as GIS software to create the maps. Microsoft Access software is used to create the database for the research area.

Visual Landscape Value

The analysis is carried out in two phases. In the first phase, a Visual Landscape Quality map is produced at the office, by using the available data from the research area according to Tab. 3. In the second phase, the land verification test of the information produced at the office is carried out. The caves are to be taken as point data for their added value.

Tab. 1: Visual Landscape Field Study Form (Modified Version of(20))

X Coordinate:455431	**Y Coordinate:** 4624919				**Date:** 07.08.2013		
Location: Bozköy Bay and Beach				**Sheet No:** e28b3 (HGK)		**Photo No:**12-	
Key Factors	**Evaluation Criteria**	**Score**	**LC**	**Key Factors**	**Evaluation Criteria**	**Score**	**LC**
Landforms	High relief, steep cliff, pointed hill, dune, valleys, canyon, cliffs	**5**	3	**Color**	Pleasing contrast with other components, vivid and rich color	**5**	3
	Medium relief, interesting erosion patterns, steep hill, single hill, narrow hill, different size	3			Contrast with components, medium intensity, rich but not dominant color	3	
	Limited relief, small round hills, levels	1			Unobtrusive color combinations and contrasts	1	
Vegetation	Interesting texture and form, vegetation with year-round or temporal diversity	**5**	5	**Rareness**	Remarkable, rare in the region, time-dependent temporary view	**5**	5
	Vegetation with limited diversity	3			Areas that look similar to others, although different	3	
	Very limited or no diversity, contrast within vegetation	1			Interesting but very common areas in the region	1	
Water	Dominant and clean steady or running water, white water cascade	**5**	5	**Cultural Changes**	Supports visual diversity while enhancing visual harmony	**2**	2
	Flowing or non-steady/non-dominant water	3			Areas where there is no visual mismatch and with little impact on visual diversity	0	
	Hardly recognized water	1			Strongly mismatching and very irregular changes	-4	

Slope Groups: Slope grouping is classified as 0–10, 10–35, and > 35 and 0–10; scored as 1, 3, and 5 points, respectively.

Protected areas: National parks, nature parks, natural conservation areas, archeological sites, and natural and archaological sites have been evaluated as protected areas and have received 5 points.

Tab. 2: Esthetical-Perceptual Assessment Field Study Form

GPS Coordinate	**X Coordinate**		**Y Coordinate**	**Date:** 07.08.2013	
	455431		4624919	**Photo No:** 13	
Location:	Bozköy Bay and Beach				
Esthetical		**Value**	**Perceptual**		**Value**
Naturality	Natural	x	**Stimulant**	Captivating	
	Semi-natural			Interesting	x
	Artificial			Ordinary	
Surrounding	Open		**Satisfaction**	Fine	x
	Semi-open	x		Content	
	Closed			Discontent	
Diversity	Diverse	x	**Safety**	Safe	x
	Monotype			Unpeaceful	
Color	Colorful	x		Frightening	
	Single Color		**Dominancy**	Land form	
	Dull			Flora	
Line	Vibrant			Water element	x
	Curved (semi-vibrant)	x		Agriculture	
	Horizontal (inert)			Settlement-Rural	
Tranquility	Dynamic	x		Settlement-Urban	
	Calm				
Perceived landscape type (survey point): Coastal and Forest Landscape					
Other types of landscapes perceived (periphery): Rural Settlement					

Rocky and Bare Areas: The areas determined through satellite images and field studies have been converted into spatial data in GIS environment and have received 5 points.

Vegetation: The landscape pattern classes are taken from Habitat class map data, and accordingly broad-leaved, coniferous, and mixed forests (D1) received 5 points; plant exchange areas (D2-D4-D6), non-irrigated orchards (T3), non-continuously irrigated mixed agriculture areas, and continuously irrigated mixed agricultural areas (T5) received 3 points; non-irrigated arable areas (T1) and continuously irrigated areas (T2) received 1 point.

CORINE: CORINE land use/cover classification and slope, stream, and lake layers obtained from existing data are used for analysis. The data included in the CORINE land use/cover classification data are reclassified and evaluated under color, rareness, and cultural regulations categories.

Water: The water data obtained from the General Command of Mapping are used. Areas to have a 100 m bumper zone around the sea and second-degree streams received 5 points.

Color: The data obtained from the habitat classification map is evaluated according to spot shapes and stained forests received 5 points, broad-leaved forests and leaved forests received 3 points, and other areas received 1 point.

Rareness and its Contribution to Landscape Character: There are no national parks, nature parks, conservation areas, and bays in the area and these areas received 5 points. The orchards, shore, beach, and sands received 3 points.

Cultural collocations: In the area, in terms of cultural collocations, there are mining areas and tunnel extraction areas for road works, and as these areas have a negative impact on the visual value, they received -4 points.

The second phase of the visual assessment studies is completed during the survey. The visual quality score range and classification of the area are created by modifying (20) (Tab. 4). The scoring made according to Tab. 3 is processed in Tab. 1 during the field studies, and the total scores of the areas are calculated. The data obtained as a result of scoring are overlapped and total scores in the range of -4 to 27 are calculated. The scores are classified according to the score range in Tab. 4 and the visual quality analysis map is created as shown in Fig. 1 (left). The scores and distributions of the points that are given visual scores in the field according to the number of resources are shown in Tab. 5. The visual quality values map created as a result of these scores is shown in Fig. 1 (right).

In the area, no point could be determined for Class A category (very low) according to the visual landscape value classification. Due to the negative impact on the environment, the ugly view in the area caused by the work on the road connecting the Bartın road and the Amasra Tunnel entrance fall into Class B category and receives 5 points (Fig. 2 left). There are 18 points in the Class C (medium) category, 7 of which received 7 points and are the most negative views in the field. Images of some are shown in Fig. 3 (middle)–(right) and Fig. 4 (left)–(middle)–(right). There are 21 points in the Class D (high) category. The remaining 30 points are classified as Class E (highest). The highest scored points in the field; Kaman bay (27 points), Akkonak Çötlük Bay (25 points), Çapak Bay view from Tarlaağzı village road (25 points), Bozköy Bay and Beach (23 points), view from Çakraz Bay to the west (23 Points), Kanatlı Village Göl Deresi Waterfall (23 points) are shown in Figs 4 and 5. As is apparent, the study area is a very important region in terms of visual landscape value. As these areas contain very sensitive ecological structures, the nature of the plan decisions should be for protection purposes.

Tab. 3: Visual Landscape Quality Map Data to Be Prepared by Utilizing the Existing Data of the Research Field (Modified Version of(20))

	5(Five) Points	3 (Three) Points	1 (One) Point
Landform			
Slope Groups	>35%	35–10 %	10–0 %
Area of Conservation	National parks, natural parks, natural protected areas, archeological sites, natural and archaeological sites		
Rocky and Bare Areas	Satellite view Bare and rocky areas		
Flora	Forest areas	Shrubland	
CORINE	CORINE: 311-D1(Broad-leaved Forests >75 %) 312-D1(Needle-leaved Forests >75 %) 313-D1 (Mixed Forests >75 %)	CORINE: 324-D2, D4, D6 (Plant Exchange Areas) 2221-T3 (Non-irrigated Orchards) 2421-T5 (Non-continuously irrigated Mixed Agriculture Areas) 2422-T5 (Continuously irrigated Mixed Agriculture Areas)	CORINE: 211-T1 (Non-irrigated Arable Area) 212-T2 (Continuous Irrigation Areas)
River Corridor Plants	The areas containing the herbal corridors around the second-degree streams that are determined through the satellite image could not be defined due to the lack of sufficient data.		
Water	Areas to have a 100 m bumper zone around the sea and second-degree streams.		

(*continued on next page*)

Tab. 3: Continued

	5(Five) Points	3 (Three) Points	1 (One) Point
Color	313-D1, D2 (Mixed Forests)	311-D1, D2 (Broad-leaved Forests) 312-D1, D2 (Needle-leaved Forests)	Agriculture Areas: 2111-T1 (Non-irrigated Arable Areas) 2421-T5 (Non-irrigated Mixed Agriculture Areas) 2422-T5 (Continuously Irrigated Mixed Agriculture Areas) 243-T6 (Agriculture Areas mixed with Natural Vegetation)
Rareness and Its Contribution to Landscape Character	National parks, nature parks, protected sites, and bays	Fruit orchards: 222 (Orchards) 331 (Shore, beach, sands)	
Cultural Collocations			Mines (-4): 131 (Mining Areas)

Tab. 4: Visual Quality Scores and Classification (Modified Version of (20))

Class	Score Range	Description	Code
A	(-4) -1	Very low	1
B	2 - 5	Low	2
C	6 - 9	Medium	3
D	10 - 14	High	4
E	15 – 27	Very high	5

Esthetical and Perceptual Landscape Value

The distribution of the esthetical and perceptual values in the study field are given in Tab. 6, and the distribution of landscape types are given in Tab. 7.

In terms of naturalness, semi-natural areas with leastwise human interference stood out with 50 resource points. In terms of encircling, the number of semi-open areas is high with 61 points. Monotype areas (58 points) mostly stood out with sea

Tab. 5: Distribution of Visual Scores by Number of Resources

Resource scores	27	25	23	21	19	17	15	13	11	10	9	7	5
Number of resources	1	2	2	8	1	11	5	10	10	1	10	7	1

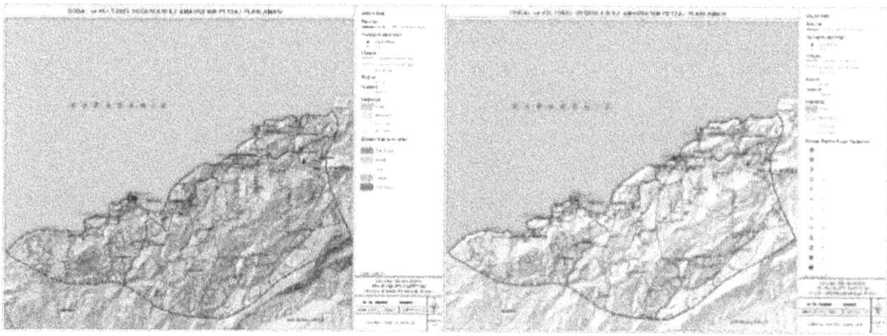

Fig. 1: Visual quality analysis map (right), visual quality scores map (left).

Fig. 2: Amasra Tunnel link road site (left), view of marble quarry from Kalaycı village road (middle), view of Hema 1 from Gömü Village (right.)

Fig. 3: South-east view from stone and debris quarry opened for road and tunnel construction on İnpiri village road (left), T1 tunnel construction site (middle), Amasra Tunnel single-tubular (right) passage.

Fig. 4: Karaman Bay (left), Akkonak Çötlük Bay (middle), Çapak Vay view from Tarlaağzı Village road (right).

Fig. 5: Bozköy Bay and Beach (left), west view from Çakraz Bay (middle), Kanatlı Village Göl Deresi Waterfall (right).

or forest cover. In terms of color, the area reveals colorful views with 53 points. When the frequency of sloping areas is examined lineally, it is observed that 35 points are curved, 28 points are vibrant, and 6 points are horizontal. In the area, tranquility is in the foreground with 41 points. Interesting areas (42 points) and captivating areas (24 points) also draw attention. In terms of satisfaction, 31 points are evaluated as content, 18 points as fine, and 20 points as discontent. While safety stands out at 42 points, 3 points have a frightening impression. In the area, while the vegetation and water elements are dominant at 21 and 17 points, respectively, the agriculture areas form a limited area with 3 points.

When examined in terms of landscape types, forest landscapes and the areas both with forest landscapes and rural settlements stand out with 8 points in the survey points. When examined in terms of surrounding areas, forest landscapes and rural settlements are the most common landscape types with 16 points. Esthetical and perceptual value maps are: Fig. 6 (left) naturalness, (right) encircling, Fig. 7 (left) diversity, (right) color, Fig. 8 (left) line, (right) tranquility, Fig. 9 (left) stimulation, (right) satisfaction, Fig. 10 (left) security and shape, and (right) dominancy.

Result

Landscapes according to their status are distinctive areas which contribute to defining characteristics such as interesting geomorphological microclimatic

Tab. 6: Distribution of Esthetical and Perceptual Values in the Area

Esthetical			**Perceptual**		
Naturality	Natural	18	Stimulant	Captivating	24
	Semi-natural	50		Interesting	42
	Artificial	1		Ordinary	3
Surrounding	Open	3	Satisfaction	Fine	18
	Semi-open	61		Content	31
	Closed	5		Discontent	20
Diversity	Diverse	11	Safety	Safe	42
	Monotype	58		Unpeaceful	24
Color	Colorful	53		Frightening	3
	Single color	16	Dominancy	Land form	5
	Dull	-		Flora	21
Line	Vibrant	28		Water element	17
	Curved (semi-vibrant)	35		Agriculture	3
	Horizontal (inert)	6		Settlement-Rural	14
Tranquility	Dynamic	28		Settlement-Urban	9
	Calm	41			

structure, etc., due to the natural, cultural, and visual landscape elements they involve. Areas that have an important place in the operation of ecological and cultural processes in the areas determined as a result of the visual analysis where the landscape functionality is high, should also be evaluated within this scope. In natural landscapes, the areas with high landscape functionality values are landscape conservation areas. They constitute the most significant landscapes in terms of conservation value, in order to maintain continuity and improvement of natural, cultural, and visual landscape.

The landscape structure of an area is closely related to the perceived visual landscape quality. Visual landscape character has a significant effect on people's recreational activity preferences. The more natural, different, diverse, harmonious, intriguing, mysterious, perspective, reassuring, and regular the landscape structure of an area is, the higher visual value it provides. In this context, the role of landscape esthetical quality in environmental management should be taken into consideration and the visual esthetical structure of the landscape should be included in planning decisions.

In order to increase the visual landscape value, it is necessary to eliminate the damages caused by the dysfunctional mining facilities in Amasra city center,

Tab. 7: Distribution of Landscape Types in the Research Area

Landscape Type	Survey Point	Surroundings	Landscape Type	Survey Point	Surroundings
Sea and Forest Landscape	1		Forest Landscape and Rural Settlement	8	16
Rock Landscape	1		Forest Landscape and Cultural Landscape		2
Urban Landscape	1		Forest Landscape and Mine Landscape	1	
Urban Settlement	5		Forest and Water Landscape	1	
Urban Settlement and Cultural Landscape		1	Forest, Agricultural-Valley Landscape and Rural Settlement	1	
Urban Settlement, Forest Landscape, and Cultural Landscape		5	Forest-Water Landscape and Rural Settlement	1	
Rural Settlement	4	2	Forest-Valley Landscape and Rural Settlement	1	
Coastline Landscape	3	1	Water Landscape	1	
Coastline Landscape and Urban Settlement	1		Agricultural Landscape and Rural Settlement	1	
Coastline Landscape and Rural Settlement	1	1	Agricultural Landscape	3	1
Coastline Landscape and Maquis Landscape	1		Agricultural and Forest Landscape	1	1
Coastline Landscape and Forest Landscape	5	5	Agricultural-Forest Landscape and Rural Settlement	3	4

Tab. 7: Continued

Landscape Type	Survey Point	Surroundings	Landscape Type	Survey Point	Surroundings
Coastline-Forest Landscape and Rural Settlement	3	1	Valley Landscape	1	1
Coastline-Forest Landscape and Urban Settlement	4	1	Valley Landscape and Rural Settlement	1	
Mine Landscape	1		Valley-Forest Landscape	2	
Cemetery Landscape	1		Valley-Forest Landscape and Urban Settlement	1	
Forest Landscape	8	24	Road Landscape and Forest Landscape	1	
Forest Landscape and Urban Settlement	2	2			

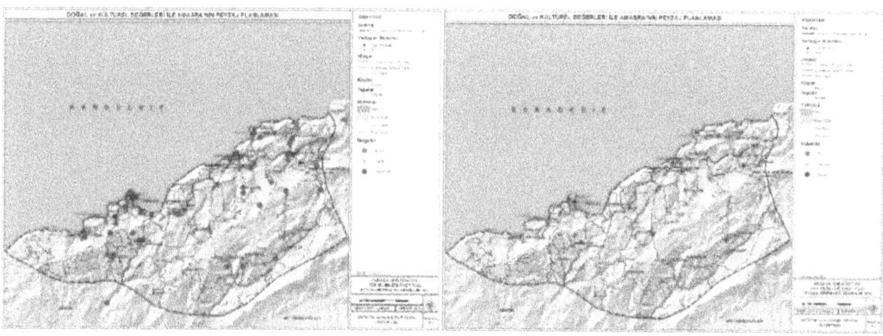

Fig. 6: Naturalness (left), surrounding (right).

Tarlaağzı, and Gömü areas, the closed marble quarries, the slope areas resulting from the road maintenance works, and the tunnel constructions along the Bartın-Amasra road, by means of landscape maintenance works. In the process of planning and design of an area, assessing the biophysical characteristics of the landscape and making planning decisions considering the existence of visual landscaping

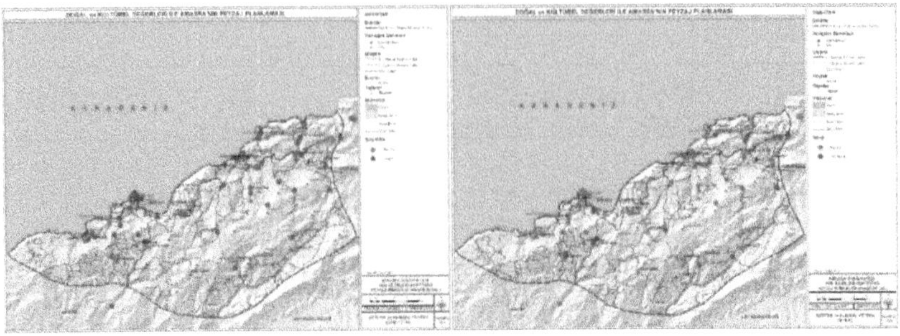

Fig. 7: Variety (left), color (right).

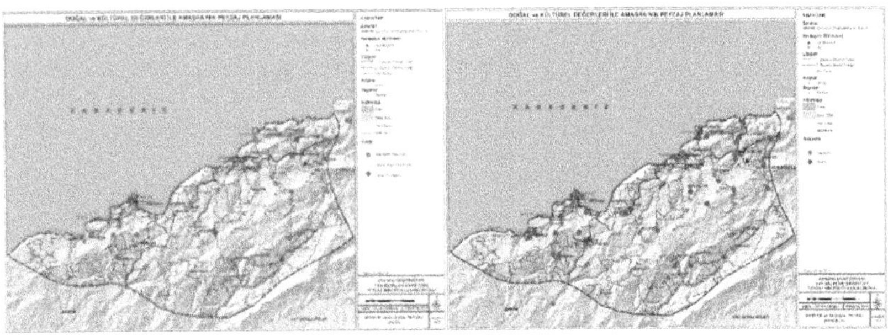

Fig. 8: Line (left), color (right).

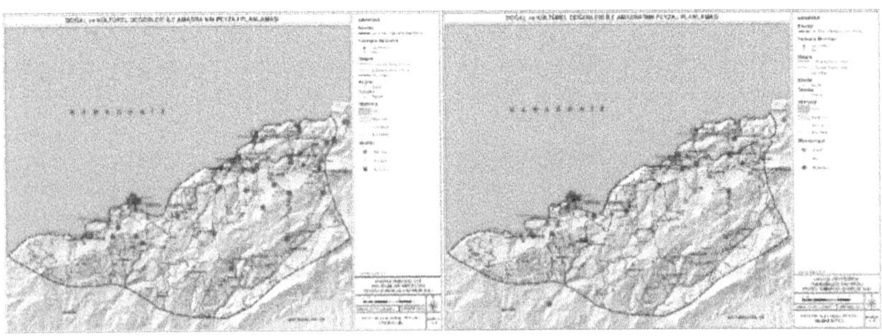

Fig. 9: Stimulant (left), color (satisfaction).

resources is important in terms of maintaining the continuity of landscape assets to be conserved. Beautiful bays, scenic viewpoints, natural vegetation, waterfalls, and historical and cultural values that have not been destroyed much in this area should be assessed in this context.

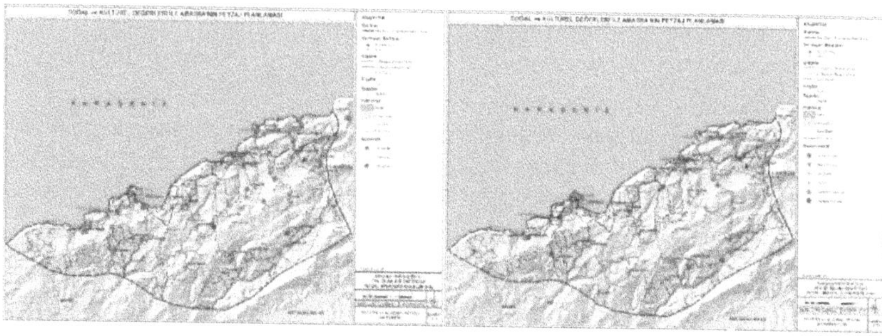

Fig. 10: Security (left), color (dominance).

Within the scope of the PhD thesis, the landscape character areas are mapped by taking the visual landscape quality assessment and esthetical-perceptual assessment criteria into consideration. This chapter shows the necessity of making a visual landscape quality assessment in order to be able to create the landscape pattern and landscape character maps within the scope of Landscape Character Analysis Assessment method.

This chapter is based on the first author's doctoral thesis entitled "Landscape Planning with Natural and Cultural Values of Amasra."

References

1 Anonymous. European Landscape Convention. Council of Europe. ETS no.176. Florence, Italy. 2000.

2 Bulut, Z. Yılmaz, H. Determination of landscape beauties through visual quality assessment method: a case study for Kemaliye (Erzincan/Turkey). Environmental Monitoring and Assessment, 2007. 141 (1–3): pp.121–129.

3 Özhancı, E. Yılmaz, H. Tekin, H. Ödüllü Peyzaj Fotoğraflarına Analitik Bir Yaklaşım. Humanitas Uluslararası Sosyal Bilimler Dergisi, 2013. 2:151–168.

4 Ak, M.M., Akçakoca kıyı bandı örneğinde görsel kalitenin belirlenmesi ve değerlendirilmesi üzerine bir araştırma. Ankara Üniversitesi, Fen Bilimleri Enstitüsü, Peyzaj Mimarlığı Anabilim Dalı, Doktora Tezi, 2010. 156s. Ankara.

5 Daniel, T.C., Whither scenic beauty? Visual landscape quality assessment in the 21st century. Landscape and Urban Planning, 2001. 54(1–4), 267–281.

6 Özhancı, E. Yılmaz, H. Rekreasyon alanlarının görsel peyzaj kalitesi yönünden değerlendirilmesi; Erzurum örneği. Iğdır Üniversitesi Fen Bilimleri Enstitüsü Dergisi, 2011. 1(2), 67–76.

7 Meitner, M.J., Scenic beauty of river views in the Grand Canyon: relating perceptual judgments to locations. Landscape and Urban Planning, 2004. 68: 3–13.

8 Palmer, J.R., Hoffman, E. Rating reliability and representation validity in scenic landscape assessments. Landscape and Urban Planning, 2001. 54: pp.149–161.

9 Çakcı, I., Peyzaj planlama çalışmalarında görsel peyzaj değerlendirmesine yönelik bir yöntem araştırması. Ankara Üniversitesi, Fen Bilimleri Enstitüsü, Peyzaj Mimarlığı Anabilim Dalı, 2007. 109 s. Ankara.

10 Kiper, T., Korkut, A., Üstün Topal, T., Görsel Peyzaj Kalite Değerlendirmesi: Kıyıköy Örneği, KSÜ Doğa Bilimleri Dergisi, 2017. 20 (3), 258–269.

11 Özgeriş, M., Karahan, F., Rekreasyonel Tesislerde Görsel Kalite Değerlendirmesi Üzerine Bir Araştırma: Tortum ve Uzundere (Erzurum) Örneği, 2015. Artvin Çoruh Üniversitesi Orman Fakültesi Dergisi, 16(1): 40–49.

12 Asur, F., Alphan, H., Görsel Peyzaj Kalite Değerlendirmesi ve Alan Kullanım Planlamasına Olan Etkileri (Derleme), 2018. YYÜ Tarih Bilim Dergisi (YYU J AGR SCI) 2018, 28(1): 117–125.

13 Yeomans, W., Visual Resource Assessment. A User Guide, 1983. MOE Manual 2. ISSN 0821-0640; 2. Ministry of Environment, Victoria B.C. Canada.

14 Wherrett, J.R., Visualization techniques for landscape evaluation. Literature review. The Macaulay land use research institute (Mluri), 1996. Aberdeen Scotland.

15 Kalın, A., Çevre tercih ve değerlendirmesinde görsel kalitenin belirlenmesi ve geliştirilmiş: Trabzon sahil bandı örneği. K.T.Ü. Fen Bilimleri Enstitüsü, Peyzaj Mimarlığı Anabilim Dalı, Doktora Tezi, 2004. 221 s. Trabzon.

16 Kaptanoğlu Ç. A. Y., Peyzaj değerlendirmesinde görsel canlandırma tekniklerinin kullanıcı tercihine etkileri. İstanbul Üniversitesi. Orman Fakültesi, Fen Bilimleri Enstitüsü, Peyzaj Mimarlığı Anabilim Dalı, Doktora Tezi, 2006. 203 s. İstanbul.

17 Wu, Y., Bishop,.I, Hossain, H., Sposito, V., Using GIS in landscape visual quality assessment. Applied GIS 2006. 2 (3): pp. 18.1–18.20. Doi: 10.2104/Ag 060018.

18 Uzun, O., Müderrisoğlu, H., (2011). Visual landscape quality in landscape planning: examples of Kars and Ardahan cities in Turkey. 2011, African Journal of Agricultural Research Vol. 6(6), pp. 1627–1638.

19 Benliay A, Soydan O (2015). Aspendos-Sillyon-Perge bisiklet güzergahı örneğinde peyzaj görsel kalitesi ve peyzaj özelliklerinin değerlendirilmesi. Artium, 3(1).

20 Şahin, Ş., Perçin, H., Kurum, E., Uzun, O., Bilgili, E., Tezcan, L., Çiçek, İ., Müftüoğlu, V., Çorbacı, Ö.L., Sütünç, S., Doğan, D., Koç, Ö., Ateş, E., Tarım, B. ve Kurdoğlu, G. 2013. PEYZAJ-44: İl Ölçeğinde Peyzaj Karakter Analizi ve Turizm/Rekreasyon Açısından Değerlendirilmesi (PEYZAJ-44). 109G074 Nolu TÜBİTAK KAMAG Projesi, Proje Sonuç Raporu.

Assist. Prof. Dr. Özge İslamoğlu

29 The Concept of Refunctioning in Terms of "What," "Why" and "How" Questions

Abstract: Architecture is a field which is influenced by social, economic and technological developments. While on one hand structures that are appropriate to new conditions are built, on the other hand structures that lose their current functions are demolished or they continue to exist with different functions. At this point, it is significant to ensure the conservation of the social, cultural and economic values of structures that continue to exist in order to form urban identity and public memory. Especially in historical buildings, reuse as a way of conservation is an important tool in protecting the values that structures carry. Within the scope of the study, the concept of reuse in historical buildings that are concrete documents of notions that form social memory and cultural values that are worth preserving have been discussed under three main titles. What is refunctioning in this context? (Why?) Why is is necessary? (Why?) How is it maintained? (How?) These are the questions that are discussed. Under the title of "What" theoretical information regarding refunctioning and historical processes are included. Under the title of "Why is it necessary?" conditions in which refunctioning is necessary have been discussed. The title "How is it maintained?" contains inferences made by selected sample structures.

Keywords: Refunctıonıng, Reuse, Conservation, Heritage, Historical building

1 Introductıon

Architecture is a field which is influenced by social, economical and technological developments. While on one hand structures that are appropriate to new conditions are built, on the other hand structures that lose their current functions are demolished or they continue to exist with different functions. At this point, it is significant to ensure the conservation of the social, cultural and economical values of structures that continue to exist in order to form urban identity and public memory. Especially in historical buildings, reuse as a way of conservation is an important tool in protecting the values that structures carry. Reuse is the name given to activities aiming to create an alternative function for a building. The term *reuse* appears in the conservation of buildings that form urban identity and in maintaining the sustainability of that identity, and it is an approach that is often preferred in contemporary conservation approach.

With this conservation approach that allows to keep the values of historical buildings alive, the building can continue to exist and economic and cultural sustainability is maintained. In order to ensure conservation in a sustainable way, it is necessary to perpetuate the authentic values of the building, and and to maintain harmony between the new function chosen for the building and current physical and cultural values.

Within the scope of the study, the concept of reuse in historical buildings that are concrete documents of notions that form social memory and cultural values that are worth preserving have been discussed under three main titles. What is refunctioning in this context? (Why?) Why is is necessary? (Why?) How is it maintained? (How?) These are the questions that are discussed. Under the title of "What" theoretical information regarding refunctioning and historical processes are included. Under the title of "Why is it necessary?" condition in which refunctioning is necessary have been discussed. The title "How is it maintained?" contains inferences made by selected sample structures.

2 What Is Refunctıonıng?

Architecture is tied to economical, social and political events, and it goes through developments over time. Architecture is in fact a reflection of society and social events. Güzer emphasizes that architecture is not only used but is a sphere of consumption at the same time that represents, attaches identities and determines the boundaries of difference, authenticity and being privileged [1]. In this context, architecture is a concept that reflects almost all the characteristics of the period it belongs to [2]. In particular, historical structures that are within the field of architecture carry a significance at this point. The reason is, beyond being buildings that are built simply to live in, they are sites [3] that represent traditions, culture, habits, public consciousness and experience. These structures have witnessed history, they carry the characteristics of that time to our present day, and contribute to the formation of urban identity. In this regard, all these values that render them concrete as expressions reflecting the history, culture, economy and lifestyles of society create the necessity for their conservation.

The conservation of historical buildings in general encompasses the set of activities that aim to avoid unwanted changes in buildings, are the symbol of social, cultural, historical and architectural values and to maintain cultural continuity and sustainability. The concept of conservation is a subject that is often discussed in the field of theory and practice. In this regard, many definitions of it have been made. According to Burden "conservation" covers the activities that aim to prevent decay, ruin, abuse and desertion of a building, as well as past records of the structure and relevant conservation measures [4]. Feilden, expresses conservation as the act of conserving natural and cultural environment and transferring them to the future [5]. According to Yaldız, conservation reflects social, economic and cultural values of societies, prevents their extinction within social changes, and thus is the act of sustaining them in a way that will cover yesterday, today and tomorrow [6]. Hasol defines conservation as taking the necessary measures in order to ensure that structures or city sections that bear historical or artistic value will continue to exist [7].

When we talk about the conservation of historical buildings, the first way that comes to mind, without doubt, is giving a different function to a building that has lost its authentic function over time. Change of function is a contemporary conservation approach in historical buildings [8]. An efficient and sustainable conservation approach is possible by enabling a building that lost its function to continue to exist with a function appropriate to our current conditions [9]. Reuse is a method that is often used in the conservation of buildings that are valuable documents of their time period as they reflect social, cultural and economic accumulation of the society and it refers to a process of transformation that aims to develop new functional requirements [10]. The use of historical buildings with respect to contemporary requirements by conserving their authentic functions is an indispensible element to the function of conservation [11].

There are many definitions regarding the concept of reuse. Bacon defines reuse as one of the conservation methods, applied in the conservation of economic and cultural identity, aiming to prevent buildings from becoming inactive and extinct [12]. Feilden stresses that reuse of historical buildings is a process that transfers concrete and abstract values, natural and cultural environments of those buildings to future generations by conserving those values [5]. Cantell defines reuse as developing new ways of utilizing the potential of a vestigial and abandoned historical building in order for it to continue to exist economically in accordanc with current requirements [13] and for Burden, reuse is, simply put, a way of making the building appropriate to emerging needs by making interventions [4]. As for Özer, it is expressed as allowing buildings to be integrated into contemporary life efficiently by attaching to them new functions in terms of meaning and use [14].

In general terms, reuse is the physical and functional obsolescence, deterioration, extinction or becoming incapable of a building because of environmental, cultural, economical, technological reasons, and therefore being unable to satisfy needs; as a result of this it is reopened to use by applying various interventions. Reuse covers the process and activities that enable the adaptation of buildings that have lost their authentic function, to current needs with the aim of conserving them.

3 Why Is Refunctioning Necessary?

The concept of keeping historical buildings alive by adapting them to the needs of current times rather than conserving them by suspending them is becoming more and more widespread [15]. In this context, allowing historical buildings to exist with new functions is a prevalent approach in conservation.

The significance of refunctioning in the conservation of historical buildings is stressed in various studies. Worthing and Bond state refunctioning as a necessity for the adaptation of historical buildings to changing life standards [16]. Madran and Özgönül emphasize that cultural property needs to be conserved through refunctioning due to many reasons such as learning about the past, benefitting from experience, taking measures for the future and transferring them to the future as documents[17]. Altınoluk states that the way to attain a meaning that influences and educates, dignifies society is possible by seeing these buildings as living structures, and this can be maintained by giving them new functions, by making them useful to society [18].

So why is refunctioning necessary? Time, nature conditions, rent concern, vandalism and many other factors accelerate the deterioration of structures [19]. Also as a result of changes in living environment over time, and changes that happen in living environment parallel to changing needs, functions of historical buildings change. These buildings that lose their function, but still bear symbolic meaning and value, need to be integrated to contemporary life by attaching to them appropriate new functions in terms of meaning and function. It is possible to explain this necessity with the values that the building to be conserved carries with it. Since these structures have values that relate to continuity; identity; symbols; document; history; aesthetics; economical, social and political reality [5, 20, 21], they need to be conserved. Because of these values they carry, conserving them through refunctioning makes significant contributions in the context of urban and public benefit.

Refunctioning of buildings is necessary due to acquired historical, cultural and economic contributions. Even though historical buildings lose their authentic function, their being reflective of social, cultural, economic characteristics and lifestyles of their society attaches to them a significance in terms of forming and perpetuating social memory. Historical buildings are witnesses of history with all the traces that they bear. In this context, historical buildings are cultural products that need to be sustained and transferred to future generations [18] and therefore they have to be refunctioned by being reconsidered **historically and culturally**. As well as in terms of conservation of historical and cultural values, refunctioning also provides economic benefits. Refunctioning emerges through the synthesis of conservation approaches and **economic** ways of thinking [22]. The reason for this is using the reserves of a present structure will be more economic that building a new one. Making use of a present structure means consuming less energy and not depleting natural resources [6]. At the same time, it transforms the building and its environment to a center of attraction and makes room for the development of the building and its environment.

Refunctioning is necessary for buildings that have deteriorated functionally but remain physically robust or those that have the potential to be reused in order to prevent them from destruction. Preventing the destruction of an architectural heritage, conserving their values, maintaining continuity and creating economic benefits is important for countries in the context of the values they possess. Therefore, refunctioning as a conservational intervention is necessary in order to intergrate historical buildings to urban life.

4 How Is Refunctioning Maintained?

What is refunctioning and why is it necessary are questions with clear answers, yet, how is it maintained, how should it be maintained, are questions with more complex answers. The reason is there are no standard intervention procedures applicable to refunctioning of buildings that are worth conserving, and every problem has strong and weak points peculiar to itself [23]. In order to properly maintain reuse, the most important step is to determine the intervention approaches [24]. Even though intervention approaches are not drawn with strict boundaries, it is necessary to specify a number of criteria for the intervention to be applied to the building that is thought to be refunctioned and then they need to be applied in a way that is peculiar to the building. In this way conservation can be done with minimum damage and in a sustainable way.

Refunctioning what the buildings represent is more important than who they serve. Since conservation of an object is trasferring the message and appearance of the conserved object, every act of conservation and the method of conservation influence the message of the conserved [22]. What is important in refunctioning is not the continuation of the function, but rather, the ability to keep the building alive [25]. In this regard, the aim in reuse is to ensure active conservation of the building, rather than satisfying the new user's needs and necessities completely [6]. Due to this, the essential thing in the application-oriented decisions regarding refunctioning is conserving the physical and cultural values of the building and choosing a function that is appropriate to it.

Reuse of historical buildings contains a number of decisions that shape the future of the building. Contrary to the process of architectural design, in refunctioning interventions that can be applied to the building are shaped according to the current structure. In refunctioning, areas that are needed for activities are limited with the spatial form and structure of the building [26]. In this regard, refunctioning is also expressed with terms like "re-programming and re-architecture" [6]. Re-programming refers to the rearrangement of existing sites in order for them to contain new utilizations, activities, events and re-architecture defines the application of new programs by utilizing the potential of architecture language [22]. That is, refunctioning should be applied with respect to boundaries and in connection with the cultural identity, namely, the first determinant of design should be the authentic architectural characteristics of the structure. The changes made should have reversible traits.

The examples discussed in this part are limited with buildings whose authentic function were industrial structures. The reason for choosing industrial structures is that, together with their mechanical equipment for production and with the structures that enable them to engage production, they are important indicators of the socio-economic past of a country, and that they point towards a overall conservation value encompassing with their stratified living patterns and spatial traces in every field of life [22] and in our present time, the fact that the conservation of "old" industrial structures due to economic, cultural and aesthetic reasons has become a phenomenon [27]. Moreover, it has been deemed appropriate as these structures are buildings that reach more users, become prominent in the publicity of the country with their refunctioned states as parks, public museums directed towards education, culture and arts. Firstly, buildings that have been refunctioned and which have a place in international literature in England, Holland and Spain

Fig. 1: Tate Modern, London [30,31].

have been included; afterwards, more detailed information is given on Rahmi Koç Museum, whose authentic function in Turkey was an industrial structure and is used today as a museum, which in 1996 received "Museum of the Year Special Award," and is a museum that I have personally experienced.

Tate Modern is one of the industrial structures that was refunctioned as a museum in London (Fig. 1). In the first years of the 19th century docks were built on the side of the River Thames for ship building, maintenance and repair; this area lost its economic vitality in 1967 and reached a state in which it could not satisfy techological needs. The Bankside Power Station in the area was transformed to Modern Art Gallery in 2000. In 1981 a contest was organized for the structure which had lost its function and Jacques Herzog and Pierre de De Meuron won this contest thanks to the conservation method and the references made to authentic function. This design won the architects The Pritzker Award and this is one of the most visited museums in the World [28,29].

Cam Framis Museum in Barcelona consists of a new structure that was designed to connect structures to each other by restoring the factory building that was abandoned and idle (Fig. 2). An old storehouse in the area was demolished, creating space between the buidings, and it was refunctioned [32].

Van Nelle Factory in its authentic function was a factory where tobacco, tea and coffee were produced (Fig. 3). The building was designated as World Heritage in 2014. In the period it was completed (1925–1931), it was defined by famous architects as the most beautiful view of the modern era. (https://tr.wikipedia.org/wiki/Van_Nelle_Fabrikas%C4%B1) Van Nelle consists of a factory building, an office building, storehouse, discovery and stock warehouse, a boiler house, workshops, cafeteria and sports areas. It is a structure planned in a green area with social facilities. The factory was complemented with production units in 1951 and exhibition centers, then after being taken over by American Standard Brands in 1989 it was sold and it was designated to be in the status of national monument [29].

In all the examples that were used by refunctioning, it is clearly seen that a conservation approach that does not do harm to industrial identity, where almost no intervention to existing architecture was made, existing material, texture, architectural elements, volume and structural elements were used.

Fig. 2: Can Framıs, Barcelona [33].

Fig. 3: Van Nelle, Rotterham [33,34].

5 From Lengerhane and Hasköy Shıpyard to Rahmi M. Koç Museum

Rahmi Koç Müzesi was constituted in Beyoğlu district of Istanbul through refunctioning of Lengerhane and Hasköy Shipyard on the waterside of the Golden Horn (Fig. 4). Lengerhane Section was opened on December 13th, 1994, and Hasköy Shipyard was opened on November 20th, 1996 [35]. The museum, which was opened with the support of Rahmi Koç, one of the renowned businessmen in Turkey, is the first important museum of Turkey that was dedicated to industry, transportation and communication. In 1996 it was deemed worthy of "The Musem of the Year Award" by the Council of Europe.

Lengerhane, whose construction date is uncertain, was built to serve as foundry to the Ottoman Fleet during the reign of Ahmet the Third (1703–1730). Lengerhane[1]

1 The designation Lengerhane comes from "Lenger," which is anchor in English, as the chain that is thrown in the water and the anchor attached to it is called by this name; therefore Lengerhane corresponds to "The place where anchor is made." (Fig. 5)

Fig. 4: Lengerhane ve Hasköy Shipyard [36].

structure is stated as "Torpedo Storehouse" in Necip maps, while in Montke map in 1852 it was designated as "Foundry" [36,37]. After using the Lengerhane structure as foundry, there is not sufficient information regarding its use at the end of the 19th century and the start of the 20th century. The structure was used by the Treasury until 1951, and later it was used as Spirit Storehouse by Tekel Cibali Tobacco Factory. Due to different usages in different periods alterations in the architectural characteristics of the structure have been made. A fire occured in 1984; in 1991 Rahmi M. Koç Industry Museology and Culture Foundation bought it and Lengerhane began to be used as Industry Museum in 1994. Within this process Lengerhane structure was evaluated as Second Degree building by First Degree Cultural and Natural Heritage Conservation Board [35,37].

Lengerhane is located in an approximately 2100 m^2 land together with the building next to it. Lengerhane structure is square plan. At the center of the structure there is a dome on four pillars and on four sides of this dome there are rooms with mirror vaults. Along with this, on four corners of the structure there are domed rooms situated on octagon pulleys. As befitting the architectural style of the period, the alternate wall technique is one row of stone and two rows of brick, the pillars carry the main dome and the coves in between are face stone. The 16 windows above the two-floor facade are arched, and the 16 rectangular windows on the ground floor have been built into arched niches, thereby setting the facade in motion.

Hasköy Shipyard, having been established in 1861 by Şirket-i Hayriye for the maintenance and repairing of ferries belonging to the company, was designed as a workshop and consisted of a few small buildings when it was first founded [38]. This establishment was called a factory in the first times, though later a wooden leigh was added to it in 1884, then another sleigh was built in 1910, also,

Fig. 5: Lengerhane Structure, pre-restoration appearance [36].

Fig. 6: Hasköy Shipard pre-restoration appearance [40].

a turning lathe, a construction workshop and carpenter's shop were established [29,39]. (Fig. 6)

In the shipyard where only repairing work was done until 1934, thereafter ship building was commenced, two city ferry routes were built in 1938. The shipyard was associated with the Directorate of State Waterways and Ports in 1945, then it was assigned to Ministry of Maritime Affairs in 1952. Then in the year 1984 the shipyard was associated with Turkey Ship Industry Inc. and thereafter performed the maintenance and repair of city route ferries and small crafts. When Şirket-i Hayriye was bought by Ministry of Transport in 1945, this shipyard was assigned to State Sea Routes and Ports Department. The institution was assigned to Maritime Bank, and from 1954 onwards was known as Directorate of Ship Building and Repair. Hasköy Shipyard was associated with Turkey Ship Industry Inc. in 1984 and it undertook the maintenance and repair of many city route ferries and small crafts. The shipyard was bought by Rahmi M. Koç Museology and Culture Foundation in 1996 [35,37].

The entrance to the shipyard that is located on the north of the Golden Horn in a 11,068 m^2 land is from Hasköy Street that carries the same name. Fourteen buildings of different size and plan characteristics that surround the land form a large U shape. The buildings are situated on a 4300 m^2 ground. Together with the intermediate levels, it covers an area of 7800 m^2. The buildings have generally been built as single-story stone buildings.

Refunctıonıng of Lengerhane Building and Hasköy Shipyard

Lengerhane structure was planned together with the small buildings and the yard surrounded by them in order to provide the museum land needed. A three-floor steel construction unconnected to the main structure was placed inside the cube-shaped building. In this way it was made possible to use the authentic architecture of the building together with contemporary building techniques. The exhibition area situated at the basement level was connected to the main building from underground with a glass ramp [29,37]. The added steel-framed mezzanine was built as a floating floor that does not connect with the stone walls. The gallery floor, which has been designed in way that allows the audience to watch the place from a different elevation, does not violate the interiors of the building.

The basement floor of the museum consists of many sections: Aviation, Restoration Workshop, Locomotive and Automobile Models, Toys, Navigation, Printing Machines, Cinema Section; on the ground floor Steam-Operated Machine Models, Steam-Operated Ship Machine Models, Hot Air and Internal Combustion Engines, Steam Rollers and Towing Machine Models, Locomotive Models and Calender Steamer Machines and on the upper floor scientific and communication tools are exhibited [36]. (Fig. 7)

The exhibition areas within Lengerhane structure were not sufficient, so Hasköy Shipyard across the structure was bought and it was restored together with its 14 abandoned buildings and one slipway; consequently the building was opened to be used as a museum in 2001. The 6670 m^2 land on the west of the shipyard was bought in 2001 to be used as car parking area and open display space.

While the main entrance of Hasköy Shipyard from Hasköy Street was preserved in the same way in restoration work, the museum entrance was provided from the land on the west of the shipyard. Since it was an old shipyard building, it has a large volume. In accordance with the volume of the building, the products exhibited here are larger in size. The reinforced concrete intermediate levels that were built into the shipyard building afterwards were removed. Over time intermediate levels were added in places due to reasons related to usage needs. In some buildings roof trusses were removed by adding intermediate levels and the buildings were elevated. Wooden hanging post trusses were used in most of the buildings [35,37]. (Fig. 8)

The ground floor of the shipyard structure consists of Underwater Section, Automobile Section, Steam Engines, Olive Oil Factory, Historical Slipway, Nostalgic Shops, Marine Steam Engine, Navigation, Rail Transportation Sections, and

Fig. 7: Lengerhane building after refunctioning (Author's personal archive).

the upper floor consists of Rahmi M. Koç Gallery, Atatürk Section, Colorful Mathematics World, Nursery Class Education Workshop, Motorcycles, Baby Carriages, Bicycles, Oxcarts/Horse-drawn Carriages/Sledges, Turning Lathes, Boats/Outboard Motors Sections [36].

The refunctioned building was not shaped with regard to the work at hand, the works were placed and classified according to the space provided by the architectural space and exhibited items were organized according to spaces in the following manner[2]. All exhibition areas formed in both structures are designed in a way that pays respect to the form and identity of the building, do not overshadow it and emphasize the building itself.

Lengerhane and Hasköy Shipyard, which were refunctioned in different ways from their authentic functions, have been functions appropriate to industrial structures as most of the exhibited item are industrial products. In this regard, if we take the matter on the level of function selection, it is observed that the

2 Since the structure is a second degree conservation building as regards the law number 2863 of Code of Protection of Cultural and Natural Properties, function changes of the building have been made in accordance with this law according to projects evaluated by the Conservation Board that the building is subject to. Consequently, changes made to this kind of registered buildings that undergo refunctioning is done within certain limits. In Rahmi Koç Museum, exhibition areas were created by placing the works according to the plan of the museum, rather than arranging the museum according to the works.

Fig. 8: Yeniden işlevlendirme sonrası Hasköy Tersanesi (Author's personal archive).

reuse approach applied conserves existing historical and architectural values of the building and pays attention to structure-function compliance.

Spatial requirements of the new function to be determined in refunctioning should be maintainable with the existing structure. In this context, it is important to be able to maintain the function diagram formed in accordance with the new function within the structure itself. It can be seen that entrance, exhibition, education, social areas, administrative division and support units that are necessary for museum structure have been created. It can be seen that functions such as exhibition areas in various spatial fictions, education areas for various age groups, social areas with a variety of concepts, etc., have been provided without disturbing the existing spatial fiction of the building.

Most of the items in the museum have been compiled from Rahmi M. KOÇ's private collection. Additionally, works that have been taken from various institutions and foundations and from persons as donation or temporarily are exhibited in the museum. The basis of the collection consists of real works and their models, scientific and mechanical objects. If we classify the tour plan of the museum as Lengerhane building and the additional building, the entrance is below street level, a narrow tunnel leads to the old building. In this tunnel, anchor, cannonball and cannon parts are displayed. The stairs that lead to two sideways connect the entrance to the upper floor.

6 Evaluatıon

The questions for which answers have been sought in this study are: What is refunctioning? Why is it necessary? and How?

Refunctioning is a contemporary conservation approach that is necessary for the conservation of historical buildings which have an important place within the field of architecture. It involves all the activities and processes engaged within this context. As for why it is necessary, this needs to be addressed with the values that these structures possess. Historical buildings are a reflection of the historical, social and cultural accumulation of the periods they belong to, and transferring these values to future generations and sustaining them is possible only through reuse of these buildings. As for the question of how refunctioning can be applied, its answer is more complex. There is no clear approach that we can infer within this context. Nevertheless, it is a matter of determining a number of intervention criteria.

The important thing here is the reversibility principle in the refunctioning of historical buildings. All refunctioning endeavors require some level of intervention to the structure. It is important to pay the necessary attention in order for these interventions not to cause loss of value in the historical building. In this context, it is necessary to avoid interventions that do damage to the formal characteristics that reflect the identity of the building. Within the frame of this principle, remaining loyal to spatial setup, volume, structural order, frontal setup and applying interventions that do not damage the authentic function of buildings, as well their cultural and historical characteristics, bears great significance.

References

1 Güzer, C. A., *Mimarlık Turizmi: Turizmin Nesnesi Olarak Mimarlık*, Mimarlık Dergisi. 2007, 336.

2 Yalçinkaya Erol Ş., Küreselleşme ve Mekanda Dönüşüm: Dışa Kapalı Siteler, Dönüşüm Konusunda Uygulamalı Bir Araştırma Galata Çalıştayı, Gür,O.Ş., Kuyumcu, Y., Ed., Beykent Üniversitesi Yayınları, İstanbul, ss.134–143, 2011.

3 Asiliskender, B., Anıt Kavramı, Kimliğin Sürekliliği ve Değişim: Gevher Nesibe Medresesi Deneyimi, *Mimarlık Dergisi*, 322. 2005, İstanbul.

4 Burden, E., *Illustrated Dictionary of Architectural Preservation:Restoration, Renovation, Rehabilitation, Reuse*, 63, 2004. New York, McGraw-Hill Press.

5 Feilden,M.,B., *Conservation of Historic Buildings Technical Studies in the Arts, Archaelogy and Architecture*, 1982, London. Butterword Scientific.

6 Yaldız, E., *Anıtsal Yapıların Kullanım Sürecinde Değerlendirilmesine Yönelik Bir Model Önerisi*, Doktora Tezi, Selçuk Üniversitesi Fen Bilimleri Enstitüsü. 2013, Konya.

7 Hasol D., *Ansiklopedik Mimarlık Sözlüğü*. 1988. İstanbul, Yapı-Endüstri Merkezi Yayınları.

8 Ahunbay, Z., *Tarihi Çevre Koruma ve Restorasyon*, 2009. İstanbul. Yem Yayınları

9 Oral, E. Ü., Ahunbay, Z., *Bursa'nın İplikçilerle İlgili Endüstri Mirasının Korunması*, İTÜ/a Dergisi. 2005. cilt 4, sayı 2,s:37–46

10 Eyüce, Ö., Eyüce., Design Education for Adaptive Reuse, International Journal of Architectural Research, Special Volume, Design Education:: Explorations and Prospects for a Better Built Environment, 2010, July–November.Volume (4), Issue(2–3), 419–428.

11 Tapan, M., *Soru ve Cevaplarla Koruma*, 2007. İstanbul, TMMOB Mimarlar Odası İstanbul Büyükkent Şubesi Yayını.

12 Bacon, K., *The Adaptive Reuse of Heritage Buildings, University of Calgary Master of Architecture*, 2001. Calgary, Alberta, https://dspace.ucalgary.ca/handle/1880/40863.

13 Cantell, S. F., *The Adaptive Reuse of Historic Industrial Buildings: Regulation Barriers, Best Practices and Case Studies*, Virginia Polytechnic Institute and State University, Submitted in partial fulfillment of the requirement for the Master degree of Urban and Regional Planning, 2005.

14 Özer, B., *Konservasyon, Restorasyon Ve Rövitalasyon Sergisi*, yapı dergisi, 1979, 31, 60–64.

15 Roders, A., R., G., M., M., P. Re-Architecture Basis: Lifespan Rehabilitation of Built Heritage. 2007. PhD (Thesis). Eindhoven: Technische Universiteit Eindhoven, 2007, Netherlands.

16 Worthing, D., Bond, S., Managing Built Heritage: The Role of Cultural Values and Significance 2nd Edition, 2016. Wiley Blackwell, London.

17 Madran, E. ve Özgönül, N., *Kültürel ve Doğal Değerlerin Korunması*, 2005. Ankara. TMMOB Mimarlar Odası

18 Altınoluk, Ü., *Binaların Yeniden Kullanımı*, 1998. Yem Yayınları, İstanbul

19 Zorlu, T., Engin, H. E., İslamoğlu, Ö., Aras, A., Büyükçam Faiz, S., Küçük, P, Efe Ziyrek, B. E., Lüleci, S., Trabzon-Sürmene Aksu Köyü Kırsal Mimari Mirasının Korunması Ve Turizm Odaklı Yeniden Kullanımı, Dicle Üniversitesi 1. Uluslararası Mimarlık Sempozyumu,,4–6 Ekim 2018, Özet Kitabı, pp.303–304, Diyarbakır, Türkiye.

20 Latham, D. *Creative Re-Use of Buildings*. 2002. Donhead Publishing, United Kingdom

21 Kurak Açıcı, F., Ertaş,.Ş., *Tourısm Inıtıated Changes on the Sustaınabılıty of Hıstorıcal Texture:Yeni Cuma Mosque Trabzon*, open house international September, 2018, Vol 43, No 3, 102–110.

22 Cengizkan, M., *Endüstri Yapılarında Yeniden İşlevlendirme: İş'i Biten Endüstri Yapıları Ne İş' e yarar?* 2006. TMMOB Mimarlar Odası Ankara Şubesi Bülteni 45. Matsa Basımevi, Ankara.

23 Broto, E., *New Concepts in Renovating*, 9, 2005. Barcelona, Structure Press.

24 Caterina, G., Pınto, R. M., Fabbricatti, K., Oppido, S., Bianchi, A., Medici, S., Torro, P., *Reusing and Managing the Real Albergo DE Poveri of Naples. Evaluation and Re-Design For İmproved Effiency, Facilities Management Maintenence Human Elements in Feilities Management Understanding Need of Our Customer Proceeding of the CIBW70*, 2004, 129–139,Hong Kong.

25 Cantacuzino, S. İ., *Re Architecture; Old Buildings/New Uses*, 1989. New York. Abbeville Press.

26 Kincaid, D., *Adapting Building for Changing Uses*, 2002. London and New York, Taylor & Francis Group.

27 Hume, J., Industrial Architecture and Structures: Their Retention and Adaptive Re-use the Scottish Case, *Council Of Europe, The Industrial Heritage: What policies, Reports of the Lyons colloquy*, 1987, 38, Strasbourg, France.

28 https://sanatkaravani.com/dunyanin-en-cok-ziyaret-edilen-muzeleri-2/

29 Gün, A., K., *Kıyılarda Yer Alan Eski Endüstri Alanlarının Değerlendirilmesi: Paşabahçe Cam Fabrikası İçin Yeniden Kullanım Önerisi*, İstanbul Teknik Üniversitesi, Fen Bilimleri Enstitüsü, Kentsel Tasarım Anabilim Dalı, Yüksek Lisans Tezi, İstanbul, 2014.

30 https://sanatkaravani.com/wp-content/uploads/2015/06/ulusal-saray-muzesi.png

31 http://www.wikizeroo.net/index.php?q=aHR0cHM6Ly91cGxvYWQud2lraW1lZGlhLm9yZy93aWtpcGVkaWEvY29tbW9ucy8wLzA3L1RhdGVfbW9kZXJuX2xvbmRvbl8yMDAxXzAzLmpwZw

32 https://www.archdaily.com/40219/cam-framis-museum-jordi-badia?ad_medium=gallery

33 https://tr.wikipedia.org/wiki/Van_Nelle_Fabrikas%C4 %B1

34 De Jonge, W., *Üründen Sürece: Rotterham' daki Nelle Fabrikaları' nın Yeniden Canlandırılması*, Mimarlık, 308, 2002.

35 Elhan, S., *Kentsel Bellek Bağlamında Sanayi Mirasının Değerlendirilmesi: İstanbul Haliç Örneği*, İstanbul Teknik Üniversitesi, Fen Bilimleri Enstitüsü, Mimarlık Anabilim Dalı, Kentsel Tasarım Programı, Yüksek Lisans Tezi, İstanbul, 2009.

36 http://www.rmk-museum.org.tr/rmk_tarihce.htm

37 Ergin, N., *Rahmi M. Koç Müzesi*, İstanbul Dergisi, Sayı:39, 2001,Sf. 136, İstanbul.

38 Akyıldız, A., "Şirket-İ Hayriyye," Islam Encyoclopedia, volume: 39; Istanbul, 2010, page: 203, Şirket-i Hayriye: Steamship Corporation that transported passengers and cargo from 1854 to1945 in Bosphorus.

39 Tutel, E., Hasköy Tersanesi, Dünden Bugüne İstanbul Ansiklopedisi,Cilt:4, Sf.12, 1994.İstanbul.

Assist. Prof. Dr. Pınar Öztürk Demirtaş

30 Interaction of Design Museums and Design Education as a Creativity Exposing Education Model: The Case of Britain

Abstract: Design is not only an economic development tool, but also an area that directs people's and societies' lifestyles and requires a creative education process. Especially primary and secondary school education is a very important learning and self-discovery process for the youth in terms of research, information gathering, problem-solving, creativity, art, design and technology. At this stage, planning of creativity and design education in schools is important. This study addresses the contribution of design and technology lectures in primary and secondary education to design awareness and design education, and the education programs organized by Design Museums. In the first part, the content and methods of design and technology course will be examined. In the second part, the educational function of the museum will be discussed and the design museum and its creativity training will be examined. Britain, where the path for the design was cleared with the Industrial Revolution, was chosen as the case study. Design education in primary and secondary education in Britain are reviewed and the contribution of the education programs organized by the London Design Museum on design education in schools is analyzed through examples.

Keywords: Education, Design education, Design museums, Creativity

1 Introduction

Today, lifestyle and attitudes adopted by the society are affected and guided by design. As a material value, design also has an important role in economic growth and global competition. Creation of an authentic design that contributes to the society requires an intense designing process. Designing is a future- and goal-oriented creative act of problem-solving and decision making. The development of this creative act is based on the knowledge and means of the person who will realize the design; namely, the designer. Therefore, in addition to the creative ability, knowledge and means that nurture and extend the design are also fundamental necessities. And such knowledge and means are possible through research and education.

Although design education is provided in colleges and universities within different professional areas such as industrial design, fashion design, interior design and graphic design, it is rooted in earlier stages of education. Especially primary and secondary school education is a very important learning and self-discovery process for the youth in terms of research, information gathering, problem-solving, creativity, art, design and technology. In many countries, lectures of "design and technology" or "art and design" are included in the school curriculum and the content of these lectures are very well thought out in developed countries. These lectures, where students can discover and develop their design skills, can be considered as an important starting point for their area of profession. In addition, these lectures have great importance in terms of increasing design awareness and raising creative individuals who ask questions and are capable of adopting different perspectives. The most decisive contribution to design education in schools is provided by design museums. Museums are more than just an exhibition of collections; they are important educational tools as well. And design museums have become prominent aspects for both contribution to education and raising social awareness on the importance of design. Education/workshop programs organized by design museums especially for pupils of primary and secondary education support the art and design education provided by these schools.

This chapter addresses the contribution of design and technology lectures in primary and secondary education to design awareness and design education, and the education programs organized by design museums. Britain, where the path for the design was cleared with the Industrial Revolution, was chosen as the case study. Design education in primary and secondary education in Britain are reviewed and the contribution of the education programs organized by the London Design Museum on design education in schools is analyzed through examples.

2 Technology and Design Education in Primary and Secondary School for Contribution to Design: The Case of Britain

Design is a concept that impacts the lives in a wide variety of fields from clothing to furniture, product packages to graphic illustrations and determines the behaviors of people.

Today, communication of design products of different scales with the users, user behavior and trends, and meeting their expectations are research topics of design [1]. Design process is a multidisciplinary field that encompasses many topics from engineering to science, psychology to mathematics [2]. In addition to the value it adds to lifestyle, design is also important for its economic aspects and Bruce Archer [3] defined design as a "problem solving action for a goal." If design is an act of problem-solving, then design research is required to detect a problem and solve it through creative solutions. "Creativity" and "design research" are two important concepts. Gordon and Poze focused on process while defining creativity and highlighted that creativity can be learned and improved [4]. Through a similar approach, Çellek also highlights that although skill may be inherited, creativity can be taught and improved [5]. Çellek states that creative abilities originate from intellectual accumulation and that research is of crucial importance in this regard [5].

Design research is a systematic research and its purpose is to determine the knowledge behind the man-made objects and systems, and their appearance, composition, structure, goals, values and meaning [3] As it is understood, the design research necessary for the act of design is a process that requires both sophisticated knowledge and method, and continues from the beginning to the end of the design. In addition to creating a product, competent designers, who are capable of managing such process, can also forge new behaviors like raising environmental awareness through the products they design [6].

Authenticity, accessibility, personalization and development of different perspectives are crucial aspects of design. Acquisition of such traits is possible through a design education that supports creativity. Technology and Design lecture in secondary school stage is a reflection of design education [7]. For this purpose, many countries now include lectures on technology and design in their national curriculum. The content of these lectures are shaped by many different areas. Design and technology lecture is an interdisciplinary field; while it embodies both design and technology, its scope is wider and includes cognitive modeling process. These lectures encourage pupils to develop capability and value judgment necessary to find effective and creative solutions to the problems presented in real life. They improve pupils' ability to use technology by focusing on design and efficiency [8]. Although technology and design education provided for pupils in schools is similar in terms of general purpose, there are different approaches among countries. According to Black, different approaches in some European countries can be summarized as follows:

- Technical Skill Approach focused on the acquisition of craft skills about processing resistant materials, food and textile, or electronics and automatic control (e.g., Finland);
- Craft Approach emphasizing the conservation of cultural and individual importance of manual skills, sense of aesthetics and traditional design (e.g., "Sloyd" tradition in Sweden);
- Technical Production Approach focused on the development of skills required by modern mass production, and its control and organization (Eastern Europe countries with traditions of former Eastern Bloc);
- Engineering Apprenticeship Approach where the school is considered as a training field for expert technicians and engineers in higher education;
- "Modern Technology" Approach where the characteristics of the work of the future are emphasized and information technology is the focal point (e.g., France);
- "Science and Technology" Approach where these two fields are considered and studied in close relation (e.g., Denmark);
- Focus on design where design is considered to be the focal point of technology (e.g., Northern Ireland);
- Problem-solving Approach emphasizing social needs within the context of "problem" definition and understanding interdisciplinary necessities for addressing the real-life problems (e.g., Scotland and the United States);
- "Practical Capability" Approach focused on engaging pupils in a personal and active manner to address real-life problems in an attempt to balance the static and receptive perception of school education; and

- Focus on the Connection between Technology and Society with an emphasis on research for technological development to initiate social change and the study of the relation between technological development and other driving forces [9].

In addition to these, "engineering apprenticeship" approach, "practical capability" approach, "technology-society connection" approach, "approach on craft skills and industrial production," "key skills approach" and "high technology approach" are also widely accepted in some European countries [9]. Despite the differences in approaches among countries, the root idea of all forms of design and technology education is to raise socially responsible individuals who can produce creative and innovative solutions using science and modern technology. Some of these individuals will choose this field as their profession and gain an important fundamental from these schools that will guide them through self-discovery.

Industrial Revolution cleared the path for design. Britain is one of the first countries that included design education in the national curriculum. The model utilized by Britain is executed through "Design and Technology" lecture that puts design content to the forefront. In Britain, where implementation of the national curriculum is mandatory, the education system consists of four key stages based on age groups for primary and secondary schools. In the British education system: key stage 1 is for pupils aged between 5 and 7 years old, key stage 2 is for years 7 to 11, key stage 3 is for years 11–14, and key stage 4 is for pupils aged between 14 and 16 years old. Evaluation of the national curriculum reveals that "Design and Technology" and "Art and Design" lectures are provided during the first three key stages, that is, between 5 and 14 years of age.

In the National Curriculum, the purpose of Art and Design lecture is defined as "Some of the highest forms of human creativity are possessed within art, craft and design. A high-quality art and design education must allow pupils to experience their own art, artisanship and design works and provide them with the necessary information and skills to execute and create. As pupils progress in this process, they acquire critical thinking ability and develop a more meticulous understanding of art and design. Pupils also should know how art and design are reflected in and shaped our history and their contribution to our nation's culture, creativity and wealth" [10].

This education process aims for pupils to learn, use and analyze drawing, painting, sculpture and other crafts and design techniques; produce creative works, form solutions through trial; learn the changes in art forms throughout the history, and gain knowledge about important artists, artisans and designers. Lecture contents in key stage 1, 2 and 3 are developed around these feats (Fig. 1).

The national curriculum defines the purpose of Design and Technology lecture as "Design and technology involves inspirational, meticulous and constructive processes. Pupils can design and create works that are able to solve real-life problems in a diverse setting while taking their own needs, desires and values as well as others'. Pupils accumulate and utilize knowledge from a wide variety of fields from mathematics to science, engineering to art. Through this process, pupils

	Key stage 1	Key stage 2	Key stage 3	Key stage 4
Age	5-7	7-11	11-14	14-16
Year groups	1-2	3-6	7-9	10-11
Core Subjects:				
English	✓	✓	✓	✓
Mathematics	✓	✓	✓	✓
Science	✓	✓	✓	✓
Foundation subjects:				
Art and design	✓	✓	✓	
Citizenship			✓	✓
Computing	✓	✓	✓	✓
Design and technology	✓	✓	✓	
Languages		✓	✓	
Geography	✓	✓	✓	
History	✓	✓	✓	
Music	✓	✓	✓	
Physical education	✓	✓	✓	✓

Fig. 1: Structure of the national curriculum (The National Curriculum in England, Department of Education, Key stages 1 and 2 Framework document, 2013).

discover how to become individuals who are able to take risks and grow into being creative, innovative and aspiring citizens. Analyzing the history and current status of design and technology, pupils can understand how they affect people's daily life and the whole world. A nation's creativity, culture, affluence and prosperity can be greatly improved through high quality design and technology education" [10].

Lewis (2000) states that the design of British technology education program is based on a mental process-oriented perspective with a purpose of providing skills such as design and problem-solving, which is in contrast with the concept and subject-oriented program approach adopted in the United States for production, construction, transportation and energy [11]. At this point, it can be concluded that the main purpose here is to raise users and designers who actively analyze their environment, improve their design culture and skills, think and question critically, detect problems, provide creative solutions, possess an advanced aesthetic understanding, know and interpret design and technology developments throughout the history while being able to utilize methods and techniques that support design process and adept at using future technologies. "Therefore, pupils must be provided with skills of developing, planning and conveying an idea in order for them to gain special behavioral goals that get more challenging in each key stage; with skills to properly use tools, equipment, materials and components necessary to create quality product; ability to evaluate processes and product, and behaviors of knowing and understanding materials and components" [12].

Design and Technology lecture in key stages 1, 2 and 3 of the British education system is developed through creative activities that promote understanding and utilization of design as a concept. These activities are not only executed within the borders of a classroom but also in related and external institutions and industries. Method of learning for the lecture is addressed in four categories: "design"—"execution"—"evaluation" and "technical knowledge." In key stage 1, pupils are encouraged to design goal-oriented innovative products for themselves and other users through basic projects. They are expected to create forms by using different materials and explain their designs with 2- and 3-dimensional models. Pupils are taught to develop information and recommendations on how to render existing mechanisms and structures more durable, strong and effective. In key stage 2, supportive and developmental education involving sketching, prototypes, computer-aided design model creation are provided. The products are designed by using a more extensive range of materials and shaping methods, and pupils are expected to build 3-dimensional models of their designs. Mechanical and electrical system information is provided as technical knowledge. For example; subjects like gears, pulley systems, cameras, robotic arms and connections are studied to understand and use the mechanical systems within the products and structures like series circuits, switches and lamps are studied for the understanding and utilization of electrical systems inside the products. In key stage 3, pupils are provided with a more advanced perspective and understanding towards design and technology. They learn how to analyze and research different cultures, to understand the needs of users, to develop innovative and functional products designs, and they are taught about various approaches like biomimicry and user-centered design, which they can utilize in their design processes. Pupils are enabled to develop their creative ideas through annotative sketches, detailed plans and 3-D models, and they learn about mathematical modeling, verbal and digital presentations and computer-based tools [10].

In primary school (stage 1–2) and secondary school (stage 3–4) of the British education system, design and technology lecture and art and design lecture are mainly based on objects with a purpose of producing creative ideas. Lectures are conducted through design subjects like electronic products, food technology, graphics illustration products, product design, resistant material technology, systems and control technology, textile technology. And this means that pupils are gradually provided with all aspects of design knowledge during the first nine years of 11-year compulsory education in the British education system.

3 Design Museums and Their Contribution to Design Education: Case of London Design Museum

Today, museums are far beyond the traditional museology; no longer just buildings for collection exhibition, museums have become life-long learning centers. In many developed countries, museums are now considered as a part of the innovation system [13].

International Council of Museums (ICOM) defines museum as "a non-profit, independent and permanent institution that is open to public and serves the society and its development with an aim to research, gather and protect materials that are witness to humanity and its environment and to share this knowledge for education and joy purposes" [14]. As can be seen from this definition, museums have different functions such as gathering, preserving, documenting, evaluating, exhibiting, scientific publication and education. In general, the overall purpose is to directly and indirectly convey scientific and cultural knowledge to all individuals of the society, who have different ages and experience. This is why the "**education function**" is of crucial importance in order for museums to communicate knowledge in the most efficient way possible.

The studies conducted with university students studying in design departments, such as industrial design, graphic design and fashion design, have shown the contribution of museums to design education and highlighted its importance [15]. Museums have become secondary institutions that support schools especially for the education of school-age children. Although schools are the first institutions that come to mind in terms of learning, in today's education perspective, it is widely accepted that many different factors should be involved in education.

Technological developments, which are fast-developing in recent years and enter into people's daily life, and research studies on learning abilities of human show that education cannot be conducted solely in a classroom environment and cannot be entirely based on textbooks. Museums have become the most utilized institutions as a different education environment by countries that adopt modern education approach [16]. In contrast with quantitative indicators used in formal education, learning experiences have turned into performance and social values in museums and this founded an intellectual interdisciplinary language. Long-lasting and effective learning in museums has put them at the center of attraction [17].

There are various museums serving in different contexts from archeology to ethnography, fine arts to history, science to children's museum. Among these, **Design Museums** are valuable institutions that increase design culture awareness, exhibit changes in design based on aesthetic and technological developments within a historical context and explain the impact of design on the development of societies (Fig. 2).

History of design museums dates back to Victoria & Albert Museum established in England in 1851. Today, there are more than 100 design museums around the world. Design Museum, which was first established inside Victoria & Albert building, was moved to an independent building in 1989 and finally relocated to a monumental building from the 1960s in 2016. London Design Museum (Design Museum) is considered a pioneer in design with a wide variety of subjects ranging from industrial design to architecture, graphic design to fashion design. Inside its permanent exhibition "Designer Maker User," the London Design Museum accommodates important designs from design history and how design was

Fig. 2: London Design Museum (Photo by Pınar Öztürk Demirtaş).

developed through technological advancements. The museum also provides an important museum experience to its visitors with interactive interactions, design and 3D printer applications (Fig. 3).

In addition to the intellectual value it adds to design, the London Design Museum also draws attention with its education programs that aim to improve design knowledge and experience of pupils especially from primary and secondary schools. In an attempt to support design education provided by British schools, London Design Museum organized workshops and education programs in line with "Design and Technology" lecture curriculum for pupils in key stages 1, 2, 3 and 4. This is a valuable example showing the functionality of design museum in terms of the development of creativity and contribution to design education (Fig. 4).

Design Museum defines itself as a campus that is in search of an answer to the problems of the world through a local-to-global perspective using its design potential. This campus encourages its pupils to think and act like a designer and presents experiential learning programs. These workshops and courses prepared with the inspiration from the global exhibition of the museum are instructed by creative professionals and design lecturers. Different applied design programs are prepared for pupils and teachers from different stages of "Design and Technology" lecture taught in schools as per the British National Curriculum. Some of these education programs under "Designer Maker User," "Stranger by Design," "Chairs and Seating," "Lighting," "Materials: Smart and Sustainable" and "Digital Workshop" topics aim to provide experiential learning for pupils through interaction with design objects and they are applied in line with the curriculum of Design and Technology lectures at different key stages. These studies enable young people to develop design ideas based on real needs. London Design Museum, which also provides courses for teachers of technology and design lectures in addition to the education programs directed at primary and secondary school pupils, has made a great contribution to design education [18].

Fig. 3: London Design Museum, "work in progress" area–drawing, 3d printing (Photo by Pınar Öztürk Demirtaş).

4 Conclusion

Design is a process of development and change that has sociological, cultural, technological and economic dimensions.It is established within people's lives in every aspect and possesses both technology and creativity. Design education is essential not just for educating qualified designers but also for raising creative individuals who question, research and can adopt different perspectives. Education that belongs to different disciplines of design is provided to designer candidates on an academic level in universities. However, earlier design education included in secondary school is very important for creating a basis.

Design and technology lectures in primary and secondary school stage are designed to encourage pupils to think creatively, reveal need-oriented solutions and experience all these processes in a tangible manner. In the British education system, which mandates a compulsory 11-year education, Design and Technology lecture is included in the first nine years and pupils are gradually provided with all aspects of design knowledge in this process. Lectures are not just taught inside the classroom but in all related institutions and spaces, and the design process is addressed in a wide framework starting from the concept of "designing" to electronic products, food technology, graphics illustration products and product design. This process attempts to provide the ability to use concepts like biomimicry and

Fig. 4: "Designer Maker User" permanent exhibition–London Design Museum (Photo by Pınar Öztürk Demirtaş).

user-centered design, which are impactful in material knowledge and experience, production, shaping, modeling and design. The support of auxiliary institutions is essential in order for this education with many inputs to reach its goal.

Design museums are very important resources and education centers in this regard. Evaluation of London Design Museum in relation to the British education system reveals that this museum provides many workshops and education programs developed according to the content of the British national curriculum for both pupils and instructors in support of "design and technology" lectures. Education programs are based on experiential learning, cover many different subjects and their intended category of pupils is defined in terms of key stages. In addition to the education programs it organizes, museum's own spaces are also developed as learning and designing medium. London Design Museum is one of the best examples as to how a design museum can effectively contribute to the design education in schools. Creation of coordinated supportive and creative workshops by reviewing the Design and Technology curriculum, and having design professionals prepare a significant portion of these workshops forges an undisputable contribution for the education of both pupils and instructors.

Extending design museums and establishing them as an education tool in support of design education will be an essential aspect in raising creative individuals and accelerating the development in the design field. Creation of education content through school-museum collaboration gains a significant importance in this regard.

References

1 Özturk Demirtas, P. and Bostanci, S.H., (2016) *Analysing of Design Approaches of Urban Furniture at City Centres Over Students' Projects*, Balkan Journal of Social Sciences, 2016, pp. 695–707.

2 Green, P., *Design Education- Problem solving and visual experience*, BT Batsford Limited, 1974, p. 9, London.

3 Archer, L. B., *A View of the Nature of the Design Research*, Design: Science: Method, R. Jacques, J. A. Powell, eds. Guilford, Surrey: IPC Business Press Ltd., 1981, p. 30–47.

4 Gordon,W. J. J., and Poze,T., SES Synectics and Gifted Education Today. New York. 1980, Thousand Oaks.

5 Çellek, T., *Sanat ve Bilim Eğitiminde Yaratıcılık*, Başkent Üniversitesi İktisadi ve İdari Bilimler Fakültesi Eleştirel – Yaratıcı ve Davranış Araştırmaları Laboratuvarı, pivolka, 2003. 8(2): pp. 3–11.

6 Bostancı S.H., Öztürk Demirtaş, P., *Influence of Urban Furniture Design on Environmental Behaviour*, Sustainable Landscape Planning and Design, ed. M. Özyavuz, Peter Lang, Frankfurt, 2017. pp. 9–18.

7 Gündüz N., and Akbulut, D.,. *"Ortaöğretim tasarım eğitiminde yaratıcılığın desteklenmesi: İşlev odaklı sayısal tasarım platformu önerisi,"* UTAK 2018, Ankara, 2018. 3: pp. 429–446.

8 Wilson, V., and Harris, M., *Creating Change? A Review of the Impact of Design and Technology in Schools in England.* Journal of Technology Education, 2004. 15(2): pp. 46–65.

9 Black, P., *An International Overview of Curricular Approaches and Models in Technology Education*, Journal of Technology Studies, 1998, Winter–Spring pp. 24–25. [cited 2018, 12 August]; Available from:https://scholar.lib.vt.edu/ejournals/JOTS/Winter-Spring 1998/PDF/black.pdf

10 The National Curriculum in England, Department of Education, 2013. [cited 2018, 20 August]; Available from: https://assets.publishing.service.gov.uk/government/uploads/system/uploads/attachment_data/file/425601/PRIMARY_national_curriculum.pdf

11 Lewıs, T., *Technology Education and Developing Countries*, International Journal of Technology and Design Education, 2000.10: pp. 163–179.

12 Walmsley, B., *Partnership-Centered Learning: The Case for Pedagogic Balance in Technology Education.* Journal of Technology Education, 2003. 14(2): pp. 56–69.

13 Rose, M. and Johnston, L., *To Design for the Future You Must Leaf through the Past: Museums as Part of Systems of Innovation.* In: Business History, Conference, 11–13 June 2009, Milan.

14 Museum Definition, International Council of Museums, ICOM, [cited 2018, 11 September]; Available from https://icom.museum/en/activities/standards-guidelines/museum-

15 Cook, B., Reynolds, R., and Speight, C., *Museums and Design Education: Looking to Learn, Learning to See,* Ashgate Publishing Limited, Farbharm, 2010.

16 Önder, A., Abacı, O., and Kamaraj,I., *The Project of "Employing the museums for educational goals:Example of Marmara,"* Pamukkale Üniversitesi Eğitim Fakültesi Dergisi,2009. (1) 25: p. 104–105.

17 Erbayi, F., *Informal Educatıon in Education and Design Workshops of Museums,* The Journal of National Education, 2017. pp. 239–253.

18 The Design Museum, Schools and Colleges, [cited 2018, 13 September; Available from (https://designmuseum.org/the-design-museum-campus/schoolsandcolleges)

Prof. Dr. Rüya Yılmaz and Assoc. Prof. Dr. Okan Yılmaz

31 Evaluation of the Concept of Green Infrastructure Systems, Planning and Management by Using Swot Analysis (Strengths, Weaknesses, Opportunities and Threats)

Abstract: In this chapter, the concept of green infrastructure systems is evaluated by using SWOT analysis. Firstly, international reports, meetings, books, seminars, scientific articles, symposiums, congresses and information published on the Internet are evaluated for literature review about the definition, scope and application of green infrastructure. In this study, the decision-management-planning stages were determined by identifying the opportunities and threats, taking into account the weaknesses and strengths of the green infrastructure determined according to the results of the swot analysis. Green infrastructure applications were created as a result of management and planning based on the results of SWOT analysis. At the end of our research in order to achieve the standard in green infrastructure works, we recommend that cerfitication systems be prepared as soon as possible like at ecobuildings.

Keywords: Green infrastructure concept, SWOT analysis, Green infrastructure systems, Landscape planning, Evaluation

1 Introduction

Nowadays, increasing pollution and environmental and natural problems such as reduction of resources with urbanization have led to efforts to create a livable environment in the whole world. As a result of these studies, the concept of sustainability has entered our lives. New strategies are being developed all over the world within the scope of sustainability concept. The relationship between the urban spaces and sustainability which has emerged to solve environmental problems in cities, to provide quality life to people, to protect natural resources and the green infrastructure which are the natural systems which are frequently used in the world in recent years have been examined. The excessive and unconscious use of natural resources in the world has created different results. The most important of this effect is that the natural resources are damaged by pollution and consequently disappearing. Many people from across the world agree that sustainable use of resources is necessary to address these natural values, which are a common property of humans, beyond politics and to solve them globally. In this case, the use of natural operating systems of ecology is important for sustainable planning. Definition of green infrastructure for the Commission of the European Communities: to provide ecological, economic and social benefits with natural solutions, a tool that helps nature understand the benefits it provides for humans (1–5).

Applicable analysis of green infrastructure is important to ensuring dissemination. Undoubtedly the biggest obstruction to green infrastructure planning work is the economy. Green infrastructure cost ought to be calculating and the benefits and value evaluation ought to be done between green infrastructure and grey infrastructure. Green infrastructure investments are needed to ensure the balance between ecology and economy that has lasted over a century in the world. Green infrastructure's planning work forms the basis of nature protection and sustainable development. The most important problem in front of the planning and implementation of green infrastructure is providing finance. The transformation of the existing grey infrastructure into green infrastructure especially in big cities will reveal very serious financial resource needs. But the green infrastructure will provide invaluable benefits ecologically, socially and psychologically to protect our world (1–5).

2 Methodology

Methodology includes a case study based on examination of online green infrastructure documents supported by literature review. Both aim to clarify the nature and characteristics of green infrastructure planning. For the literature review of this study, articles and books published both locally and internationally in relation to the green infrastructure and green infrastructure planning were examined. Although there were many other sources, similar examples and published literature were preferred. In the study, literature on green infrastructure systems has been examined and evaluated. As a result of this review and evaluation, SWOT analysis was established.

A SWOT analysis is a structured planning method used to evaluate the strengths, weaknesses, opportunities and threats involved. SWOT analysis is a structured araştırma method used to evaluate the strengths, weaknesses, opportunities, and threats in green infrastructure system. The internal and external environment of the green infrastructure is expressed by the concept of strategic planning:

- Strengths characteristics: advantage of subject,
- Weaknesses characteristics: disadvantage of subject,
- Opportunities elements: exploit to its advantage,
- Threats elements: trouble to its disadvantage (Fig. 1.) (6–11).

3 Green Infrastructure Concepts, Benefits, Planning Hierarchy

The definition of the multi-function concept has been made in detail by the European Commission. The European Commission explains the functions of green infrastructure in four main terms, these are:

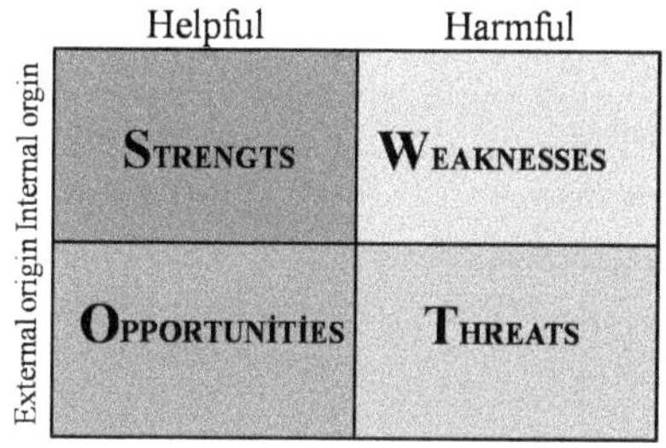

Fig. 1: Scheme of SWOT analysis (6–10).

- Conservation of ecosystems and maintaining biodiversity;
- Improve the current functioning of the ecosystem and ecosystem services;
- Encourage social well-being and health; and
- Shoring the development of a green economy, ensure the sustainable management of land and water.
- Then describe the green infrastructure properties that support these functions. For example, managed agricultural land through sustainable means and ecological corridors, greenways, ecological steppingstones, buffer areas contribute to the function of conservation of ecosystems and maintaining biodiversity.
- Restored habitats, water availability and wetlands, trees in the city and permeable pavement contribute to the function of improving the current functioning of the ecosystem and ecosystem services. Wetlands, public parks and urban vegetation contribute to the function of encouraging social well-being and health.
- For the last function; Shoring the development of a green economy, ensure the sustainable management of land and water all the features listed above contribute with the functions their provide (1–5, 12–14).

The green infrastructure covers a wide range of planning areas from micro-to-macro. In this context, the scale and typology of the green infrastructure are given in Tab. 1. (15).

The benefits of green infrastructure are summarized under environmental, social, climate change adaptation and biodiversity conservation. Fig. 2. summarizes the benefits of green infrastructure (12–17).

Green Infrastructure Planning Hierarchy

Green infrastructure planning is addressed in the areas of landscape planning and landscape design. In the planning: it is one of the most basic areas of environmental planning and it provides a rational basis for directing land use change in general. Green infrastructure is generally understood as landscape planning. However, landscape planning is a major activity that often deals with issues beyond

Tab. 1: The Scale, Typology and Examples Area of the Green Infrastructure (15)

Scale	Typology	Examples
Micro	Street trees	Urban trees, green road verges
Building	Green roofs and walls	Green roofs, living/green walls, vertical gardens, rooftop gardens, planter boxes and handing baskets on balconies, bird boxes and roost sites
Site	Amenity spaces	Informal recreation spaces, communal and private green spaces in and around housing and business developments, domestic gardens
	Derelict lands	Abandoned land, temporary greening of vacant sites
	Water management spaces	Sustainable urban drainage schemes, flood storage areas, swales, rain gardens
Neighbourhood	Park and gardens	Urban parks, country parks, formal gardens, village and town greens, playgrounds
	Land used for urban agriculture	Allotments, community gardens, city (urban) farms, permaculture, urban orchards
	Civic spaces	Squares and outdoor market spaces, other hardsurfaced areas and designed for pedestrians
	Institutional grounds	Land surrounding schools, hospitals, universities, cemeteries and churchyards
	Outdoor sports facilities	Tennis courts, bowling greens, sports pitches/ovals, golf courses, athletics tracks, playing fields
Settlement	Green corridors	River and canal banks, aisle ways , rights of way, paths and greenways, railway cuttings, embankments, infrastructure corridors
	Natural and semi-natural spaces	Woodlands, scrublands, wetland, coastal habitats, bare rock habitats, lakes and other open water bodies

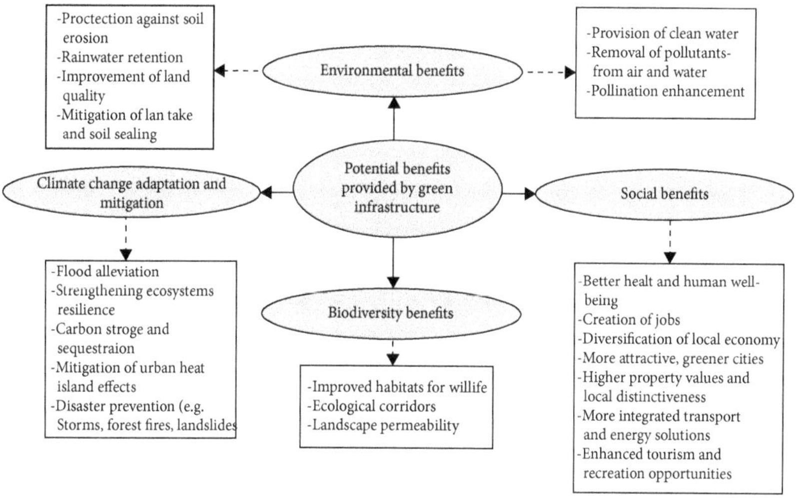

Fig. 2: Benefit of green infrastructure.

the focal point of green infrastructure planning (e.g., estimation of environmental impacts, risk assessment). Therefore in the Fig. 1, there is green infrastructure planning within the scope of landscape planning. It shows as dashed line because it is relatively underdeveloped compared to the other "planning sub-areas" shown in Fig. 3 as a subdomain. However, it is within the complex planning hierarchy, such as green infrastructure. The simplest way a part of landscape planning (Fig. 3) is generally intertwined with all planning types (15).

Green infrastructure top scale rural planning is connected with sustainable development, environmental protection, economic planning, regional planning, urban planning, social planning, etc. (Fig. 3). It should be included in these plans and contribute to the achievement of these plans.

4 SWOT Analysis (Strengths, Weaknesses, Opportunities and Threats)

In this study, strengths, threats, weaknesses and opportunities are explained by SWOT analysis method. Surveys to identified wareness, the gathering of data and the evaluation of cultural and natural values were combined by the SWOT evaluation (6–11, 18).

Detected issues in concept of green infrastructure, collected of data, literature for the assessment of green infrastructure (1–5, 12–17, 19, 20) were combined with results of SWOT analysis.

Fig. 3: Planning with linkages to other planning subfields (15 with chaning).

Strengths

- Green infrastructure like forests, plants and other flora stabilize the soil, prevent erosion, enhance the land's capacity to store water, and moderate air and soil temperatures,
- Reduction of green house gases (Sequestering carbon, reduction in or alternatives to green house gases emitters),
- Green infrastructure (e.g., wetlands) can protect coastlines against storm and flooding,
- Environmental benefits (provision of clean water, removal of pollutants from air and water, pollination enhancement, protection against soil erosion, rainwater retention, increased pest control, improvement of land quality, mitigation of land take and soil sealing),
- Wetlands, public parks and urban vegetation contribute to the function of encouraging social well-being and health,
- Green infrastructure reduces the risks and changes in air and water quality resulting from climate change,
- It can create buffer zones for habitats and species,
- It makes a major role in providing sustainable urban drainage,
- Reduces flood risk by controlling water flow,
- Enables people and animals to live in healthier environments,
- Increases sporting opportunities for healthy life,
- Psychologically makes people feel more comfortable,
- Balances the temperature in the city,

- Increases the richness of natural and cultural resources,
- Green infrastructure such as rain gardens, green roofs and walls capture water flow runoff from impervious cover before it reaches overburdened sewer systems,
- Green infrastructure application can reduce particulate pollution by absorbing and filtering particulate matter,
- Reduce public infrastructure cost,
- Provides protection of biodiversity,
- Green infrastructure provides substantial opportunities for the conservation of pollinating insects within urban and rural landscapes,
- Disaster prevention and mitigation (protecting an area against floods, storm damage, landslides, earthquakes, fires, droughts and mitigation of disaster impact, among others),
- Improves roof gardens and urban green spaces,
- Enables people and animals to live in healthier environments,
- Increases sporting opportunities for healthy life,
- Psychologically makes people feel more comfortable,
- Balances the temperature in the city,
- Roof gardens, vertical planting, etc., increases green areas within the city.

Weaknesses

- Lack of legal regulations on the application of green infrastructure,
- Lack of budget for green infrastructure applications,
- Lack of planning strategies in green infrastructure,
- Inadequate grey infrastructure,
- Unplanned understanding of local people,
- Insufficient green infrastructure project conditions,
- Inefficient orientation, promotion, advertisement and encouraging support,
- Lack of qualified staff and technical equipment in green infrastructure,
- The impossibility of green infrastructure pratices in four seasons,
- Lack of green infrastructure planning project,
- Insufficient introduction of green infrastructure activities,
- Inefficient communication among communities and local authorities.

Opportunities

- Roof gardens, vertical planting, etc., increase green areas in the city,
- Parks, urban vegetation can promote societal well-being and health,
- Provides restoration of degraded habitats,

- To enhance, conserve, and restore biodiversity by inter alia increasing spatial and functional connectivity between natural and semi-natural areas and improving landscape permeability and mitigating fragmentation,
- To maintain, strengthen and restore the good functioning of ecosystems in order to ensure the delivery of multiple ecosystem and cultural services,
- Improves quality of life,
- Good aesthetics (provide attractive sceneries, decorations, etc.),
- Providing opportunities for cognitive development, awareness, school excursions, and scientific research among others,
- Improved habitats for wildlife,
- Ecological corridors (provides uninterrupted green area),
- Provides recreation opportunity (visit to natural ecosystems, ecotourism, outdoor sports, play and relaxation),
- Growing interest to protection natural and cultural resources,
- Green infrastructure such as forest conservation and restoration averts the need for new water filtration plants to maintain clean water flows to city residents,
- The financial support of the European Union, etc., to alternative local green infrastructure projects,
- To make best use of the limited land resources,
- Contribution of green infrastructure as entrepreneurial activity,
- To enhance the societal and cultural link with nature and biodiversity,
- To acknowledge and increase the economic value of ecosystem services,
- To create incentives for local stakeholders and communities to deliver them.

Threats

- Lack of legal regulations,
- Lack of planning strategies in green infrastructure,
- Insufficient funds for supporting green infrastructure research,
- Lack of assurance of the balance between protection and usage,
- Weakness of inspection in building and construction,
- Lack of experience related to green infrastructure design and planning,
- Pollutants of the environment,
- The natural environment being under threat due to anthropogenic effects,
- Reduction of natural areas due to rapid land use changes,
- Inadequate care taking of natural, historic and cultural assets,
- Pressure on natural resources of uncontrolled wastes.
- Low request against green infrastructure,
- Community and local decision-makers are unaware of the benefits of green infrastructure,
- Initially, a high budget is needed for green infrastructure investments (especially restoration of degraded areas).

5 Results and Recommendations

In this chapter, the decision-management-planning stages are determined by identifying the opportunities and threats, taking into account the weaknesses and strengths of the GI determined according to the results of the SWOT analysis. Results and recommendations related to GI applications were created as a result of management and planning based on the results of SWOT analysis.

As a result, first of all, public policies and strategies are needed to promote green infrastructure activities. In particular, public and public institutions may introduce basic conditions for the development of green infrastructure systems through legal arrangements (e.g., making it compulsory to have a roof garden in public buildings). In our research, it is seen that green infrastructure practices are given different importance in every country.(Even in the USA and EU member states, a comprehensive planning network has not been established at the desired level.) The meetings should be held to inform local communities about strategies for the short- and long-term benefits of green infrastructure. Needs are to be established of local, regional, national and even European Union funding for green infrastructure systems. To implement the strategy of green infrastructure works, there should be sustainable development taking into account national or local targets; basic institutional, regulatory and economic aspects (concepts, guide connection rules, standards, taxes, rules and regulations).

References

1 Forest Research. *Benefits of green infrastructure.* Report to Defra and CLG. 2010, Forest Research, Farnham.

2 Ahern, J. *Greenways as a planning strategy.* Landscape and Urban Planning. 1995. (33): 131–155.

3 Ahern, J. *Greenways as Strategic Landscape Planning: Theory and Application. Doctoral thesis.* ISBN 90-5808-605-4; 2002. Wageningen University, The Netherlands.

4 Benedict, M.A., and Mc Mahon, E.D. *Green Infrastructure: Smart Conservation for the 21st Century.* Renewable Resources Journal. 2002, Autumn edition. pp. 12–17.

5 Science for Environment Policy - European Commission March 2012. *The Multifunctionality of Green Infrastructure,* 2012. [cited 2018, 02 June]; Available from: http://ec.europa.eu/environment/index_en.htm.

6 Pahl, N., and Richter, A. *SWOT Analysis. Idea, Methodology and a Practical Approach,* p: 92 GRIN Publishing, 2009, UK, ISBN-10: 3640303032, ISBN-13: 978-3640303038.

7 Pickton D.W., and Wright, S. *What's swot in strategic analysis? Briefings in Entrepreneurial Finance,* 1998. (7), 2: pp. 101–109 https://doi.org/10.1002/(SICI)1099-1697(199803/04)7:2<101::AID-JSC332>3.0.CO;2-6

8 Rachid G., and El Fadel, M. *Comparative SWOT analysis of strategic environmental assessment systems in the Middle East and North Africa region,* Journal of Environmental Management, 2015. (125): pp. 85–93, https://doi.org/10.1016/j.jenvman.2013.03.053

9 Paliwal, R. *EIA practice in India and its evaluation using SWOT analysis Environmental Impact Assessment Review,* 2006. (26), 5: pp. 492 – 510 [cited 2017, 10 June]; Available from: https://doi.org/10.1016/j.eiar.2006.01.004

10 Vonk G., Geertman, S., and Schot, P, *SWOT Analysis of Planning Support Systems Environment and Planning A, 2007.* (39): pp. 1699 – 1714, https://doi.org/10.1068/a38262.

11 Nikolaou, E.I., Ierapetritis D., and Tsagarakis, K.P. *An evaluation of the prospects of green entrepreneurship development using a SWOT analysis, International Journal of Sustainable Development &* World Ecology, 2011. 18:1, 1–16, DOI:10.1080/13504509.2011.543565, [cited 2018, 30 April]; Available from: concrete/http://dx.doi.org/10.1080/13504509.2011.543565

12 Fábos, J.G., *Greenway planning in the United States: Its origins and recent case studies. Landscape and Urban Planning,* 2004. (68) 2–3: pp. 321–342.

13 Weber, T., Sloan, A., and Wolf, J. 2006. *Maryland's Green Infrastructure Assessment: Development of a comprehensive approach to land conservation. Landscape and Urban Planning* (77): pp. 94–110.

14 Leah Hollstein, *Handbook on green infrastructure: Planning, design and implementation,* edited by Danielle Sinnett, Nick Smith, and Sarah Burgess, Journal of Urban Affairs, 2018. 40:1, 155–156, DOI: 10.1080/07352166.2017.1292082.

15 Sinnett, D., Smith, N., and Burgess, S. *Handbook on the Green infrastructure: Planning, Design and Implementation, 2015.* Edward Elgar Publishing Limited, ISBN-13: 978-1783473991, USA.

16 Martin Mwirigi M'Ikiugua, QianNaa, W., and Kinoshitaa, I., Procedia - *Social and Behavioral Sciences,Potential benefits provided by green infrastructure European Commission,* 2013. Berlin.

17 Mell, I. C., *Green Infrastructure: concepts and planning,* FORUM E-journal 8, 2008: pp. 69–80, 1354–5019 - 2008 – 04 [cited 2018, 11 July]; Available from: https://research.ncl.ac.uk/forum/v8i1/green%20infrastructure.pdf

18 Yuan, H., *A SWOT analysis of successful construction waste management* Journal of Cleaner Production, 2013. (39): pp. 1–8, [cited 2018, 5 June]; Available from:https://doi.org/10.1016/j.jclepro.2012.08.016

19 Çetinkaya, G. *Kentsel peyzaj planlamada yeni yaklaşımlar: Yeşil altyapı ve yeşil koridorlar. 5. Peyzaj Mimarlığı Kongresi "Dönüşen Peyzaj,"* 14–17 Kasım 2013, Çukurova Üniversitesi, Adana, Türkiye.

20 *Çetinkaya, G,.* Kentsel Peyzaj Ekolojisinin Sürdürülebilirliği İçin Yenilikçi Bir Yaklaşım: Yeşil Altyapı ve Planlama Politikası, *İdealkent,* 2014. (5) 12, Sf 218 -245

Assoc. Prof. Dr. Seda H. Bostancı

32 Cognitive Studies in Urban Design

Abstract: The main idea of environmental psychology is understanding the perception, cognition and the way of visual thinking mechanisms of inhabitants, and observing their experiences and behaviors in urban spaces. Accordingly, this approach enables urban designers to create positive places and design environmental and human-friendly public spaces. The aim of this chapter is to describe the cognitive dimensions of urban design and present recent cognitive applications for urban design. This research has roots from cognitive mapping to neuroscience. Current debates focus on how our visual world has changed in recent years. Innovations related to visual simulations from virtual reality to augmented reality are the contemporary design issues. These subjects are newly positioning in urban design. In this chapter, it has been determined that while the new technologies affect the human cognitive process and urban design, the designer should seek a balance between nature, environment, human expectations, and innovation, and consequently establish a holistic relationship among them.

Keywords: Urban design, Environmental psychology, Neuroscience, Perception, Cognition, Sense of place, Cognitive mapping, Visual thinking

1 Introduction

The cognitive process can be defined as the operation of information processing of the messages that the mind receives from environmental conditions and its biological structure, based on vital priorities. Among the stages of this process, filtering, the perception of the environmental information through the sensory channels and selection of useful information constitute the initial stage. Then some complex operations such as abstraction, taking place in instant memory and storing in permanent memory as a result of experiences are carried out. Loewenstein (2013) states that contrast relationships of the objects are emphasized, moving objects are spotted and boundaries are sharpened in the information channels of the mind for the survival and well-being of the organism. Accordingly, he concludes that the brain has formed a depiction of the world with its modified regular flow through its filter [1]. Therefore, it can be concluded that the mind can create a simulated living space with its ability to abstraction by picturing a fictional world suitable for itself through its sense organs. Eagleman (2011) stated that the brain is able to distinguish only what it can focus on the real image but it can't recognize the rich details of the world [2]. According to Arnheim (1969), abstraction is the most common characteristic of perception and visual thinking [3]. For example; visual deficiencies, such as the inability to see infrared rays, enhance vital adaptation, and thus allow the ability to combine visuals and create a meaningful whole, in other words, they create gestalt. Gestalt means shape, figure, and integrity.

In urban design, it is very valuable to learn how natural and designed spaces make users feel. The research dimension of this knowledge is the aesthetic quality and the research area is environmental psychology. In aesthetic evaluations of a particular region, implementation of surveys using photographs exclude most of the sensory perception parameters. Prioritization of moving objects in the cognitive process leads the studies in this area to researches such as video screenings or virtual space simulation experiments [4]. Experimental studies, which do not have an outdoor connection, cause most sensory-cognitive factors to be excluded, such as vitality, discovery, and fear. Based on these approaches, researchers gravitate toward studies that measure greater real experience. In order to measure the experience, studies such as Photo-Projective Method (PPM), which analyses the experiences of participants such as taking photographs by themselves, are carried out [5]. Urban image studies on transcranial magnetic stimulation (rTMS) [6] and functional magnetic resonance imaging (fMRI) techniques [7,8] offer new opportunities to understand the effects of the aesthetic perception and the cognitive process of the visual characteristics of cities on brain physiology. Cognitive studies, including how sensory experience and perception affect a particular urban image in memory and how it affects behavior patterns, have been among the intriguing issues of urban design discipline since the 1960s. Starting from the 2000s, the introduction of imaging techniques in the field of environmental psychology has brought a new technical dimension to this area.

Researches on the cognitive process and environment interaction include data collection, analysis, and implementation of two interrelated processes in general. The first one of these processes is the study of brain structures of humans and animals. Accordingly, it is about explaining how the transition from sensory communication and perception to cognitive process occurs with biological and behavioral indicators. In general, it focuses on understanding the fundamentals of decision-making systems. The research interests span the medical sciences and biology, as well as a wide range of subjects including psychology, economics, architecture, and urban design. In urban design, understanding the mechanism of the human brain in the feelings of aesthetics, safety, and comfort creates a guiding effect for design goals. The second one includes implementations based on artificial intelligence and machine learning by neuroscience. In this field, studies on the modeling of the cognitive process of the human mind by physics, probability, informatics, computer, electronic engineering are carried out. The second process influences a wide range of social sciences from psychology to urban design. At the same time, the implementation of the second process depends on an in-depth analysis of the first process.

The objective of this chapter is to understand the cognitive process of human minds' place recognition and make a discussion about the effects of this information on the urban design field. After discussing the place and importance of perception, cognition, and experience in environmental psychology, spatial projections of the effects of technical developments on the human mind will be discussed from the perspective of urban design discipline. This study basically includes the evaluations related to the cognitive processes in the interactions of people, defined as the first process, with space. Therefore, digital information processes such as computer-aided visualization techniques, genetic algorithm, computer vision, drone technology, artificial vision, paramedic design, visual recognition, neural network, and big data are excluded.

2 Perception, Cognition, and Experience in Environmental Psychology

The most fundamental question that environmental psychology tries to analyze is that Whyte (1980) and Nasar (1989) emphasized in their researches [9,10]; "Why do people want to visit certain places and spend a long time there, and why would they prefer to move away and avoid some places even if they sometimes have the same function from the preferred place?" Surveys and interviews are often inadequate in studies aimed at answering such questions and understanding the human decision-making and preference system. Because some of the participants of the experiment are unconscious or deliberately giving answers that do not reflect the truth. Behavioral observations and experimental aesthetic studies [11] are more effective in analyzing space-user interaction. For example, eye-catching and brain imaging analysis techniques give important clues about people's subconscious preferences. Understanding the process of formation of these patterns of behavior is the basic starting point for conceptualizing how perception, cognitive process, gestalt, visual thinking and experience, and internal expectation models are formed in the mind. While Wertheimer evaluated the organization of perceptual forms, he made assessments in terms of gestalt on how proximity, similarity, and simplicity facts were perceived and how the mind established figure-ground relationships [12]. The basis of gestalt psychology is how the human mind perceives the figure-ground relations from the whole image. Gestalt psychologists have determined that the forces operating between certain perceptual stuations are processed in different parts of the brain and that this information is combined in the cognitive process. Consequently, they reveal the piece as a whole relationship between figure-ground in the representative world [13]. As far as this process is concerned, it can be deduced that the bridge between visual perception and the cognitive process is established based on gestalt psychology.

The mind stores information that it can filter from the past, retrieves the repetitive information from the memory repository, and calculates the probability of what happens in the future from their frequency [1]. Humankind's ancestors led their mental cycles which they acquired in order to survive in savannah to other areas after they began to live in safer environments. Thus, they have gained place attachment-based behavior patterns such as enjoying the environment they are in, giving meaning to space, reflecting their cultural values to space, connecting with space, and transforming their surroundings from space to place. The behavioral patterns of the human mind formed with its interaction with the environment are connected with both social and spatial environments. Lewin, considered among the pioneers of the field of environmental psychology, defined behavior patterns with the potential for change that can be produced through social interactions between individuals and groups [14]. Rapoport, while talking about the culture filter of behavioral patterns, found that the design interventions in the living space of different cultures could harm the life of that society more than benefits [15]. When Alexander mentions that life patterns and patterns of behavior are shaped by a pattern language [16], on the other hand, Norberg-Schulz explains the importance of the genius loci concept and the unique content integrity of particular urban space in the long term [17]. Phenomenological approaches related to how the place attachment [18] is felt from the relationship between transferring from a space to a place and the experience, and the concept of topophilia [19] to the concept of sense of place [20], constitute the basis of environmental psychology. Based on all these approaches, environmental psychology investigates what the design interventions actually cause in the process of the positive or negative spatial experience [21].

Kaplan et al. state that a definable meaning, a desire to research, providing a relaxing and enjoyable experience of a particular urban environment are the main determinants of its preferability [22]. This approach shows that the positive connection and communication built with the environment depend on having the necessary motivation to start the process by arousing curiosity and interest. As a result of the processing of the data transmitted by the sensory organs acquired by perception, it is a process of understanding by transmitting the space information of the cognitive process to the memory channels. Consequently, a transformation process (or an adaptation process) from the space to the place occurs in lifestyle habits with the increase of experience in the environment.

3 Cognitive Mapping in Urban Design

In the human mind and environment interaction, it is seen that the urban space offers many messages and the responses to these messages start the new information-energy and communication processes by paying its thermodynamic cost, as Loewenstein states [23]. This approach refers to the loss of energy that can no longer be reproduced since energy is consumed for each cognitive process in the decision-making system, from remembering a particular phone number to walking, due to the semi-open biological system of humans. Thus, according to entropy, which is expressed as the second law of thermodynamics, every decision and its implementation has thermodynamic costs that should be met by the organism. It may be appropriate to interpret this as the experience-based decision-making response of the behavioral outcome in the cognitive process. Faria et al. state that cognitive mapping has a system that allows the inclusion of qualitative factors into the decision-making framework [24]. Thanks to cognitive mapping, there is an opportunity to observe what is perceived from the urban environment by the user and which message is interrupted in the process. In the theory of information, Shannon formulates this interruption by describing it as the noise problem between the encoder and the decoder [25]. Even though Loewenstein criticizes the reductionist aspect of Shannon's communication theory by saying "the formula only grasps the amount of information but does not deliver its quality and meaning," this approach remains current as the first starting point in order to simply understand the process [1]. However, Shannon published an article condemning the use of his formula out of its context in many areas from social sciences to psychology [26]. At this point, Schrödinger stated that in biology, genetics, and physics intersections, this communication process would be very complicated to define in the subatomic particle and quantum wave dimensions [27].

The research techniques, such as attitude surveys and cognitive mapping in urban design help designers to understand how inhabitants comprehend different locations and their movement experiences [28].

From the time Tolman initiated the subject and Lynch developed it in relation to environmental psychology in urban design, highly innovative technologies have been added to today's cognitive mapping approach [29,30]. Systems such as navigation system, vehicular satnav systems, and specialist navigation apparatus have become simplified and accessible for everyone to use by GPS-enabled smart-phones [31]. As a consequence, modern techniques to help understand how users feel while walking in urban areas, such as smart-watch applications that allow users to record and monitor their route and heart rate, have contributed brand-new data to the area of cognitive mapping. However, an important issue should also be discussed: "To what extent can people be manipulated and can stay natural in the existence of so much orientation and observation?" For example, Harari mentions that technologies like GPS deteriorate the ability of people to navigate since people can drive carelessly into a lake or a closed road whilst following the GPS from a screen [32]. Thus, while trying to examine and evaluate the behavior of people with various techniques, a process can be experienced in which people become inactive with excessive technology and cannot exhibit their natural behavior.

Although the behavior and sensory reactions of the users in a particular space can be measured, user's drawings of what remains in their minds as a result of their spatial experiences are still crucial in environmental psychology studies. The reason for this is that when users produce cognitive mapping drawings, they reveal indicators related to the visual information that has been stored and the functioning system of the memory part of the mind. The users can comprehend and express the functioning of the state of their external environment relations and visual perception system combined with experience, which can be defined as visual-spatial awareness. In this way, experts get clues about what is more decisive and preferable in an urban space.

Fig. 1: Postcards from different countries from the early 1900s (From Hilmi Bostancı's collection).

4 Change in the Perception, Cognition, and Experience in Visual World from 1900 to 2000

In the early 1900s when digital technologies did not become widespread, cinema, painting, and visual arts were the main representations of the visual world. In those years, visual descriptions and postcards were accepted as the main documents in order to understand the unknown urban spaces. Still descriptions of grandparents' neighborhoods where they have been living can be subject to areas such as historical research and design exhibitions [33]. Postcard and stamp collection were considered as popular hobbies in the 1900s. Postcards considered as objects and metaphors are the public images of cities where the visual features of cities are represented [34,35].

Fig. 1 shows postcards from different countries from the early 1900s. In these postcards, the historical structures, squares, and coastal landscapes of the cities are noteworthy. Images from 2000s such as historical buildings, night views, bays, river views, and bridges are seen in Fig. 2. In general, postcards include urban centers, skyline and nature views, and historical buildings. Additionally, historical personalities, landmark features, statues, city landscapes at night, cultural-local

Fig. 2: Postcards from different countries from the 2000s.

features, festivals, and tourism destination centers are among the postcard themes representing identity values. Postcards illustrate the transformation process of cities as visual documents. In the postcards of well-preserved historical cities, it is seen that the building and street textures have not changed as well as some details such as cars and people's clothing and street furniture. In today's world, when it's easy to take pictures with mobile phones, postcards still keep a certain amount of interest.

Postcards are defined as public images while photos can be expressed as personal images. Taking a photo of a landscape often shows that people are enthusiastic about turning this photo into a memory. The examination of the landscape photographs people take has contributed to the field of aesthetic evaluation. The photo-projective method (PPM), which is among the reflective methods, includes experimental research on the photographs taken by people. Using this method, Yamashita investigated how the water in the river landscape in Japan was perceived and evaluated by adults and children. In the study, residents were asked to take photographs of their surroundings and record each scene by adding explanations in the space. The participants were given a camera and were asked to take 50 photos of the river and its surroundings and take notes about the scenes.

Among the results, it was found that water elements in landscaping attracted more attention to children, and there were significant differences in space preferences between children and adult groups [5].

Human behavior towards sharing the photos that they take as well as their preferences for taking photos are the subject of study in the field of urban design. A photograph of urban scenery shared through social media can be considered as an indicator of personal aesthetic experience. Among the widely used social media applications which visual compositions are at the forefront is Instagram. The Instagram world provides important information about people's interactions with their environment and their visual thoughts. However, this information contains so much data that it is one of the big databases. Visual data processing in the context of big data is very difficult. However, various interactive studies on the Instagram information of a specific sample group can solve the big data problem. At the same time, new techniques are being developed to examine this data. For example, Karduni et al. have developed "Urban Activity Explorer" as a new prototype for a planning support system that uses visual analysis to understand mobile social media data. The system consists of connected visualizations that include spatial, temporal, and topical data, and enables the discovery of multiple scenarios [36]. Technologies in smartphones can analyze inhabitants as perimeter sensors. It allows people to move, produce, and share rich spatial data useful for new types of collaborative governance schemes [37]. As a result of the processing of this data by various methods, urban planning and design studies are carried out on the use of urban areas [36–38].

In the past, experiences of places that arouse the feeling of the sense of place were verbally told and remembered. In today's world, while the number of places photographed and shared through social media are increasing, the experiences and memories that make up the content of the visuals are forgotten. This situation causes the lack of experience.

5 Technical Innovations for Understanding New Visual World: From Virtual Reality to Augmented Reality

Today's human can use techniques that go beyond imagination triggers such as dream, literature, and cinema. In order to get away from the real world, people find opportunities to create and live alternative world experiences ranging from virtual space experiences to augmented reality.

Virtual reality simulations, which allow the experience of the new space before the design projects are realized, have made a great contribution to urban design. Virtual reality has also become an effective means to promote existing cities. Crang states that virtual projects such as Digital Amsterdam, Helsinki Arena 2000, can communicate with the city and enable the participation to cities even if it is virtual [39]. Virtual reality brings the sensation of touch, smell, etc., to bring users closer to the feeling of the real experience. In addition, it can bring the users a kind of touristic experience and memories with applications like postcards in virtual reality games, such as virtual postcards applications [40].

Urban designers work with virtual simulations that show people how to find a specific pattern or how interesting spaces are directed, how the perception of distance is developed, and creation a pre-design test environment [4,41–43].

It can be detected by combining online and offline worlds with the interface that can add imaginary elements within the real space created with devices such as Google Glass. This experience is defined as augmented reality. The goal of augmented reality is to go over the limit between machine and body by direct interaction between brain and computer via biometric sensors [32]. It has become possible to interact with virtual objects with the fingertip in augmented reality (AR) environments [44]. Augmented reality gave a new point of view to the studies in the field of urban design by giving an opportunity of creating an innovative user-space and machine interacted interface. Urban designers have got some technical opportunities in order to experience more design possibilities and simulate the behaviors that people can show in these designs in a real space [45]. Thus, it will be able to make great progress in preventing failures in urban design.

In addition to the many distinctive features of augmented reality from the virtual space, the most fundamental is that it can be experienced within the real world and create dangerous effects such as hallucinations. Besides, since the real spaces where people will be able to see things that are not real will make the cities more interesting, it will allow the outdoor space become more attractive again.

6 Contribution of Neuroscience in Understanding Cognitive Process in Urban Design

From computer-aided interactive applications such as eye-tracking to brain imaging techniques, technological advances are taking place in the field of neuroscience. In the mid-1900s, this process offered new opportunities to prove the experiments and predictions regarding the functioning of the brain. The development of the field and the opportunities it offers have begun to affect a wide range of fields such as social sciences, economics, marketing, architecture, urban design, as well as medical, engineering, and technical sciences.

Eagleman [2] states that the visual system focuses on the interpretation of visuals in the mind that the perception foregrounds for some specific internal reasons, but not the real ones. The eye-tracking technique was first used by Yarbus who used an eye-tracking device [2]. In his study, he asked the participants to examine the people on the painting in order to examine the eye movements on the perception of complex objects. This study showed that participants focused on eye movements towards people in the painting [46]. Eye-tracking analyses are also used for landscape perception and urban aesthetics. With this approach, the researches on the landscape, it was found that the visual openness was effective in the landscape observation of strong and detailed photographs [47]. Eye-tracking enabled some interactive applications including selective and capturing the perspective approaches [48]. These analyses allow us to understand how visual images of cities, such as video and photos, are followed by the eye in the field of urban design. The stable and moving visual images of cities show what the eyes are focused on.

Studies to understand where and how aesthetic perception occurs in the brain revealed the area of neuroaesthetics [49,50]. In these researches, findings show that there might be a direct connection between the limbic parts which trigger the feelings of "feature binding" and "rewarding" in the functioning of the brain [51]. Sensory measurement and brain imaging techniques are used to investigate how the mind is affected by the interaction with art. Calvo-Merino et al. [6] applied a repetitive transcranial magnetic stimulation (rTMS) on brain parts over dance movements of humans. The aim of the study was to determine the parts where the aesthetic sensitivity of the brain was formed. This research is an experimental study on how aesthetic sensibility is formed in the brain, accompanied by dance movements, music, and rhythm. It demonstrates the importance of the aesthetic experience and its connection with other sensory organs [6].

Neuroscience makes important contributions to understanding the biological and physical properties of the aesthetic aspects of people's artistic expressions and creative processes. It shows that seeing senses, emotions, memories, and imagination have a meaning and characteristics far beyond a technical process [52]. Neuroscience offers new opportunities for people to understand how their brain activity is shaped and their interactions in the body system.

Through advances in the field of neuroscience, people are able to measure the states of brainwave and autonomic nervous systems as they move in urban space. People's willingness to use mobile phone applications allows these systems to be used in the measurement of urban movements and behaviors. Thus, it provides a lot of psychological data to understand how the urban system works and what effects it has on people's perception and cognitive process. With this data collection and analysis methods, the ability to discover the reactions of the users against the simulations to be realized in the city is quite developed [53]. For example, researches on green areas showed that green area images could change the physiology of people, as well as their happiness, and autonomic nervous systems showed strong symptoms of relaxation responses [53–55]. However, there are also findings that virtual spaces containing green area images have positive impacts [53]. These developments create effects in the field of urban design that will change the perspective of nature and technology dialectics. However, the preservation of natural areas and the creation of spaces that will let a natural sense of space within the city will continue to be important among the universal and unchanging targets of urban designers.

There are studies using functional magnetic resonance imaging (fMRI) techniques to understand how the brain is affected by the architectural styles of buildings. Choo et al. determined that the neural connections of human classification of visual categories were not only affected by perceptual features but also by meaning and culture, such as architectural styles and expertise of architects of buildings with fMRI [8]. Marchette et al. [7] used multivoxel pattern analysis of fMRI data to understand how the effects of landmarks on the locating strategies were coded by the brain. Based on the technical explanations of this study, it has been found that in the biological structure of the brain, specific objects or topographic elements supporting the identity code perform a triple collaboration processing the spatial and conceptual information and perceptual details of the brain [7]. This information coincides with the findings regarding the understanding of all the parts of gestalt psychology before the possibilities for these imaging techniques were available. Studies on brain imaging techniques performed by urban design and neuroscience have recently begun. As the studies done in this field diversifies, the unknowns in the process of interaction and communication with the urban space will be clarified.

7 Conclusion

The success of urban design implementations is closely related to the way of people's interaction with a specific physical environment and the comprehension of this interaction's contributions to their well-being situations. When inhabitants show the sense of place behaviors in a particular public space, important clues can be found on the success of the design process of this area. Besides, being able to get user comment about pre-design through simulation may prevent the designs that may have negative results after the design. New technologies such as photo-projective method (PPM), eye-catching technique, and fMRI have been added to experimental studies and qualitative research techniques in which environmental behavior is examined. Thus, new approaches are being developed for the cognitive interpretation of people's communication with space.

In addition to the benefits of predicting the decision-making system offered by technology, there are some drawbacks of becoming more directive. For example, navigation and GPS can impair people's ability to navigate. So this affects the place attachment ability. People can become dependent on technological devices to interact with the environment. This can cause people to start to lose their natural behavior over time. Thus, measurements can be misleading because it evaluates the manipulated user, not the natural mood of the user.

This study discusses the effects of the concepts of mind, cognitive process, and neuroscience, which are of interest to every discipline, to the field of environmental psychology and urban design. The research shows the diversity and complexity of the technologies that make up the sub-branches of this process. From the postcards of the 1900s to the virtual tours of the 2000s and the augmented reality experiences, innovations that will change the visual thinking system from the beginning to the end have emerged.

In the future, studies can be carried out that combine different techniques such as visual thinking, virtual reality, augmented reality. Neuroscience can add new dimensions to this kind of studies. In the area of cognitive mapping, a wide variety of researches are made about real and virtual spaces. In addition to these studies, different implementations that deal with the third dimension experience, such as the ability to remember the city skyline and estimate the distance of a building by looking at the skyline, can be examined. Augmented reality practices on reviving the historical features of city skylines can be planned. Probability calculations for future projects and alternative skyline prediction studies can be designed through augmented reality. Neuroscience-based research can be carried out to understand how users who examine these past and future urban skylines feel.

Understanding the mechanism of human mind may increase the level of satisfaction with urban design practices. However, designers have important professional missions, such as conserving historical and natural values, rather than ensuring the happiness of people. In the light of this information, how will urban designers behave as the observer-protective and how much will they act as guiding-intrusive? As far as possible, they can try to design natural habitats in cities or carry out activities to protect natural areas from the threat of mass urbanization. On the other hand, they can aim to respond to increasing the demands of technology, innovation, and consumption. Thus, they can design consumer spaces that are constantly engaging, intriguing, directing and encouraging photo-sharing by being sensitive to market conditions. Although the designer wants to make a balance between these two, the results of these processes are still uncertain. Here, the basic self-features of urban designers, such as biographies, political thoughts, values, personality, educational lives, and professional experiences can be determinant. The conditions and governance behaviors of countries mostly influence urban areas above all these determinants. The holistic neuroscientific analyses of how multivariate cognitive processes such as life expectancies, visual, natural, environmental and political tendencies, physiological, and psychological needs lead the design decisions of designers can lead the future of urban design. The complex process of the human mind determines the fate of the living spaces from the past to the future in design, space, inhabitants and designer interaction. New technologies enable designers and users to create experimental spaces with more possibilities. But without holistic mental accumulation, it is not possible to make a meaningful future for the visual world by adding new values to the values of the past.

References

1 Loewenstein, W.R., *Physics in Mind: A Quantum View of the Brain.* 2013, Basic Books: New York.

2 Eagleman, D., *Incognito: The Secret Lives of the Brain.* 2011, New York City: Pantheon.

3 Arnheim, R., *Visual Thinking.* 1969, Berkeley: University of California Press.

4 Chon, J., and Scott Shafer, C., *Aesthetic Responses to Urban Greenway Trail Environments.* Landscape Research, 2009. **34**(1): pp. 83–104.

5 Yamashita, S., *Perception and Evaluation of Water in Landscape: Use of Photo-Projective Method to Compare Child and Adult Residents' Perceptions of a Japanese River Environment.* Landscape and Urban Planning, 2002. **62**(1): pp. 3–17.

6 Calvo-Merino, B., Urgesi, C., Orgs, G., Aglioti, S.M., and Haggard, P., *Extrastriate Body Area Underlies Aesthetic Evaluation of Body Stimuli.* Experimental Brain Research, 2010. **204**(3): pp. 447–456.

7 Marchette, S.A., Vass, L.K., Ryan, J., and Epstein, R.A., *Outside Looking in: Landmark Generalization in the Human Navigational System.* Journal of Neuroscience, 2015. **35**(44): pp. 14896–14908.

8 Choo, H., Nikrahei, B., Nasar, J.L., and Walther, D. *Neural Decoding of Architectural Styles from Scene-Specific Brain Regions.* Journal of Vision, 2015. **15**(12): pp. 520–520.

9 Whyte, W.H., *The Social Life of Small Urban Spaces.* 1980, Washington, DC: The Conservation Foundation.

10 Nasar, J.L., *Perception, Cognition, and Evaluation of Urban Places.* Public Places and Spaces: Human Behavior and Environment ed: I. Altman, and E.H. Zube, 1989, Plenum Press, pp. 31–56.

11 Berlyne, D.E., *The New Experimental Aesthetics.* Studies in the New Experimental Aesthetics, 1974. pp. 1–25.

12 Wertheimer, M., *A Brief Introduction to Gestalt, Identifying Key Theories and Principles.* Psychol Forsch, 1923. **4**, pp. 301–350.

13 Köhler, W., *Gestalt Psychology Today.* American Psychologist, 1959. **14**(12): pp. 727–734.

14 Lewin, K., *Behavior and Development as a Function of the Total Situation.* Manual of Child Psychology. ed: L. Carmichael, 1946, Hoboken, NJ, US: John Wiley & Sons Inc. pp. 791–844.

15 Rapoport, A., *The Meaning of the Built Environment: A Nonverbal Communication Approach.* 1990, University of Arizona Press.

16 Alexander, C., *The Nature of Order: The Process of Creating Life.* 2002, Taylor & Francis.

17 Norberg-Schulz, C., *Genius Loci: Towards a Phenomenology of Architecture,* 1980, Academy Ed., London.

18 Low, S. M., and Altman, I., *Place Attachment.* Place Attachment ed: I. Altman, and S.M., Low, 1992, Boston, MA: Springer, pp. 1–12.

19 Tuan, Y.F., *Space and Place: The Perspective of Experience.* 1977, University of Minnesota Press.

20 Steele, F., *The Sense of Place.* 1981, CBI Publishing Company, Inc.

21 Gifford, R., *Introduction Environmental Psychology and Its Methods, Research Methods for Environmental Psychology.* ed: R. Gifford, 2016, John Wiley & Sons.

22 Kaplan, R., Kaplan, S., and Ryan, R., *With People in Mind: Design and Management of Everyday Nature*. 1998, Island Press.

23 Loewenstein, W.R., *The Touchstone of Life*. 1999, Oxford: Oxford University Press.

24 Faria, P.A., Ferreira, F.A., Jalali, M.S., Bento, P., and António, N.J., *Combining Cognitive Mapping and MCDA for Improving Quality of Life in Urban Areas*. Cities, 2018. **78**, pp. 116–127.

25 Shannon, C.E., *A Mathematical Theory of Communication*, The Bell System Technical Journal, 1948. **27**(3): pp. 379–423.

26 Shannon, C.E., *The Bandwagon*. IRE Transactions on Information Theory, 1956. **2**(1): pp. 3.

27 Schrödinger, E., *What Is Life? The Physical Aspect of the Living Cell and Mind*. 1944, Cambridge University Press, Cambridge.

28 Salama, A. M., Al-Maimani, A., and Khalfani, F., *Understanding Inhabitants' Spatial Experience of the City of Doha through Cognitive Mapping*. Open House International, 2013. **38**(4): pp. 37–46.

29 Tolman, E.C., *Cognitive Maps in Rats and Men*. Psychological Review, 1948. **55**(4): pp. 189–208.

30 Lynch, K., *The Image of the City*. 1960, MIT Press.

31 Kullmann, K., *The Mirage of the Metropolis: City Imaging in the Age of Digital Chorography*. Journal of Urban Design, 2018. **23**(1): pp. 123–141.

32 Harari, Y.N., *21 Lessons for 21th Century*, 2018, Popular Science.

33 Eldem, E., *Bankalar Caddesi: Osmanlı'dan Günümüze Voyvoda Caddesi*. Osmanlı Bankası Bankacılık ve Finans Tarihi Araştırma ve Belge Merkezi, 2000, İstanbul.

34 Larsen, L., and Swanbrow, L., *Postcards of Phoenix: Images of Desert Ambivalence and Homogeneity*. Landscape Journal, 2006. **25**(2): 205–217.

35 Lydecker, K., *Constructing Memory: Picture Postcards of New York City*. The Hopkins Review, 2011. **4**(2): pp. 229–270.

36 Karduni, A., Cho, I., Wessel, G., Dou, W., Ribarsky, W., and Sauda, E., *Urban Activity Explorer: Visual Analytics and Planning Support Systems*. International Conference on Computers in Urban Planning and Urban Management, 2017, July. Springer, Cham, pp. 65–76.

37 Guerrero, P., Møller, M.S., Olafsson, A.S., and Snizek, B., *Revealing Cultural Ecosystem Services through Instagram Images: The Potential of Social Media Volunteered Geographic Information for Urban Green Infrastructure Planning and Governance*. Urban Planning, 2016. **1**(2): pp. 1–17.

38 Giridhar, P., Wang, S., Abdelzaher, T., Ganti, R., Kaplan, L., and George, J. *On Localizing Urban Events with Instagram*. In INFOCOM 2017, IEEE Conference on Computer Communications, pp. 1–9.

39 Crang, M., *Public Space, Urban Space and Electronic Space: Would the Real City Please Stand Up?* Urban Studies, 2000. **37**(2): pp. 301–317.

40 Book, B., *Traveling through Cyberspace: Tourism and Photography in Virtual Worlds.* Conference Tourism & Photography: Still Visions - Changing Lives in Sheffield, 20–23 July 2003, UK.

41 Darken, R.P., Allard, T., and Achille, L.B. *Spatial Orientation and Wayfinding in Large-Scale Virtual Spaces: An Introduction.* Presence, 1998. **7**(2): pp. 101–107.

42 Cubukcu, E., and Nasar, J.L. *Influence of Physical Characteristics of Routes on Distance Cognition in Virtual Environments.* Environment and Planning B: Planning and Design, 2005. **32**(5): pp. 777–785.

43 Drettakis, G., Roussou, M., Reche, A., and Tsingos, N., *Design and Evaluation of a Real-World Virtual Environment for Architecture and Urban Planning.* Presence: Teleoperators and Virtual Environments, 2007. **16**(3): pp. 318–332.

44 Buchmann, V., Violich, S., Billinghurst, M., and Cockburn, *A. FingARtips: Gesture Based Direct Manipulation in Augmented Reality.* Proceedings of the 2nd International Conference on Computer Graphics and Interactive Techniques in Australasia and South East Asia, ACM. 2004, June. pp. 212–221.

45 Seichter, H., *Benchworks - Augmented Reality Urban Design.* Computer Aided Architectural Design Research in Asia (CAADRIA 2004), ed: H.S. Lee and J.W. Choi, 2004, Seoul, Korea, pp. 937–946.

46 Yarbus, A.L., *Eye Movements during Perception of Complex Objects.* Eye Movements and Vision, 1967, Boston, MA.: Springer.

47 Dupont, L., Antrop, M., and Van Eetvelde, V., *Eye-Tracking Analysis in Landscape Perception Research: Influence of Photograph Properties and Landscape Characteristics.* Landscape Research, 2014. *39*(4): pp. 417–432.

48 Duchowski, A.T., *A Breadth-First Survey of Eye-Tracking Applications. Behavior.* Research Methods, Instruments & Computers, 2002. **34**(4): pp. 455–470.

49 Zeki, S., *Inner Vision: An Exploration of Art and the Brain.* 1999, Oxford: Oxford University Press.

50 Di Dio, C., and Gallesse, V., *Neuroaesthetics: A Review.* Current Opinion in Neurobiology. 2009. **19**(6): pp. 682–687.

51 Ramachandran, V.S., and Hirstein, W., *The Science of Art: A Neurological Theory of Aesthetic Experience.* Journal of Consciousness Studies, 1999. **6**(6–7): pp. 15–51.

52 Gallese, V., *The Power of Images: A View from the Brain-Body.* 2018, Frienze University Press, pp. 70–79.

53 Ellard, C., *Cities and Their Psychology: How Neuroscience Affects Urban Planning.* 2014. The Guardian.

54 Ulrich, R.S., *View through a Window May Influence Recovery from Surgery.* Science, 1984. **224**(4647): pp. 420–421.

55 Thompson, C.W., Roe, J., Aspinall, P., Mitchell, R., Clow, A., and Miller, D. *More Green Space is Linked to Less Stress in Deprived Communities.* Landscape and Urban Planning, 2012. **105**(3): pp. 221–229.

Assoc. Prof. Dr. Şengül Yalçınkaya

33 Construction of Memory through Architecture: Current Approaches

Abstract: Given the relationship between memory and space, with globalization, architecture is today used as a tool for remembering and forgetting. The dominant force of the time shapes the physical environment according to its own system of thought, manifesting itself in a multiple scale ranging from household to urban. This study shows how the power of memory over remembering and forgetting is manipulated through architecture. The resulting mechanism of destruction and reconstruction is analyzed through the Trabzon Tekel Building. People and communities are adversely affected by the intensive destruction and reconstruction around them.

Keywords: Memory, Remember, Forget, Space, Architecture, Globalization

1 Introduction

Memory is what people and communities can hold onto in the face of ephemerality, speed, change and transformation. The primary source of memory formation is space. People and communities are adversely affected by the intensive destruction and reconstruction around them. Memory is formed, destroyed and sometimes distorted through spaces; therefore, the effect of current conditions on the relationship between architecture and memory is an important issue to be addressed. The sensitivity of the situation turns the architectural activity into a paradox. In the relationship between architecture and memory, on the one hand, the individual and society are guided by architecture while, on the other hand, they need space to exist.

Every era is marked by its own conditions affecting the way of life, production, consumption and therefore, the space. In this sense, the modern era differs markedly from its predecessor and shapes the present. This study primarily addressed the time-related change in the relationship between memory and space.

Secondly, this study focused on how remembering is constructed over space and what spaces of memory, which have emerged with the disappearance of real places of memory, actually signify? The object of the relationship between architecture and recollection through iconic buildings and concept spaces is to keep the past alive in the present and to stick in people's minds, albeit temporarily. Recollection requires an object of reminiscence, the most important among which are spaces.Forgetting is essential for memory. However, what to forget and what to remember are key questions. This chapter focuses on the thought and power behind the disappearance of architectural structures. Especially today, cities are facing serious destruction. This chapter describes the impact of this destruction on the individual and society through examples from Turkey.

The last section addressed "destruction and reconstruction," which is one of today's widespread architectural activities, through a local example. We first focused on the significance of the Trabzon Tekel Building for collective memory. We presented designers' perspectives and views regarding the significance of the building in collective memory and what traces were to remain intact in its reconstructed version. The winning projects of the competition were evaluated in terms of collective memory.

2 Memory in Modern Times

Having begun with the scientific revolution of the 16th and 17th centuries, the modern era differs markedly from its predecessor. Modernity is based on the concepts of "progress" and "pace," and always seeks the new and identifies with it, resulting in the duality of destruction and reconstruction in the world. Encompassing continuous change and progress, modernity removes the notion of permanence. According to Octavio Paz, modernity "is cut off from the past and continually hurtling forward at such a dizzy pace that it cannot take root ..." [1]. Similarly, Harvey saw "the transitoriness, the fleeting, and the fragmentary as a necessary condition through which the modernizing project could be achieved" [2]. With his remark "Both/And supplanted by Either/Or," Berman suggests that the modern world is based on expendability and criticizes that many things have been abandoned for the new [1]. Modernity's impulse to seek the new comes into play on space in architecture, and the notion of permanence is replaced by the mechanism of destruction.

Modernity effectively uses spaces in forgetting and remembering. In fact, the elimination of the old and seeking of the new in modernity is in contradiction with the structure of memory consisting of familiarity, experiences, and marks and traces. The understanding of the "now" and production of the "new" in modernity, on the one hand, pave the way for forgetting, while on the other hand, they create spaces that are reminiscent of things that must not be forgotten.

Globalization has brought a different dimension to the economic, social and technological developments that started with modernization. The search for the new in modernity has been replaced by consumer culture, which actively shapes the processes of remembering and forgetting, and turns everything into a commodity. In this culture, commodities are made while other things are made to be forgotten. Everything is consumed at a fast pace and every item remembered and forgotten are meant to increase consumption.

The fast-paced life in cities and loss of sense of permanence and time in spaces have changed the meaning and value of the relationship between spaces and people. Cities have now turned into concrete jungles of motorways, horizontally and/or vertically large buildings, satellite districts and majestic malls. This has forced people to share spaces without discernible borders with strangers. Besides, immediately consumed experiences and constant reconstruction of spaces weaken the bond that people share with their surroundings [3]. The growth of new human who is extremely active, the emergence of new everyday life practices and the new urban fiction that promotes these practices trigger forgetting. The sense of cities resembling each other more and more, the feeling of disconnectedness and ephemerality and the lack of familiarity cause memory loss.

At this point, the disappearance of real places of memory has given way to the emergence of spaces of memory, leading to an effort to compensate for the resulting discontinuity by spaces. Today, libraries, archives, monuments and museums are used to maintain traces of the past [4]. These spaces convey the knowledge of the past all the way to the next generation, and therefore, play an important role in the creation and handing down of collective memory. In this way, the relationship between time and history for the modern individual is re-established through spaces.

3 Construction of Memory through Architecture: Current Approaches

Given the relationship between memory and space, with globalization, architecture is today used as a tool for remembering and forgetting. The dominant force of the time shapes the physical environment according to its own system of thought, manifesting itself in a multiple scale ranging from household to urban. This chapter shows how the power of memory over remembering and forgetting is manipulated through architecture. The resulting mechanism of destruction and reconstruction is analyzed through the Trabzon Tekel Building.

3.1 Memory Constructed through Architecture

Spaces play a key role in the creation and handing down of collective memory. Each society needs spaces that are the symbol of its identity and the basis of its memories [5]. Spaces of memory can, therefore, be thought of as a reaction to the phenomenon of forgetting, which is an inevitable consequence of modernization. Spaces of memory bring the past to the present and ensure that the present survives in the future.

Spaces are an important tool that gives a sense of continuity. The irresistible urge for constant change and seeking the new induced and introduced by modernization has paved the way for the disparagement of the past and the destruction of sites of memory. New ways have been sought to preserve the past in the face of its gradual disappearance, which has resulted in the emergence of spaces of memory [6]. Spaces of memory serve the function of alleviating the alienation felt in the physical environment and the fear and loneliness that accompany it. The goal to be achieved by spaces of memory is to form a collective memory and to preserve it.

Museums and monuments are prominent spaces of memory. Architectural structures sometimes undertake the task of memory, and sometimes host the elements of memory. Architectural structures do this by commemorating a prominent figure or a historical event or by retaining the traces of a historical process. Oftentimes, these structures are removed from their current context and converted into museums. In other words, architectural structures become memory interfaces where elements of reminiscence are displayed.

On the other hand, memory is also used to make sure that the present prevails in the future. In general, the "thing" that is supposed to be remembered or forgotten is presented in a space or as the space itself by the authority. Architecture is transformed into a spectacle and an indicator that encompasses collective and political memory. The use of architecture for political purposes dates back to ancient times. Political authority in any given society plays an active role in the formation of cities, especially public structures [7]. Architecture is the most effective way for a sovereign to leave a mark in history and to fulfill his/her wishes. Today, the political power in Turkey promotes the construction of physical environments with traditional, local and Islamic references. In addition to the political power supporting the revival of the past, especially in public buildings, globalization and popular culture are the other driving forces that shape the physical environment.

Popular culture is a commercially driven artificial culture reinforced and legitimated by the discourse of modernity [8]. In general, continuity is achieved not by permanence, but by change. Popular culture based on consumption makes use of different aspects of architecture to be permanent in memory for a while. In some cases, an image created for a brand is applied simultaneously in all venues of that brand throughout the world. For example, Starbucks have almost the same interior set design all over the world. Although this leads to a mnemonic deviation, the brand creates the effect it wants to create and increases its recognition.

Cities are also interested in branding. They highlight their unique and authentic characteristics and promote their landmarks to attract tourists and investments, to increase recognition and to stand among their rivals. They also focus on the design and construction of iconic buildings, the most prominent examples of which are the Sydney Opera House, Hotel Renaissance, The Royal Ontario Museum, the Museo Soumaya and the Capital Gate (Fig. 1). These structures are designed to be not only imaginable and symbolic but also original and new. Difficult images are chosen in the design of iconic buildings so that they stick in the mind [9]. On the other hand, iconic buildings only aim to be visually permanent, their number increases every day and their focus is always on consumption. It is, therefore, difficult for them to stand the test of time.

3.2 (De)constructed Memory through Architecture

Architectural structures are not purely physical objects. They exist not only with their own assets but also through their interaction with the space on which they are built. They begin to partake in the system with what they gain from and contribute to the space they take up in the system. They begin to communicate with

Fig. 1: Examples of iconic buildings [10].

the public the moment they start to be constructed. With their physical, psychological, social and temporal characteristics, these structures bring a new dimension not only to the space but also to people. They take on a new meaning with what they offer for new experiences and what they represent. The destruction of architectural structures imbued with such a plethora of meaning is, therefore, more than their physical destruction. In other words, destruction means not only the physical disappearance of a building but also the loss of its meaning, value and experience.

In the modern era, destruction runs through the idea of opening space for the new. Modernity is a period of time where the built environment is destroyed by people. This causes the reproduction of the present and the destruction of the past. This takes place as the destruction of the built environment in cities. Today, destruction is changing its meaning and increasing its pace and scope. The basis of this notion is the continuation of consumer society that has come into existence with postmodernity. The "new" in cities turns into a commodity, and each action is taken to promote consumption [11]. This causes rapid changes in the appearance of cities leading to the disappearance of spaces that memory is located, and hence, resulting in the weakening of memory.

It should be kept in mind that urban architectural structures are elements of reminiscence. Transmitting messages, witnessing, and forming interfaces, architectural structures help, directly or indirectly, to remember. They also take on a new meaning within the urban setting. They witness historical events from the time they were built until today, play an important role in memory-making process, reflect the architectural characteristics of the period they were built, or simply symbolizes authority. According to Rossi, architectural structures, besides their functional and physical existence, are reproduced in time by meanings ascribed to them by communities and by images created in minds [12]. Therefore, what is destroyed and why it is destroyed becomes an important issue in interventions in urban settings.

Nowadays, the rapid change and indiscriminate destruction in urban settings destroy collective memory as well. Halbwachs suggests that people witnessing the destruction of structures that make up their physical universe also lose a part of themselves [13]. Ganic, on the other hand, emphasizes that the feeling of absence defined by the new physical condition that arises with the termination of the physical continuity of an architectural structure comes into conflict with the image of that space in collective memory [14]. Artun sees the destruction of some buildings in urban space as a kind of defeat [15]. The destruction of architectural structures set forth by a particular discourse is the destruction of the spectacle of those structures and their representation in collective memory. Destruction in urban space is a multidimensional phenomenon and has different objectives and repercussions. At this point, the impact of interventions in urban areas on the urban population should also be considered.

The constant change in the physical, social and cultural environment, consumption habits, and complexities and pace of modern life facilitate collective forgetting [16]. Especially symbolic structures in urban settings are abandoned and eventually destroyed by various reasons such as functional deficiency, physical obsolescence, economic hardships, political concerns, etc. In urban settings, large areas are destroyed under the name of urban transformation and given a completely different identity. Boyer (1994) states that when individuals or communities establish a special relationship with any structure in an urban setting, that structure should be preserved [17]. As the new replaces the old, that is, as elements of memory in an urban setting are destroyed, the bond between the city and its inhabitants weakens.

İller Bankası Merkezi Genel Müdürlük Building, which was once the symbol of the Republic and Development in Turkey, Atatürk Kültür Merkezi, which was the first opera house of Turkey and Mecidiyeköy Tekel Liqueur Factory, which was important in terms of art, architecture and history and was one of the first structures of the Republic, have been demolished (Fig. 2). Another striking example

Fig. 2: Recently demolished registered buildings [18].

is that numerous city stadiums and gyms [Kayseri Atatürk Stadium (2008) Rize Atatürk Stadium (2009), Ali Sami Yen Stadium (2011), BJK İnönü Stadium (2013), Konya Atatürk Stadium (2014), İlkadım Stadium (2014), Antalya Atatürk Stadium (2016), Diyarbakır Atatürk Stadium (2016), Hüseyin Avni Aker Stadium (2017), Elazığ Atatürk Stadium (2018) and Abdi İpekçi Gym (2018)] have been demolished throughout Turkey in 10 years (Fig. 3). These structures that hosted hundreds of major sporting and cultural events were demolished for various reasons. It is not only the physical disappearance of those structures but also the gradual disappearance of their meaning and significance from collective memory. Given the fact that sports bring people together, the destruction of those stadiums and gyms is a significant loss for collective memory.

Forgetting begins with the destruction of the founders and bearers of collective memory. Physical environments are also elements of memory based on habits, associations and experiences ascribed to them. The more the urban setting is

Fig. 3: Recently demolished sports facilities [19].

transformed, the more the collective memory is damaged. People lose their connection with their memories as a result of destruction in urban settings. In some cases, destruction-induced trauma is overcome by nostalgia. Depending on the needs of the day, nostalgia helps to recall some of what has already been lost.

4 Localization of Destruction-Reconstruction in Globalization Process: Trabzon Tekel Building

This section discusses the destruction of the Trabzon Tekel Building, which was located in the city center of Trabzon and had a significant role in industrial memory and urban memory. The earliest architectural drawings of the building date back to 1948. The building was constructed in two parts in 1951 and 1965. The first part consisted of two storys to the southern front, five storys to the north of this block and three storys to the east of the courtyard. In the second part, three-story and five-story blocks were added to the west and north of the courtyard, respectively [20]. The main entrance to the south of the building had the characteristics of the Second National Architecture Movement (Figs. 4,5).

Having been constructed as a tobacco factory and warehouse, the building was used only as a warehouse and stood idle for a certain period of time. The building was registered in 1998 [21]. The reduction of tobacco cultivation in the region and loss of the original function of the building sparked debates on the building. The demolition of the idle building became an issue in 2002 and the

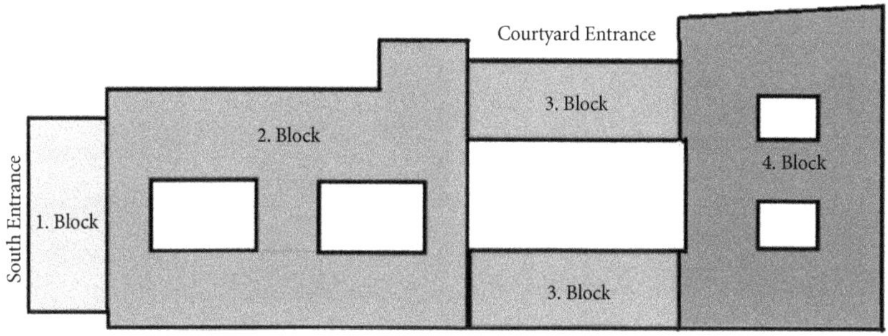

Fig. 4: Construction blocks.

national competition in 2006 aimed to revive it. However, in the end, the building was completely demolished, and a shopping mall was built there in 2008 and a town hall in 2011.

The building was both an industrial heritage and an authentic structure in terms of its location. It was located in the center of Trabzon within the historical site. On the other hand, the city is located in an industrially underdeveloped part of Turkey. The building provided job opportunities for the inhabitants and was especially a place of work for women of low-income families. With its contribution to economic life, the building assumed a central role in the identity of the city. It was located on a network of communication and transportation connecting the center to the periphery. It made a visual impression not only on the workers but also on city-dwellers. Not only did it not overshadow the historic characteristics of its surroundings but also enhanced the aesthetic appeal with its color and texture.

A competition on a national scale was held to reintroduce the idle building to the city life. Within the scope of the Trabzon Old Tekel Building Revitalization Architectural Project, the significance of the building in urban memory was emphasized by the competition specifications and project design. However, the competition specifications included only the decision of the Trabzon Cultural and Natural Heritage Preservation Regional Board with regard to the significance of the building in urban memory. After registration, the decision of the board in 2005 stated that "the significant function undertaken by the building is expected to prevail in the future and in the urban memory, and to help the city to have a new vision [23].

In the report of the winning project, the building was divided into four blocks to make an emphasis on its significance in urban memory and on the sections planned to be preserved. The evaluation regarding Block 1 was that "it has the characteristics of the II. National Architecture Movement Era and an important place in urban memory." The evaluation regarding Block 3 was that "the 3-storey block has cork flooring, which is a conical cap flooring system widely used in

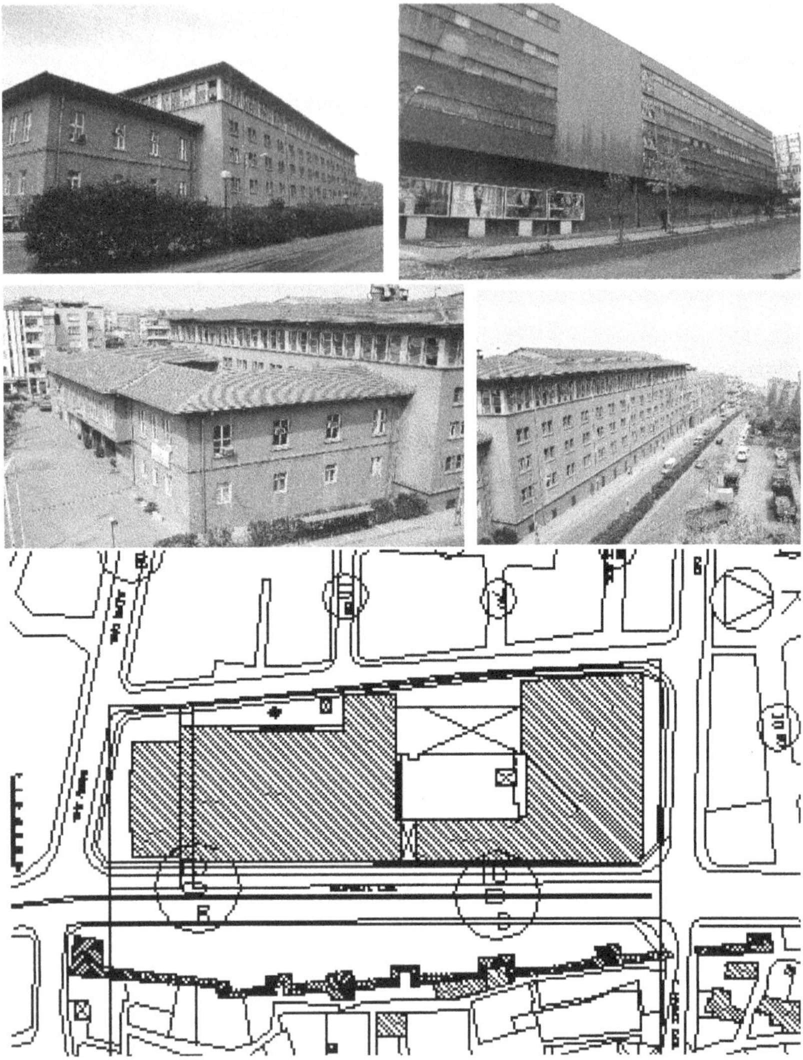

Fig. 5: Drawings and images of the Old Tekel Building [22].

Anatolia at that time" [24]. In the design process, the decision was to preserve Blocks 1 and 3 completely and the facade of Block 2, and to completely demolish and rebuild Block 4.

In the report of the project that ranked second in the competition, the evaluation was that "Blocks 1 and 2 from South to North have the characteristics of II.

National Architecture Movement Era in terms of architectural style while Block 3 belongs to a more recent period and is more ordinary in terms of architectural style [25]. During the design process, the decision was to preserve Blocks 1 and 2 completely and to completely demolish and rebuild Block 3. The project that ranked third stated that "it has been decided that the massive impact created by the building on people's perceptions must be preserved for the continuity of urban memory even if the function of the building had changed [26]. For this, the project preserved the building facade as it was.

In the project that received the honorable mention in the competition, "the interior space in Block 2 and the tobacco tanks with haunched columns were found to be exciting [27]. The outer shell of the two-stoey section of the first part was preserved. In the project that received the second mention, it was stated "that Trabzon Old Tekel Building has the characteristics of the industrial buildings of the 2nd Republican Period" [28]. For this, the Block to the south and north was preserved. The project that received the third mention emphasized that "the first part of the building projects the image of the building, and therefore, it should be preserved" [29].

While all projects decided on the preservation of the Block to the south of the building, designers focused, in general, on different aspects of the building and found them to be worth preserving at different levels (Fig. 6).

While the section of the building to the south was supposed to be preserved according to the result of the competition, the whole section was demolished during reconstruction. Some parts of the building were replicated [30]. The competition deviated from its original objectives and ended up with the destruction of the whole building. Today, there is a shopping mall and a town hall where the building used to be (Fig. 7). However, the shopping mall did not attract the expected attention and a significant part of it remained unoccupied. All in all, an important site of production for the city has been turned into a site of consumption. The building has been completely demolished let alone some part of it be physically preserved. The social, cultural and architectural traces of the building have been destroyed resulting in a multifaceted loss of memory.

5 Conclusion

We explained the relationship between architecture and memory through remembering and forgetting. We focused on the elements underlying this relationship that has brought a different dimension to today's conditions and on the way architecture has mediated it. Today, what to remember and what to forget during the formation of individual and collective memory is determined within the framework established by actors. Through spaces that it creates, architecture mediates this process directly or indirectly. The desired collective memory is presented to the community through spaces of memory. However, the rapid and deliberate change in the physical environment triggers forgetting. The point to note is that these interventions on the physical environment have negative impacts on people who need spaces to exist.

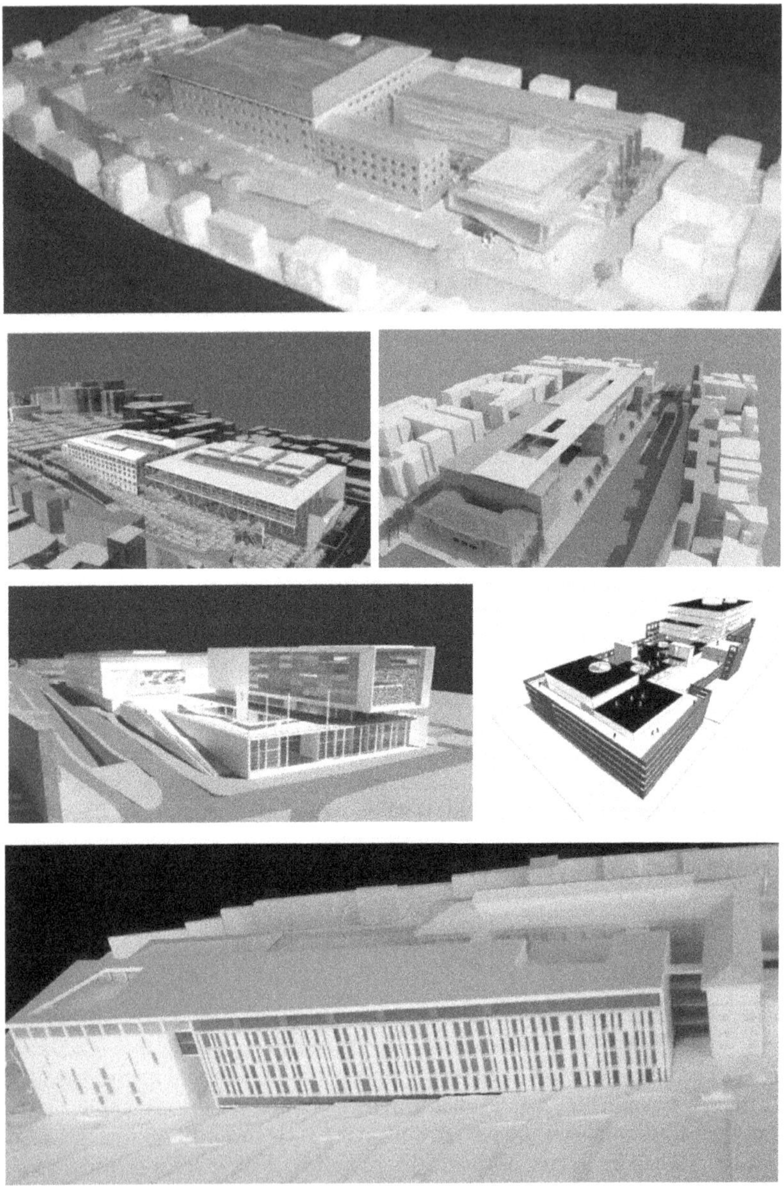

Fig. 6: Projects submitted for the revitalization of the building [31].

Fig. 7: Images of the project submitted for the demolished Tekel Building [32].

References

1 Berman, M.,. Katı Olan Her Şey Buharlaşıyor, (Trans. Altuğ Ü., Peker, B.) İletişim Publications, 1999, İstanbul: p. 57.

2 Harvey, D. Postmodernliğin Durumu: Kültürel Değişimin Kökenleri, (Trans. Savran, S.,), 2003, Metis Publications, İstanbul: p. 6.

3 Connerton, P. Modernite Nasıl Unutturur, (Trans. Kelebekoğlu, K.) Sel Publishing, 2012, İstanbul: pp. 99–102.

4 Kurak Açıcı, F. and Konakoğlu, Z., N. Kültürel Mirasın İzlerini Kent Müzelerinde Sürmek: Trabzon Müzeleri, Journal of History Culture and Art Research, 7(3), 2018: p. 669.

5 Assmann, J. Kültürel Bellek. (Trans. Ayşe Tekin). Ayrıntı Publications, 2001, İstanbul: p. 47.

6 Nora, P. (2006). Hafıza Mekanları. (Trans. Özcan M.E). Dost Kitabevi Publications. Ankara: p. 17.

7 İslamoğlu Ö. and Usta G., Sosyo-Kültürel Bir Mekan Olarak Kentin Biçimlenişinde Siyasal Eğilimlerin Rolü-Rize Örneği, Uluslararası Yapı ve Yaşam Congress, 20–21 March 2009;Bursa, Turkey: pp. 123–131.

8 Özdemir, İ.M., Hasançebi, Ö., Araz, A., Dönüşüm Paralelinde Popüler Kültürün Yarattığı Aynı Mekanlar, XIX. Yapı-Yaşam Congress, 22–24 March 2007, Bursa, Turkey: pp. 453–462.

9 Yılmaz, A., Ikonik Bellek Mimarlığı: Imge Olarak Yapı, Anı Olarak Mekan, *Mimarist* , 2009, 33: pp. 62–67.

10 Icon buildings, [cited 2018, 10 November]; Available from: https://en.wikipedia.org/wiki/Capital_Gate#/media/File:Capital_Gate.jpg; https://uncouthreflections.com/2013/06/29/eyesore-of-the-month-carlos-slims-museo-soumaya/; https://www.ensonhaber.com/galeri/dunyanin-en-guzel-kultur-merkezi-haydar-aliyev

11 Çetken, P., Sil Baştan Yaşayan Kent ve İnsanları, Bir Varmış Bir Yokmuş: Toplumsal Bellek, Mekân ve Kimlik Üzerine Araştırmalar, KÜY, 2017, İstanbul: p. 94

12 Rossi, A. Şehrin Mimarisi (Trans: Gürbilek,N.,), Kanat Kitap, 2006, İstanbul.

13 Halbwachs, M. (1992). On Collective Memory. (Trans. L. A. Coser,). Chicago: University of Chicago Press.

14 Ganiç, K., Kentsel Yapıtlar Üzerinden Bir Mekânsal Bellek Okuması: Atatürk Kültür Merkezi (AKM), 2016, İTU, Institute of Natural Sciences, Postgraduate, İstanbul.

15 Artun, A. 09/11. Yıkıntı Estetiği. Sürrealist Eylem Türkiye (SET), 2011, http://www.aliartun.com/yazilar/911-yikinti-estetigi/

16 Tuğrul, S. AVM'li Hatırlama ve Unutma... *Moment Dergi*, Hacettepe Üniversitesi İletişim Fakültesi Kültürel Çalışmalar Dergisi 1, 2014 (2): 16–33 ISSN: 2148-970X

17 Boyer, M. C. The City of Collective Memory: Its Historical Imagery and Architectural Entertainments. Cambridge, 1994: MIT Press.

18 Destroyed registered buildings [cited 2018, 10 November]; Available from: http://www.arkiv.com.tr/proje/iller-bankasi-belediyeler-bank-genel mudurluk-binasi/6873; http://v3.arkitera.com/tools/watermark.php?src= UserFiles/Image/ig/Gundem/AKM_imaj/102.jpg; http://www.mimarist.org/file/2017/04/danistay.jpeg

19 Destroyed sports buildings [cited 2018, 10 November]; Available from: http://www.milliyet.com.tr/avni-aker-in-kaderini-basbakan-trabzonspor-1900561-skorerhaber/; https://www.galatasaray.org/haber/futbol/1933 ten-2011e-ali-sami-yen/34471; https://www.emlaktasondakika.com/haber/kent-bolge-haberleri/besiktas-inonu-stadyumunun-yikimi-icin-henuz-izin-cikmadi/57408;http://www.haskoinsaat.com/proje/istanbul-abdi-ipekci-kapali-spor-salonu-kule-kolonlari-ve-tribun-temelleri.

20 Topaloğlu, G. Bir Yok-Mekân Öyküsü: Trabzon Tekel Binası, 2013, KTU, Institute of Natural Sciences, Postgraduate Trabzon.

21 Sancar K., and Yalçınkaya, Ş., Trabzon Tekel Binası'nın Ardından... ", 7. International Sinan Symposium, 2011, 28–29 Nisan 2011, Edirne, Turkey.

22 Özen, H., & Sert, A. Karadeniz'de Unutulan Endüstri Mirası, Gazi Üniv. Müh. Mim. Fak. Der, 2006, Vol. 21, No 3: p.499–508.

23 Trabzon Monopoly Building Competition Specification [cited 2018, 10 November]; Available from: http://www.mo.org.tr/yarismadocs/Trabzon-TekelSartname.pdf

24 Trabzon Old Tekel Building Revitalization Architectural Project Competition, 1st prize[cited 2018, 10 November]; Available from: http://v3.arkitera.com/yp108-trabzon-eski-tekel-binasi-yeniden-canlandirma-mimari-proje-yarismasi.html?year=&aID=821.

25 Trabzon Old Tekel Building Revitalization Architectural Project Competition, 2 nd prize[cited 2018, 10 November]; Available from:: http://v3.arkitera.com/yp108-trabzon-eski-tekel-binasi-yeniden-canlandirma-mimari-proje-yarismasi.html?year=&aID=837

26 Trabzon Old Tekel Building Revitalization Architectural Project Competition, 3st prize[cited 2018, 10 November]; Available from: from: http://v3.arkitera.com/yp108-trabzon-eski-tekel-binasi-yeniden-canlandirma-mimari-proje-yarismasi.html?year=&aID=788

27 Trabzon Old Tekel Building Revitalization Architectural Project Competition, 1st mention [cited 2018, 10 November]; Available from: http://v3.arkitera.com/yp108-trabzon-eski-tekel-binasi-yeniden-canlandirma-mimari-proje-yarismasi.html?year=&aID=788

28 Trabzon Old Tekel Building Revitalization Architectural Project Competition, 2nd mention [cited 2018, 10 November]; Available from: http://v3.arkitera.com/yp108-trabzon-eski-tekel-binasi-yeniden-canlandirma-mimari-proje yarismasi.html?year=& aID=820

29 Trabzon Old Tekel Building Revitalization Architectural Project Competition, 3rd mention [cited 2018, 10 November]; Available from: http://v3.arkitera.com/yp108-trabzon-eski-tekel-binasi-yeniden-canlandirma-mimari-projeyarismasi.html?year=&aID =822

30 Asasoğlu A.O., Kuloğlu N., Öztürk A., Bayrak D., Bir Tüketim Nesnesi Olarak Mekân: Trabzon'da Üç Yer, TRANSITION in/at/on/and/over/under SPACE, Kuloğlu N., Öztürk SD., Polat A., Bayrak D., Ed., 2017, Altan Özyurt Typography, Ankara: pp.126–138.

31. Trabzon Old Tekel Building Revitalization Architectural Project Competition, Winning projects, [cited 2018, 10 November]; Available from: http://v3.arkitera.com/yp108-trabzon-eski-tekel-binasi-yeniden-canlandirma-mimari-proje-yarismasi.html

32 Varlıbaş shopping mall and Trabzon town hall [cited 2018, 10 November]; Available from: http://www.varlibasavm.com/galeri/alisveris-merkezi; http://galeri3.arkitera.com/var/albums/Arkiv.com.tr/Proje/Mimari-Tasar% C4%B1m-Fikirleri/Trabzon-Belediyesi-Hizmet-Binas%C4%B1/DSCF5348.JPG

Assoc. Prof. Dr. Sennur Akansel

34 The Effects of River on the Urban Development and Water Structures of Edirne

Abstract: Rivers, as natural thresholds, have redirector role on urban planning and Edirne is one of the examples of these cities. Tunca and Meriç Rivers have been determinative in the process of urban formation of Edirne. In the connection to outer world, these rivers have either restrictive or binding roles. Hadrianopolis as located along the river side in ancient Rome, after dominated by Ottoman Empire as a castle city, locations like Gazi Mihal, Yıldırım and Sarayiçi are also along the river side as new settlement places.Rivers were the thresholds in the extension of Edirne in the Ottoman era. Edirne is a city rich in water, it is known that, political power has also contributed to build Beyazıt complex along the water side where patients get cured by sounds of water. Otherwise, villas which had been located on the riverside in the Ottoman era point out the water-city structure of Edirne in the Ottoman time period.Nowadays, although most of the architectural values gained by water have been lost, it can be documented that Edirne was a water-city.

Keywords: Historical settlement, Water structures, Edirne

Introduction

Water is one of the indispensable elements of human life such that it is the most important element for the human life's continuity following air. Water also affects the life style of a human being. On the other hand, it's one of the important input datas of architectural design and urban planning. It has an effective role in the process of the passing from nomadic life to settled life, the formation of the trade routes of civilizations and the connection of the cultures. Reduction of the water resources is one of the global problems that may cause inter-social wars. Water is the most necessary and effective element of human life and main component of biological, physical and psychological living environment. Mankind needs water as a tool to regulate the environment and as his life source. People prefer mostly water edges for recreational activities. The power of water's visual impact makes people to rest on the edges of the lakes, the banks of the rivers, the benches on the seaside. There are physical, visual reasons as well as psychological reasons for going to the water's edge. However, people do not always live near natural water resources. In the process of history, humanity has taken water as an important element of environmental design and it has built in the civilizations. In Japanese garden architecture designs, it is noteworthy that water expresses the ocean, rivers, lakes and marshes (1) Besides being decorative, designs developed around water in Islamic architecture are reflected in the design of the environment as a functional element that gives a coolness to the environment (2).

Rivers, as natural thresholds, have redirector roles on urban planning and Edirne is one of the examples of these cities. In the process of urban formation of Edirne, Tunca and Meriç Rivers have been determinative. In the connection to outer world, these rivers have either restrictive or binding roles. Hadrianopolis as located along the river side in ancient Rome; after being dominated by the Ottoman Empire as a castle city, locations like Gazi Mihal, Yıldırım and Sarayiçi are also along the river side as new settlement places. Yildirim District is located on the north side of the Tunca River and its southern ridge is adjacent to the river. Saraçhane is located on the south side of the river opposite the Yıldırım District and its northern ridge is adjacent to the river. In addition, the southern parts of Kaleiçi, Darphane, Ayşe Kadın and Kirişhane districts are adjacent to the Tunca river. Kıyıkis is the only place that has no relation with neighborhood. In this settlement, when the city morphology is examined, it can be said that the city uses to maximum level the geography of the place where the river is and the commercial centers tend to return near the water places. The reason for this is the need to use the attractive force of the city's water and waste disposal requirement.

Both Karaağaç —which is a kind of suburb of Edirne—and Tunca and Meric Rivers—which connect the central parts of the city—have been the definitive natural element for urban development. These rivers, with the historical bridges on them and as an recreative attraction places of the city, are giving the image of "the city of the bridges" to Edirne. By containing the natural and cultural heritage, today, Edirne has a significant place with its landscape and architecture. And it is usual to see the designs, as a data of cultural and natural heritage, in the formation of such a wealthy city like Edirne. Edirne, surrounded by water, is almost a peninsula. As the determining factor of the location of the city, water defines urban planning on macro scale and building design on micro scale. Nowadays, although most of the architectural values gained by water have been lost, it can be documented that Edirne was a water-city.

1 Historical Settlement of Edirne

In order to name the city as the historical settlement, it should include a unique city structure, important architectural features and an ongoing social life (3). Considering this point of view, it should not be overlooked that physical spaces are in a continuous process of change by taking into account the fact that the occurrences of each time zone must be old (4). Historical circles reflect the socio-cultural life of the period in which they belong. In these environments, the preservation of these places, which have a bridge between the past and the future, where the historical, architectural, social, economic, technological, cultural, etc., indicators are present, and it is very important in terms of ensuring cultural continuity. Historical circles are particularly highlighted and noded in the regions where they are found. The traditional markets of established residential areas close to the center usually form the important part of city(5).

The history of Edirne (Hadrianapolis) on the Via Egnatia, which was built by the Roman Republic in the 4th century BC, dates back to the 1st century BC. The first city, where the city of Edirne is located today, was established by the Odris the old Thracians and they named it as ODRISIA. Roman Emperor Hadrian gave to Edirne his name by giving status to the city due to its strategic importance (6). In the Roman period, Edirne was called as Hadrianapolis, and it was plundered many times during the Crusades after the Roman sovereignty of the Roman Empire (Byzantine). The city, which joined the Ottoman lands by I. Murat in 1361, served as the capital for 90 years until the conquest of Istanbul. Edirne with Istanbul had the best examples of the architecture in Ottoman Empire on a creative level: religious buildings, bridges, old bazaars, caravanserais with a unique architectural fiction. The first known settlement of Edirne is the Kaleiçi, which was formerly surrounded by city walls.(Phot 1) This neighborhood can be defined as the area bounded by Tunca in the west and south, Mumcular Street in the north and Saraçlar Avenue in the east. When the city entered the Turkish domination "KALE" carried the cities qualifications. Turks did not touch the Kaleiçi settlement in accordance with the agreement made in Bursa, and established their own neighborhoods in the north-south and east sides of the walls. The settlement, which is called as closed city after Turkish sovereignty Edirne has gained an open city plan diagram like Bursa. With Kaleiçi, the neighborhood nowadays called Yıldırım was located in the district of AİNA. Non-Muslims lived inside Kaleiçi, while Muslim inhabitants lived outside the castle. Kaleiçi, which hosted the big bazaars, lost its works in the fires in 1700, 1903, 1912 (Map 1-2-3).

Most of the architectural works of the city, which entered the Ottoman sovereignty in 1361, can be found in the 15th and 16th century.

The main architectural works defining the historical core in Edirne are as follows:

Phot 1: Edirne Castle Gravure (Fazıl Bülent Kocamemi Archive)

Map 1: Edirne Kaleici (Edirne Municipality Archive)

Phot 2-3-4: Old Mosque (http://www.edirnevdb.gov.tr/kultur/eskicami.html)

Phot 5-6-7: Bedesten (http://turkisharchaeonews.net/object/bedesten-edirne)

Old Mosque 1414 (It was completed during the time of Çelebi Sultan Mehmet. İts Architect Hacı Alaaddin)(Phot 2-3-4)

Bedesten 1422 (built by I.Mehmet)(Phot 5-6-7)

Three Balconies Mosque 1447 (Sultan II.Mehmet)(Phot 8-9-10)

Rüstempaşa Caravanserai 1561 (built to Mimar Sinan by Rüstem Pasha)(Phot 11-12-13)

Arasta 1569 (built by Sultan III.Murat in 1574-1595 by Architect Davut Agha)(Phot 14-15-16)

Phot 8-9-10: Three Balconies Mosque (www.google.com.tr/search?q=üç+şerefeli+camii)

Phot 11-12-13: Rüstempaşa Caravanserai (http://www.edirnekulturturizm.gov.tr/TR-76567/rustempasa-kervansarayi.html6-31.)

Phot 14-15-16: Arasta (https://www.google.com.tr/search?tbm=isch&q=edirne+arasta)

Selimiye Mosque 1575 (Built by II.Selim to Mimar Sinan)(Phot 17-18-19)

Edirne lost its old city texture in the end of 15th century after earthquakes and fires occurred. The Ali Paşa bazaar, which was built adjacent to the fortification walls, and the bazaars in the historic center determined the commercial identity of the region. (Map 5)

The railway, educational buildings, new trade centers, military barracks, educational buildings, etc., started the process of change in Edirne at the end of 19th century. At the end of this century, the historical core of Edirne was preserved and its new layers were preserved (Map 6). The architectural works of Edirne in this century:

Harbiye Barracks 1871, Men's Art School 1879, Town Hall 1898–1905

The city, which has been growing in the direction of Istanbul with the shape given by the topography in the 20th century, has become a university city with

Phot 17-18-19: Selimiye Mosque (https://www.google.com.tr/search?q=selimiye+camii+edire)

Phot 20-21-22: Harbiye Barracks. Men's Art School. Town Hall (https://www.google.com.tr/search)

university structure. Ayşe Kadin's Quarter was the first settlement point of the university and became a residential area of new housing areas (Map 7).

Map. 2–7 (Özdeş, G., pp: 26–31) Edirne, which grows rapidly in the east direction, is the direction on attraction of the residential areas with the Güllapoğlu campus of Trakya University located at the entrance of Edirne-Istanbul headway.

Small and medium-sized industrial facilities from the bar, such as Industrial Site, Oil Factory, Flour Factory, large industrial structures based on agriculture, Shopping Centers, Trakya University - Karaağaç campus, Sarayiçi campus, Ayşe Kadin's campus, Macedonia campus, Balkan campus as centre campus and the candidate to become a Technology Development area has grown toward east side of Edirne through the Intercity Bus Terminal with the collaboration of Prof. Dr. Ahmet Karadeniz Campus.

2 Architectural Court of the Edirne City Communicating with Water

Arda, Meric and Tunca rivers in the fiction of the confluence place the city of Edirne and Rivers which are components of the natural environment, has been limiting in the west direction.

The fact that Kaleiçi is situated on the banks of the river in Edirne and is limited to the city walls offers the opportunity to open the riverside streets of Edirne to the rectal areas.

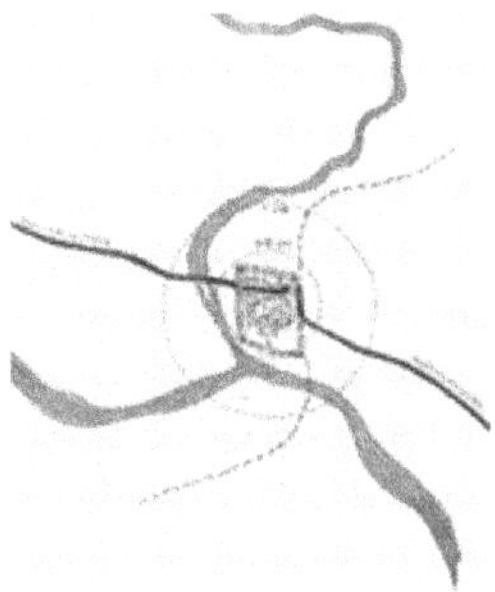

Map 2: End of II century (Özdeş, G. 1951. Edirne Foundation Plan Preparation Notes, Istanbul, , p: 26-31

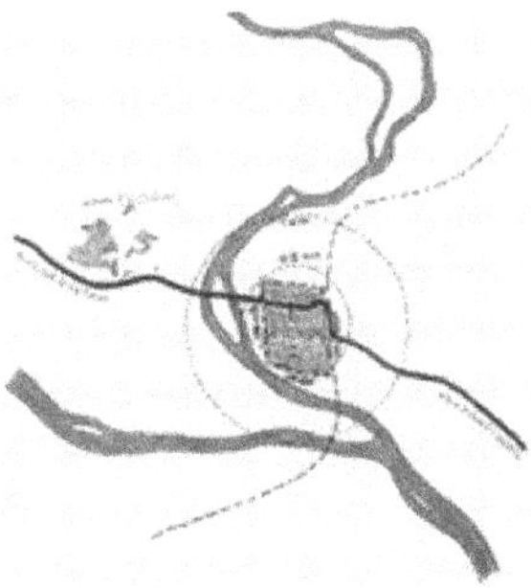

Map 3: Start of XIV century (**Özdeş, G. 1951. Edirne Foundation Plan Preparation Notes, Istanbul, , p: 26-31**

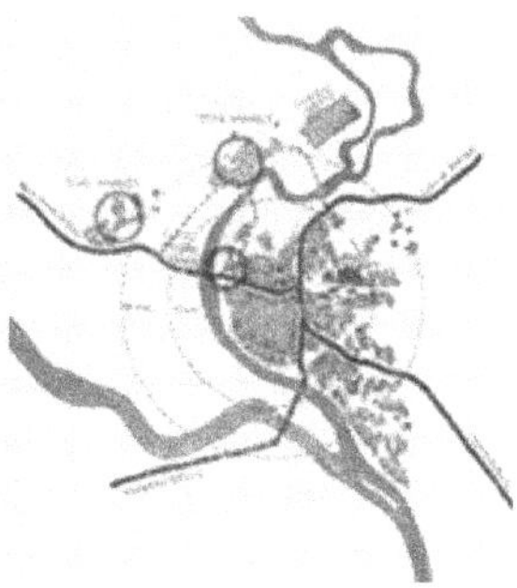

Map 4: End of XV century about 50 hectares about 100 hectares about 350 hectares (**Özdeş, G. 1951. Edirne Foundation Plan Preparation Notes, Istanbul, , p: 26-31**

Map 5: Ed of XVII century (**Özdeş, G. 1951. Edirne Foundation Plan Preparation Notes, Istanbul, , p: 26-31**

Map 6: End of XIX century (**Özdeş, G. 1951. Edirne Foundation Plan Preparation Notes, Istanbul, , p: 26-31**

We see that the palace, Kasr, religious buildings and flour factories, which have been in contact with the Ottoman period, were built on the river.

Although these buildings, which are in the nature of a mansion, have not survived to the present day, it is possible to read the traces of the literature as a document of the past life.

Known from these buildings are:

Ekmekçizade Ahmetpaşa Beach Palace (Around Kanipaşa Mosque)
Mihal Beyzadeler Beach Resort (in Bülbül Island)
Köprülü Beach Palace (on the coast of Meriç, near Kırkavak Gardens)
Bostancibasi Kırkayak Sinanpaşa Beach Palace (Near Sarachhane Bridge)
Hoca Vani Efendi Beach Palace (In the vicinity of the Old Bridge at Fındıklıyan's Mill)

Karaağaç is a suburb of Edirne and has taken its place as a promenade. It is possible to reach the center of Edirne from Karaagac through the bridges built on the Meric

Map 7: Middle of XX century about 850 hectares about 750 hectares about 600 hectares **(Özdeş, G. 1951. Edirne Foundation Plan Preparation Notes, Istanbul, , p: 26-31**

and Tunca rivers. Bridge architecture has been the return of settlement near the river in Edirne. The reconstructive spatial organizations on both sides of the river, where the bridges on the river are connected to the shore, continue to live today and show a unique relationship with the river (7).

The rivers in Edirne are dynamic and the water flow can lead to heavy floods due to Bulgaria's opening of the floodgates. The water level of the river bed with the alluviums carried by the river Meriç and the water level rises during the rainy periods. In order to minimize the future floodplain, it is known that in the past, there is a wide canal on the side of the present Gazimihal Bridge which is connected to the Yildirim and that the excess water is directed to the interior lands. With the control of the water movement, it was possible not only to evaluate the floods in the land but also to prevent ecological damage. In the palace, Saray-ı Cedid s Kum Square (second Inner Square) is the stream where the arsenal bridge connects the Justice Kasr-I and the creek, where the Sultan Süleyman Bridge, which connects the palace with the old palace site, is another element that protects the palace settlement from floods. When we examine, it is observed that the side surfaces of the stream evoke the quay and the surface of the river is smoothed and paved with stones.

Bridges built on waterways define the connection in a city and play a role in providing transportation and recreation opportunities. Mansions are built on the water shore, promenade areas, breathtaking thresholds of the city. In the past in Edirne, we have learned from the sources that the coastal architecture on the banks of the river was formed but they could not survive.

Edirne is a city that establishes a relationship with water. There are café-restaurant-tea gardens and promenade places on both shores of the city. Although it is not written in the past, it is possible to perceive that a systematic development occurred in the Ottoman city, both in historical buildings and in the neighborhood construction.

2.1 Sultan II. Bayezid Complex

The Sultan II. Bayezid complex was one of the most important health care, social, cultural and religious buildings of Sultan Mehmet II's reign. He was the son of Sultan Mehmed the Conqueror. It was built on the banks of the Tunca River in the Yeni Imaret district. Edirne Palace is 1 km away from it. The most important part of this complex is the Darüşşifa (Hospital) section, which is currently used as the Health Museum. Darüşşifa was a multifaceted hospital in its first years of life and later became the center where mental patients were treated. In this hospital, which was accepted as one of the first examples of the central plan hospitals, patients were treated with music, water tone and fragrances besides the medical knowledge of the period. It served continuously in Edirne, until 1878.

The complex, which was abandoned for many years, was allocated to Trakya University in 1984 and the university started a comprehensive restoration period. The Darüşiffa Department of Sultan II. Bayezid Complex was converted into a Health Museum by Trakya University in 1997. This museum, which was organized in a thematic style, showcased the past and attracted much attention in 2004 as winning the Council of Europe Museum Award. In 2007, the museum also won the Best Presentation Award in Europe among the award-winning museums. In 2007, Trakya University organized the madrasah section of the Külliye as a continuation of the Health Museum and in this section was held the 15th century's medical education (8).

Located in the center of Darüşşifahane and with a pool in the middle filled with water, it was used in the treatment of patients, not only with water but also with aromatic plants. It is seen that the rivers around the walls of Edirne limit the western and southern flank of the city and that the neighborhoods are constructed in the north and east also the neighborhoods fills the empty areas, and it is seen that the city is built on the Yeni Imaret side or on the way to Istanbul. It is important to note that the hospital structure is built at the edge of the water, indicating that the advantages of the environment at the treatment facility are presented to the structure with a systematic distinction (9).

2.2 Demirtaş Kasri

Edirne was the capital city of the Ottoman monarchs who had not lost its importance even after the conquest of Istanbul. The mansions have been built in relation with the river and have had the buildings of kasr and palace. Fourth Mehmet Hunter Han's Work Demirtaş Pavilion has been the residence of foreign guests and guests—"Edirne'de Saray-ı Cedid-i Amire ve Kasrları." It is possible to read the facade features of Kasr in the engraving (10). Kasr has a symmetrical facade structure. There is an outwardly protruding outcropping of the ground floor in the water direction. This is designed as open exit on the ground level. It was probably the entrance of the guesthouse from the waterway as a pier. On the upper floor of this area is a closed space. This room has windows on the

Phot 23-24-25-26: II.Bayezid Complex (Akansel S.,v.d., 2008 p. 20–23)

front and side walls in water direction. Kasr, which has wooden frame construction system, is similar to the window processes of masion buildings seen in Bosphorus in Istanbul. Especially on the lower floor, there are wide floor windows, which are covered by the overhead windows. With the windows close to the water, both enlightenment and ventilation have been provided. The light reflected in the water, especially in the ceiling of the interior makes the place rich in fiction. Kasr has chimneys with metal material that is used for roofing. It is also observed that the plants around the building are integrated with the architecture of the building. The fact of the richness of the architectural organization tells that the buildings do not exceed the tree lengths and they are in the position of being in the coasts, the connection of the scaffolds shows us the contact with water.

2.3 Caged Kasr in the Coast of Chicken Forest

It is known that the palaces located on the coasts of Meriç are the mansions and each plain has six flat ridges. There are cages in the windows of the chicken forest on the coast because of the caged Kasr, the engravings in the building, single-storyy, with a roof on the top, the ceiling in the middle of the roof with an orientalist volume evokes the application of the ceiling and the realm takes its place. This Kasr has an entrance to the pier with stairs. The ratio of windows encountered in traditional architecture was applied as two modules in opening of windows.

This arrangement, which allows light to be reflected from the water to be perceived from the interior, is made with the need to have the structure at the water's

Phot 27: Demirtaş Pavilion (İsmet Osmanoğlu personal archive)

edge. The building on the shore of the Chicken Forest, which includes the hunting pavilion, shows the pier that has access to the waterway (11).

2.4 Kasimpaşa Camii'ne Yakin Tunca Sahilinde Bir Yali

The coasts of Tunca were built from Saraçhane Bridge to Kasımpaşa Cape. It is known to use these embankments with six flat embellished embroidered boats as well as embroidered thrusters in the surrounding farms to go by road. The tradition of opening up to highland in Turkish architecture has left traces in the city life, and the resorts have been built on this way. Edirne is also the architectural reflection of this life requirement on the coast of Tunca near the Kasımpaşa Mosque. This floor, which was raised with feet on a stone dock, was largely empty. The architectural arrangement on the sea is an important example of the regional and environmental conditions. It is constructed under the structure of the shaded structure. On the upper floor, there are architectural features on the quay where the rooms are located on our traditional dwelling house, with the windows of the chimney, the stylistic chimney, the metal roofing and the roof with the broken top. At the dockside of the pier there is a gush pool with the water technology of that day. The ground floor of the plain on the quay reflects an organization that allows the use of summer. In addition, the water level of the docks hardens the floor, the boat areas for boats were also ready and prepared (12). There are six flat boats on the river and excursion trips in the writings of Dr. Rıfat OSMAN (13) (Phot 30).

Phot 28: Caged Pavilion at the Chicken Forest Coast (Duymaz R. Topaloglu Y. 2015, p.123)

Phot 29-A: Yalı in Tunca Beach close near Kasımpaşa Mosque (Duymaz R., Topaloğlu Y., 2015,s.125)

2.5 Palace-I Cedid Amire's Sand Square in Tunca Coast

Sarayiçi has been appreciated throughout history in Edirne. Located on the shore of Tunca, the square was used for rest during the Byzantine period, the birds lived in the area, as is expressed by Evliya Çelebi in his books. Due to the diversity of these birds, the promenade is called Chicken Forest and is still present today. Located close to the chicken forest "Saray-ı Cedide-i Amire," the shores of the branches of Tunca are equipped with docks, while the flooring of some of them is furnished with marble. Small and elegant villas and pavilions have been built,

Phot 30: The example of the river boat used by Edirne Tahirrevancari and used in Edirne during the time of Hunter Sultan Mehmet (Duymaz R, Topaloglu Y. 2015 s.131)

Phot 31: Palace Square at the Tunca Coast (Second Inner Square) of Cedid-i Amire (Duymaz R., Topaloğlu Y., 2015,s.145)

such as Bulbul Pavilion, Iftar Pavilion, Kafesli Mansion, Değirmen Pavilion, Libra Pavilion, Şikar Pavilion, Tedbil Pavilion.

The military and political situation in the time of bridges, Fourth Mehmet's fondness for hunting, has brought the building of the Kasr's in Edirne. The people of Edirne were enriched with the bazaars, and they increased the knowledge of the people of science and wisdom. In these gardens, it is known that the children of the population are presented with candies sent from the harem (14).

Phot 32: View of the palace of Edirne by Bostancibasi and Justice Pavilions (Osman, Dr.Rifat., p. 166)

2.6 Pavilion of Justice

The bridge over one of the tributaries of one of the branches of the Tunca is taken from the KUM Square to the area of the Justice Pavilion. Located near the water, this barn indicates that the political power was built near the water in the decision-making zone. This Pavilion with tower building is located on the quadrangular prism. The venue where the meetings are held is on the top floor of the tower and is separated from the bottom by a floor erase. The top of this floor is covered with a pyramidal roof cover (15)(Phot 32).

2.7 Edirne and Its Rivers

Meriç, born from the high mountains of the Balkans, meets Arda and Tunca rivers in Edirne and is poured into the Aegean Sea by taking the waters of the Ergene River in the south. The history, economy and culture of Edirne are important for these bridges as well as for the rivers. The important structures of the city, together with the bridges, are the works defining the identity of the Ottoman capital (14). The spatial pattern of Edirne related to rivers necessitated the linking of the sides of the city with bridges. This phenomenon revealed remarkable examples of bridge architecture in the city.

2.7.1 Bridges on the Edirne Rivers

Fatih (Armory) Bridge; Three arched (Armory) Bridge (1452) in the palace connects Kumkasri square to the Palace of Justice. The coastal shore of the bridge connecting both sides is docked (Phot 33-34-35).

Gazimihal Bridge:16 arched bridges on 17 legs were built between 1261–1282. Although the plate on the Hamidiye Bridge is written, it is known as Gazimihal. İt is located where Arda, Meriç and Tunca rivers intermingle. Evliya Çelebi states that the bridge is large enough for three cars to pass side by side and has large

Phot 33-34-35: Fatih Bridge https://www.google.com.tr/search?q=fatih+köprüsü+edirne

Phot 36-37: Gazimihal Bridge (https://www.google.com.tr/search?q=gazi+mihal+kö prüsü+edirne)

pavements consisting of large stones. In 1429, Çelebi Sultan Mehmet repaired, who wrote the history of Edirne (16)(Phot 36-37).

Meric Bridge: Mecidiye Bridge (1842–1847), built during the reign of Abdülmecit, is located on the Meriç River. Ahmed Pasha is known as founder of Ekmekçizade (17)(Phot 38-39-40).

Bayezit Bridge: The bridge, which was built by Mimar Sinan in 1567 has foursmall arches in the middle (18)(Phot 41-42).

Sarachane Bridge: The bridge which was built by Şahabettin Pasha in 1451 has 10 arches (19). The bridge that provides the connection between Edirne and Saray-ı Cedid was built to provide connection to the 2nd Murad villa built before the construction of the new palace in Sarayiçi. The bridge is 120 m long, 5 m wide and 10-eyed (Phot 43-44-45).

KANUNI BRİDGE: (Palace Bridge) is a four-arched bridge, built by Mimar Sinan (1553–1554) (20)(Phot 46).

TUNCA BRİDGE: (Ekmekçizade Ahmed Pasha Bridge) was built between 1608 and 1613 by the architect Mehmed Ağa (21)(Phot 47-48-49).

3 Result

Throughout Edirne's history Arda, Tuna and Meriç rivers are so important. After Edirne's conquest, city's Kaleiçi and it's surroundings were constructed. Under the Ottoman sovereign, Edirne was constructed with a open-city plan instead of a

Phot 38-39-40: Meriç Bridge (https://www.google.com.tr/search?q=meriç+köprüsü +edirne)

Phot 41-42: Bayezıd Bridge (https://www.google.com.tr/search?q=beyazıt+köprüsü+ edirne)

Phot 43-44-45: Sarachane Bridge (Culpan, 1975.,P:16.)

close-city plan, what other castle cities were constructed with. Thanks to this plan, streets' construction was more systematic. During Edirne's historical settlement having Kaleiçi and it's surroundings were constructed more properly, it gave birth to discovering new lands and Tunca coast and Aina were home to new constructions. During Edirne's construction rivers were both restrictive and incentive. Thanks to this, rivers gave birth to excellent bound between city and river itself.

Throughout the history, organization near rivers required knowing river's eco-system and more proper approach. If these requirements were not met, the city was

Phot 46: Kanuni Bridge (https://www.google.com.tr/search?q=Kanuni+köprüsü +edirne)

Phot 47-48-49: Tunca Bridge (https://www.google.com.tr/search?q=tunca+köprüsü)

exposed to major disasters and these disasters both took lives and destroyed things. Hence, lands larger than the coast are planned and constructed. It must be known that in today's constructions precautions are taken due to adverse events in the past. Thus, evaluating the river's motion and motion's evaluating within the land can prevent ecological destruction. Placed both of the river's coast, seawalls are precautions. Because Edirne's rivers are so active, it is required to take precautions for a flood. (It is known that river floods happened in Edirne in the past.) If we look at the city's history, placed in the old palace's land, the second stream protects the palace from floods. From time to time soil barriers were solutions to protect land and cultivated area from floods.

Rivers, when we look at the city's texture and history, besides being a threat, created links, provided travelling and recreation. Bridges constructed on the river, mansions built near coast and pavillions are datas of stream's opportunity to the city.

Surrounded with streams, Edirne's city organization was created with environmental data link. The streams gave birth to authentic architecture and some parts of those are missing: imaged old written artifacts and gravures. Those authentic buildings were definitely achievement of regional and environmental conditions. Created in a natural environment, it gave opportunity for people to live in peace. These regional and environmental conditions not only gave birth to authentic architecture but also protected the cultivated area and organization of recreative lands.

Survival struggle during the past regional and environmental conditions gave birth to authentic architectures. When planning a construction, old experiences of the land must be known. It must be known regional and environmental, natural or unnatural social environment. Only newly constructed lands which make diffusion with their surroundings will be known with their sensitivity of surroundings.

References

1 Young, D.M., 2005, The Art of the Japanese Gardens, Tuttle Pub., Tokyo, Japan.

2 Evyapan, G.A.,Tokol, A.S.,2000, Peyzaj Tasarımı Ders Notları - Landscape Design Lectures, Orta Doğu Teknik Üniversitesi Mimarlık Fakültesi, Mimarlık Bölümü Yayınları, Ankara.

3 Papageorgıoui, 1971, Continuity and Change, Preagers Publishers, London, s:27

4 Akansel S., Kaprol, T., Varlı E., 2009 "Osmanlı Başkenti Edirne'de Tarihi Çekirdeğin Oluşumu ve Tasarım Dili" 3.Uluslararası Sinan Sempozyum Kitabı, Edirne

5 Bilsel A., Bilsel G., Aksulu I., 1994 'Kentsel Koruma Planlamasında Projelendirme Sürecinde Sorunlar" 2.Kentsel Koruma ve Yenilemeler Kolokyumu, M.S.Ü., İstanbul, s.66

6 Mansel A.M. 1993, "İlk çağda Edirne" Edirne, Edirne'nin 600. Fetih yıldönümü Armağan Kitabı, TTK.,s.21–37

7 Cengiz B., "Urban River Landscapes" Peyzaj Mimarlığında Gelişmeler Editör: Özyavuz M., s.566,583., http://dx.doi.org/10.5772/56106

8, 9 Akansel A., Kaprol T., Zülfikar C., Minez B., Varlı E., Şengül E,. 2008 "Sultan II.Bayezid Külliyesi","Sağlık Müzesi" Edirne Trakya Üniversitesi Yayınları 126, s.20–25

10 Duymaz R., Topaloğlu Y., 2015 'Tarih Sahifeleri:Edirne'nin Ecnebi Misafirlerinin Kabulleri Hakkında' Edirne Abideleri Yazıları, Tosyavizade Doktor Rıfat Osman'ın Milli Mecmua'daki Yazıları-(Evleri, Konakları, Bahçeleri, Nehirleri Köprüleri) Trakya Üni.Yayınları N0.168, s.73–87

11 Duymaz R., Topaloğlu Y.,a.g.e., s.119–126

12 Duymaz R., Topaloğlu Y., a.g.e., s.125

13 Duymaz,R., Topaloğlu Y.,a.g.e.,s.130–131

14 Duymaz R., Topaloğlu Y., a.g.e., s.143–148

15,16,17,18,19,20,21 Akansel A., v.d., 2008, s.50–53

Assoc. Prof. Dr. Suat Çabuk

35 Innovations in Campus Planning: The case of Karabuk University

Abstract: A myriad of ideas discussed in the design concept today will be influential in shaping the university campuses of the future. Next-age campuses will entail a variety of usages and forms within the framework of mission and visions. Nevertheless, such a radical transformation is at a slower pace than expected due to traditional approaches and unordered decisions. In this chapter, our experiences on campus planning will be elaborated specifically for Karabük University established in 2007.

Keywords: Campus planning, Urban planning, Campus architecture, Education building

1 Introduction

In the aftermath of the Industrial Revolution, the radical shifts witnessed in the Western world transformed all of the states and social institutes in the 20th century. One of the areas that experienced the increasing impact of this transformation was the university domain.

The university concept that progressed based on churches and monasteries during the Middle Ages climbed to a new rank in 1088 upon the establishment of Bologna University. Subsequently Paris, Salerno, Oxford, Cambridge and miscellaneous other universities followed lead [1]. In the architectural design of universities in 1264, "Oxford Merton College" was accepted as the base model [2]. Colleges were replicates of monastery settings due to their inner courts and rigid architectural style. College edifices served to meet students' and lecturers' needs via classrooms, studios, housing, dining and praying areas. It is attested that during those ages universities took up an area equal to the size of several wards within cities.

As an effect of the Age of Enlightenment the unprecedented advancement of science in the 18th and 19th centuries transformed universities into modern education and research foundations. Established in 1810 in pursuant to the advice of Wilhelm von Humboldt (1767–1835) Berlin Humboldt University was the very first university that integrated in its complex the standard, modern academic freedom thereby emphasizing researcher and social roles that higher education was expected to play. Nonetheless this higher-education model of German origin failed to offer a novel and creative perspective in architectural style [2].

In the United States Harvard University established in 1636, William Mary University established in 1693, Yale University of 1701 and Virginia University of 1819 exemplified the higher-education model with a variety of aspects. Designed by the US president, Thomas Jefferson Virginia University [3] was the preliminary modern and integrated campus and laid the ground for a number of modern universities.

Campus is a concept that refers to the premises in American universities. The word relates to the architectural expression of ideas from American Higher Education and social structuring [3]. In English language the notion of "campus" generically points to colleges and universities entailing a large number of academic, housing and administrative buildings settled on the unoccupied and expansive wild lands at the outskirts of cities. Campus-style universities are premises isolated from the corruption of urban life and can integrate all essential nonacademic activities such as housing, substructure, recreation, sports, social, economic and cultural activities within its own composition [3], [4].

After the Second World War, university campus settlements started to be liberated and also became degraded due to additions of mismatching structures to the original and traditional edifice. It was witnessed that in that process wide courtyards in particular or extensive grass areas were either shrunk or disjointed.

Currently, because of communication and information technologies' growth as well as the rapid development in Internet, there is fastened interpersonal-information sharing at different times and in various locations. Higher education is one of the domains in which these changes have left its footprints. In the post 1990s, virtual university education and distance education were the proposed models. It is evidenced that thanks to greater advancement of Artificial Intelligence in the future it will be viable to popularize innovative educational settings.

In Turkey, after the Proclamation of the Republic, until the formation of university models spread to İstanbul and Ankara the dominant came from Europe and specifically the German university system up to the 1950s. Mimar Sinan University of Fine Arts, İstanbul University, İstanbul Technical University and Ankara University were city universities not having large campus sites.

In the aftermath of 1950, US-style spacious campus universities in the city peripheries emerged. The most exemplary examples are the Middle East Technical University (METU-Ortadoğu Teknik Üniversitesi) founded in Ankara in 1956 and Atatürk University founded in Erzurum in 1957. These were USA funded establishments and the first administrators of these universities were USA citizens. These schools were governed by a Board of Trustees formed by nine personnel who spoke English language well. Based on department system rather than a chair system, their mission was to train United States like-minded engineers and architects qualified to work for US companies stationed in the Middle East. Atatürk University was the first in Turkey with a campus site exclusive to itself. Competitions for campus architecture projects were organized for Atatürk University in 1955 and for METU in 1961.

Regardless of being old or modern, all state universities in Turkey seek to possess their own campus. Despite its sixty years of knowledge on campus planning and application it is not yet viable in Turkey to see a successful campus-shaping process. Originally well-planned campuses of the 1950s and 1960s were deformed due to uninformed interventions and additions. In Turkey is no integrated policy to manage the planning of university campuses. Since there is uncertainty regarding the associate, bachelor's and master's programs, as well as departments to open in the establishment phase of universities, it is most often the case to cope with lack of sufficient campus areas and structures. In the end, the physical formation of campuses fails to attain an aesthetical unison and resources allocated to that mission are simply wasted away.

Within the context of this chapter, our experiences on campus planning will be elaborated specifically for Karabük University established in 2007. Campus plans and applications of Erciyes University and Nevşehir University, of which planning processes we are aware, became our guideline identifying the core principles in the campus plans of Karabük University.

2 Karabük City and Selection of the Site for the University Complex

Until 1932 Karabük was an unheard quarter located in the Öğlebeli Village of Safranbolu District. Karabük gained popularity after the erection of the "Karabük Station" in the Karabük Post alongside the Ankara-Filyos Railway line that is 8.5 km distance from Safranbolu [5]. Upon the arrival of the first train in 1934, the constructional and financial development of the region was initiated [6]. Administrators of the Early Republican Period held the belief that economic development could be achieved via industrialization. As of 1934 the first Five-Year Development Plan was put into effect, after completing industrial facilities included in this plan within a relatively short span of five years. The buildings were put into operation [7]. Karabük was the selected site to build an Iron and Steel Plant being the last industrial facility to be erected also with the highest operational costs. Upon completing all the legal, financial and substructural preparations, the plant was opened on April 3, 1937, and the very first blast-furnace of the factory was fired in the ceremonial opening on September 9, 1939 [8].

With the opening of the Iron and Steel Plant, the transformation from a village of 13 houses to a small settlement and then to a city took place. The Karabük Municipal Organization was founded on June 25, 1939. In 1941, Karabük became a subdistrict of the Safranbolu District. Census figures of 1940 indicate that there was a population of 6,825 people in Karabük. This figure increased to 10,782 in 1945. In the same year, the number of factory workers was recorded as having risen to 4,181 [9]. The rapid industrialization process allowed the opportunity to predict the level of upcoming urbanization and thus it was resolved to establish a modern quarter to settle the officials, technicians, engineers, and workers of the factory [10]. In 1938, the planning work of this quarter, named Yenişehir, was commissioned to then-renowned French Architect & City Planner Henri Prost. Although the Yenişehir Quarter was constructed in accordance with an orderly framework, it was also evident that there were a number of unplanned housing models within the territory of the municipality between 1939 and 1948. Rapid population growth and emergent slums nearby the factory resulted in the spatial expansion of Karabük. The settlement irregularly extended along the direction of the Araç Stream and Filyos River. In the following years, the city plan was redrawn five times (1948–1968–1980–1988–1996) but all of them failed to prevent the irregular urbanization of Karabük [6].

Karabük became a city on June 6, 1995. Although Karabük University was established in 2007, the university site was already drawn in development plans in 1993 [11]. Today, the campus site is located on the east of the Karabük-Safranbolu highway between the Kılavuzlar and Öğlebeli quarters on Balıklar Rock terrain. The Araç Stream limits the northern and western borders of the university premise. The eastern border is identified with a forested land while its southern border is drawn by the Öğlebeli Quarter and Karabük-Ankara intercity highway.

In the development plan the *"University Site"* legend an area of 39.2 hectares of land is indicated. Having earned the status of University in 1978, a majority of academic departments are currently functional in Erciyes University; it has already completed the physical shaping to large extent and there is also a Technopark within its territory of 500 ha land and it is among the top-10 list of Turkish universities. Therefore allocated area for Karabük University is substantially small for a university.

This failure evidences that it was initially unplanned to build a new university in Karabük and as a reflection of the common practice at that time; the site was dedicated to erect a department and college only wherever possible. Furthermore, the selection of the university site was not based on scientific investigations but rather under the prevailing conditions of the age and it was left to the decision of prominent local leaders. Except for the land for the Faculty of Technology, there was no prior geological survey performed on this site. In addition, there are limited means of transportation to this peninsula- shaped site. A certain partition of the area specified in the development plan could not be expropriated until 2006.

Sabotaged by such adversities, it was as late as 1994 that the first academic building, the Faculty of Technical Education (23,600 m^2), could eventually be built on the campus site. Afterwards a dormitory complex was erected on the eastern side of this structure to meet the accommodation needs of the university students.

3 Older Plans of Karabük University

There was a lack of plan or campus design for the university site between 1994 and 2007. Chaired by Architect Prof. Dr. Hüseyin Yurtsever, the first campus plan of Karabük University was drawn up in September 2007. This plan was designed for an approximate land area of 46 hectares. The plan was still small in spatial aspect and the designs were formed to meet the target of 12,000 students. Focusing mainly on the principle of functionality (resolutions to meet the demand for housing, study, recreation and transportation) this project suggested a dense structuring due to the limited area on hand (Fig. 1). The positioning of the Rectorate and Faculty of Engineering buildings was also fixated in accordance with the resolution of this

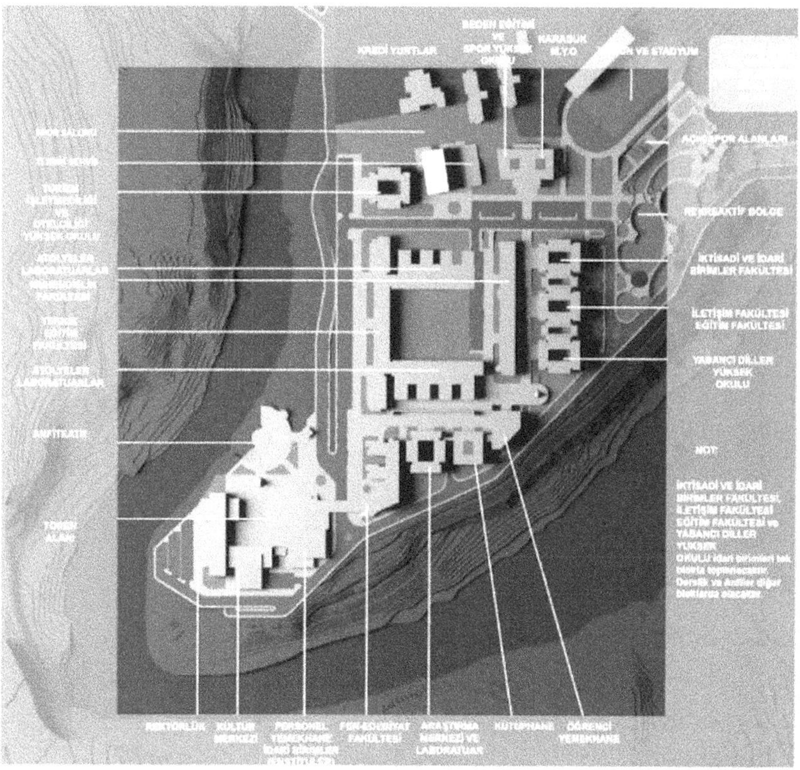

Fig. 1: Date: 2007, Karabük University Campus Project, Prof. Dr. Hüseyin Yurtsever Team.

plan. However, other resolutions taken in the plan failed to be effectuated. In the following stages geological surveys continued to be unimplemented.

For Karabük University Campus, the second project was devised in December 2009 under the leadership of the City Planner and Urban Designer Assoc. Prof. Dr. Suat Çabuk. In this plan, based on present structuring, a medium-size university campus was designed to host 20,000 students. Before devising the plan a geological survey was conducted in the university-extension area. This survey particularly noted that the university extension area could not offer solid ground for the structuring. Thus, it was required that slopes and bevels in the construction phase be avoided and that it was essential to position the buildings in required distances.

The campus size reached the size of 119 hectares land area and the compact plan envisaged for the 2007 project was forsaken due to geologic concerns, so a

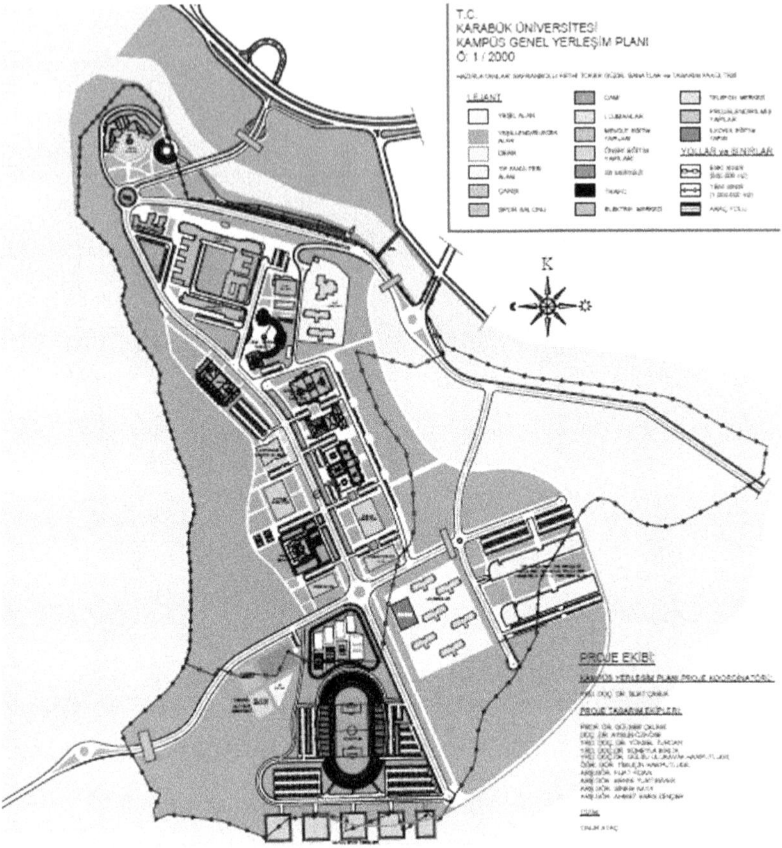

Fig. 2: Date: 2009, Karabük University Campus Project.

lightly dense housing plan was suggested. In the 2025-projected settlement plan of the university campus site, it was resolved that based on the existing conditions of land and available transportation means the best solution would be a "cross-shape or inverse T" diagram (Fig. 2). In addition, by contacting the project owners of Research-Development, Faculty of Education, Health College, Kindergarten, Stadium and Bridge of which project preparations had started, it became viable to integrate with the campus plan. However, during the approval stage of the campus plan, the site selection for the campus bazaar and mosque was spared to an administrator's resolution.

In the 2009 plan of the campus, a 22 m-wide main road extending from the northern direction bridge divided the administrative area from the academic area.

This road was included inside the campus from the western part of the Faculty of Science and by drawing a parallel with the Faculty of Economics and Administrative Sciences, building was extended towards the southwestern borders. On the eastern direction of the main road, academic units and on the western direction campus bazaar, library, cafeteria, Research and Development and similar facilities were positioned. Underlying factors in building these facilities were lower imputed values and occupying a smaller land surface.

On the southern direction of the campus, three regions formed the sub-area. In the first sub-area a sports complex was built. The main structure of the complex was the stadium and its capacity was for 20,000. Additionally for the first time, it was aimed in this project to solve the problem of building a school of sports within the premises. In addition to the stadium a great many open-sports areas were settled. Forecasting that the current substructure center would fail to meet the needs, it was suggested to build a new substructure center on the western direction of the stadium.

The second sub-area was dedicated to the public-housing site stationed on the eastern side of the stadium. The public-housing site was designed as a tiny neighborhood unit and it was foreseen in this plan to meet all its needs (school, trade, etc.) within its own body. The third sub-area was the health complex. The main complex building was the Hospital of the Medical Faculty and predicting that the building would maintain close contact with the city, it was settled in an independent region directly interconnected with the city. One of the main entrance gates of the campus was positioned on the main road accessing this complex.

4 The Final Master Plan of Karabük University

One of the envisioned decisions of the 2009 Karabük University Campus Plan could be put into effect while some could only be implemented after certain modifications. The main road in the 2009 project, the substructure channel, Residential Development Unit and stadium were some of the realizable decisions. Although the Faculty of Technology was built on the area reserved for the Health College, it was impossible to transfer the Health Facility land pointed in the development plan to the university. Therefore the Central Classroom Building was constructed in the area set for the public-housing site. A large parking lot was built in the space designed to include a Kindergarten, cafeteria and bazaar. As per the framework of a contract signed with the Public Housing Administration (TOKİ), the public-houses of the campus were constructed as 220 houses on 500 meters in the southern direction. With the installation of natural gas lines in the campus, it was resolved to heat each building independently instead of operating the central heating system. Hence, there was no longer need for a second substructure center located in the western part of the stadium.

Total number of students, academicians and administrative personnel envisaged for the year 2025 in the campus project became a reality as early as 2012. In 2007, the total size of the university campus site equaled 39.2 hectares of land and

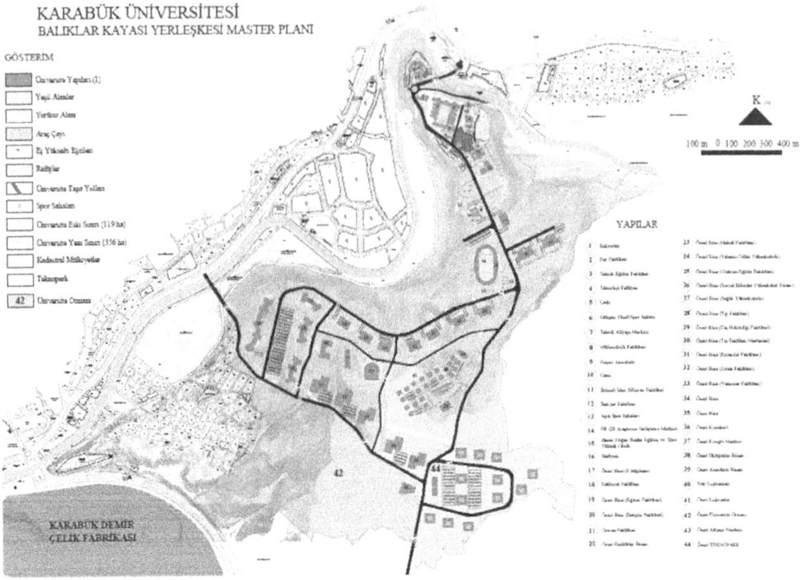

Fig. 3: Date: 2012, Karabük University Campus Master Plan.

although the size was extended to 119 hectares in 2009, it was still not capacious enough to meet the new objectives. In the subsequent stage, novel design activities were commenced for the campus in March 2012. In June 2012, after the revised plan was approved by the Karabük Municipality, the size of the campus site rose to 356 hectares. As an effect of the spatial expansion of the campus, it was resolved to dwell on a "campus master plan" technique instead of project approach.

In the Karabük University Campus Master Plan (Fig. 3) completed in August 2012, some of organizational approaches were specified with respect to nine core principles. They are as specified below [11]:

- Utilizing the aesthetical potential (the union between the green and blue areas, scenery, aqua sports, silhouette, etc.) offered by the Araç Stream that largely draws the borders of the university campus.
- In the master plan seven function areas were identified. These areas were respectively; an administrative area, academic area, equipment area, recreation area, housing area, Technopark and health area.
- In order to uncover the ecologic potential of the campus it was aimed to allocate a wider space for gardens and faculty courtyards.
- As a precaution against risks of disaster, in the original design, placing too heavy and densely situated blocks was avoided. Around 40–50 meters of distance was

freed between the structures towards the end of ensuring a homogeneous distribution overall in the entire campus site.
- The master university plan expansion diameter was 2 km, hence in addition to pedestrian transportation it was aimed to allot a place for motor vehicles being used as a means of public transportation.
- To spread the modern architectural style in the entire campus, it was aimed to renovate some facades of a number of old buildings.
- To make use of sustainable energy resources and allow the entire campus to self-provide its own energy demands.
- Technopark and the Research and Development departments bolstering industry–university coordination, the leading of which is KARDEMİR industry.
- With the motto "a campus that is active for 24 hours a day" events were to be organized that focus on sports, art and culture.

In the master plan the total number of students, lecturers and administrative personnel to use Karabük campus was computed to be 35,000. In the campus plan, it was envisaged to spare 46.5 hectares of land for technopark and 71.1 hectares of land for forestry. Since these lands were not included into the scope, 68 m^2 of land was spared per person. Gross density predicted for the campus site is 147 person/ha.

Although the master plan was completed, construction works were halted as the Karabük Revised Development Plan was annulled because despite the fact that University Site had its own master plan, there were no development rights. In 2014, according to the university campus master plan, revised development plans of the university site were forged by Assoc. Prof. Suat Çabuk submitted to the Karabük Municipality, and was effectuated upon the approval of the plan by the Municipality Board (Fig. 4).

This campus project that was drawn three times for a period of seven years failed to assist achieving a rapid growth in a controlled order. Primarily every spot of the campus was designed in line with the principle of maximum benefits with minimum costs in the substructures of road and technology. Campus diagram was positioned onto a single wide road and technical substructure axis. Between the erected structures there are homogenous spaces. In the design and application of structures the geologic disadvantages of campus surface were taken into account.

5 Problems and Novel Approaches in Campus Planning in Turkey

Experimenting the campus planning of Karabük University and our personal campus planning experiences of earlier years unveiled that in Turkey there are some serious issues related to university campus planning and applications. These problems are outlined below:

- Absence of a comprehensive and updated map of the campus site.
- Failure to conduct comprehensive geological surveys on the entire site, prior to designing campus plans.

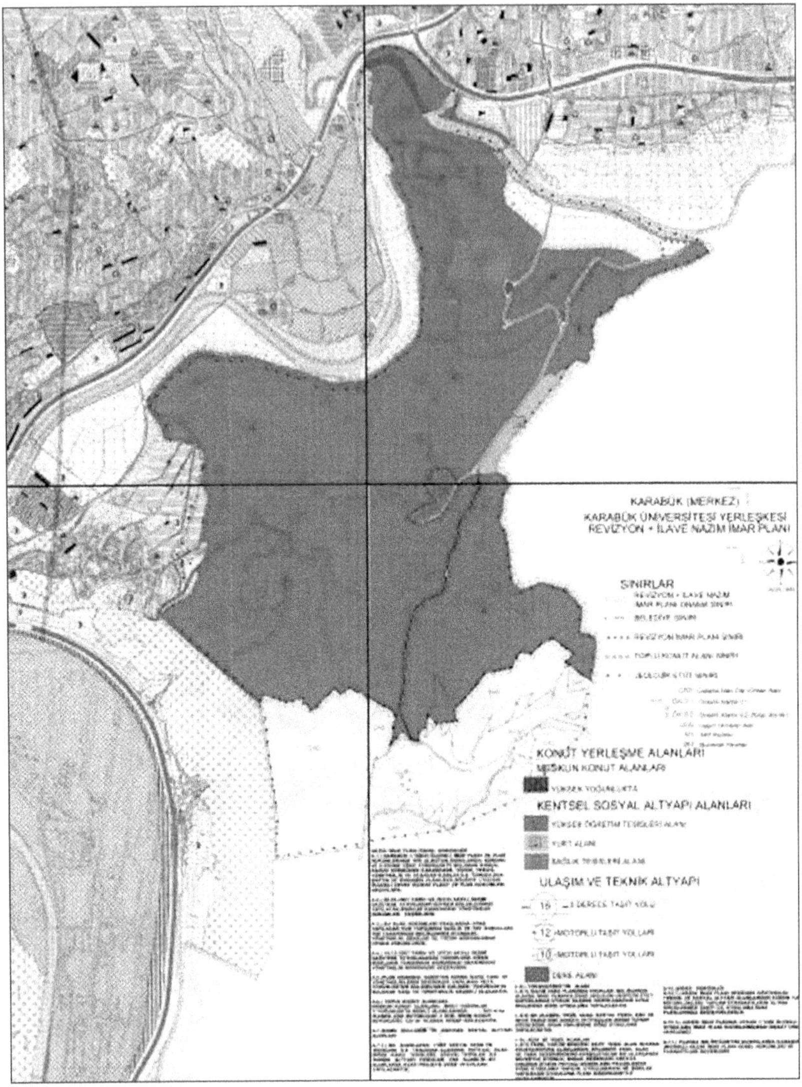

Fig. 4: Revised Development Plan of Karabük for the Karabük University Campus.

- In the city development plan there is a common failure to dedicate sufficient area for university site.
- Setting the campus plan within the borders of a restricted area.
- Unwillingness of selected or appointed university rectors to order the preparation of a campus plan.
- If a campus plan has already been prepared, there is a general tendency not to implement the resolutions or implementing them only after conducting radical modifications.
- As a result of the uncertainty in the projected academic departments and other academic units, there is failure to reflect accurately on the plans for an estimated number of buildings as well as total number of students and lecturers that the university will host.
- Despite the radical transformations in campus plans, the failure to update the plan.
- Emergent aesthetical problems because of the failure to prepare concordantly the plan and architectural projects of the structures.
- Within the framework of procurement laws, the fact that building companies having offered the lowest cost, built the structures with poor-quality materials albeit complying with the company rules.
- A general insufficiency in the totality and quality of technical staff employed in construction units of the university to put the campus plan into effect.
- Insufficient budget allocated by the university for construction work.
- In addition to the campus plan devising a substructure center project and campus substructure plan concomitantly.
- Based on the campus plan and campus substructure plan failure to forge a campus landscaping project [12].
- Failure to form a campus geo-information system [13].
- Failure to devise campus plans and campus architectural projects that are grounded in sustainability and smart campus approaches.

6 Conclusion

A myriad of ideas discussed in the design concept today will be influential in shaping the university campuses of the future. Next-age campuses will entail a variety of usages and forms within the framework of mission and visions. Nevertheless, such a radical transformation is at a slower pace than expected due to traditional approaches and unordered decisions.

The campus planning and application experience of Karabük University is a model of moving beyond traditional design and application patterns in Turkey and entails in itself an accumulation of background knowledge. It was witnessed that on accounts of systemic problems, budget deficiencies, arbitrary decisions of administrators, and failure to accurately operate the system that expected success could not be secured and will not be. Even the slightest change in the administrative team at the university is likely to spoil the campus plan and application.

Within that context avoiding the flaws mentioned in the fifth part, accurate operation of the process in campus planning and in university campus planning, organizing contests on the latest hot topic like smart-campuses and sustainable campuses would potentially offer huge benefits.

References

1 Aytaç, K., *Avrupa Eğitim Tarihi.* 1992, Marmara Üniversitesi İlahiyat Fakültesi Vakfı Yayınları, No: 58: İstanbul.

2 Hashimshony, R., and Haina, J., *Designing the University of the Future,* Planning for Higher Education, 2006, **34**(2): 5–19.

3 Turner, P. V., From Independent University Collections to a Wissenstheater: An Ambitions Project at the Humboldt University of Berlin, Museologia, 1990, **2**: 81–88.

4 Muthesius, S., *The Postwar University: Utopianist Campus and College,* 2000, Yale University Press: New Haven.

5 Çabuk, S. *Cumhuriyet Dönemi Karabük Kentinin Biçimlenişi ve Kamu Mekanları,* Kamusal Alanların Mekânsal Organizasyonu, Ed. H. S. Çelikyay, 2017, Bartın Üniversitesi Yayınları: Bartın.

6 Çabuk, S., Demir, K. and Gökyer, E., *Cumhuriyet'in Yeni Kenti Karabük ün 1937–1988 Dönemi Mekânsal Gelişimi ve Şehir Planları,* Karabük Üniversitesi Sosyal Bilimler Enstitüsü Dergisi, 2016, Özel Sayı 2: 20–39.

7 Soylu, P.Ü. ve Yaktı, Ö., *Devletçiliğe Yönelmede Bir Köşe Taşı: Birinci Beş Yıllık Sanayi Planı.* History Studies, 2012, Enver Konukçu Armağanı Sayısı: 365–379.

8 Kütükçüoğlu, M., *Milli Şef Dönemi'nde Karabük.* Yüksek Lisans Tezi. 2009, Marmara Üniversitesi Türkiyat Araştırmaları Enstitüsü: İstanbul.

9 Kalyoncu, H., *Cumhuriyet Kenti Karabük.* 2007, Karabük Belediyesi Yayınları: Karabük.

10 Karabük Belediyesi, *Karabük 85,* 1985, Karabük Belediyesi Yayını, Ufuk Matbaası: İstanbul.

11 Çabuk, S., *Karabük Üniversitesi Balıklar Kayası Yerleşkesi 2012 Master Planı,* 2012, Karabük Üniversitesi Yayınları No: 2: Karabük.

12 Dönmez, Y., Türkmen, F., and Çabuk, S., *Üniversite Yerleşkelerinin Planlarına İlişkin Öğrenci Görüşleri,* Uluslararası Hakemli Tasarım ve Mimarlık Dergisi, 2015, **6**: 1–14, Doi Number: 10.17365/TMD. 2015614142.

13 Özyavuz, M., Şişman E.E., and Korkut A., *Namık Kemal Üniversitesi Yerleşke Bilgi Sisteminin Oluşturulması,* Tekirdağ Ziraat Fakültesi Dergisi, 2009, 6(3): 227–234.

Assoc. Prof. Dr. Şebnem Ertaş and Assist.Prof.Dr. Elif Sönmez

36 Determination of Visitor Expectations in the Scope of Sustainable Tourism: The Case of Konya/Sille

Abstract: Culture tourism is a tourism activity which is fed by cultural treasures. Konya and its surroundings is an important tourism destination in Turkey with its tourism resources. With its 5000 years old history, its cultural diversity and its rich cultural heritage, Sille as a valley of history and culture, is an important tourism corridor for Konya. This article has been prepared depending on data collected through a survey study with 125 visitors who visited Sille in 2015. It was aimed at identifying resource values for Sille with the obtained data, through the determination of the expectations of domestic and foreign tourists to the area in order to enliven tourism at Sille and by attracting attention on its tourism potential.

Keywords: Culture tourism, Sustainable tourism, Visitor expectations, Konya/Sille

1 Aims and Background

In the past decade, it can be seen that various balanced and healthy approaches have emerged in order to prevent the negative developments which have been experienced in tourism globally. Among these approaches, the concept of "sustainable tourism" comes forward (Avcıkurt, 1996). According to Goodall (1992), sustainable is based on the approach of meeting the needs of visitors in a way that will improve the living standards of locals and protect environmental and cultural values (Eser, 2011). According to Müller (1993), it is defined as "tourism that has environmental and social compliance" (Eser, 2011). Müller (1993) expressed the concept of environmentally and socially compliant sustainable tourism with a pyramid they named as "the magic pentagon." Based on this pyramid, while the economic prosperity of the local community in the region is dependent subjective health and meeting the numbers of visitors to the greatest extent, the preliminary conditions for this are "protection of active natural resources" and "active culture" (URL-1, 2017). Future generations were stated as the next aim.

In the long run, culture tourism serves the purposes of sustainable tourism much more compared to mass tourism (Özdöl and Yürik, 2007). Sustainable culture tourism is related to the travels that people make in order to see culturally rich places, to learn about and observe diminishing life styles and to collect cultural values as memories. Thus, it is provided that the use of resources is continuous and balanced (Avcıkurt, 2003). For this reason, cultural tourism is inherently the most dominant activity in the scope of tourism which is fed by sustainable cultural treasures (Demiroğlu et al., 2008). Culture tourism whose basic element of attraction is the cultural values of a region includes tourism that takes place in areas like museums and theatres where, especially in historical cities, the lifestyles of people and the art objects they have created are displayed. In addition to these elements, touristic attractions which are evaluated in the scope of culture tourism include all phenomena and objects from art creations, handicrafts, traditional lifestyles to tools used by local societies and their belief systems specific to that country or region (Kahraman and Türkay, 2006).

According to the International Culture Tourism Regulation accepted by the ICOSMOS (1999) at the 12th general assembly carried out in Mexico, expressing the importance of cultural heritage to visitors has also been emphasized as an important goal in the management of cultural heritage. In the regulation, the guiding principle "Protection and Tourism Planning done for Cultural Heritage Zones should provide the visitor experience to be remarkable, satisfying and pleasurable" is included. In accordance with this principle, the planning done for tourism activities should meet the requirements and desires of the visitors (Cracolici and Nijkamp, 2008; Eser et al., 2010). People are becoming increasingly interested in seeing culturally rich places and learning about the cultural values of a region by observing the continually diminishing life styles. For this reason, a sound relationship should be established between visitors' expectations and the settlements with tourism potential. In this way, the determination of the supply and demand of both national and foreign tourists would aid the development of culture tourism.

In recent years, natural areas have become a preferred recreation area for people looking to escape their busy urban lives (Cetin and Sevik, 2016). According to the World Tourism Organization, 37% of international tourists travel for cultural purposes (du Cluzeau, 2000; Cetin and Sevik, 2016). In Turkey, a policy of moving tourism from coastal areas towards the central regions has been followed in recent years. In this context, the natural, historical and cultural richness and diversity of the Anatolian geography have brought Turkey into an ambitious position in culture tourism. While culture tourism provides significant opportunities to protect and maintain the cultural heritage of Turkey, it also turns the attention towards new investment and business opportunities and options that will increase the power of development and competition especially for historical cities (Çekül, 2012).

This study was conducted in the residential area Sille in the province of Konya where tourism activities have not yet intensified but there is a potential towards such a position. The data obtained for this purpose are important in terms of determining the touristic supply and demand potential of Sille, which was genuinely analyzed in terms of tourism, and creating strategies for this potential to form a significant corridor for the province of Konya in Turkey. This way, it was aimed to determine and analyze the current tourism activities and expectations of domestic and international visitors who visited Sille.

2 Material and Method

The study area was selected as the residential region Sille, which is an old settlement located 8 km north of the Konya city center. Sille has a potential to become a

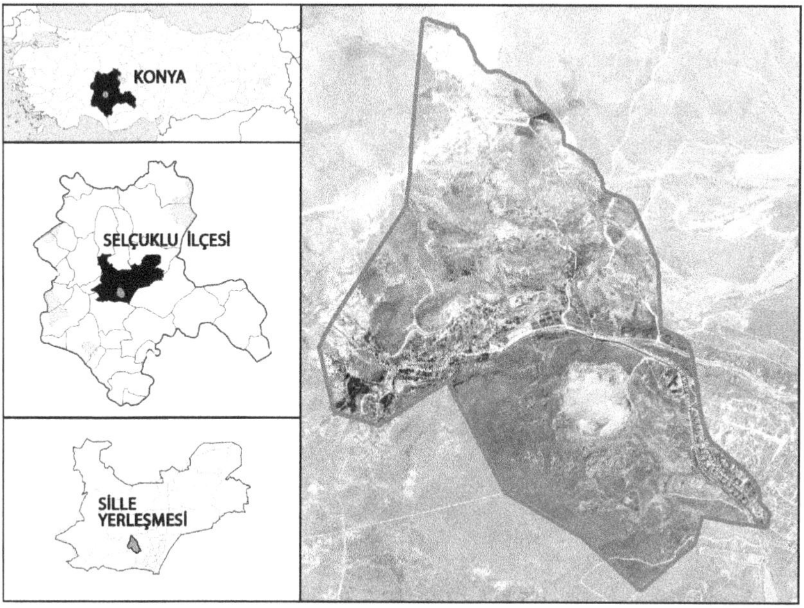

Fig. 1: Location of Konya/Sille (Sönmez, et al., 2017)

highly significant destination in terms of culture tourism with its significant touristic supply resources within the province of Konya (Fig. 1).

Throughout its 5000-year history it has hosted many civilizations mainly Greeks, Seljuks and the Ottomans. At Sille whose history goes back to very old ages and which is located on the Silk and Spice Roads seen as important commercial centers, where Muslims and other religious groups (Christians, Jews, etc.) have lived together has caused the settlement to become a center of commerce with its diverse culture (Erdem et al., 2010) (Fig. 2).

Before the Republic, Sille was a city containing 16 villages that were socially, culturally, economically and civically developed, and it had a high population density. After 1940, it became a residential area which acted as a source of migration towards metropolitan cities such as Istanbul, Ankara and Bursa due to economic reasons and increasingly lost its population (Aklanoğlu, 2009). While it was a municipality until 1980s, after 1989 it became a district of Selcuklu Province (Fig. 3).

Sille, which has hosted many buildings from Middle Ages to the Republic era, is a valley of history and culture. Mosques, churches, cemeteries, waterways, bridges, fountains, baths, houses and streets hosted by Sille are very important values which have witnessed a rich history. It is an important region, which has lived with these values to our day and which continues to preserve its social life at present like in the past (Fig. 4).

Fig. 2: Sille in 1925 (Aklanoğlu, 2009).

Fig. 3: Sille in 2016.

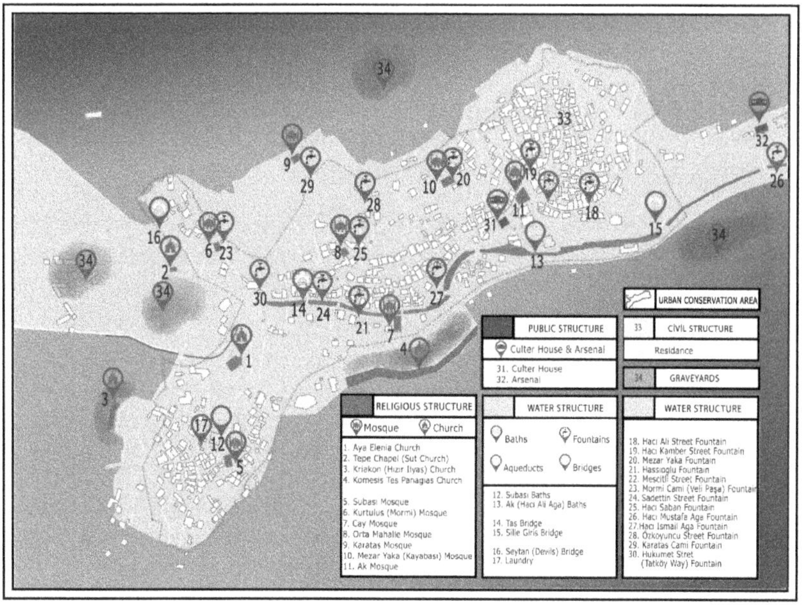

Fig. 4: Sille cultural assets in the urban conservation area.

In the study, to determine the attitudes and desires of domestic and foreign tourists of Konya/Sille settlement, the on-site data collection technique of "survey" with standard forms was used. With the survey study conducted with the participation of domestic and foreign visitors, the resource values of the region, the visitor profile, the frequency and duration of their visits, the activities they carry out, their problems with the area and their expectations were determined and evaluated.

During the survey study which was carried out in a face-to-face manner, the participants were selected randomly. In the determination of the number of participants the sample size formula was used (Kalıpsız, 1981). While the visitors to Konya is 2,313,293 people, the number of visitors to Sille is 91,264 people according to the Selcuklu Municipality museum visitors records. About 38% of these visitors are foreign visitors. The survey study was carried out in June–July 2015 in summer season and in October–November 2015 in winter season in line with the suggestions of tour operators with a total of 125 people, 75 being domestic visitors and 50 being foreign visitors.

While preparing the survey for the visitors, Kiper's (2006) study was used. In the survey study carried out with visitors it was aimed at collecting data on the profile of domestic and foreign visitors (age, sex, educational background, etc.);

their purpose of visit to Sille, their demand/desires/tourism-recreation tendencies; motivating factors for their selection of Sille and their foresights regarding Sille. While selecting the motivating factors for visitors preference for the area, the scale from Kara Lea Wolfe's (2002) study was taken as an example.

The data collected from the responses of domestic and foreign visitors were analyzed using SPSS V. 22 statistical program.

3 Findings

According to the evaluation of survey participant tourists' demographic characteristics (Tab. 1):

- When the age condition was evaluated, 11% were found to be between the ages of 18–24, 10% were found to be between the ages of 25–34, 21% were found to be between the ages of 35–44, 30.4% were found to be between the ages of 45–54, 20.8% were found to be between the ages of 55–64 and 5.6% were found to be 65 years of age or more.
- While in terms of domestic tourists visitors of all age groups were found, in terms of foreign tourists the visitors were 35 years of age and more, and especially a concentration was observed in the ages between 45–54 years.
- When evaluated generally, it could be said that a majority of the participants in the survey had a high education level.
- The 54.4% of survey participants were comprised of males and 45.6% were comprised of females. When we evaluated the tourists to the area as domestic and foreign separately, while in terms of domestic tourists female participation ratio was higher, in terms of foreign tourists, male participation ratio was higher.
- When evaluated in terms of marital status, it was seen that 72.8% of survey participants were married and 27.2% were bachelors.

It was found that 89.6% of domestic and foreign visitors had visited the area for the first time and especially all of the foreign visitors had visited Sille for the first time. A majority of these foreign visitors come with daily tours; and a majority of domestic visitors come to the area for daily visits together with friends and family. During these daily visits the visitors buy souvenirs, handcrafted knitting and agricultural products.

98.4 % of the visitors have stated that they would definitely visit the area again if they had the opportunity and that they would recommend to their family and friends to visit Sille and they would recommend it to those who enquire about Sille. For this reason, it could be said that the tourism potential of Sille will show a tendency to increase in the future.

In order to determine the factors motivating survey participants for preferring Sille, 11 suggestions were presented; and the participants were asked to rate these suggestions for motivating factors in terms of importance (Tab. 2). Five different ratings were presented to the participants (1: not important at all, 2: not important, 3: neither important nor unimportant, 4: important, 5: very important). When the motivating factors were evaluated in terms of importance for the total population

Tab. 1: Demographic Characteristics of Domestic and Foreign Tourists

		Domestic		**Foreign**		**Total**	
		Frequency	**%**	**Frequency**	**%**	**Frequency**	**%**
Age	18–24	14	18.7	-	-	14	11.2
	25–34	13	17.3	-	-	13	10.4
	35–44	21	28.0	6	12.0	27	21.6
	45–54	13	17.3	25	**50.0**	38	**30.4**
	55–64	8	10.7	18	36.0	26	20.8
	65 +	6	8.0	1	2.0	7	5.6
Total		75	100.0	50	100.0	125	100.0
Gender	Male	36	48.0	32	**64.0**	68	**54.4**
	Female	39	**52.0**	18	36.0	57	**45.6**
Total		75	100.0	50	100.0	125	100.0
Education	Primary	7	9.3	6	12.0	13	10.4
	High	24	32.0	16	32.0	40	32.0
	Undergrad	14	18.7	-	-	14	11.2
	Graduate	30	**40.0**	28	**56.0**	58	**46.4**
Total		75	100.0	50	100.0	125	100.0
Married Status	Single	21	28.0	13	26.0	34	27.2
	Married	54	**72.0**	37	**74.0**	91	**72.8**
Total		75	100.0	50	100.0	125	100.0

of visitors, it was found that visitors come to the area to see new places, to observe traditional architecture, to come together with friends and family, to obtain general knowledge, to be intertwined with nature, to get rid of personal daily hardships, to taste new dishes, to meet new people, to bring difference to the routine daily life, to pass solitary time and to visit religious sites.

Among purposes of visitors to prefer Sille, while for domestic visitors motivating factors such as seeing new places, traditional architecture, and to come together with family and friends are more effective; for foreign tourists seeing new places, traditional architecture and visiting religious sites are motivating factors.

Among the responses regarding what first takes survey participants attention in Sille, with 66.4 %, the traditional architecture is the first factor. This is followed by historical attractions with 23.2 % and unique nature and views with 7.2 %.

Samples of the rich architectural heritage in Sille are the most attractive elements for the visitors; for this reason, they could be listed among the most important factors for the development of cultural tourism at the area (Tab. 3).

Tab. 2: Motivating Factors for Domestic and Foreign Tourists to Prefer Sille

			Strongly Disagree	**Disagree**	**Neither Nor**	**Agree**	**Strongly Agree**
	Frequency		%	%	%	%	%
See/ experience new things/ places	Domestic	75	-	-	1.3	4.0	**94.7**
	Foreign	50	-	-	-	-	**100.0**
	Total	125	-	-	0.8	2.4	**96.8**
Traditional Architectural Construction	Domestic	75	-	-	-	8.0	**92.0**
	Foreign	50	-	-	-	-	**100.0**
	Total	125	-	-	3.2	5.6	**91.2**
Religious purposes	Domestic	75	17.3	16.0	28.0	12.0	26.7
	Foreign	50	-	-	8.0	2.0	**90.0**
	Total	125	10.4	9.6	19.2	10.4	50.4
Try new foods	Domestic	75	-	1.3	16.0	34.7	48.0
	Foreign	50	-	-	6.0	8.0	**86.0**
	Total	125	-	0.8	9.6	28.0	**61.6**
Be close to nature	Domestic	75	-	-	9.3	37.3	53.3
	Foreign	50	-	-	-	18.0	**82.0**
	Total	125	-	-	8.0	26.4	**65.6**
Forget about personal worries/ troubles	Domestic	75	-	-	5.3	37.3	57.3
	Foreign	50	-	-	6.0	10.0	**84.0**
	Total	125	-	-	6.4	29.6	**64.0**
Meet new people	Domestic	75	-	1.3	6.7	38.7	53.3
	Foreign	50	-	-	8.0	18.0	**74.0**
	Total	125	-	0.8	4.0	34.4	**60.8**
Be away from work/daily routine	Domestic	75	-	-	4.0	45.3	50.7
	Foreign	50	-	-	-	28.0	**72.0**
	Total	125	-	-	3.2	37.6	59.2
Spend time alone	Domestic	75	2.7	8.0	9.3	37.3	42.7
	Foreign	50	-	-	2.0	26.0	**72.0**
	Total	125	1.6	4.8	5.6	33.6	54.4
Gain general knowledge	Domestic	75	-	-	4.0	34.7	**61.3**
	Foreign	50	-	-	-	28.0	**72.0**
	Total	125	-	-	2.4	31.2	**66.4**

Tab. 2: Continued

	Frequency		Strongly Disagree	Disagree	Neither Agree Nor	Agree	Strongly Agree
			%	%	%	%	%
Be with friends and family	Domestic	75	-	-	-	28.0	**72.0**
	Foreign	50	-	-	-	26.0	**74.0**
	Total	125				27.2	**72.8**
	Foreign	50	-	-	-	-	-
	Total	125	-	-	-	-	-

Tab. 3: Features Which First Attract the Attention of Tourists in Sille

	Domestic		Foreign		Total	
	Frequency	%	Frequency	%	Frequency	%
Traditional Architectural Construction	67	**89.3**	16	32.0	**83**	**66.4**
Local Life and Culture	4	5.3	-	-	4	3.2
Unique Nature and Scenery	1	1.3	8	16.0	9	7.2
Historical Attractions (Churchs. Ruins etc.)	3	4.0	26	52.0	29	23.2
Agricultural Landscape	-	-	-	-	-	-
Other	-	-	-	-	-	-
Total	75	100	50	100	125	100

Three most loved basic features of Sille stated by survey participants are traditional architectural structure with 97.6 %, historical attractions with 70.4 % and handicrafts with 58.4 % (Tab. 4).

At Sille, where different cultures live in harmony, history and culture are the primary tourist attractions. In this context, the major features which attract tourists at the Sille settlements are its traditional architecture, houses, churches and historical sites.

Regarding the negative features at Sille, 70.4 % of the survey participants pointed out to high prices. The other negative features that came forward were respectively inadequate accommodation facilities with 33.6 %, inadequacy of touristic activities with 28.8 %, neglect of historical resources and their under maintenance with 24.8 %, low service quality with 16.8 %, transportation problems with 9.6 %, inadequate catering facilities with 8.8 % and the negative attitude of local

Tab. 4: Most Loved Features by Domestic and Foreign Tourists

	Domestic		Foreign		Total	
	Frequency	%	Frequency	%	Frequency	%
Traditional Architectural Construction	73	97.3	49	98.0	122	**97.6**
Local Life and Culture	24	32.0	25	50.0	49	39.2
Local Cuisine	21	28.0	32	64.0	53	42.4
Unique Nature and Scenery	20	26.7	34	68.0	54	43.2
Historical Attractions (Churchs, Ruins etc.)	43	57.3	45	90.0	88	**70.4**
Agricultural Landscape	6	8.0	-	-	6	4.8
Easement of Access	-	-	5	10.0	5	4.0
Accommodation Possibilities	2	2.7	3	6.0	5	4.0
Food and Beverages Opportunities	-	-	2	4.0	2	1.6
Attitudes of Local Community	5	6.7	6	12.0	11	8.8
Handicrafts	32	42.7	41	82.0	73	**58.4**
Service Quality	5	6.7	8	16.0	13	10.4
Other	-	-	-	-	-	-

people with 8.0 %. Considering other ratios 7.2 % of the participants stated various problems such as the existence of too many cafes at every corner, pollution caused by too many businesses, inadequacy and costliness of accommodation (Tab. 5). That accommodation facilities receive a low ratio is an admirable feature and that it is seen as a negative feature is an indication that the accommodation facilities are inadequate.

As a response to the survey question regarding what kind of tourism activities could be carried out in Sille, visits and stays at houses with traditional architecture were stated by 66.4 %; festivals were stated by 63.2 %, touristic activities by 53.6 % (nature walks, biking tours, etc.) as activities that could be organized. These are followed by handicraft exhibitions with 32.8 % and garden walks which could allow tourists to pick up fruit and vegetables with 24.8 % (Tab. 6). In this context, it could be thought that for activities to be planned for the region, primarily accommodation facilities and touristic activities should be organized.

In the survey participants' responses to the question "What should be done in Sille to enliven tourist?", the most important activity was found to be increasing publicity with 95.2 %. This emphasizes the need for publicity and promotion for the

Tab. 5: Negative Features Expressed by Domestic and Foreign Tourists at Sille

	Domestic		Foreign		Total	
	Frequency	%	**Frequency**	%	**Frequency**	%
Insufficient accommodation possibilities	27	**36.0**	15	**30.0**	42	**33.6**
Insufficient food and beverages opportunities	8	10.7	3	6.0	11	8.8
Insufficient tourism activities (running. horse-riding. walking etc.)	28	**37.3**	8	16.0	**36**	**28.8**
High Prices	54	**72.0**	34	68.0	**88**	**70.4**
Neglected Ancient and Historical Attractions	25	**33.3**	6	12.0	**31**	**24.8**
Low Service Quality	17	22.7	4	8.0	21	16.8
Negative Attitudes of Local Community	4	5.3	6	12.0	10	8.0
Problems about the transportation	5	6.7	7	14.0	12	9.6
Other	8	10.7	1	2.0	9	7.2

creation of new touristic demand at Sille. Especially foreign tourists have pointed out that restoration works (50.0%) and increased accommodation facilities (48.0%) should be provided. Domestic tourists, on the other hand, have stated that the number of festivals should be increased (40.0%) (Tab. 7).

4 Results and Discussion

Tourism is a large-scale activity around that world which involves more than a billion people per year today. Every year, the numbers of people, countries and regions involved in these activities are increasing, and the demands of participants vary based on the conditions of the date. These demands, which have changed in time in the world and in Turkey, have led to the development of alternative types of tourism. Culture tourism is the top form of tourism among the developed alternative types of tourism.

In this study which covered European Union countries, it was found that more than half of the tourists who travelled for cultural purposes had higher education degrees (EU average: 21 %) and they had occupations with high income in general. According to the same study, the ratio of those who described their travel activities as cultural was two times higher among those with higher education degrees in

Tab. 6: Evaluation of Domestic and Foreign Tourists Regarding Activities That Could Be Carried Out in Sille

	Domestic		**Foreign**		**Total**	
	Frequency	%	**Frequency**	%	**Frequency**	%
To visit and stay in traditional architectural house	57	**76.0**	26	**52.0**	83	**66.4**
To organize trip in agricultural land and collect the vegetables. fruits etc.	17	22.7	14	28.0	31	24.8
To organize festivals	56	**74.7**	23	46.0	79	**63.2**
Exhibition about handcrafts	28	37.3	13	26.0	41	32.8
Tourism activities (nature walking. bicycle tour. trekking etc.)	37	49.3	30	**60.0**	67	**53.6**
Other	-	-	-	-	-	-

comparison to other education levels. Again, tourists with higher levels of education visited cultural attractions two times more than other tourists (Richards 2001; Öter and Özdoğan, 2005:129). These results show that culture tourism concerns a highly qualified market with high spending capacity. Turkey should prioritize culture tourism in its touristic product diversification policies. In order to improve this type of tourism, it will be reasonable to know people who travel for cultural purposes more closely (Öter and Özdoğan, 2005:129).

Considering the domestic and international visitor potential, it was seen that all ages of domestic visitors visited the region. This was because one-day visits are popular. International visitors were mostly in the middle-aged group. This was mostly because it is an alternative corridor for groups that visit Konya and the Aya Eleni Church in the area. It was observed that most visitors did not prefer to stay in the region due to the shortcomings in accommodation in the region. If we want to attract more domestic and international tourists to the area, it is necessary to increase the means of accommodation throughout the area (Düzgüneş and Saraç, 2018).

Generally speaking, it is possible to improve culture tourism in Sille only by preservation and utilization of historical and cultural properties by restoration works, improvement of handicrafts, and completion of infrastructure for accommodation, dining, etc. Thus, in order to improve the image of the region, cultural

Tab. 7: Evaluations of Domestic and Foreign Tourists Regarding Things to Be Done at Sille to Enliven Tourism

	Domestic		**Foreign**		**Total**	
	Frequency	%	**Frequency**	%	**Frequency**	%
To focus on advertisement	74	**98.7**	45	**90.0**	119	**95.2**
To constitute an association for organizing tourism	11	14.7	8	16.0	19	15.2
To prepare an education program for local community and local employee	17	22.7	10	20.0	27	21.6
To finance public and private organizations making infrastructure and superstructure for tourism	18	24.0	9	18.0	27	21.6
To increase restoration attempts	17	22.7	25	**50.0**	42	33.6
To improve service quality	17	22.7	5	10.0	22	17.6
To supply coordination in between public and private organizations to develop tourism	2	2.7	6	12.0	8	6.4
To increase tourist accommodation by working together with travel agencies	5	6.7	24	**48.0**	29	23.2
To increase festivals	30	**40.0**	12	24.0	42	33.6
Other	-	-	-	-	-	-

resources should be utilized, and the region should be made more attractive. It was observed that expectations included artistic activities (concerts, exhibitions), festivals, hiking, thematic activities like bicycle tours, and exhibition of products by improvement of regional handicrafts.

Therefore, the supply and demand data obtained in the scope of the study towards domestic and international visitors will contribute to the development of culture tourism in Sille. Additionally, this study and similar studies will have to be replicated in certain time intervals to monitor the development of culture tourism in a region.

5 Conclusions

Considering and examining the results of the study, the issues that require attention are the following:

- The problem of transportation for the Sille destination should be solved for visitors (public transportation vehicles) and quality of service should be increased.
- In order to provide opportunities for domestic and international tourists visiting the region, accommodation, cafes, restaurants and cultural spaces (exhibition, workshops, etc.) should be established.
- The restoration work for the cultural property in the historical residential area of Sille should not be conducted for single properties in particular, but it should be planned for the whole residential area.
- Tour operators should adjust the time they spare for the Sille destination in Konya-themed tourism activities in a way that will satisfy the customers.
- Visitors who attend the tours should not be forced to go and see displays of carpets, candles and pots.
- Local shopkeepers should avoid behaviors of selling things by harassing tourists as much as possible.
- Local governments should provide training for the shopkeepers in Sille about communication with visitors and sales.
- It is important that security services in the Sille destination and historical places should be provided properly.

References

1 Aklanoğlu, F., *Geleneksel Yerleşmelerin Sürdürülebilirliği ve Ekolojik Tasarım: Konya-Sille Örneği*, Doktora Tezi, Ankara Üniversitesi, Fen Bilimleri Enstitüsü, 2009. Ankara.

2 Avcikurt, C., *Turizm Sosyolojisi*, 2003. Ankara: Detay Publication.

3 Avcikurt, C., *Turizm ve Surdurulebilirlik*, Adnan Menderes Universitesi, Kusadası Belediyesi, 1st International & National Tourism Congress, 1996. 29th November-3th December.

4 Çekül Vakfı - Tarihi Kentler Birliği, *"Sürdürülebilir Kültür Turizmi İçin Kamu-Yerel-SivilÖzel İşbirliği"*, Çekül Vakfı-Tarihi Kentler Birliği Yayınları Kılavuz Kitapçıklar Dizisi 1, Ebat Ofset, 2012. Gaziant [cited 2016, 03 March]; Available from: http://www.tarihikentlerbirligi.org/wp-content/uploads/S%C3%BCrd%C3%BCr%C3%BClebilirK%C3%BClt%C3%BCrTurizmi_2012-Ekitap.pdf

5 Cetin, M. and Sevik, H., *Evaluating the recreation potential of Ilgaz Mountain National Park in Turkey*, Environmental Monitoring and Assessment, 2016. 188: 52.

6 Cracolici, M. F. & Nijkamp, P., *The Attractiveness and Competitiveness of Tourist Destinations: A Study of Southern Italian Regions*, Tourism Management, 2008. 30, 336–344.

7 Demiroglu, O. C., Dincer, M. Z. & Izgi, T. M., *Revitalization of Cultural Heritage Assests within the Context of Sustainable Tourism Development: A Proposal for the Rejuvenation of Prinkipo Palace*, International Tourism Conference, 2008. 382–403.

8 Du Cluzeau, C. O., *Le Tourisme Culturel*, 2000. Paris: Presses Universitaires de France.

9 Düzgüneş, E. & Saraç, E., *Evaluation of Urban Transformation Areas in Terms of User Satisfaction: The Case Study of Zağnos Valley (Trabzon/Turkey)*, Environmental Monitoring and Assessment, 2018. 190:11.

10 Erdem, R., Yildirim, H., Ciftci, C., Dulgerler, O. N., Cibikdiken, A. O., Levend, S. & Erdogan, A., *Sille, A Conservation Development Plan and After (Present Situation)*. J. Fac. Eng. Arch. Selcuk Univ, 2010. 25, 2.

11 Sönmez E, Sadiklar Z., Kulak Torun F., Torun A., "Sille Kent Tarihi Aksının Kent Kimlik Bileşenleri", Kent Mobilyası ile Anlatılan Tarih Konya/Sille, ERTAŞ Ş., Ed., Detay Anatolia Akademik Yayıncılık, ANKARA, ss.1-16, Ağustos 2017

12 Eser, S., Dalgin, T. & Ceken, H., *Cultural Tourism as a Sustainable Tourism Type: The Case of Ephesus*, Aegean Geographical Journal, 2010. 19/2, 27–34.

13 Eser S., *Sürdürülebilir Turizm ve Efes Örneği*, Yüksek Lisans Tezi, Muğa Üniversitesi, Sosyal Bilimler Enstitüsü, 2011. Muğla.

14 Goodall, B., *Environmental Auditing for Tourism*, Progress in Tourism, Recreation and Hospitality Management, 1992. London.

15 ICOMOS, *International Cultural Tourism Charter: Managing Tourism at Places of Heritage Significance*, 1999. Translation: Z. Ahunbay, 2011. (http://www.icomos.org.tr/?Sayfa=Tuzukler1&dil=tr)

16 Kahraman, N. & Turkay, O., *Turizm ve Cevre*, 2006. Ankara: Detay Publication.

17 Kalipsiz, A., *Istatistik Yontemler. Istanbul Universitesi Orman Fakultesi Yayınları*, Istanbul Universitesi Yayın No. 2837, 1981. Istanbul.

18 Kiper, T. *An Assessment of the Landscape Potentıal of Safranbolu Yorukkoyu in the Frame of Rural Tourısm*, PhD Thesis, University of Ankara, 2006. Ankara

19 Müller, H., *Çevre Bilincine Erişmiş ve Sosyal Açıdan Uyumlu Turizm, Turizm ve Gelişme*, Türkiye'nin Şansları ve Riskleri, Türk-Alman Kültür Işleri Kurulu Yayın Dizisi, 1993. (4)-27, Ankara.

20 Ozdol, S. & Yurik, E., *The Cultural Tourısm and Prehıstorıc Sıtes in Turkey: The Example of Catalhoyuk*, Cesme Ulusal Turizm Sempozyumu, Izmir, 21–23th November, 2007. 721–732.

21 Richards, G., *Cultural Tourists of a Culture of Tourism: Developments in the European Cultural Tourism Market.* Edited by, J. Butcher, Innovations in Cultural Tourism, Tilburg: ATLAS, Association for Tourism and Leisure Education, 2001. pp. 1–9.

22 Öter, Z & Özdoğan, O. N., *Kültür Amaçlı Seyahat Eden Turistlerde Destinasyon İmajı: Selçuk-Efes Örneği*, Anatolia: Turizm Araştırmaları Dergisi, 2005. Cilt 16, Sayı 2, S. 127–138, Güz.

23 URL-1, http://www.İto.Org.Tr/Dokuman/Sektor/1-31.Pdf, Accessed: 15 November, 2017.

24 Wolfe, K. L., *Investigating Seeking and Escaping Aspects of Tourists' Motivations And Specific Barriers of Travel: A Study of Two National Park Sites*, Doctor of Philosophy, Kansas State University, 2002. Manhattan, Kansas.

Asst. Prof. Şehriban Eraslan

37 Future Landscape Constructions: Green Infrastructures (GI)

Abstract: GI is described as a tool for providing ecological, economic, and social benefits through natural solutions, helping us to understand the advantages that nature offers human society and to mobilize investments that sustain and enhance these benefits. This explicitly relates to the exclusive use of expensive "gray" infrastructure, which typically only fulfills a single function, such as drainage or transport, whereas nature often provides multiple solutions that are also cheaper, more robust, and more sustainable both economically and socially. GI systems are practical integrations of built and ecological systems that incorporate all natural, semi-natural, and constructed green spaces within, around, and between built areas, to replace or augment more traditional gray infrastructure. Cities need to upgrade and expand their "green" infrastructure—the network of open space, woodlands, wildlife habitats, parks, and other natural areas, which sustain clean air, water, and natural resources and enrich their citizens' quality of life.

Keywords: Green infrastructure, Urban, Landscape management

1 Introduction

Planning methods and understandings combining present green spaces are now becoming common as a way to decrease the amount of concrete and asphalt, or in other words the gray infrastructure, inside urbans around the world.

The understanding of green infrastructure (GI), different from the concept of green space, is a system created to protect and direct not the parks, natural areas, and recreational areas that are not connected to each other and the urban but rather ecological benefits for people and the environment in such a way as to create green corridors among natural area systems and other open spaces. GI is a system of network planning rather than planning segmental areas and therefore it is based on the agency and mobility of humans and other living beings. While green spaces are self-sustainable structures, it is necessary to actively protect GI.

Implementing GI features in urban areas creates a greater sense of community, strengthens the links among voluntary actions undertaken by civil society, and helps combat social exclusion and isolation. They benefit the individual and the community physically, psychologically, emotionally, and socio-economically. GI creates opportunities to connect urban and rural areas and provides appealing places to live and work in (1).

The idea that nature is also an infrastructure is not new, but it is now more widely understood to be true. Nature can be harnessed to provide critical services for communities, protecting them against flooding or excessive heat, or helping to improve air and water quality, which all underpin human and environmental health. When nature is harnessed by people and used as an infrastructural system, it is referred to as GI (2).

A Green Sponge for a Water-Resilient City: Qunli Stormwater Park (3)

GI might be, for example, a network of decentralized storm water management practices, such as green roofs, trees, rain gardens, and permeable pavement that can capture and infiltrate rain where it falls, thus reducing storm water runoff and improving the health of surrounding waterways.

The ability of these practices to deliver multiple ecological, economic, and social benefits or services has made GI an increasingly popular strategy in recent years. In addition to reducing polluted storm water runoff, GI practices can also positively impact energy consumption, air quality, carbon reduction and sequestration, property prices, recreation, and other elements of community health and vitality that have monetary or other social value. Moreover, GI practices provide flexibility to communities faced with the need to adapt their infrastructures to a changing climate.

Many cities have taken on the challenge of managing storm surge, storm water runoff, water conservation, and water pollution reduction, increasingly through the use of GI. That challenge has become even more urgent with the advent of global climate change and the more frequent and intense storms that have accompanied it (4).

GI is described as a tool for providing ecological, economic, and social benefits through natural solutions, helping us to understand the advantages that nature offers human society and to mobilize investments that sustain and enhance these benefits. This explicitly relates to the exclusive use of expensive "gray" infrastructure, which typically only fulfills a single function, such as drainage or transport, whereas nature often provides multiple solutions that are also cheaper, more robust, and more sustainable both economically and socially (5).

For example, GI can be used to decrease the flow rate of rainwater being mixed into sewage systems and, through there, into lakes, rivers, and streams due to the retention and absorption capacity of the plant cover and the soil. In such a case, increased carbon storage, high quality of air, decreased heat islands in the urban, and increased natural habitats and recreation areas can be considered among the benefits of GI. Furthermore, green spaces create an identity for sites, urban areas, and city surroundings where people live and work by contributing to cultural and historical areas (6).

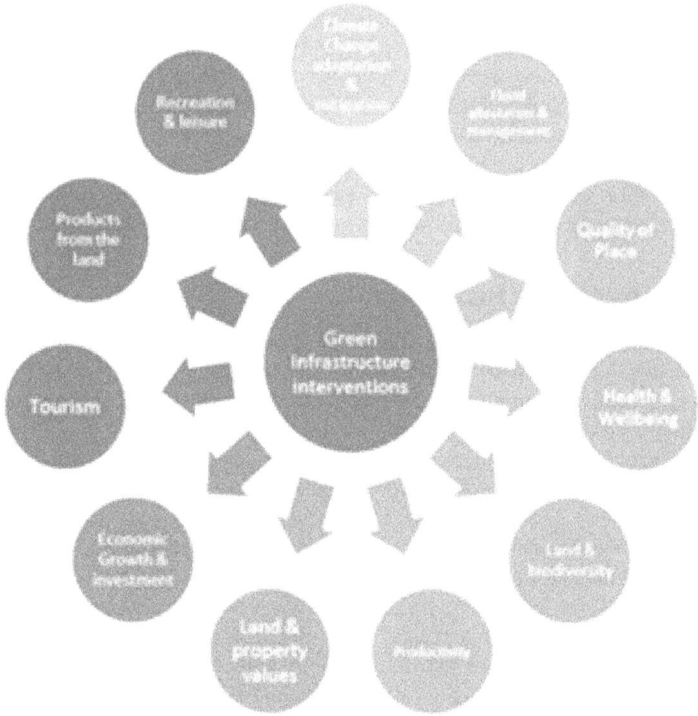

GI provides multiple functions:

Green Infrastructure Inventions (7)

In urban environments, GI covers everything from parks to street trees and green roofs to bioswales—anything that helps absorb, delay, and treat storm water, mitigating flooding and pollution downstream. GI also creates oxygen, sequesters carbon, and creates wildlife habitats (8).

Landscape architects are involved in regional master planning, conservation and restoration efforts, urban design, and park and green roof design and construction. At all scales, they bring a critical eye for social and artistic value to the design process.

2 Defining Green Infrastructure

GI is a notion that is appearing more and more frequently in land conservation and development discussions around the world. It means different things to different people depending on the context in which it is used. For example, some people refer to trees in urban areas as GI because of the "green" benefits that they provide, while others use GI to refer to engineered structures, such as water treatment facilities or green roofs that are designed to be environmentally friendly (9).

GI can be broadly defined as a strategically planned network of high quality natural and semi-natural areas with other environmental features, designed and managed to deliver a wide range of ecosystem services and protect biodiversity in both rural and urban settings. More specifically, GI, being a spatial structure providing benefits from nature to people, aims to enhance nature's ability to deliver multiple valuable ecosystem goods and services, such as clean air or water (10).

Zidell Yards Green Infrastructure (11)

GI can be also considered a conceptual framework for understanding the valuable services that nature provides to the human environment. At regional or national levels, interconnected networks of park systems and wildlife corridors preserve ecological functions, manage water, provide wildlife habitats, and create a balance between built and natural environments.

GI Techniques

There are many design choices for GI installations. More variety in the plant palette can increase bioswale function, enhance biodiversity (for instance, creating pollinator pathways), and create more vibrant urban places (12). These include:

- Green Roofs and Walls
- Rain Harvesting
- Downspout Disconnection
- Planter Boxes/Rain Gardens
- Vegetated Swales
- Green Parking
- Permeable Pavements
- Green Streets & Highways
- Pocket Wetlands
- Bio-retention/Detention Areas

- Riparian Buffers
- Wildlife Habitats and Corridors

3 Green vs. Gray Infrastructure

GI systems are practical integrations of built and ecological systems that incorporate all natural, semi-natural, and constructed green spaces within, around, and between built areas, to replace or augment more traditional gray infrastructure.

There are a variety of benefits of choosing GI over traditional gray infrastructure (e.g., concrete pipes), including:

- It is more attractive, effective, and multifunctional.
- It provides habitat for beneficial wildlife.
- It can reduce maintenance and eliminate mosquito breeding habitats.
- It can incorporate existing park features—both natural elements and man-made infrastructure.
- Using it can reduce infrastructure costs compared to gray infrastructure.

GI can always be counted as "parks," but when properly designed, constructed, and managed, GI can be a park, especially under broader definitions (4).

The term *GI* was selected to emphasize its difference from traditional conservation practices and the need to change several popular perceptions about green space planning and protection (9).

- Whereas green space is often viewed as something that is nice to have, the notion of GI implies something that we must have. Protecting and restoring our nation's natural life-support system is a necessity, not an amenity.
- Whereas green space is often thought of as isolated parks, recreation sites, or natural areas, GI emphasizes interconnected systems of natural areas and other open spaces that are protected and managed for the ecological benefits they provide to people and the environment.
- Whereas green space is often viewed as self-sustaining, GI implies something that must be actively maintained and at times restored.

4 Scales

GI occurs on all scales (2).

- While it is often closely associated with green storm water management systems, which are smart and cost-effective, it is really bigger than that.
- GI can be a centerpiece of smart regional and metropolitan planning, ensuring that communities have a livable environment, with clean air and water, for generations to come. GI can be designed to address the needs of wildlife, which are increasingly threatened by climate change, by providing systems of corridors

or greenways to enable movement through human settlements. Those corridors are often beautiful places that people want to live close to, as well.
- GI also comprises park systems and urban forests. The message here is that trees are a critical piece in green infrastructural systems and should not be discounted in favor of other technologies.
- Constructed wetlands are another way to harness nature to manage water locally and provide wildlife habitats.
- Finally, at site-scale, smart communities are using GI for transportation systems (green streets) and for green roofs, which can bring the benefits of nature to the built environment.

Potential Components of a Green Infrastructure (10)

On a larger scale, GI includes preserving and restoring natural landscape features (such as forests, floodplains, and wetlands), and reducing the amount of land covered by impervious surfaces. On a smaller scale, GI practices include green roofs, permeable pavement, rain gardens, vegetated swales, planters, and stream buffers. Others suggest that true GI is not engineered or "built," but is "natural" and in its simplest form consists of trees, plants, and soil (4).

5 Structure of GI

GI encompasses a wide variety of natural and restored native ecosystems and landscape features that make up a system of "hubs" and "links" (13).

Hubs

GI hubs come in all shapes and sizes and serve different purposes. For example:

- *Reserves:* Large protected areas, such as national wildlife refuges or state parks, serve as primary sites for conserving biological diversity and natural resources such as fisheries and fresh water. Reserves can also protect important historical and cultural sites, and can provide nature-based recreation opportunities.
- *Managed Native Landscapes*: Large publicly owned lands, such as our system of national and state forests, managed for resource extraction as well as natural and recreational values.
- *Agricultural Preservation Districts*: A number of states have recognized the importance of protecting large blocks of contiguous working farmland. This minimizes conflicts between farmers and suburban homeowners while providing farm machinery and seed dealers with the critical mass of farmland they need to stay in business.
- *Regional Parks and Preserves*: Less extensive hubs of regional significance, like the forest preserve system in Cook County, Illinois, provide ecological benefits and conserve biological diversity as well as offering important compatible resource-based recreational opportunities.
- *Cultural/Historic/Recreational Sites*: Community parks or cultural/historical sites that provide recreational opportunities, help protect and interpret a community's heritage, and can often serve as an origin or destination for a recreational trail.
- *Trailheads*: Selected ecological, recreational, or cultural/historic sites with appropriate visitor services that serve as points of origin or destination linked by trail corridors. Trailheads can occur within rural natural areas and working landscapes or within urban areas ranging from large metropolitan areas to small communities. Trailheads serve as human hubs within greenways systems.

Links

Many different types of landscape links can go into creating a system of green space. For example:

- *Landscape Linkages*: Large protected natural areas that connect existing parks, preserves, or natural areas and provide sufficient space for native plants and animals to flourish while serving as corridors connecting ecosystems and landscapes. Landscape linkages can also provide space for the protection of historic sites and opportunities for human use like hunting, fishing, canoeing, and hiking.
- *Conservation Corridors*: Less extensive linear protected areas, such as river and stream corridors, serve as biological conduits for wildlife and, in many cases, also provide opportunities for compatible outdoor, resource-based recreational activities.

- *Greenbelts:* Protected natural lands or working landscapes, such as the one around Boulder, Colorado, serve as a framework for development while also preserving native ecosystems and/or productive farms or ranchland and directing urban and suburban growth.
- *Trail Corridors*: Designated routes, such as rail-trails and greenways, provide access to and appreciation of the values of natural areas and other green spaces, present diverse resource-based outdoor recreational opportunities, and enhance the understanding of historical sites and cultural diversity. Trail corridors include linear urban open spaces that accommodate moderate to intense recreational use for residents and visitors.
- *Utilitarian Corridors*: Linear features, such as powerline and pipeline rights-of-way and canals, can provide a way of connecting recreational, cultural, or natural system features. Addison, Texas, for example, has created an outstanding park system using utility rights-of-way, while communities in New York have used the Erie Canal as the backbone for a series of linear park and recreation areas.

6 Benefits and Functions Provided by Green Infrastructure

GI brings multiple benefits to both rural and urban areas. With rapid global population growth and increasing urban and suburban density, green spaces are sometimes seen as a luxury. However, the presence of high-performing green spaces, or GI, provides real economic, ecological, and social benefits. Integrating GI into the built environment must be a priority.

The benefits of GI are numerous. From the broadest environment benefits to site-specific ones, GI is an effective and cost-efficient tool for absorbing and sequestering atmospheric carbon dioxide (CO_2), filtering air and water pollutants, stabilizing soil to prevent or reduce erosion, providing wildlife habitat, decreasing solar heat gain, lowering the public cost of storm water management infrastructure and providing flood control, and reducing energy usage through passive heating and cooling (14).

GI systems help protect and restore naturally functioning ecosystems and provide a framework for future development. In doing so, they provide a diversity of ecological, social, and economic functions and benefits: enriched habitat and biodiversity, maintenance of natural landscape processes, increased recreational opportunities, improved health, and better connection to nature and sense of place (9).

GI enhances recreational value, creates attractive park features, enhances social and environmental equity, reduces maintenance, improves drainage, helps to educate the public, improves water quality, benefits the overall environment, and provides economic benefits (15).

With careful design, green spaces can manage runoff and provide a range of co-benefits. Integrated planning of GI and park systems helps to cost-effectively provide multiple benefits and contributes to more livable communities.

These benefits can be environmental (e.g., conserving biodiversity or adapting to climate change), social (e.g., providing water drainage or green space), and economic (e.g., supplying jobs and raising property prices). As such, GI has the potential to offer win-win solutions by tackling several problems and unlocking the greatest number of benefits within a financially viable framework (16). GI can therefore be a highly valuable policy tool to promote sustainable development and smart growth by meeting multiple objectives and addressing various demands and pressures (7).

The Benefits of Applying GI Principles Can Be Summarized within Four Main Groups (10)

Environmental Benefits

- Provision of clean water/water flow regulation (17)
- Water purification (7)
- Removal of pollutants from air and water
- Air quality regulation
- Pollination enhancement
- Protection against soil erosion
- Soil structure and quality
- Rainwater retention
- Increased pest control
- Improvement of land quality
- Mitigation of land take and soil sealing
- Soil structure and quality (18)
- Erosion protection (7)

Sustainable urban drainage system (SUDS) is integrated with the street system (19).

Social Benefits

- Better health and human well-being
- Creation of jobs
- Diversification of local economy
- More attractive, greener cities
- Higher property values and local distinctiveness
- More integrated transport and energy solutions
- Sustaınable transportatıon (20)
- Enhanced tourism and recreation opportunities

Wildlife Bridge Over Motorway (10)

Climate Change Adaptation and Mitigation Benefits

- Flood alleviation
- Strengthening of ecosystem resilience
- Carbon storage and sequestration (21)
- Mitigation of urban heat island effects
- Disaster prevention (e.g., storms, forest fires, landslides)
- Air quality regulation (22)

Biodiversity Benefits

- Improved habitats for wildlife
- Pollination (23)
- Ecological corridors
- Coastal protection (24)
- Landscape permeability

7 Conclusion

In order for a city to be livable and prudential, healthy, sustainable, and clean, it is first necessary to set up infrastructure systems and reinforce the present infrastructure. GI highlights the importance of integrative consideration pairing up rather than separating the green and gray colors in a city.

An ecological life where nature and city are interbedded can be pursued with a natural infrastructure system. It has been observed that applied GI systems are creating places with high visual quality as landscape items along with positively contributing to the ecosystem.

While such a city embraces solutions supporting the GI and nature, it considers being in cooperation with nature, not reconciling with it.

GI systems are complex. For example, urban vegetation must be chosen for durability and resilience, character, growth habit, and aesthetic value. Paving materials used in these systems must be permeable and also funnel water into bioswales or bioretention areas to treat and absorb storm water. Landscape architects must understand how to incorporate ecological functions into urban areas while also designing spaces for a diverse group of users and programs.

Greening initiatives isolated within themselves are beneficial; however, they do not have transformative effects. While these initiatives meet the needs of a specific area, difficulties on the regional scale storm water management, improvement of air quality, protection of biodiversity or walking/bicycle networks, etc., must be developed as infrastructure connected to a network and systematized like in water, electricity, transportation, and digital services.

Landscape architects create master plans for parks, trail networks, wildlife habitats, recreation, and gathering spaces and draft design details from road systems down to visitor centers, transportation systems, and interpretive art and signage (25).

Landscape architects serve as dual advocates for both human and non-human communities. Landscape architects work with ecologists, conservation biologists, and other scientists to map existing and potential habitat patches and corridors, and then propose designs that protect critical habitats, connect fragmented green spaces, and weave together built communities and natural systems into a broader system that maximizes benefits.

Landscape architects often develop specialized knowledge in these areas: hydrology and riparian habitat for river ways, horticulture for native habitat planting, and ecology for systems (26).

Cities need to upgrade and expand their "green" infrastructure—the network of open space, woodlands, wildlife habitats, parks, and other natural areas, which sustain clean air, water, and natural resources and enrich their citizens' quality of life.

However it is defined, GI is quickly becoming a major tool in designing and building sustainable cities, and increasingly is seen as a way to both improve park design and have GI itself function as a park.

As we work to create sustainable, resilient cities, GI, with appropriate planning, will be a way to create new, well-funded, multi-functional public parks and open spaces, large and small.

References

1 European Commission, The Multifunctionality of Green Infrastructure. [cited 2018, 18 September]; Available from: http://ec.europa.eu/environment/nature/ecosystems/studies.htm.

2 American Society of Landscape Architects [cited 2018, 25 June]; Available from: https://www.asla.org/greeninfrastructure.aspx

3 American Society of Landscape Architects 2012 General Design Award of Excellence. A Green Sponge for a Water-Resilient City: Qunli Stormwater Park. Haerbin City, Heilongjiang Province, China. Turenscape and Peking University, Beijing.

4 Benepe, A., Parks as Green Infrastructure, Green Infrastructure as Parks: How Need, Design and Technology Are Coming Together to Make Better Cities. New York City. [cited 2018, 17 April]; Available from: https://www.thenatureofcities.com/2013/04/17/parks-as-greeninfrastructure-green-infrastructure-as-parks-how-need-design-and-technology-are-coming-together-to-make-better-cities/

5 European Commission, Communication from the Commission to the European Parliament, the Council, the European Economic and Social Committee and the Committee of the Regions 'Green Infrastructure (GI) Enhancing Europe's Natural Capital' (COM (2013) 249 final of 6 May 2013) [cited 2014, 25 January]; Available from: http://eurlex.europa.eu/LexUriServ/LexUriServ.do?uri=COM:2013:0249:FIN:EN:PDF

6 European Environment Agency [cited 2018, 20 October]; Available from: https://www.eea.europa.eu/downloads/98f1261382d143be874e6d29b7f029fa/1527156406/yesil-altyapi-dogaya-dayali-cozumlerle.pdf

7 European Environment Agency, Spatial Analysis of Green Infrastructure in Europe Technical Report 2/2014. ISSN 1725-2237.

8 American Society of Landscape Architects [cited 2018, 30 June]; Available from: https://www.asla.org/ContentDetail.aspx?id=43535

9 Benedict, M. ve McMahon, E., Green Infrastructure: Smart Conservation for the 21st Century, 2001. Sprawl Watch Clearinghouse Monograph Series, Washington.

10 European Commission, Building a Green Infrastructure for Europe. European Commission Building a Green Infrastructure for Europe. 2013. Luxembourg: Publications Office of the European Union 2013–24 pp. doi: 10.2779/54125. [cited 2018, 15 September]; Available from: http://ec.europa.eu/environment/nature/ecosystems/docs/green_infrastructure_broc.pdf

11 Green Works [cited 2014, 1 January]; Available from: https://greenworkspc.com/blog/2013/11/06/zidell-yards-greeninfrastructure-scenarios-wins-asla-oregon-honor-award

12 Wolf, K. L., Stormwater Report. [cited 2014, 2 April]; Available from: http://stormwater.wef.org/2014/04/water-wellness/

13 McMahon, E. T., "Green Infrastructure," Planning Commissioners Journal, Number 37, Winter 2000.

14 American Society of Landscape Architects, Green Infrastructure: Overvıew [cited 2018, 30 June]; Available from: https://www.asla.org/ContentDetail.aspx?id=43532)

15 Environmental Protection Agency, Green Infrastructure in Parks: A Guide to Collaboration, Funding, and Community Engagement, June 2017, EPA 841-R-16–112.

16 European Commission, The Multifunctionality of Green Infrastructure. Science for Environment Policy, In-depth Reports, March 2012. [cited 2013, 30 December]; Available from: (https://docs.google. com/viewer?docex=1&url=http://ec.europa.eu/environment/nature/ecosystems/docs/Green_ Infrastructure.pdf

17 Ennos, R., Quantifying the Environmental Benefits of Trees in Towns. [cited 2018, 8 October]; Available from: http://www.emonfur.eu/public/pub_files/Efuf/presentazioni/venerdi/Ennos_-_Milan_Talk.pdf

18 Green Infrastructure Research Group, [cited 2018, 20 August]; Available from: http://thegirg.org/future-research

19 CABE Space, Street Design and Place making: Case Study 4, [cited 2014, 7 July]; Available from: http://www.cabe.org.uk/files/streets-upton.pdf

20 American Society of Landscape Architects [cited 2018, 22 June]; Available from: https://www.asla.org/sustainabletransportation.aspx#GreenStreets

21 Grid Arendal [cited 2018, 12 May]; Available from: http://www.grida.no/publications/rr/natural-fix/page/3724.aspx

22 Environmental Protection Agency [cited 2018, 3 July]; Available from: http://water.epa.gov/infrastructure/greeninfrastructure/gi_why.cfm

23 European Environment Agency [cited 2018, 14 May]; Available from: http://www.eea.europa.eu/soer/synthesis/synthesis/chapter3.xhtml

24 European Commission [cited 2018, 14 October]; Available from: http://ec.europa.eu/environment/integration/research/newsalert/pdf/IR3.pdf

25 American Society of Landscape Architects [cited 2018, 30 June]; Available from: https://www.asla.org/ContentDetail.aspx?id=43533

26 American Society of Landscape Architects [cited 2018, 30 June]; Available from: https://www.asla.org/ContentDetail.aspx?id=43534

Assoc. Prof. Timur Kaprol

38 Functional Transformation: 2 Cities, 2 Buildings Bursa: Veysel Karani Sevgi Köyü (2003, 2018) Edirne: Trakya University – Prof. Dr. Ahmet Karadeniz Campus Technical Sciences Vocational School (2012, 2018)

Abstract: Data on how buildings are transformed and used throughout human history have survived. From the remains, it is understood that the functions of buildings changed in ancient times during the multifunctional period. The economic inequalities that occur after the changes in the economy, the location and economic values of the buildings cause transformations. In this transformation, the location of the building, its economic value, its continuity and sustainability in the new use are important. This study, which focuses on the use of public and education, deals with the continuity of social life with the continuation of the physical qualities of buildings which have lost their living conditions after development and rapid development. The purpose of this chapter is to establish a system of criteria that scientifically investigates the design and application of the transformation of the new function needed when the building does not conform to the necessity of the spatial organization performing its function.

Keywords: Functional transformation, Social environment, Making environment, Natural environment, Architectural design, Reuse

1 Introduction

Data on how buildings were transformed and used throughout the human history have reached the present day. It can be understood from the remnants that functions of buildings have changed during the antiquity in the polytheistic era. One can encounter in the excavations that there had been more than one civilization on the same topography in different eras and that each civilization utilized the buildings constructed prior to them according to their cultural needs, which is the functional transformation of a building.

If function is to be considered as an organism, the organism possesses biological installations that provide for the vital necessities. The subcomponents that provide for the vital necessities of these biological installations appear both as individuals and create their own morphologies when assembled together. The defining function of each morphological subject can also change, and this suggests a new definition in its morphology. The fact that in natural sciences some creatures provide the continuance of the lineage by changing their sex contributes to the sustainability of nature. Biological transformation can take place.

In human psychology, an emotion or desire pushed into the subconscious may rise to the conscious in an opposite manner or another form thus providing transformation. As a result of civilizations affecting each other with both wars and interaction, religious buildings, among historical buildings, are frequently subjected to this transformation. For instance, the transformation of Hagia Sophia Church into a Mosque after the conquest of Istanbul; along with the transformation in the Post-Republic era in Turkey and the population exchange, while the non-Muslims migrating to Greece from Anatolia and the Muslim population coming from the Balkans, religious buildings can particularly be seen to have transformed. Transformation of a church into a mosque, just as some mosques having been turned into museums in Balkans, took place not only as a result of the needs of public but also because of economic disparities.

Economic disparities that occurred after changes in economy, the position of buildings, and their economic values cause transformations. The position of the building, its economic value, maintaining its nature in the new usage, and its sustainability are significant in this transformation. Transformation in architecture is the transformation of a series of certain buildings and their peculiarities into other buildings and peculiarities on a structural scale.

Reutilization of buildings is apparent throughout human history. It is necessary in reutilization-transformation that data about the spirit of a place should be embraced in the design on both urban and structural scales. The typology and the structuring of the design and a simultaneous reading of types are important in terms of sustainability.

The aim of the chapter is to establish a criteria system which scientifically investigates the design and practices in which transformations of the new function necessitated are applied in case the spatial organization that fulfils the function of the building does and does not comply with the necessities of the new function.

A structure can undertake different functions throughout its life. The transformation of the design requires the function, which is almost an organism, to react to different usages. Transformation of a shopping mall into a hospital or school, and a research facility into a faculty again requires a construction that will react to transformation during the design process of the architect, and prioritizes flexibility in usage. A single building in the techno-city differs in use according to the departments in it. A spatial organization peculiar to research for mechanical and chemical industry can be equipped with properties that will serve a field of medicine. Not only similar usages, but also different and extreme usages may be necessary. The designer may be asked of spatial organizations that will cover all of these challenging features together.

2 Bursa Osmangazi Veysel Karani Sevgi Köyü Construction Design (2003)

2.1 Objective of the Project

The high amount of the number of children who polish shoes, sell bagel or tissues, and clean windshields on the intersections among the dangerous traffic in the densely populated and commercial areas in Bursa is of significance. These children, mostly in the compulsory elementary education era, constituted a marginal sector with a demand of their own creation which is not a result of the supply that economy requires. The reason why children who are made to work on the street, whose education, psycho-social and physical development is under all kinds of threat and risk, has been found out in scientific studies carried out by medical specialists, and the reason lies in the families, and social and economic conditions that affect families. Likewise, immigrations are highlighted as a reason by the same specialists. It has also been pointed out that unemployment, poverty, formation of unhealthy residential areas, maladaptation to the city, resolution in social values, inclination to crime, adopting bad habits in the post-immigration era appear in children who live and work on the streets. In the scientific field research carried out by Child and Youth Centre that was opened under Bursa Section of Social Services in 2001, it can be seen that families of the children who work on the street are highly uneducated, have jobs that are not permanent and without social security, work in temporary jobs if they can find, or are unemployed. The data of this research state that the issue of the children who are made to work on the streets in Bursa stems from the immigration to the city from the rural, and the families make their children work because of issues such as lack of education and unemployment. This problem that stems from the social and economic condition that family is in can be solved through planned and programmed social service practices that will cover the families as well. With reference to this, it can be asserted that the following social service facilities are a necessity for Bursa.

1. A child and youth center to encourage children who work on the street not to continue this work, and promote their education and healthy development.
2. A social center that will prevent, educate, develop, guide, and rehabilitate, with an aim to solve the social and economic issues of families.
3. A guest children center to provide children who live on the street with temporary and emergency shelter and reintroduce them to the society.

2.2 Design Process

With a group of architects who are the members of the Union of Chamber of Turkish Engineers and Architects, before getting to the design stage of the project that is prepared free of charge and as a public service, there has been made a preliminary study carried out by a group of Faculty of Medicine associates led by Prof. Dr. Suna TANELİ of the Faculty of Medicine, Uludağ University, presented to Bursa governorship. Architect Ebru BİLGİLİ, Architect Hatice DÜLGEROĞLU, Architect Oya AYDIN, Architect Nizamettin KAYA, Architect Dr. Timur KAPROL, and Architect Recep ÇOLHAN participated in the architectural design process.

Social service specialists, psychologists and psychiatrists, and architects participated in the work group of Bursa Sevgi Köyü. The aim of the project has been to construct social service facilities that will solve the problems of children and

Ground Floor Basement Floor First Floor

Fig. 1: Floor plans of Bursa Sevgi Köyü Community Center.

families who are in need of social services and social assistance. In these facilities known as “Bursa Sevgi Köyü,” three main service units and supporting units that will serve the main ones commonly have been designed (Fig. 1,2,3).

Building Programs of the Main Service Units of Bursa Sevgi Köyü

Supporting units that will be open to common use in Bursa Sevgi Köyü: specialists have adviced architects to build these units in the same garden as separate units, their common needs should be met together, and activities peculiar to there should be organized in units.

Exhibition and Retail Gallery: it is thought as a gallery that will exhibit, sell, and earn an income from the products produced in work ateliers by individuals who receive service from Bursa Sevgi Köyü. The unit is required to be planned so as to provide service to the general public in the campus it is in.

Medical Unit: it aims to provide medical and emergency services to the children and families who benefit from Bursa Sevgi Köyü.

Game, Sports, and Show Areas: children who benefit from social services need areas to play games, do sports, and exhibit shows in the garden. Functional areas that are needed are as follows 1) An outdoor play ground with equipment that includes game sets such as slides, swings, jungle gym, etc. 2) A sports area that includes volleyball, basketball, and football areas.

3) A show space that includes an amphitheater that can accommodate 100 people.

Reception: a necessary place to guide visitors and control the passes to the campus.

3 Transformation of Bursa Osmangazi Veysel Karani Sevgi Köyü into Bursa Immigration Center

3.1 Reason of the Transformation

The centers which are divided into three as Child and Youth Center, Guest Children, and Community Center have developed the social life actively. As a result

Fig. 2: Perspective of Bursa Sevgi Köyü Community Center.

Fig. 3: Location plan of Bursa Veysel Karani Sevgi Köyü.

of the growth in developing Turkey, because the collectively designed buildings began to be inadequate, the society put new and comprehensive designs into practice that focus on services to children and youth. Development has brought about changing needs to the social life. Economic development and growth act as a locomotive for social needs. Industrialization, population growth, work life, and developments that take place during the economic growth can introduce some social issues and problems that need to be solved. Thus, community centers in which equipped specialist personnel will be serving to solve these problems are needed. Scientific developments provide for the construction of technically well-equipped and advanced spaces. Public services are provided to solve problems and guide the society. Authentic centers that are constructed out of their own functions of these spaces and include spatial organizations are needed. Architectural typologies of these buildings which are advancing and specializing in their areas are forming, and this formation introduces macro solutions on an urban scale and micro scale solutions for the building. As a result of the developments, community centers with enhancing volumes cannot fit in their present area and a new spatial construction is needed. In this case, buildings which lost their function but are still of economic rentable value remain empty. This situation that the 21st century and rapid urbanization will introduce to us very frequently will require 21st century architects to refunction these areas and buildings which lost their authentic function but still possess a rent value. In fact, this formation has happened throughout the history and refunctioning an old structure has been the duty particularly of the architects and designers who work in protection and restoration. Protecting the modern era buildings and transferring them to the future has become a separate field of expertise in our age. A building or set of buildings which lost their function are planned to be refunctioned and continue its life in a dynamic manner in social life.

3.2 Design Process

The building complex constructed under the name Bursa Sevgi Köyü was planned together as a result of the then economic portrait although the functions in it require separate areas of expertise. The level we reached with the economic growth has enabled for the construction of separate children's village today. This complex is composed of three centers. The first, Child and Youth Center, aims to include children and teenagers who live with their family and work on the street in community life and support them with vocational education. Another one is Guest Children Center which saves our children who became drug addicts that began to be widespread in the late 20th century, and rehabilitate and prepare them for life. The third one, Community Center aims to provide unemployed adults and parents with employment. These three centers became a complex that raises the level of social life by rehabilitating the vulnerable structure of the community.

As the 100th anniversary of the Republic is approaching, after the political and economic events, and wars, people immigrated to other countries as a result of such dramatic events. Direct and indirect immigrations to Turkey, and the fact that Turkey is in the immigration route that runs from the Middle East into Europe and therefore is subjected to immigration, have made immigration centers a necessity in our country.

Bursa, a very important industrial city, has been a city that receives immigration throughout the century, starting from its conquest by the Turks until today.

Fig. 4: Photograph of the transformation of Bursa Sevgi Köyü Complex into immigration center.

The employment opportunity the industry gives can be an attraction both in the country and abroad nowadays. Centers with new spatial organizations are needed in order to solve the problems of immigrants who entered out country illegally. Two of these buildings which have lost their function but will meet the mentioned needs have been refunctioned as temporary accommodation and shelters separately for families and men. The third building is planned to by the technical staff of Bursa Provincial Directorate, Ministry of Environment and Urban Planning of Turkey to serve within the complex and its construction is being prepared by specialists. The shafts and wet areas of the buildings do not change in the transformation projects; the necessary interior space equipment is enhanced and the interior space design is made to be more comfortable (Fig. 4,5,6).

4 Edirne, Prof. Dr. Ahmet Karadeniz Campus Technical Sciences Vocational School

4.1 Objective of the Project

Trakya University Prof. Dr. Ahmet Karadeniz Technical Campus is far away from the Edirne urban area and therefore from urban services. Its visual and physical relation to the main campus is broken and it is almost like an opposite neighbor marginalized by the partition of the TEM highway (Fig. 7). Apart from a building which was designed for the Faculty of Architecture and Engineering in 1990s but left to the Engineering Faculty after the Architecture Department was moved to the city center and the building of Food Engineering, it does not have a constructed environment with which it can interact (Fig. 7,8).

With the design of the techno-city in the campus that can be used as the engineering and technical campus of the Trakya University, a suitable physical

Fig. 5: One of the buildings of Bursa Sevgi Köyü transforming into an accommodation building of the immigration center.

Fig. 6: Photograph of the alteration work which indicates that only the interior space is being transformed and the outer space is maintained in the building which carries remarks of traditional architecture in its design.

construction gave way to collaboration between the university, industry, and public. The campus has a capacity to be a science cuckoo in the production of information, an education center in technological development, and a zone for universal knowledge and production.

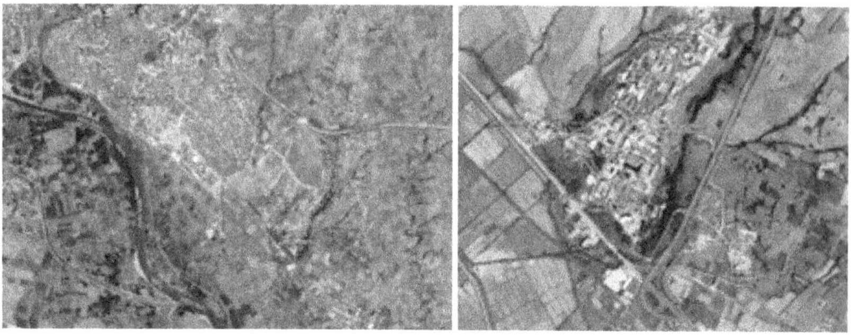

Fig. 7: The location plan indicating the Trakya University Prof. Dr. Ahmet Karadeniz Campus and Güllapoğlu Campus.

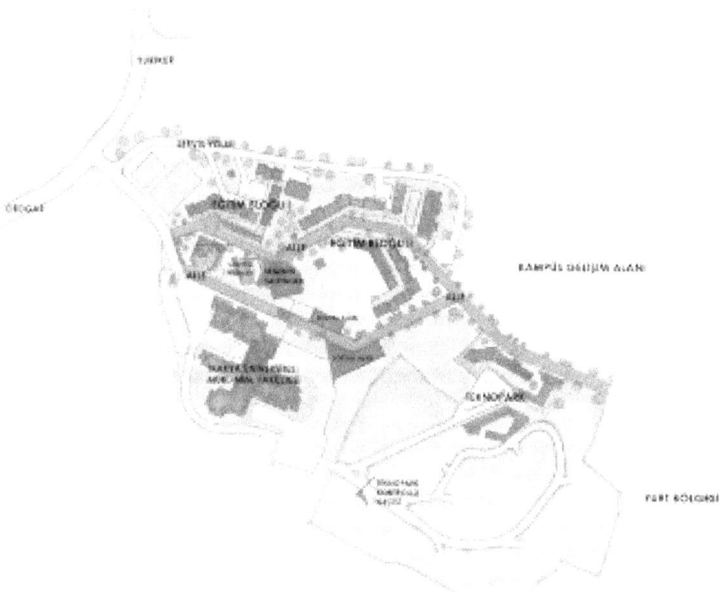

Fig. 8: Layout plan of the Trakya University Edirne Technical Sciences Vocational School Campus Design.

4.2 Design Process

The design team of the campus includes the present academic staff of Architecture and Landscape Architecture departments of the Faculty of Architecture, the architect member of the Department for Construction, and the academic staff of Technical Sciences Vocational School, and the proposal was prepared with members coming from a variety of professions. By taking the opinions of all the departments that will be situated in the campus during the architectural design process, the plan has been presented to the relevant departments both in the formation of building program stage and after the design process. As a result of these opinions, the project has been evolved to meet the needs in terms of both architectural organization and structural construction at development stage (4).

Architectural Design Team: From the Academic staff of Trakya University, Architecture Department Asst. Prof. Dr. Sennur AKANSEL, Asst. Prof. Dr. Timur KAPROL, and Inst. Gülay DALGIÇ; in architectural design, **Trakya University Landscape Architecture Department** Asst. Prof. Dr. Candan ZÜLFİKAR; **in landscape design,** Hüseyin ÇİFEL, Burcu DEDE, and bachelor degree students of Architecture Department worked as assistants (4).

Trakya University Department for Construction President Ayhan DİKİCİ, **Architecture Department academic staff** Asst. Prof. Dr. Sennur AKANSEL, Landscape Architecture Department academic staff, Asst. Prof. Dr. Candan ZÜLFİKAR, Asst. Prof. Dr. Beyza ŞAT GÜNGÖR, Technical Sciences Vocation School staff, Asst. Prof. Dr. Hüseyin İNCE, Inst. Gülay DALGIÇ, Şehit Ressam Hasan Rıza Vocational School of Fine Arts academic staff Inst. Aycan ÖZENÇ, and topographical engineer Bülent ÖGET worked in the project commission (4).

Program of the Buildings in the Design

Edirne Technical Sciences Vocation School is constituted of two education blocks, administration building and a social complex. In the complex, which includes open sports areas nearby, there have been regulations that will enable physical activities along with technical and scientific education (4).

Spatial organization of three departments has been designed in the first education block. These are Electronics and Energy Department which includes Electric Division and Gas Installation Division; Electronics and Automatization Department which includes Technology of Electronic, Biomedical Device Technology, Radio and Technologies, and Mechatronics Divisions; Department of Material and Material Processing Technologies which covers Furniture Decoration Division (4).

The second education block accommodates seven departments: Machinery and Metal Technologies Department which covers Machinery Division; Computer Technologies Department which covers Computer Programming Division; Visual and Auditory Technologies and Media Production Department which includes Media Technologies Division, and Radio and Television Programming Division; Textile, Clothing, Footwear, and Leather Department which includes Clothes

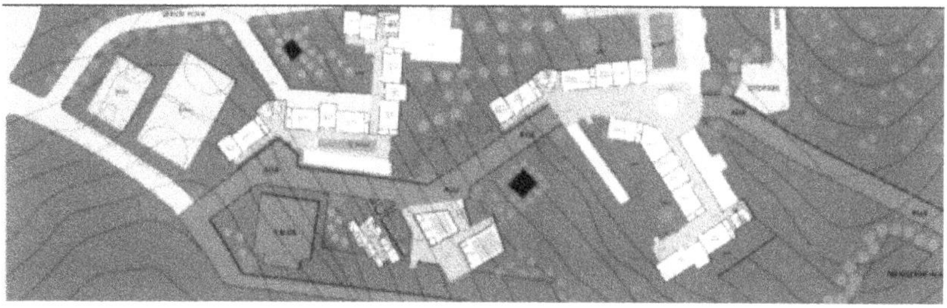

Fig. 9: Edirne Technical Sciences Vocation School education blocks and the floor plan that includes the administration and social facility buildings.

Production Technologies Department and Textile Programming Division; Architecture and City Planning Department which includes Mapping and Cadaster Division and Building Inspection Division; Design Department which includes Graphic Design Division; Worker's Health and Labor Safety Department, and the academic staff of these departments along with the ateliers which are included in the building design. The total construction area of this building is 9450 m^2(Fig. 9)

4.3 The Aim of the Design

The promoted project, as well as being a new project, it is also an effort to unify the present/living area that is already in the campus on different elevations, to reinstate the existing one, bring together the education, social, and technological areas, and create a common design language. In this context, as the proposed project obtained the ALLE reference and was included in the present structure of the engineering faculty, this produced the interconnection between the present and proposed buildings, and provided an expansion in the manner of a cluster city diagram with the possibility of new constructions the proposal tends to host.

The new hypothetical interconnection of ALLE is; the students of Engineering Faculty and Vocational School both interact with each other and provide the circulation who enable a diffusion from the inside-out and outside-in. ALLE not only assumed the duty of intermingling the vocational school and bachelor degree students but also acts as a bridge of technology. Education blocks and techno-park area are unified with the help of this *meeting and coming together* platform. ALLE is terminated with urban equipment spaces which starts with the school and where social needs are met in terms of decisions made for the upper scale accommodation where the student dormitories are. It has almost constituted the basis of a future pedestrian boulevard. Also, this strong foundation is predicted to be the reference for a hypothetical future (4).

The façades of the buildings that face the Alle are closed and with steel frame construction, and winter gardens that enable for summer and winter use, and the

Fig. 10: ALLE proposal and which serves as the main theme of the Trakya University Prof. Dr. Ahmet Karadeniz Campus Technical Sciences Vocational School Idea Project and its visuals.

Fig. 11: Visual of the construction stage of Trakya University Prof. Dr. Ahmet Karadeniz Campus Technical Sciences Vocational School.

air conditioning of the building were also taken into consideration. In Edirne, where there is continental climate conditions, qualified spaces were included in the campus design in which students, apart from their education, come together especially in winter months, prepare workshop activities, do atelier work in their spare times, and also students from different departments can assemble the information they learned in their lessons and use them as a step in their work life, preparing a foundation for information production and sharing (Fig. 10,11).

5 Transformation of Edirne Technical Sciences Vocational School into the Engineering Faculty

5.1 Reason of the Transformation

Trakya University is continuously growing and the present departments have spatial needs as both the number of the students and faculty members are rising. Prof. Dr. Ahmet Karadeniz Campus is separated from the Güllapoğlu Campus by the

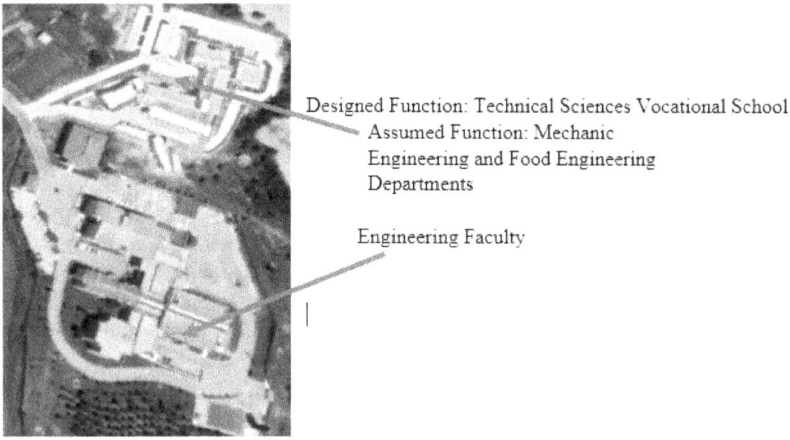

Fig. 12: Trakya University Prof. Dr. Ahmet Karadeniz Campus Technical Sciences Vocational School (Mechanic Engineering and Food Engineering Department building). Source: Google Earth – accessed at 14.08.2018.

highway. The two campuses, which are connected with a subsidiary road, have an indirect means of travel. In summary, as Prof. Dr. Ahmet Karadeniz Campus includes the new techno-park area and is projected to be the new technology campus, engineering is naturally expected to grow in technical terms. The campus, being near the bus station and also has social equipment units and accommodation spaces, presents possibility for expansion both with the transportation network and closeness to the bus station. Not only the growing of the departments and emergence of new science fields but also the area position that enables new construction necessitates new designs. The Engineering Faculty which was designed in 1990 accommodates Computer Engineering (1991), Electric-Electronic Engineering (2012–2013), Genetics and Bioengineering Department (2014), Food Engineering (2008), Civil Engineering, and Mechanical Engineering (1982) departments. As a result of scientific variety which covers day and night education, the present building has become insufficient and consequently Edirne Technical Sciences Vocational School building was planned to accommodate engineering departments and a functional transformation took place. As a result of the function in the first education block that will include technological structuring about education and engineering, there was not a necessity to change the project design; the present technical spaces were used without any alteration in the present spatial organization. The building designed for the Technical Sciences Vocational School is expected to be completed to accommodate Mechanical and Food Engineering Departments (Fig. 12).

6 Conclusion

Upon examining the constructed environments the societies build in history, it can be seen that these environments were constructed to meet social needs. Constructed within either traditional building systems or with unorthodox construction systems, all were constructed in order to answer the needs of human life. Whether the construction is made by the wishes of the society or the political power, the aim is the same for all: obtaining living spaces. While assembling them, all these buildings were constructed in terms of their authentic texture, and their connection to the natural, social, and constructed environment. The urban texture formed by the assembly of these buildings has constituted an authentic morphology peculiar to the area. On a micro scale, the building defines the function it has and the architectural typology. When there is a new function or a function change proposed on a single building scale that will not disturb the existing environment, alterations can be accepted that will not change the former use of the building nor harm the structure. However, new function proposal can bring about a different function on a campus scale. In this case, an architectural approach should be adopted by taking the former spatial organization into consideration. Particularly the outcomes of the function transformation are analyzed during the design process and proposals of additional buildings are visualized through architectural expressions and then decided (6,7).

The constant expansion of cities raises the rent value of the building in the city center. New settlements that are built to ease people's lives in terms of their needs become attractions for the surrounding areas to be constructed, and these areas are also left among the center of urban texture. Therefore, public buildings, which have lost their function over time, become attractive locations for reutilization both because of the length of their physical life and the high rent value of their parcel, as well as their central location. Bursa Sevgi Köyü undergoes a function transformation as a result of such a development.

Edirne Technical Sciences Vocation School, on the other hand, was designed as a part of the technological settlement, but Trakya University later decided to utilize this school in terms of the needs of Engineering Faculty in its settlement policy. Functional transformations took place in areas such as laboratories and studios for different branches. Differences in the primary proposals can frequently be observed. Architects should foresee a transformation during the construction process or a while after it was used as they make a design for such buildings, and prioritize sustainability and transformability in spatial production. The construction of a building requires a large economic investment, and it stays on the location for a long time. The function may be of short-term just as it may continue for a long time. Industrial and production-oriented cities particularly like Bursa are inclined to change and transform with the missions they assumed.

With the advanced technological materials in our time, building construction methods differ, and new buildings are constructed which are long lasting, demountable, receive additions, or transform. Flexible design necessitates an architect to prioritize change, transformation, and sustainability concepts in his design. In order to use a space in more than one function, it needs to enable transformation, and accommodate disassembly in case of fatigue of materials. It is also necessary to use research for and use in architecture materials like bioplastic that develop the standard production, lower the cost and at the same time raise the duration, promote research and development (6,7).

Changes that take place in the social life can bring about a reshaping and assuming a different shape in architecture. According to Leupen et al. (1997), transformation can be defined as "assuming a shape other than the self, get another form, to reshape," rather than transfer. In biology, it is defined as "a change that appears in an organ as its function changes." Psychology, however, defines transformation as "an emotion or desire pushed into the subconscious to elevate to the conscious in an opposite appearance or in a different form." In the light of these, in architectural production which is the activity area of the architect, there can occur unprecedented situations during the design or construction stages apart from economic and social reasons, and the location of the parcel of the building, as well as its economic value. In this case, the architect should assume a very important mission in our age and produce designs that comply with the living spaces which do not harm the nature necessary to humans and give it the opportunity to live, by guiding his design with a methodology he produced and through multidisciplinary studies. Thus, particularly by accommodating function transformations, the architect can assume the role of producing spatial organizations that are in harmony with the social, natural, and constructed environment, and enable sustainability, flexibility, and transformability by using initiative in situations when transformation is needed. Therefore, when the design is realized, it can present a new whole with its environment, and become significant in how it promotes the given context and the existing elements, while the architect is renowned for his approach, and the innovative and environmentalist design.

References

1 Leupen B., vd. (1997) **Design and Analysis**, Van Nostrand Reinhold, USA, pp. 13–17.

2 Anon. (2003) "**Bursa Sevgi Köyü- Çocuk, Gençlik ve Toplum Merkezi**" Bursa Governorship, Bursa.

3 Anon.(2001) "**Bursa ilinde göç ile ilgili bilimsel araştırma**" Bursa İl Sosyal Hizmetler Müdürlüğü, Çocuk ve Gençlik Merkezi, Bursa.

4 Kaprol T., Akansel S., (2018) "Trakya Üniversitesi Prof.Dr.Ahmet Karadeniz Yerleşkesi Fikir Projesi," Mimarlık ve Dekorasyon Dergisi, Sayı 274–3, pp. 80–84, İstanbul ISSN:1301-6741

5 Anon (2018) Bursa Valiliği Çevre ve Şehircilik İl Müdürlüğü, Proje İşleri Arşivi

Internet sources:

6 Anon., "Dönüşüm" *Güncel Türkçe Sözlük*, Türk Dil Kurumu, http://www.tdk.gov.tr/index.php?option=com_bts&arama=kelime&guid=TDK.GTS.5b6c2a53295370.71822395, accessed: 9.08.2018.

7 https://mimaritasarimveelestiri.wordpress.com/2012/05/21/kosullandirilmis-fiziksellik-etkisinde-ortaya-cikan-mimarlik/accessed:9.08.2018https://gayrimenkulturkiye.com/2016/12/08/donusen-ofislere-butuncul-bakis/

Assist. Prof. Dr. Tuba Rastgeldi Doğan

39 An Investigation of Indoor Air Quality in Old and New Campus Buildings: Harran University

Abstract: While most people spend 70–90 % of their time in indoor spaces (home, office, etc.), students spend a large part of their time in educational buildings. The quality of the air that is inhaled in indoor environments is determined based on the standards as it directly affects human health. As the quality decreases in indoor environments, it can cause chronic diseases such as asthma and bronchitis together with acute disorders such as weakness, difficulty in concentration and headache. Universities are research-based public institutions that serve as the pioneers for educational institutions. In this study, PM_{10}, $PM_{2.5}$, CO_2, temperature and relative humidity measurements were made in an old-structured campus (Eyyübiye Campus–built in 1988) and a newly built campus (Osmanybey Campus–built in 2003– construction still continues), both of which belong to Harran University. It has been found that indoor air quality values (PM_{10}, $PM_{2.5}$, CO_2) exceeded the standards in the formerly built Eyyübiye Campus when compared to the WHO and ARSHAE standards, and CO_2 level in the Osmanbey campus exceeded the limits in the cafeteria between 11.00 and 15.00 because of lack of ventilation. It has been determined that the temperature was high in both campuses, and the relative humidity was within an acceptable range. The reason why indoor air quality levels were high at the Eyyübiye Campus can be that the buildings are prefabricated, are 30 years old, the ventilation is inadequate and the ceiling height is 2.5 m, which is very low. In this chapter, it has also been discussed what can be done in the short and long term to improve the indoor air quality in the buildings where the values are high.

Keywords: İndoor air quality, Particulate matter, Carbon dioxide, Campus buildings, Harran University

1 Introduction

Air pollution is a growing problem worldwide and is also an important risk factor attributed to climate change, global warming, and people's various health problems. Effects of air pollution on health from both indoor and outdoor environments have been a major concern because of the high risk of exposure even at relatively low air pollutant concentrations. As a result of the damage caused by air pollution to the lungs and respiratory system, it is estimated that more than two million deaths will occur in the world every year [1]. The causes of these deaths include fine particulate matter (PM) and ozone [1–2]. Although there is so much research on external air pollution, studies on the impact of this pollution on indoor air are more limited.

Most of us know that outside air pollution can affect our health, but indoor air pollution can also seriously affect our health. As a matter of fact, it is a necessity to minimize the influences of indoor pollutants in buildings, particularly because it has been reported by the Environmental Protection Agency (EPA) that there will be severe increases in the levels of indoor pollutants which will be higher than those of outdoor levels [3]. Therefore, the studies show that indoor air pollution is one of the top five factors in terms of public health. The quality of the indoor air of the houses, offices or other public or private houses, which we spend most of our time in, can be seen as one of the main determinants of a healthy life and the well-being of each individual [4].

Universities constitute a significant part of the society as a result of the knowledge they provide and the use of knowledge in practice for the quality of city life [5]. Universities also increase sustainability through such internal ways as sustainability policy, campus sustainability and environmental initiatives also through such external ways as universities' role in the region [6], ensuring that the necessary Indoor Environmental Quality (IEQ) leads to less frequent and less severe illnesses [7]. As schools and universities are more interested in having more idea about their buildings in terms of their measurements and performances, the IEQ can be said to have more influence for the users' educational facilities. A number of policies have been developed by the architecture, engineering and construction industries in order to enhance the conditions of the faculties and their users. In recent years, the concept of "sustainable buildings" has started to include Indoor environment performance as a significant element and several studies have been conducted. Moreover, a number of universities have started to make use of sustainable building practices as a result of the rising interest in green building. Similarly, there have developed standardization systems like Leadership in Energy and Environmental Design (LEED) and Building Research Establishment Environmental Assessment Method (BREEAM) for a more systematic practice [3]. In this sense, there are several considerations to be dealt with in reference to the sustainable and efficient alternatives for the users of educational facilities [5]. Inadequate internal environment in schools and campuses can be attributed to four main reasons: 1) Inadequacy or inadequate operation and maintenance of ventilation systems, 2) Uncomfortable and poorly cleaned indoor surfaces, 3) The effect of number of students on the occupancy rate of the classrooms, 4) The influence of the indoor environment due to the poor quality of the outdoor air [8].

In order to develop these systems, it is necessary to investigate and study the current situation from the point of the indoor air quality of the campuses. In this study, the internal air quality of the buildings which are frequently used by academic staff, employees, university students and visitors, and are located in the two campuses of Harran University, the Eyüpüyye Campus and the Osmanbey Campus was investigated. The PM_{10}, $PM_{2.5}$, CO_2, temperature and relative humidity values of the classrooms, class corridors, cafeterias, library buildings in campuses were analyzed separately and in comparison to each other. The measured values were compared to the WHO and ARSHAE limit values. In addition, the possible sources and reasons of high values in the studied areas are discussed in this chapter.

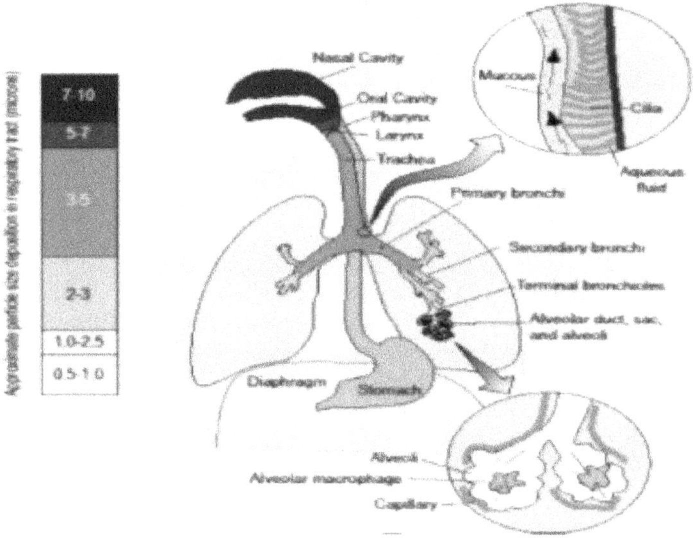

Fig. 1: Simplified diagram of the human respiratory system. The approximate particle size deposition that may occur in each section is color coded and identified in the legend [13–15].

1.1 Particulate Matter and Health

Airborne PM includes a heterogeneous mixture of solid and liquid particles in air that differs continuously in size and chemical composition in space and time [9]. Although there are several ways of defining or classifying PM, aerodynamic diameter is considered to be one of the most important features to describe its ability of transport in the atmosphere and/or inhaling ability by means of a respiratory organism. EPA has been regulating particles mainly in two size categories depending on their predicted capacity to penetrate into the lung as either (i) coarse particulate matter with an aerodynamic diameter of 10 μm or (ii) fine particulate matter with an aerodynamic diameter of 2.5 μm [10].

The PM has a breathable size and can go to the most important part of the lungs, which is the alveoli (Fig. 1). Particles with the most impact on human health have been found to be those less than 10 μm in diameter. These particles can follow the respiratory tract starting from the nasal passages to the alveoli, deep within the lungs as they can be penetrated easily. Particles with the sizes between 5 and 10 μm are often deposited in the tracheobronchial tree. On the other hand, those between 1 and 5 μm are accumulated in the respiratory bronchioles and the alveoli where gas exchange occurs [11]. These particles have the potential of influencing

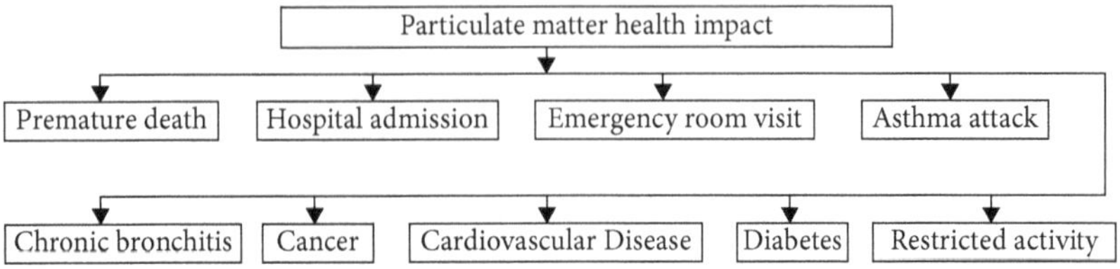

Fig. 2: Health impacts of PM exposure [17].

gas exchange within the lungs and can even penetrate the lung. Ultimately, these particles will go into the blood stream which can lead to some health problems [12].

Particles that can be so deeply entrained in our body cause a variety of illnesses. [16] reported in 2009 that transfer of atmospheric particulate matter in the size of desert dusts to Turkey, Sanlıurfa, is associated with an increase in the number of patients with respiratory problems. Fig. 2 is the result of the study conducted by Kim et al., in 2015.

1.2 CO_2, Temperature, Relative Humidity and Health

Indoor CO_2 concentrations are primarily dependent on the occupancy level and outdoor air supply rate [18]. Humans produce and exhale CO_2; therefore, a greater occupancy coupled with lower ventilation rates—intended to reduce energy consumption—give rise to higher and often harmful CO_2 concentrations indoors [19]. The American Society of Heating, Refrigerating and Air-conditioning Engineers (ASHRAE) recommends a maximum indoor CO_2 concentration of 1000 ppm [20].

Many studies have shown that inadequate ventilation in classes in elementary schools reduces students' comfort, performance and participation, and increases the prevalence of adverse health symptoms [21]. Elevated CO_2 can also reduce the cognitive performance of students in schools, while long-term, regular exposure has been associated with increased absenteeism, weight gain, and obesity [18,22–23]. Another issue for indoor environments is low relative humidity (RH). An RH below 30% has been found to lead to eye irritation and skin dryness, with an RH below 10% leading to dryness of the nasal mucous membrane. Moreover, low RH can trigger influenza transmission, increase indoor ozone concentration and produce static electricity [24–25]. However, high RH (> 60%) too can lead to some cases by encouraging fungal/mold growth and contributing to the deterioration of building materials [26]. Ensuring indoor levels to be between 40% and 60% can prevent a great number of negative effects on health [27].

2 Material and Method

2.1 Introduction of the Campus and Measurement Points

Harran University has been known as a very old university from a historical point of view but was founded in 1987. Harran University possesses two campuses. The

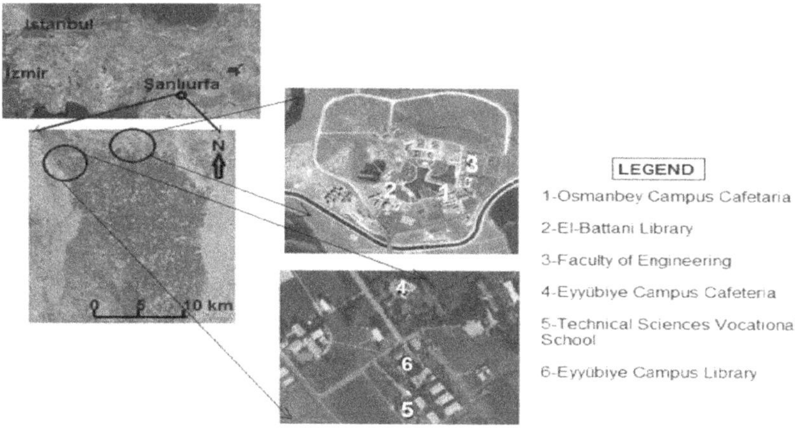

Fig. 3 : Osmanbey and Eyyübiye Campüs measurement locations in Harran University.

educational activities first started in various public buildings, but the first campus was moved to Eyyübiye Campus in 1987 as part of the World Bank Industrial Training Project. Located on an area of 752,521 m^2 in the 5th kilometer of Şanlıurfa, Eyyübiye Campus now has 1 faculty, 2 vocational schools (12 departments in the social sciences, 10 departments in technical sciences) and 7296 students are now studying at the university. Osmanbey Campus, which is the second campus, was established in 2003 and currently several facilities are constructed for faculty buildings. 17,107 students are studying at 11 faculties and 3 institutes of Osmanbey Campus which is located on a 1500 ha land in Şanlıurfa. The area used in our study consists of classrooms where students spend most of their time, corridors to which the classes are connected, libraries and cafeterias where they go to eat (Fig. 3).

The study was focused on indoor environments (classes, corridors connected with classes, libraries and especially the cafeterias where they go to eat) where students, academic staff, employees and visitors spend about 90% of their time. The buildings of Eyyübiye Campus were built in 1988 and the building is prefabricated. As it is now 30 years old, no ventilation system has been constructed; the number of windows is less and the ceiling height is 2.5 m. The Osmanbey Campus was built in 2003 and has a ventilation system as it is 15 years old.

At Eyyübiye Campus, a. The library has a ceiling height of 2.5 m and an area of 80 m^2 (Fig. 4a) which is ventilated only through windows, **b.** Classes are 40 m^2 and the number of students is usually 50 (Fig. 4b), **c.** The corridors connecting the classrooms are 12.5 m^2 and **d.** The cafeteria has an area of 200 m^2.

At Osmanbey Campus, a. The El-Battani library is 620 m^2 (Fig. 4c), **b.** Classrooms are 100 m^2 and the number of students is 40 (Fig. 4d), **c.** Corridors have an area of 72 m^2 and **d.** The cafeteria has an area of 393 m^2.

Fig. 4: **a.** Eyyübiye Campus Library **b.** Eyyübiye Campus Technical Sciences Vocational School Class **c.** Osmanbey Campus El-Battani Library **d.** Osmanbey Campus Engineering Faculty Class.

2.2 Devices Used in the Measurement

In order to evaluate the indoor air quality of the buildings on the campuses, measurements were made at 8.00–16.00 hours, during which a student spends his/her time on campus. Termo Scientific pDR 1500 personal DataRam devices were utilized for PM_{10} and $PM_{2.5}$ measurements in the course of 2018 with periods of 15-minute sampling for each parameter. The device consisted of PM_{10} and $PM_{2.5}$ headers, and it was calibrated before each different environment. In addition, Testo 440 device, which has the capacity of measuring CO_2 concentration in the range of 0–5000 ppm with an accuracy of ± 1 ppm CO_2 levels, was used to measure temperature and relative humidity in the campuses. Measurements were taken at a minimum of 3 points depending on size in classrooms, corridors, cafeterias and libraries, and the averages were calculated.

3 Results and Discussion

The efficiency in learning and the physical and mental well-being of students are significant and are greatly influenced by the indoor air quality (IAQ) in university dormitories [28]. In this chapter, the indoor air quality of two different campuses Osmanbey and Eyyübiye at Harran University in 2018 were investigated

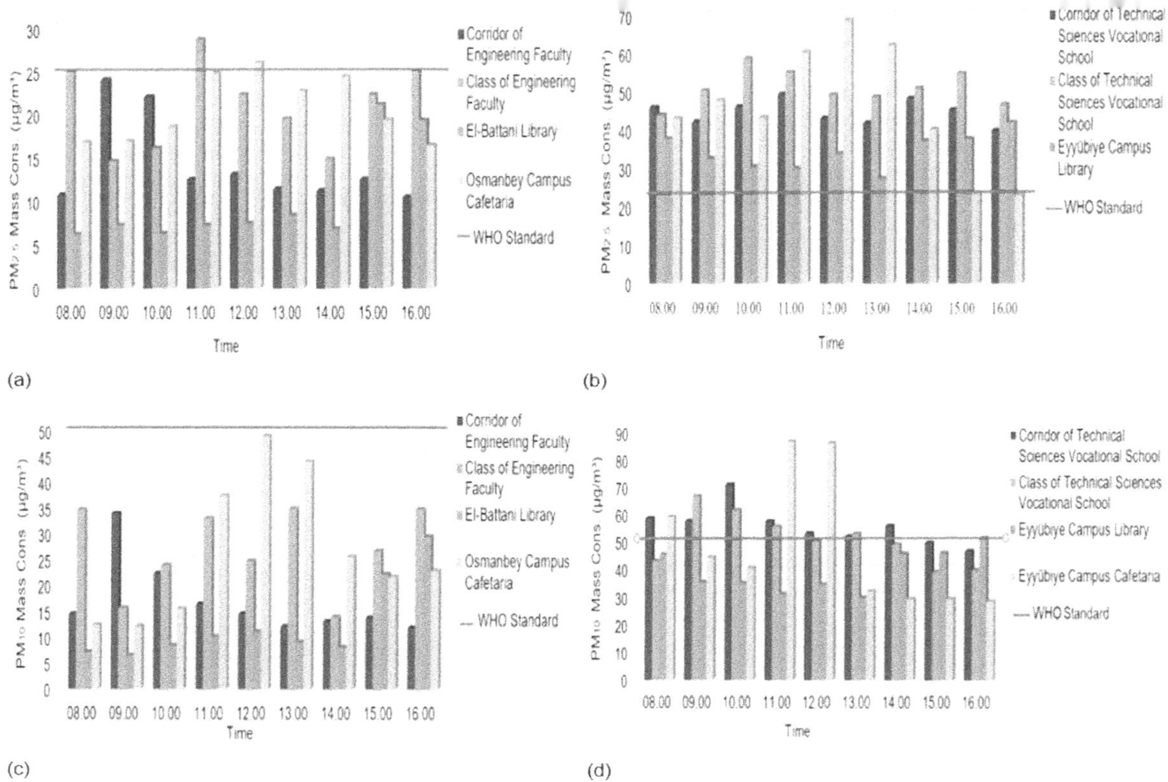

Fig. 5: **a.** Osmanbey Campus $PM_{2.5}$ concentration, **b**. Eyyübiye Campus $PM_{2.5}$ concentration, **c.** Osmanbey Campus PM_{10} concentration, **d.** Eyyübiye Campus PM_{10} concentration, **e.** Osmanbey Campus CO_2 concentration, **f.** Eyyübiye Campus CO_2 concentration, **g**. Osmanbey Campus temperature value, **h.** Eyyübiye Campus temperature value, **ı.** Osmanbey Campus relative humidity value, **i.** Eyyübiye Campus relative humidity value.

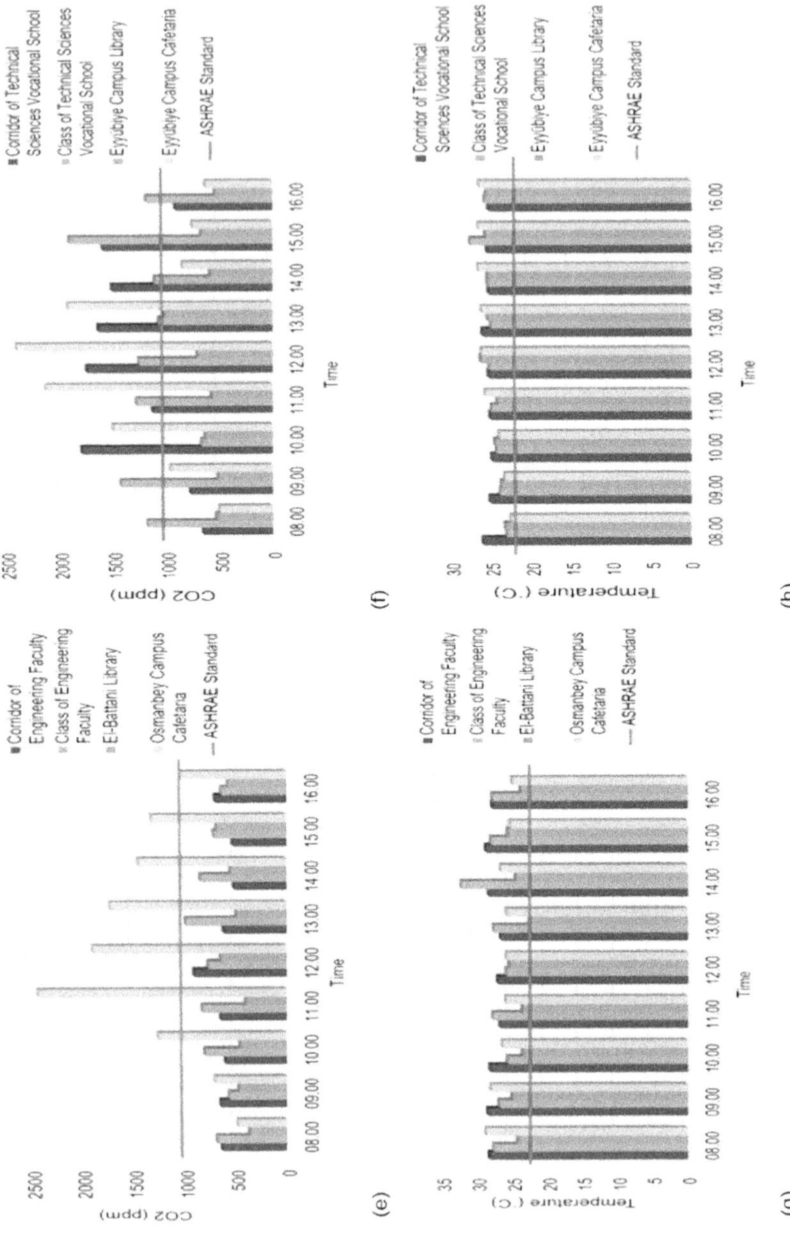
CO2 (ppm)
Temperature (°C)
Time
08.00
09.00
10.00
11.00
12.00
13.00
14.00
15.00
16.00
Corridor of Engineering Faculty
Class of Engineering Faculty
El-Battani Library
Osmanbey Campus Cafetaria
ASHRAE Standard
Corridor of Technical Sciences Vocational School
Class of Technical Sciences Vocational School
Eyyübiye Campus Library
Eyyübiye Campus Cafetaria
(e)
(f)
(g)
(h)

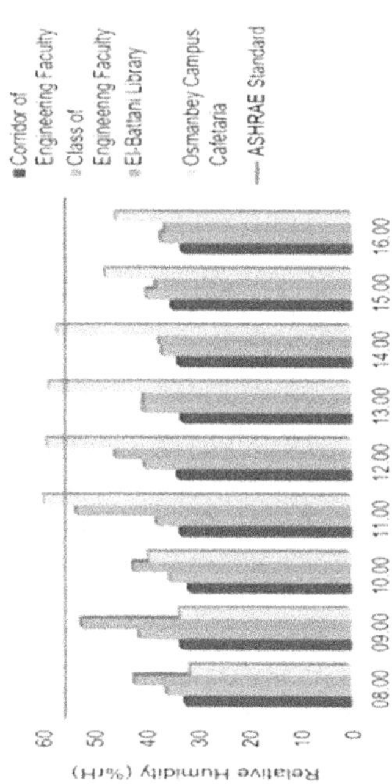
Relative Humidity (%rH)
0
10
20
30
40
50
60
08.00
09.00
10.00
11.00
12.00
13.00
14.00
15.00
16.00
Time
Corridor of Engineering Faculty
Class of Engineering Faculty
El-Battani Library
Osmanbey Campus Cafetaria
ASHRAE Standard

(l)

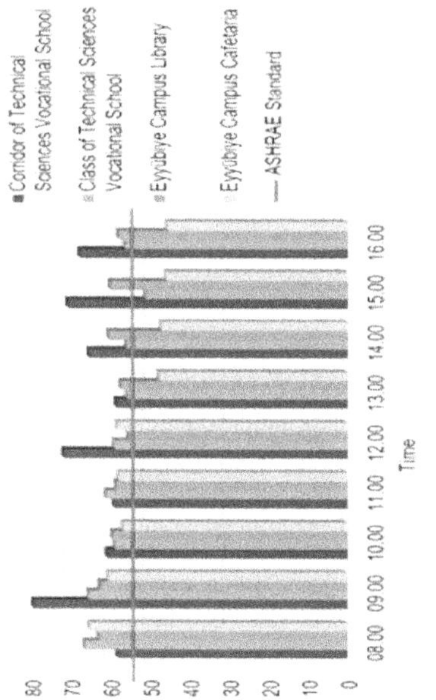
Relative Humidity (%rH)
0
10
20
30
40
50
60
70
80
08.00
09.00
10.00
11.00
12.00
13.00
14.00
15.00
16.00
Time
Corridor of Technical Sciences Vocational School
Class of Technical Sciences Vocational School
Eyyübiye Campus Library
Eyyübiye Campus Cafetaria
ASHRAE Standard

(l)

in comparison to WHO standard values of 50 µg/m³ for PM_{10}, 25 µg/m³ for $PM_{2.5}$, 1000 ppm for ARSHAE standards CO_2, 20 °C –25.5°C for temperature, 30–60 % for relative humidity [29–30] (Fig. 5).

Osmanbey Campus **Eyyübiye Campus**

- The PM_{10}, $PM_{2.5}$ and CO_2 values were high in the classrooms of Eyyübiye Campus due to the small size of classrooms, buildings at the age of 30, inadequate ventilation and poor cleaning (Fig. 5 b, d, f). There have been a number of studies dealing with the indoor air quality in classroom. [31] in their study conducted in 2009 investigated indoor air quality in classroom environments to identify the effect of cleaning and ventilation on PM. It was concluded in their study that greater level of particulate matter is a result of cleaning which is not done frequently and inadequate ventilation. Burdova et al., in their PM research conducted in the University of Slovakia in 2016, state that all of the values were above 100% mainly due to the use of solid fuel for the heating houses [32]. Aziz et al., in their study conducted in 2015 in Pakistan, investigated the relationship between $PM_{2.5}$ inside and outside the classrooms in the University of the Punjab. The average concentrations pertaining to $PM_{2.5}$ in Classrooms I, II and III over the course of 24 hours were found to be 282 µg/m³, 75 µg/m³ and 673 µg/m³. On the other hand, outdoor levels for Classrooms I, II and III were 324 µg/m³, 121 µg/m³ and 998 µg/m³, respectively. The ambient air and occupancy level was reported in the study to have a significant influence on $PM_{2.5}$ levels in classrooms, and the values were found to exceed the WHO limits [33]. In her study, Rastgeldi Dogan (2018) emphasizes that atmospheric desert dusts can influence the air quality of the indoor environment in Sanlıurfa, Turkey [34]. [35] have found out that the internal air quality values of the Batman University's central campus in Turkey are higher than the West Raman campus, and they report that this may be caused by traffic. It was found in the study conducted by Issa et al. that teacher and staff absenteeism in green Canadian schools improved by 2%–7.5 %, whereas student performance improved by 8–19 % in comparison to traditional schools [7]. It has been shown in a number of studies that the comfort, performance and attendance of the students are negatively affected by insufficient ventilation and that this increases the occurrence of negative health conditions [21,36].
- The $PM_{2.5}$ values in the Eyyübiye campus library exceeded the WHO standard during the time period between 8.00–16.00. This is considered to be due to the fact that the library is a 33-year-old prefabricated building, with no ventilation system, and there is a lot of dust due to the open shelf system (Fig. 4a, 5b).
- CO_2 is a parameter that can increase in indoor air quality depending on the human density. In both campus cafes, it is believed that the high amount of CO_2 between 11.00 and 14.00 is because it is the mealtime and there is an excessive number of students during that period (Fig. 5e, f). However, in Eyyübiye Cafeteria, the WHO standards of PM_{10}, $PM_{2.5}$, temperature and relative humidity

were also exceeded (Fig. 5b, d, h, i). This is due to the fact that the Eyyübiye Cafeteria has an area of 200 m^2 and the ventilation is insufficient during this time. It has been reported in recent studies investigating the classrooms of schools in Denmark and abroad that classroom air quality is often associated with carbon dioxide (CO_2) concentrations, which are significantly above the level of 1000 ppm, the recommended amount for buildings [37–38]. Classrooms where ventilation is provided through opening the windows manually have been found to be environments where this situation occurs more often especially during the heating season [39–42]. In his study on hospital environment, [15] has found that CO_2 concentration is exceeded in some clinics due to excessive number of people.

- It is thought that the reason for the high values of PM_{10}, $PM_{2.5}$ and CO_2 in the corridors of Eyyübiye campus is the lack of windows in the corridors, the old prefabricated structures of the buildings and the small size of the area, which is 12.5 m^2 (Fig. 5 b, d, f).
- The temperature has been found to be higher than ARSHAE standards in both campuses. This is due to the hot climate of the region and lack of isolation systems in the buildings (Fig. 5 g, h). Children's abilities to conduct tasks involving mental skills such as concentration, addition and completing sentences have been found to be influenced by changes in room temperature (e.g., 77°F to 68°F). It has been reported in several studies that cooler temperatures and modest relative humidity influence more than others [43].
- Relative humidity at both campuses was found to be not too much below the ARSHAE standard and was considered appropriate (Fig. 5 ı, i).

As a result, it has been determined that all indoor air quality values are better in Osmanbey Campus than in Eyyübiye Campus. The factors that were taken into account for the evaluation of indoor air quality are, old and new buildings, ventilation status, frequency of cleaning, ceiling height and number of people in the building.

4 Recommendations

Indoor air quality is the most important parameter affecting human health. Educational institutions, which are considered to be the most important public areas in the world, have responsibilities for ensuring these standards. The fact that the Eyyübiye Campus buildings are 33 years old and have a prefabricated structure has negative effects. The demolition and rebuilding of these buildings can be expensive. Students and academic staff using the buildings should be instructed to adopt the habit of opening windows manually in buildings until the completion of this process. As it is easier for diseases to spread in indoor envrionment, these places should be cleaned more frequently. In order to ensure internal thermal comfort, air conditioning systems must be installed and annual maintenance must be carried out without delay.

Acknowledgments

I would like to thank administration of Harran University for their support, Department of Construction for sharing structural details of the building, my graduate students Merve Nur DİNÇ and Melek YILDIZ for their assistance with the measurements and my PhD advisor Prof. Dr. Mehmet Ali ÇULLU for his guidance and support for map drawings.

References

1 Shah, A.S.V., J.P. Langrish, H. Nair, D.A. McAllister, A.L. Hunter, K. Donaldson, D.E. Newby, and N.L. Mills, *Global association of air pollution and heart failure: a systematic review and meta-analysis*, The Lancet. 2013. 382, pp. 1039–1048.

2 Chuang, K.-J., Y.-H. Yan, S.-Y. Chiu, and T.J. Cheng, *Long-term air pollution exposure and risk factors for cardiovascular diseases among the elderly in Taiwan*, Occup Environ Med. 2011. 68, pp. 64–68.

3 El Asmar, M., A. Chokor, and I. Srour, *Are building occupants satisfied with indoor environmental quality of higher education facilities?*, Energy Procedia. 2014. 50: pp. 751 – 760.

4 Almeida, S.M., Canha, N., Silva, A., Freitas, M.C., Pegas, P., Alves, C. and Pio, C.A., *Children exposure to atmospheric particles in indoor of Lisbon primary schools.* Atmospheric Environment. 2011. 45: pp.7594–7599.

5 Stein, P.P. and A.N. Rodrigues da Silva, *Barriers, motivators and strategies for sustainable mobility at the USP campus in São Carlos, Brazil*, Case Studies on Transport Policy. 2018. 6: pp. 329–335.

6 Dagiliūtė, R. and G. Liobikienė, *University contributions to environmental sustainability: Challenges and opportunities from the Lithuanian case*, Journal of Cleaner Production. 2015. 108: pp. 891–899.

7 Issa M., J. Rankin, M. Atallah, and A. Chritian, *Absenteeism, performance and occupant satisfaction with the indoor environment of Green Toronto Schools*, Indoor Built Environment. 2011. **20**(5): pp. 511–523.

8 Matic, B., U. Rakic, V. Jovanovic, S. Dejanovic, and N. Djonovic, *Key factors determining indoor air PM10 concentrations in naturally ventilated primary schools in Belgrade, Serbia.* Slovenian Journal of Public Health. 2017. **56**(4): 227–235.

9 World Health Organization (WHO) Health effects of particulate matter. Policy implications for countries in eastern Europe, Caucasus and central Asia.WHO Regional Office for Europe, Copenhagen (2013).

10 Esworthy, R., *Air Quality: EPA's 2013 Changes to the Particulate Matter (PM) Standard*, Prepared for Members and Committees of Congress. 2013. Amsterdam.

11 Löndahl, J., A. Massling, J. Pagels, E. Swietlicki, E. Vaclavik, and S. Loft, *Size-resolved respiratory-tract deposition of fine and ultrafine hydrophobic and hygroscopic aerosol particles during rest and exercise*, Inhalation Toxicology. 2007. **19**(2), pp. 109–116.

12 Fu, M., F. Zheng, X. Xu, and L. Niu, *Advances of study on monitoring and evaluation of PM2.5 pollution*, Meteorol Disaster Reduc Res. 2011. 34: pp. 1–6.

13McClellan, R.O., *Particle interactions with the respiratory tract*, Particle–Lung Interactions. 2011. p. 823.

14 Morman, S.A. and G.S. Plumlee, *The role of airborne mineral dusts in human disease*, Aeolian Research. 2013. 9: pp. 203–212.

15 Rastgeldi Doğan, T., *Investigation of indoor air quality in a hospital: A case study from Şanlıurfa, Turkey*, Journal of Natural Hazards and Environment. 2019. **5**(1): pp. 101–109.

16 Rastgeldi Dogan, T., C. Saydam, M.İ. Yeşilnacar, and M. Gencer, *In-cloud alteration of desert-dust matrix and its possible impact on health: a test in southeastern Anatolia, Turkey*, European Journal of Mineralogy. 2010. 22 (5): pp. 659–664.

17 Kim, K.H., E. Kabir, and S. Kabir, *A review on the human health impact of airborne particulate matter*, Environment International. 2015. 74: pp. 136–143.

18 Zhang, X., P. Wargocki, and Z. Lian, *Physiological responses during exposure to carbon dioxide and bioeffluents at levels typically occurring indoors.* Indoor Air. 2017. 27: pp.65–77.

19 Satish, U., M.J. Mendell, K. Shekhar, T. Hotchi, D. Sullivan, S. Streufert, and W.(B).J. Fisk, *Is Co2 an indoor pollutant? Direct effects of low-to-moderate Co_2 concentrations on human decision-making performance.* Environ Health Perspect. 2012. 120: pp.1671–1677.

20 Torpy, F.R., M. Zavattaro, and P.J. Irga, *Green wall technology for the phytoremediation of indoor air: A system for the reduction of high Co_2 concentrations.* Air Qual Atmos Health. 2017. 10: pp.575–585.

21 Bakó-Biró, Z., D.J. Clements-Croome, N. Kochhar, H.B. Awbi, H., and M.J. Williams, *Ventilation rates in schools and pupils' performance*, Building and Environment. 2012. 48: pp. 215–223.

22 Nieuwenhuis, M., C. Knight, T. Postmes, and S.A. Haslam, *The relative benefits of green versus lean office space: three field experiments.* Journal of Experimental Psychology: Applied. 2014. **20**(3): pp.199–214.

23 Vehvilainen, T.Y., H. Lindholm, H. Rintamaki, R. Pääkkönen, A. Hirvonen, O. Niemi, and J. Vinha, *High indoor Co_2 concentrations in an office environment increases the transcutaneous Co_2 level and sleepiness during cognitive work.* Journal of Occupational and Environmental Hygiene. 2016. 13: pp.19–29.

24 Zhang, H., and H.Yoshino, *Analysis of indoor humidity environment in Chinese residential buildings*, Building and Environment. 2010. 45: pp.2132–2140.

25 Abusharha A.A, and E.I,Pearce,*The effect of low humidity on the human tear film*. Cornea.2013. 32: pp.429–434.

26 Frankel, M., G. Beko, M. Timm, S. Gustavsen, E.W. Hansen, and A.M. Madsen, *Seasonal variations of indoor microbial exposures and their relation to temperature, relative humidity, and air exchange rate*. Applied and Environmental Microbiology. 2012. 78: pp.8289–8297.

27 Arundel, A.V., E.M. Sterling, J.H. Biggin, and T.D. Sterling, *Indirect health effects of relative humidity in indoor environments*, Environmental Health Perspectives.1986. 65: pp.351–361.

28 Yang, J., X. Wu, K.T. Chan, and X. Yang, *Measurement and analysis of indoor air quality in the university dormitories*, International Conference on Bioinformatics and Biomedical Engineering. 2009. 3rd International Conference on (p. 1–4). IEEE. Beijing, China.

29 ASHRAE, (2003), ASHRAE Handbook CD, 2001 Fundamentals, Chapter 8: Thermal Comfort, Atlanta, USA

30 WHO, (2000), *WHO air quality guidelines for Europe*, WHO Regional Publications Eur Ser No. 91. Regional US EPA, 1997. National Ambient Air Quality Standards

31 Heudorf, U., V. Neitzert and J. Spark, *Particulate matter and carbon dioxide in classrooms – The impact of cleaning and ventilation*. International Journal of Hygiene and Environmental Health. 2009. 212: pp.45–55.

32 Burdova, E.K., S. Vilcekova, and L. Meciarova, *Investigation of particulate matters of the University Classroom in Slovakia*, Energy Procedia. 2016. 96: pp.620–627.

33 Aziz, K and Ali, Z and Nasir, ZA and Colbeck, I *Assessment of airborne particulate matter (PM<inf>2.5</inf>) in university classrooms of varrying occupancy*. Journal of Animal and Plant Sciences, 2015. **25** (3): pp.649 – 655.

34 Rastgeldi Doğan, T., M.İ. Yeşilnacar, And M.A. Çullu, *Seasonal investıgatıon of atmospherıc desert dust affectıng sanlıurfa usıng Modıs Satellıte And Hysplıt model data*, Sigma Journal of Engineering and Natural Sciences. 2018. **36** (3): pp.905–916.

35 Argunhan, Z. and A.S. Avci, Statistical evaluation of indoor air quality parameters in classrooms of a university, Advances in Meteorology. 2018. p.10.

36 Gaihre, S., S. Semple, J. Miller, S. Fielding, and S. Turner,*Classroom carbon dioxide concentration, school attendance, and educational attainment*. Journal of School Health. 2014. **84**(9): pp.569–74

37 Energistyrelsen. 2014. Bygningsreglement 2010 (Danish Building Code 2010), BR10. http://bygningsreglementet.dk.

38 ISO Standard 15251. 2007. Indoor Environmental Input Parameters for Design and Assessment of Energy Performance of Buildings Addressing Indoor Air Quality, Thermal Environment, Lighting, and Acoustics. Geneva, Schwitzerland: International Organisation for Standardisation.

39 Clausen, G., J. Toftum, and B. Andersen. 2014. *Indeklima iklasselokaler—Resultater Masseeksperiment (Indoor environment in classrooms—Results)*, Danish Science Factory & Technical University of Denmark.

40 Rosbach, J., E. Krop, M. Vonk, J. van Ginkel, C. Meliefste, S. de Wind, U. Gehring, and B. Brunekreef, *Classroom ventilation and indoor air quality—Results from the FRESH intervention study*. Indoor Air .2016. 26: pp.538–45.

41 Stabile, L., M. Dell'Isola, A. Frattolillo, A. Massimo, and A. Russi, *Effect of natural ventilation and manual airing on indoor air quality in naturally ventilated Italian classrooms*. Building and Environment. 2016. 98: pp.180–9.

42 Toftum, J., B.U. Kjeldsen, P. Wargocki, H.R. Menå, E.M.N. Hansen, and G. Clausen, *Association between classroom ventilation and learning in Danish schools*. Building and Environment. 2015. 92: pp.494–503.

43 Wargocki, P. and D.P. Wyon, *The effects of moderately raised classroom temperatures and classroom ventilation rate on the performance of schoolwork by children*. HVAC&R Research. 2007. **13**(2): pp.193–220.

Prof. Dr. Tuğba Kiper, Prof. Dr. Aslı Korkut and Sümeyra Betül Tayfur

40 Determination of Visual Landscape Quality of Campus Areas: Tekirdağ Namik Kemal University Example

Abstract: In this study, a visual impact assessment study was carried out in the sample of Tekirdağ Namık Kemal University with the prediction that "Determination of the quality of the visual landscape in the physical locations of campus areas will be effective." For this purpose, 11 photographs were evaluated visually using five-point Likert scale based on eight parameters by a group of 72 observers. The study was supported by surveys based on questionnaires, observations and related literature. The findings were evaluated by the tables in which the percentile average of the scores the observers have given to the evaluation criteria for each photo. As a result of the evaluation, various suggestions were developed for the physical layout of the area by determining the preferred areas with visual appeal in the main campus of Tekirdağ Namık Kemal University.

Keywords: Campus areas, Visual landscape, Visual landscape quality, Campus areas

1 Introduction

Campus areas have many functions including educational and scientific activities, social and cultural life, shopping and housing. Accordingly, they consist of academic buildings (classrooms, laboratories and research areas), administrative buildings (rectorate, deanery, etc.), cultural areas (libraries), residential buildings (dormitories, lodgings) and recreational areas (cafeterias, facilities, green spaces, etc.) [1–5]. Campuses form a whole with their structures, open spaces and users. They are shaped by natural (climate, topographic structure, flora) and cultural (transportation situation, economic structure, etc.) characteristics of the city they are located in. Hence, they form small models of cities and give little clues about them. In this respect, campus areas need to be planned and designed in a functional, environmental and visual sense. The preservation of the environmental character, the indoor-outdoor relationship between the spaces, the continuity of the urban texture, the provision of accessibility and data such as symbolization, formalism, perceptibility, visuality and aesthetics should be considered in the design of the outdoor spaces of the university campuses [6]. Pouyaa and Bekhbahani [7] state that every landscape owns exclusive sentiment, notions and connotation that enhance sensation of the place and its character [7]. In his study, Abu-Ghazzeh [8] also stated that the outer spaces and structures should have an aesthetic appearance and a common character since visual preference is based on visual perception [8]. Visual landscapes are an inherent piece of cultural awareness of the habitat, people's loyalty to the area and their standard of living [9, 10]. Visual aesthetic experience in civilized territories has significance for creating a convenient and fulfilling experience for the society. A concrete example of a good visual image of urban space will support the rise of affirmative recognition and understanding that urges society to respond well to urban space [11]. Landscape visual assessment is an essential element of landscape architecture, landscape planning and different applications that involve decision making [7]. As reported by Uzun and Müderrisoğlu [12] and Scott [13], determining the visual and environmental characteristics of the landscape is a pioneering subject of landscape planning and at the same time it has various usage possibilities in landscape studies [12,13]. Griffith [14] pointed out that effective campus planning should evoke a sense of "space, purpose, order and quality" of the campus visually. In this context, it is very important to determine the visual landscape quality in the physical space settlements of campus areas [14]. For this purpose, a visual impact assessment study was carried out in Tekirdağ Namık Kemal University with the prediction of that "Determination of the quality of the visual landscape in the physical locations of campus areas will be effective" and the answer was searched for the question "What are the areas that are visually attractive and preferred in Tekirdağ Namık Kemal University Main Campus?" The study was supported by surveys based on questionnaires, observations and related literature. In this context, various parameters for visual quality were determined. Eleven photographs were evaluated visually by a group of 72 observers using five-point Likert scale based on eight parameters and the most and the least preferred images were defined at the level of these parameters. The findings were evaluated in the tables with the percentage averages of the scoring given by the observer group for each photograph. As a result of the evaluation, various suggestions were developed for the physical layout of the area by determining the preferred areas with visual appeal in Tekirdağ Namık Kemal University Main Campus.

2 Materials and Methods

Tekirdağ Namık Kemal University Main Campus is located between 27 ° 34'38.95"-27 ° 35'57.24" East longitudes and 41 ° 1'10.12 "– 40 ° 59'17.83" northern latitudes. It is positioned on a land of 262.3 ha about 7 km from city center on the side of Atatürk Forest at the entrance of the city following the İstanbul-Çanakkale road. The campus includes living center, hotel, indoor sports hall and stadium in addition to rectorate building, academic units and dormitories. In this context, the campus allows for activities such as sightseeing, walking, sitting and resting, as well as educational and research activities (Fig. 1).

Visual landscape analysis method was used in the context of this study. Visual character assessment is a fundamental analysis method when considering landscape as an aesthetic asset [15]. Visual landscape analysis contains research that perceptually evaluates the characteristics of spaces and the interplay between its visual components [16]. It covers the appraisal of a range of physical, aesthetic and psychological aspects [17]. In this regard, the aim of visual quality assessment is

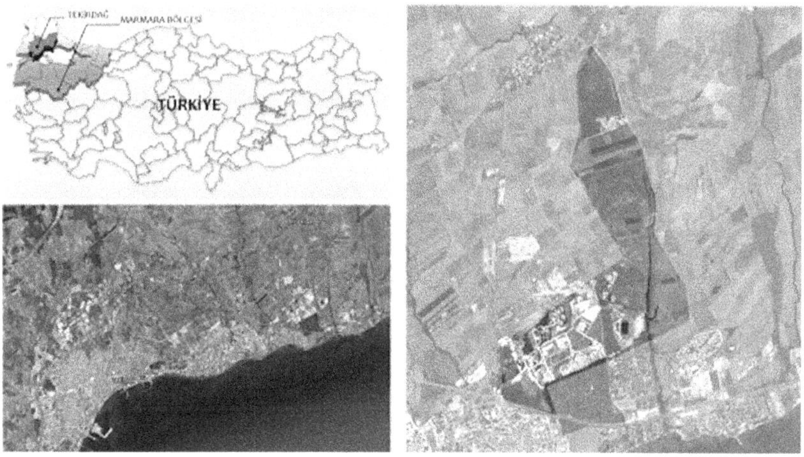

Fig. 1: Location of the study field.

to define a landscape aesthetically and to establish a common denominator of the aspects and tangible elements of landscape affecting the preferences of the people living in a certain area in order to consider when starting the exact planning and design [18,19]. Visual landscape quality is based on observers' evaluation of landscape qualities (visual quality, landscape quality, landscape beauty) by scoring the landscape scenes presented with photographs in terms of the parameters determined [20]. The main objective of visual quality is to enable people to evaluate the landscape they see in terms of visual appeal [21]. In this context, 11 photographs taken at different points of Tekirdağ Namık Kemal University Main Campus, which are thought to be essential for determining the visual landscape quality, were used (Fig. 2). In the selection of photographs, images that offer a wide range of view with many landscape elements have been influential. The 11 selected photographs were evaluated by the observer group of 72 people within the scope of the parameters created by using in similar studies. In determining the evaluation parameters, the works of [22–26] were used. Eleven photographs were evaluated with five-point Likert Scale (ranging from 1–5 points) in terms of parameters consisting eight adjective pairs (Tab. 1) by the observer group. Scores based on the use and availability of the determined parameters in the relevant photograph are defined as 1 "very negative," 2 "negative," 3 "medium," 4 "positive" and 5 "very positive." The average percentages of the scoring given by the observer group for the evaluation criteria are shown in the tables. The highest and the lowest scored criteria are presented for 11 photographs individually by calculating the percentage values for each parameter used in each photograph and analyzing the rating of the observer group.

Fig. 2: Photographs used in visual landscape analysis and shooting points.

Tab. 1: Parameters Used in Visual Landscape Analysis [22–26]

Neglected/Well-maintained	The image and the elements within the view to be maintained
Displeasing/Pleasing	Image interpretation is not confusing or difficult
Incompatible/Compatible	The absence of foreign elements that are not integrated with the remaining landscaping elements in the image
Repulsive/Appealing	The multitude of appealing views from the visual aspect
Distractor/Concentrator	Absence of distracting negative objects in the image
Not Reliable/Reliable	The elements of the image do not evoke risks or hazards; offer welcoming, safe and secure appearance
Static/Dynamic	The image has a large number of disparate elements
Irregular/Regular	The elements that make up the landscape are in order

Tab. 2: Evaluation Results for Image 1

Evaluation Criterion	**Evaluation (Rating) Score**				
Neglected/Well-maintained	**1**	**2**	**3**	**4**	**5**
Displeasing/Pleasing	14	11.3	35.2	36.6	15.2
Incompatible/Compatible	7.0	18.3	39.4	28.2	7.0
Repulsive/Appealing	**4.2**	**25.4**	23.9	35.2	11.3
Distractor/Concentrator	4.2	19.7	40.8	28.2	7.0
Not Reliable/Reliable	4.2	14.1	43.7	28.2	9.9
Static/Dynamic	8.5	11.3	21.1	**31.0**	**28.2**
Irregular/Regular	4.2	19.7	38.0	19.7	18.3
Neglected/Well-maintained	5.6	14.1	22.5	33.8	23.9

The criterion that has the highest score for the determination of the visual landscape quality is the "**dynamic**" effect, while the lowest score is the "**repulsive**" effect.

3 Results

In the study, images from Namık Kemal University Main Campus and the average percentage of the ratings for the evaluation criteria scored by the observer group were given in the form of tables (Tabs. 1, 2, 3, 4, 5, 6, 7, 8, 9, 10, 11).

4 Conclusions

In this study, which was conducted to determine the visual landscape quality of Tekirdağ Namık Kemal University Main Campus, 11 photographs were evaluated by the group of 72 observers within the scope of the parameters consisting eight adjective pairs. As a result of this evaluation, the image with best visual quality is the photograph F9 while the lowest quality image is determined as the photograph 7. In the images, the areas that have large amount of greenery where the natural and artificial landscape elements are in harmony with each other and the

Tab. 3: Evaluation Results for Image 2

Evaluation Criterion	Evaluation (Rating) Score				
	1	2	3	4	5
Neglected/Well-maintained	1.4	12.7	38.0	23.9	23.9
Displeasing/Pleasing	**2.8**	**22.5**	23.9	35.2	15.5
Incompatible/Compatible	1.4	14.1	25.4	43.7	15.5
Repulsive/Appealing	1.4	19.7	46.5	18.3	14.1
Distractor/Concentrator	4.2	16.9	42.3	23.9	12.7
Not Reliable/Reliable	-	12.7	28.2	35.2	23.9
Static/Dynamic	5.6	9.9	26.8	29.6	28.2
Irregular/Regular	5.6	12.7	19.7	**35.2**	**26.8**

The criterion that has the highest score for the determination of the visual landscape quality is the "**displeasing**" effect, while the lowest score is the "**regular**" effect.

areas where the view is perceived clearly are found to be high in terms of visual landscape quality. The areas where natural and artificial landscape elements are irregular, the existence of green areas is insufficient and the architectural unit is dense and does not conform to the environment are found to be low in terms of visual landscape quality.

The most preferred images are given in the Tab. 13 in respect to parameters. According to this, in terms of maintenance, compliance, regularity and reliability parameters, image 9 has received the highest score. The image is preferred due to the harmony of the visual elements with each other, the well-defined location, the integrity of the buildings and the green texture and the central location close to the entrance.

Tab. 4: Evaluation Results for Image 3

Evaluation Criterion	Evaluation (Rating) Score				
	1	2	3	4	5
Neglected/Well-maintained	4.2	8.5	32.4	33.8	21.1
Displeasing/Pleasing	4.2	16.9	31.0	35.2	12.7
Incompatible/Compatible	5.6	11.3	28.2	35.2	19.7
Repulsive/Appealing	**5.6**	**16.9**	31.0	31.0	15.5
Distractor/Concentrator	4.2	16.9	45.1	18.3	16.9
Not Reliable/Reliable	2.8	11.3	33.8	29.6	22.5
Static/Dynamic	4.2	15.5	28.2	31.0	21.1
Irregular/Regular	2.8	11.3	22.5	**36.6**	**26.8**

The criterion that has the highest score for the determination of the visual landscape quality is the "**regular**" effect, while the lowest score is the "**repulsive**" effect.

Image 5 is the most preferred in terms of pleasantness and appeal parameters. When the image is examined, it is seen that the structural elements form integrity of the green area, the urban settlement layout and the water element.

In terms of distraction parameter, image 4 is the most preferred and image 2 is the most preferred in terms of dynamism parameter. There are no distracting negative objects in the image 4. The presence of the intersection point of the monument and the square has created a concentrating effect in the image 2.

Seating-resting places in the vicinity of social facilities in the campus as well as large grass surfaces near the administrative and academic buildings and areas arranged with plants offer relaxing environments; however, they are not able to meet the user crowd's requirements. In particular, vehicles parked on the roadside and the movement of the vehicles; the continuation of construction works for academic, administrative and cultural units; insufficiency of planting activities;

Tab. 5: Evaluation Results for Image 5

Evaluation Criterion	Evaluation (Rating) Score				
	1	2	3	4	5
Neglected/Well-maintained	2.8	9.9	32.4	35.2	19.7
Displeasing/Pleasing	5.6	16.9	39.4	26.8	12.7
Incompatible/Compatible	4.2	16.9	32.4	29.6	19.7
Repulsive/Appealing	**8.5**	**19.7**	31.0	25.4	15.5
Distractor/Concentrator	4.2	16.9	32.4	39.4	16.9
Not Reliable/Reliable	5.6	12.7	25.4	**35.2**	**22.5**
Static/Dynamic	4.2	15.5	40.8	22.5	21.1
Irregular/Regular	8.5	8.5	32.4	23.9	26.8

The criterion that has the highest score for the determination of the visual landscape quality is the "**reliable**" effect, while the lowest score is the "**repulsive**" effect.

and the uneven distribution of structural and vegetal elements weaken the visual impact.

As a result, Tekirdağ Namık Kemal University Campus is an important area for Tekirdağ city and its users with its location, natural and cultural landscape features and its structure and green area texture. This area needs to be made visually more attractive. For this purpose, the environmental perception must be strengthened by considering the combination of buildings and green spaces with a holistic approach and supporting the pedestrian-oriented uses in terms of quality and quantity.

Tab. 6: Evaluation Results for Image 5

Evaluation Criterion	Evaluation (Rating) Score				
	1	2	3	4	5
Neglected/Well-maintained	1,4	16,9	35,4	29,6	16,9
Displeasing/Pleasing	**5,6**	**22,5**	14,1	**45,1**	**12,7**
Incompatible/Compatible	8,5	9,9	29,6	36,6	15,5
Repulsive/Appealing	1,4	12,7	32,4	40,8	12,7
Distractor/Concentrator	5,6	15,5	28,2	35,2	15,5
Not reliable/Reliable	4,2	14,1	25,4	36,6	19,7
Static/Dynamic	4,2	21,1	33,8	26,8	14,1
Irregular/Regular	7,0	15,5	40,8	23,9	12,7

The criterion that has the highest score for the determination of the visual landscape quality is the **"pleasing"** effect, while the lowest score is the "**displeasing**" effect

Tab. 7: Evaluation Results for Image 6

Evaluation Criterion	Evaluation (Rating) Score				
	1	2	3	4	5
Neglected/Well-maintained	9,9	14,1	40,8	21,1	14,1
Displeasing/Pleasing	2,8	25,4	33,8	23,9	14,1
Incompatible/Compatible	7,0	15,5	36,6	32,4	8,5
Repulsive/Appealing	5,6	18,3	46,5	21,1	8,5
Distractor/Concentrator	**7,0**	**23,9**	40,8	21,1	7,0
Not reliable/Reliable	4,2	18,3	28,2	**29,6**	**19,7**
Static/Dynamic	4,2	25,4	35,2	25,4	9,9
Irregular/Regular	7,0	21,1	25,4	36,6	9,9

The criterion that has the highest score for the determination of the visual landscape quality is the "**reliable**" effect, while the lowest score is the "**distractor**" effect

Tab. 8: Evaluation Results for Image 7

Evaluation Criterion	Evaluation (Rating) Score				
	1	2	3	4	5
Neglected/Well-maintained	5.6	22.5	32.4	29.6	9.9
Displeasing/Pleasing	**8.5**	**23.9**	26.8	31.0	9.9
Incompatible/Compatible	5.6	22.5	33.8	32.4	5.6
Repulsive/Appealing	2.8	23.9	47.9	15.5	9.9
Distractor/Concentrator	8.5	19.7	45.1	16.9	9.9
Not Reliable/Reliable	9.9	19.7	26.8	26.8	16.9
Static/Dynamic	7.0	19.7	31.0	26.8	15.5
Irregular/Regular	4.2	19.7	32.4	**28.2**	**15.5**

The criterion that has the highest score for the determination of the visual landscape quality is the "**regular**" effect, while the lowest score is the "**displeasing**" effect.

Tab. 9: Evaluation Results for Image 8

Evaluation Criterion	Evaluation (Rating) Score				
	1	2	3	4	5
Neglected/Well-maintained	2.8	9.9	39.4	33.8	14.1
Displeasing/Pleasing	4.2	16.9	31.0	39.4	8.5
Incompatible/Compatible	2.8	7.0	36.6	39.4	14.1
Repulsive/Appealing	**4.2**	**19.7**	33.8	26.8	15.5
Distractor/Concentrator	4.2	9.9	35.2	36.6	14.1
Not Reliable/Reliable	1.4	18.3	28.2	32.4	19.7
Static/Dynamic	1.4	16.9	33.8	23.9	23.9
Irregular/Regular	4.2	18.3	22.5	**32.4**	**22.5**

The criterion that has the highest score for the determination of the visual landscape quality is the "**regular**" effect, while the lowest score is the "**repulsive**" effect.

Tab. 10: Evaluation Results for Image 9

Evaluation Criterion	**Evaluation (Rating) Score**				
	1	2	3	4	5
Neglected/Well-maintained	1.4	5.6	33.8	36.6	22.5
Displeasing/Pleasing	1.4	16.9	35.2	33.8	12.7
Incompatible/Compatible	1.4	11.3	19.7	**47.9**	**19.7**
Repulsive/Appealing	.	15.5	38.0	33.8	12.7
Distractor/Concentrator	2.8	14.1	42.3	22.5	18.3
Not Reliable/Reliable	7.0	9.9	19.7	36.6	26.8
Static/Dynamic	5.6	12.7	28.7	21.1	32.4
Irregular/Regular	**1.4**	**21.1**	12.7	33.8	31.0

The criterion that has the highest score for the determination of the visual landscape quality is the "**compatible**" effect, while the lowest score is the "**irregular**" effect.

Tab. 11: Evaluation Results for Image 10

Evaluation Criterion	**Evaluation (Rating) Score**				
	1	**2**	**3**	**4**	**5**
Neglected/Well-maintained	5.6	11.3	47.9	21.1	14.1
Displeasing/Pleasing	**7.0**	**32.4**	14.1	25.4	21.1
Incompatible/Compatible	4.2	22.5	35.2	26.8	11.3
Repulsive/Appealing	4.2	16.9	33.8	26.8	18.3
Distractor/Concentrator	4.2	19.7	38.0	19.7	18.3
Not Reliable/Reliable	4.2	15.5	28.2	31.0	21.1
Static/Dynamic	5.6	21.1	28.2	25.4	19.7
Irregular/Regular	4.2	21.1	26.8	**36.8**	**21.1**

The criterion that has the highest score for the determination of the visual landscape quality is the "**regular**" effect, while the lowest score is the "**displeasing**" effect.

Tab. 12: Evaluation Results for Image 11

Evaluation Criterion	**Evaluation (Rating) Score**				
	1	**2**	**3**	**4**	**5**
Neglected/Well-maintained	2.8	7.0	36.6	**43.7**	**9.9**
Displeasing/Pleasing	4.2	11.3	42.3	32.4	9.9
Incompatible/Compatible	2.8	11.3	36.6	35.2	14.1
Repulsive/Appealing	1.4	9.9	50.7	23.9	14.1
Distractor/Concentrator	2.8	18.3	33.8	32.4	12.7
Not Reliable/Reliable	-	15.5	33.8	33.8	16.9
Static/Dynamic	**8.5**	**21.1**	23.9	31.0	15.5
Irregular/Regular	7.0	9.9	31.0	35.2	16.9

The criterion that has the highest score for the determination of the visual landscape quality is the "**well-maintained**" effect, while the lowest score is the "**static**" effect.

Tab. 13: Highest Rated Images B ased on Parameters

Evaluation Criterion	**Image No**
Neglected/Well-maintained	F9
Displeasing/Pleasing	F5
Incompatible/Compatible	F9
Repulsive/Appealing	F5
Distractor/Concentrator	F4
Not Reliable/Reliable	F9
Static/Dynamic	F2
Irregular/Regular	F9

References

1 Korkut, A.B., Çilek, M.S. (2005). Üniversite yerleşkeleri peyzaj planlama ilkeleri, II. Ulusal Üniversite Yerleşke Planlaması ve Çevre Düzenlemesi Çalıştayı, 9–10 Haziran, Kahramanmaraş.

2 Eminağaoğlu, Z., Arslan Muhacir, S. (2018). Artvin Çoruh Üniversitesi kampüs alanlarının planlama ve tasarım ilkeleri kapsamında değerlendirilmesi. Artium, 1, 38–43.

3 Erkman, U. (1990). Büyüme ve gelişme açısından üniversite kampüslerinde planlama ve tasarım sorunları. İ.T.Ü. Mimarlık Fakültesi Baskı Atölyesi, İstanbul.

4 Aydın, D. (2003). Üniversite Kampüs tasarımı üzerine düşünceler. S.Ü. İletişim Fakültesi Uygulama Gazetesi, Sayı 50, Konya.

5 Erçevik, B., Önal, F. (2011). Üniversite kampüs sistemlerinde sosyal mekan kullanımları. MEGARON, 6(3):151–161.

6 Atabeyoğlu, Ö. (2014). Sosyal bilimler meslek yüksekokulu kampusu peyzaj tasarım ve uygulama çalışması. Artium, 2(1), 85–101.

7 Pouyaa, S., Behbahania, H.I. (2017). Assessment of the visual landscape quality based on the subjectivist paradigm to design the memorial garden. Universty of Süleyman Demirel Turkish Journal of Forestry, 18(3): 171–177.

8 Abu-Ghazzeh, T.M. (1999). Communicating behavioral research to campus desıgn factors affecting the perception and use of outdoor spaces at The University of Jordan. Environment and Behavior, 31 (6), 764–804.

9 Van Eetvelde, V., Antrop, M. (2009). Indicators for assessing changing landscape character of cultural landscapes in Flanders (Belgium). Land Use Policy, 26, 901–910.

10 Kienast, F., Frick, J., Van Strien, M.J., Hunziker, M. (2015). The swiss landscape monitoring program–A comprehensive indicator set to measure landscape change. Ecol. Model. 295, 136–150.

11 Santosa, H., Ernawati, J., Wulandari, L.D. (2018). Visual quality evaluation of urban commercial streetscape for the development of landscape visual planning system in provincial street corridors in Malang, Indonesia. Friendly City 4 "From Research to Implementation for Better Sustainability" IOP Publishing IOP Conf. Series: Earth and Environmental Science. 1–13.

12 Uzun, O., Müderrisoğlu, H. (2011). Visual landscape quality in landscape planning: examples of Kars and Ardahan cities in Turkey. African Journal of Agricultural Research. 6(6): 1627–1638.

13 Scott A., (2002). Assessing public perception of landscape: the LANDMAP experience, Landscape Research, 7 (3): 271–295.

14 Griffith, J. (1994). Open Space Preservation: An imperative for quality campus environments, The Journal of Higher Edıcation, vol. 65(6), pp. 645–669.

15 Keleş, E., Atik, D., Bayram, G. (2018). Visual landscape quality assessment in historical cultural landscape areas. European Journal of Sustainable Development, 7 (3): 287–300.

16 Malkoc, E., Kılıçcaslan, Ç., Özkan, M.B. (2010). Visual landscape analysis of urban open spaces: a case study of the coastline of Gocek Settlement, Mugla, Türkiye. Indoor Built Environ, 19 (5):520–537.

17 Cañas I., Ayuga, E., Ayuga, F. (2009). A contribution to the assessment of scenic quality of landscapes based on preferences expressed by the public. Land Use Policy, 26, 1173–118.

18 Kaya, S., Başar, H., Can, T., Müderrisoğlu, H. (2016). Evaluation of visual landscape quality Konuralp Campus in Duzce University. Düzce Üniversitesi Orman Fakültesi Ormancılık Dergisi, 12 (2):123–142.

19 Kiper, T., Korkut, A., Üstün Topal, T. (2017). Implementation of visual landscape analysis regarding eco-tourism planning ın Turkey. Oxidation Communications, 40, 1007–1020.

20 Daniel, T.C. (2001). Whither scenic beauty? Visual landscape quality assessment in the 21st century. Landscape and Urban Planning, 54 (1-4), 267-281.

21 Polat, Z., Acar, C. (2009). Peyzajda neden "görsel kalite analizi" yaparız? Artvin Çoruh Üniversitesi Orman Fakültesi Dergisi, 10 (2), 19–29.

22 Fuante de Val, G., Atauri, A. J., Lucio, J. V. (2006). Relationship between landscape visual attributes and spatial pattern indices: A test study in Mediterranean- climate landscapes. Landscape and Planning, Volume 77, Issue 4, 393–407.

23 Clay, G.R., Smidt, R.K. (2004). Assessing the validity and reliability of descriptor variables used in scenic highway analysis. Landscape and Urban Planning, 66, 239–255.

24 Özhancı, E., Yılmaz, H. (2011). Rekreasyon alanlarının görsel peyzaj kalitesi yönünden değerlendirilmesi; Erzurum Örneği. Iğdır Üni. Fen Bilimleri Enstitüsü Dergisi, 2, 67–76.

25 Kaplan, A., Coşkun Hepcan, Ç. (2004). Ege Üniversitesi Kampüsü "Sevgi Yolu"nun görsel (etki) değerlendirme çalışması. Ege Üniv. Ziraat Fak. Derg., 41 (1), 159–167.

26 Kiper T., Korkut A., Üstün Topal, T. (2016). Kıyıköy örneğinde ekoturizm planlamasına yönelik görsel peyzaj analizinin uygulanması, NKUBAP.00.18. AR.14.10 Nolu Araştırma Projesi.

Assoc. Prof. Dr. Ülkü Duman Yüksel and Ezgi Ceyda Büyüktaşkın

41 Applying the Theory Challenge: Measuring the Environmental Quality in Turkey

Abstract: Monitoring the quality of the environment in cities is considerably important in terms of present and future generations that can live in a healthy and sustainable environment. The main objective of this study is to evaluate the environmental quality of urban settlements in Turkey. The results of the study show while the highest value of the environmental quality of urban areas in Turkey is 76.5 out of 100, the lowest value of environmental quality is 40.9. While the provinces of Tokat, Eskişehir, Çorum, Çanakkale and Kilis have the highest environmental quality values, the provinces of Muğla, Muş, Ardahan, Ordu and Yalova have the lowest environmental quality value. In this chapter (1) current environmental pressures and problems in Turkey have been analyzed, (2) problematic areas and regional differences in terms of environmental quality have been revealed, (3) a tool for local governments, central government and the public to monitor the quality of the environment has provided. In this chapter also, areas with priorities and issues are demonstrated in terms of environmental protection and sustainable development by indicating the quality of the environment of urban settlements in Turkey.

Keywords: Urban environment, Sustainability, Environmental quality, Indicators, Turkey

1 Introduction

1972 published report entitled as "The Limits to Growth" underlined the fact that world's natural resources are limited and (some of them) non-renewable and when it was expressed that a growth in an ignored environment would cause unsustainable level of environmental problems in the future [1], and thus, environmental issues began to be discussed intensively. After the publication of the Brundtland Report in 1987 the interest in environmental issues increased, not only worldwide but also at local level.

Today, 46 years after the publication of the Report, 55% of the world population live in cities, and it is projected that 68% of the world population will be urban in 2050 [2]. As the world continues to urbanize, the impacts of this growth to the environment cannot be ignored. As techno-ecosystems, cities are unsustainable independently and they sustain themselves by consuming natural resources, energy and ecosystem services mostly provided by other areas in their hinterland and imposing environmental pollutants such as wastes, greenhouse gas emissions and pollutants in and beyond their boundaries. Therefore expansion of the cities means more stresses on ecosystems, air, water, land and physical and social infrastructure.

Increasing quality of urban environment is very important in terms of living in a healthy and balanced environment for generations of today and future. Therefore, the management of the environment in urban settlements is very important. Geray [3] refers to importance of evaluation of natural resources required for a healthy and balanced life of human, plant and animal; determination of the problems caused by input-output differences in the production; transportation and consumption processes in the cities; the search for the solutions of the problems. He emphasizes the importance of planning, coordination, communication and control processes for the implementation of these solutions which are among the duties and responsibilities of local governments.

2 Quality of the Urban Environments

Environmental quality is a measure of the health of the environment itself (including the plants and animals it supports), and of the effects it has on the health, comfort and psychological state of the people that inhabit it [4]. According to Marans and Counter [5] "an environment of high quality conveys a sense of well-being and satisfaction to its population through characteristic that may be physical, social or symbolic."

The concept of sustainability, which started to be discussed and discussed in different scales and sectors in developed and developing countries, contributed to the questioning of the balance between urbanization, industrialization and environment, and Verhoef and Nijkamp [6] (2008) state, after that the avalanche of sustainability studies has also called for a renewed interest in urban environmental quality.

On the other hand, the quality of the urban environment is defined as an essential component of quality of life [5], and there is a linear relationship between the two concepts.

As environmental quality is related to many issues such as sustainability and quality of life and public health, urban environmental quality issues are highly interdependent [5]; for example, the environmental impacts of poor urban sanitation are closely related to the land use patterns and public health in cities. Domestic and industrial sources and mainly motorized traffic are responsible for pollutant emissions and noise which decisively affect the people in cities. They affect human health both directly (air, noise, etc.) and indirectly (climate change, biodiversity, etc.). As environmental quality is related to many issues such as sustainability and quality of life and public health, urban environmental quality issues are highly interdependent [7]

It is difficult to identify and evaluate the quality of the urban environment as "optimal." The basic principles should be minimum resource consumption and minimum production of waste from environment point of view. But the correlation among the quality of the urban environment, economic and socio-cultural structures should not be ignored. While the quality of some environmental quality components increases due to economic development (such as water, sewage), some of them are declining due to economic development (such as air quality, water, soil).

Elements of the Urban Environmental Quality

Air, water, land, plants, animals, the built environment, recreation, aesthetics are the elements of the urban environmental quality [4]. It is a heterogeneous group, partly measuring the scale of the problem, and partly the effectiveness of planning decisions and distribution of public services in handling it. The heterogeneity also exists in the approach of the perception of the environment too.

Air

Air pollution adversely affects the quality of life and public health especially in urban centers, therefore, it is "a major public health concern and focus of regulatory activity" [8]. As air pollution decreases the quality of life in cities, but also poses a risk to public health, today, air pollution is in the eighth row among the mortality risk factors, and it is responsible for 2.5% of deaths in developing countries [9].

The impacts of air pollution on human health in urban areas vary depending on amount of the contaminants in the air and duration of exposure to these substances. Correctly identified and known air pollution levels in the settlements is important in terms of identifying and implementing policies aimed at preventing and solving the pollution.

The EPA has set National Ambient Air Quality Standards for six principal pollutants, which are called "criteria" air pollutants. These pollutants are carbon monoxide (CO), lead (Pb), nitrogen dioxide (NO_2), ozone (O_3), particle pollution (PM) and sulfur dioxide (SO_2) [10].

Water

Water is a natural resource that is vital for people. The extent of water resources (their amount and distribution) and their condition (physical, chemical, and biological attributes) are critical to ecosystems, human uses, and the overall function and sustainability of the hydrologic cycle [11].

Secretary-General Kofi Annan on the World Day for Water in March 2002 announced that [12] "an estimated 1.1 billion people lack access to safe drinking water, 2.5 billion people have no access to proper sanitation, and more than 5 million people die each year from water-related diseases—10 times the number killed in wars, on average, each year" and emphasized "even where supplies are sufficient or plentiful, they are increasingly at risk from pollution and rising demand."

The point and diffuse sources of water pollution are manifold, and their temporal and spatial impacts on water quality range from short-term local to long-term global [13]. The sources of water pollution and reasons of water quality decrease are also similar to other environmental pollution in urban areas and urbanization—population growth, urban wastewater, sealing of land surfaces are the most relevant pollutant sources.

The water contaminated by liquid wastes which result from industrial activities contains chemicals that constitute a major health threat to people living in the rural areas as well as in the urban regions. Urban waste waters discharging into rivers without treatment in wastewater treatment plants causes organisms to pass into the water, which is hazardous to human health; pollution of surface waters is a threat to human health. With similar reason, it is necessary to prevent plants and agricultural waste from contaminating surface water.

Land

Land use has generally been considered as a local environmental issue, but it is becoming a force of global importance [14]. In urban areas, after construction, land turns into impermeable surfaces that disrupt ecological cycle and cause their transformation. The characteristic and type of land use determines the quality of the environment in terms of the people as well as other components of the urban ecosystem.

Soil has versatile functions and in urban areas its ecological service, especially the ability to buffer and purify pollutants, is very much needed. However, due to various intensive human activities, soils in urban areas are often subject to fundamental changes, even serious degradation. Their degradation is essentially a process of sacrifice because they provide ecological services, such as being a sink to all kinds of pollutants, at the cost of their own quality. Therefore, as a result, they often are subject to variety of environmental problems.

Infrastructure

Infrastructure is the foundation of a system or organization. Urban infrastructure are facilities that are built for electricity, gas, drinking water, sewage, all kinds of transports, communications, treatment services which are necessary to fulfill and develop vital functions of the settlements. Completed and effective infrastructure service contributes to the quality of life in terms of public health as well as environment.

Due to the infrastructure services of sewerage systems, waste treatment systems and the amount of waste related to the amount of population and urban growth; infrastructure development falling behind population growth and urbanization pose a threat to the livability and environmental quality.

3 Studies on Environmental Quality in Urban Areas

Before describing the studies on urban environmental quality, the need for urban indicators should be explained. In many parts of the world, urban phenomena and processes of urbanization remain poorly documented, understood, and measured. Many cities lack a sustained or systematic appraisal of urban problems, such as loss of public space, environmental impact, and land consumption [15]. Monitoring cities' social, economical, physical and institutional changes/developments, loss and damages from environmental point of view, developing and evaluating urban indicators become a vital issue for healthy, livable and sustainable cities.

Tab. 1: Environmental Management Indicators of Habitat Urban Indicators Program[16]

Subject	Indicator
Settlement	Urban population growth
	Planned settlements
Water	Price of water
	Water consumption
Urban Pollution	Wastewater treated
	Solid waste disposal
	Regular solid waste collection
Disasters	Disaster prevention and mitigation instruments Houses in hazardous locations
Transport	Travel time
	Transport modes
Support Mechanisms	Local environmental plans

There is an urgent need to build environmental indicators at all levels of government (1) to collect useful information on urban environmental conditions and trends; (2) to analyze this information to improve environmental conditions through improving basic services and other urban infrastructure and implementing urban policies and programs.

UN-HABITAT Urban Indicators Program

UN-HABITAT has been a pioneer organization in the collection of urban indicators. In 1991, it initiated the Housing Indicators Program, focusing on monitoring shelter performances. It then became Urban Indicators Program in 1993 in order to focus on a larger range of urban issues. This program which created roots of many national indicator programs focused Habitat Agenda goals in five key urban issues, and environmental management is one of these key issues (Tab. 1)

EPA's Report on the Environment

In 2001, U.S. Environmental Protection Agency (EPA) begun an initiative to assemble an extensive set of environmental indicators and presented these indicators in its Draft Report on the Environment Technical Document, released in 2003. Since then, EPA has revised, updated, and refined this Report in response to scientific developments and feedback [17].

The report is organized around five main chapters: air, water, land, human exposure and health, and ecological condition related to EPA's five strategic goals (Clean

Tab. 2: Environment Indicators of CDB Index [17]

Goal	Indicator
Environment	Levels of air pollution concentrations
	No. and type of noise complaints
	Current levels of household liquid waste disposal
	Current levels of household solid waste disposal
	Current percentage of wastewater subjected to some form of treatment
	Current percentage of BOD removed from wastewater
	No. and extent of disasters over past 10 years

Air, Clean and Safe Waters, Healthy Land, Healthy Communities and Ecosystems, Stewardship and Compliance) [11]. Most of the EPA's environmental indicators are by national and regional levels, not city level.

Cities Data Book

The Cities Data Book: Urban Indicators for Managing Cities was designed firstly for policy makers and urban managers of cities in the Asian and Pacific region in 1999 and published in 2001 from the Asian Development Bank. The study intended to establish city-level information resource persons and a network of city managers who are prepared to devote some resources to using urban indicators and benchmarking in their work, and to trigger reforms in their city environment.

The City Development Index ranks cities in the development spectrum and combines city product with infrastructure, waste management, health and education indicators [17] (Tab. 2).

Earth Day Network – Urban Environment Report Study

Earth Day Network's Urban Environment Report is a study which explores the environmental state of America's largest cities. In 2004 EDN formed an alliance with civic and community organizations to mobilize citizens around local environment issues. As an outgrowth of Campaign for Communities programs EDN has completed a research project, the Urban Environment Report [18]. Its goal was to inform and thereby promote civic participation and engagement in order to bring about change in environmental policies and practices on a community level. Subject Areas of this report was vulnerable population, toxics and waste, air quality, drinking and surface water, quality of life (SITES-Safety, Income, Transportation, Education, Sprawl), parks and recreation opportunities, human and public health, and global warming climate change.

The China Urban Sustainability Index

The China Urban Sustainability Index funded by the Urban China Initiative, is a report into the sustainability of nearly 200 Chinese cities.

An indicator set was developed from the 2011 China Sustainability Index as well as the China Urbanization Index, using the indicator framework set out in the 2011 CSI [19]. The analysis deploys 23 metrics, which cover four categories; economy, society, resources, and environment (Tab. 4). About 185 cities, of varying sizes and at different stages of development were ranked by their level of sustainability from 2005 to 2011 [20].

Environmental indicators: emphasis is placed on environmental indicators through a weighting system. The strength of the CSI Indicator set is that it is a tool to quantify urban growth and development, rather than a static benchmarking tool [19].

4 Case Study: An Analysis of the Environmental Quality in Turkey

Enforcement of environmental law began in the early 1980s, since which time there has been increasing regulation in the environmental sphere. In the process of European Union harmonization, many environmental issues were discussed, the required legal regulations in this matter developed, implementations have been initiated[21].

Turkey has been experiencing environmental pressures due to population growth, industrialization, and rapid urbanization. Some cities undergoing rapid growth are suffering from deficient infrastructure and services, severe environmental degradation, increasing traffic congestion, and poor air quality.

Environmental Indicator Reports and Provincial State of the Environment Reports are the main sources of the National State of the Environment Reports that are published annually. But in these reports each dimension of the environment for each city or area is examined solitarily and thus, these reports cannot go beyond inventory studies.

The main objective of this study is to evaluate the environmental quality of urban settlements in Turkey.

Methodology of the Study

It's a benchmark study for Turkey's environmental quality and for the study, a list of indicators has been established to determine the quality of the environment of Turkey by examining international studies. At this stage, accessible and reliable data were considered in Turkey and Minx et al.'s [22] approach of "Urban Metabolism" is based on it. The approach that addresses the city as an urban metabolism describes city as source as well as sink [22] (Fig. 1). In this approach that handles the city as input-output relationship; the more the city and its hinterland consumed as source and produced waste, environmental quality and livability decline in so much.

Tab. 3: Environmental Indicators of Urban Environment Report by Subject Area (Indicators in City Level Are Presented) [18]

Subject Area	Indicator
Toxics & Waste	Recycling Rate
	% Municipal Solid Waste Recycled by state
	% Municipal Solid Waste to Waste-to-Energy
	% Change in Municipal Solid Waste Generated
Air Quality	Sulfur Dioxide (SO_2) Released
	Nitrogen Oxides (NOx) Released
	Carbon Dioxide (CO_2) Released
	Mercury Released, Pounds
	High Ozone Days, annual # of, Weighted Average
	Short-term Particle Pollution Weighted Average, 24 Hour
	Year-Round Particle Pollution – Pass/Fail
	High Ozone Days – Grade
	Short-Term Particle Pollution Grade
	Year-Round Particle Pollution - Pass/Fail
	EPA Ambient Carbon Monoxide (CO) (8-hr ppm)
	EPA Ambient Lead (PB) Q Max
	EPA Mean Ambient Nitrogen Dioxide (NO_2 ppm)
	EPA Larger Particle (PM_{10}) 24-hr (ug/m^3)
	EPA Mean Ambient Sulfur Dioxide (SO_2 ppm)
	EPA Days over 100 AQI (Warning Level)
Drinking and Surface Water	Drinking Water Quality
	Water Withdrawn
	Water Pollution Summary
	Water Health & Testing Summary
	Water Violation Summary

In the study, for the purpose of determining the quality of the environment, 4 subject areas and 16 indicators were determined. Air quality, water quality, waste and land use are four subject areas in the study. Indicators concerning to the four subject fields in different units normalized and converted into dimensionless value by taking into consideration their maximum and minimum values and the relationship (positive-negative) with the quality of the environment. The arithmetic

Tab. 4: Environment Indicators of the China Urban Sustainability Index

Cleaness	Air Pollution	Concentration of SO_2, NO_2, PM_{10}
	Industrial Pollution	Industrial SO_2 discharged per unit GDP
	Air Qualified Days	Days of air qualified equal or above level II
	Waste Water Treatment	Wastewater treatment rate
	Household Waste Management	Domestic waste treated
Built Environment	Urban Density	Persons per km^2 of urban area
	Mass Transit Usage	Passengers using public transit
	Public Green Space	Area of green space
	Public Water Supply	Public water supply coverage
	Internet Access	Household access to Internet

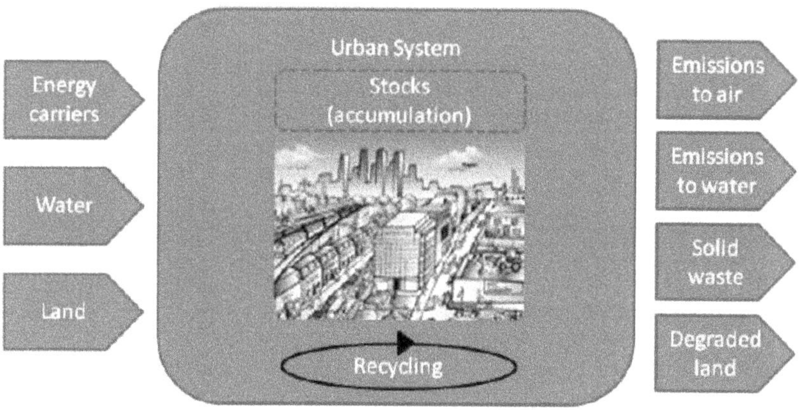

Fig. 1: Schematic Representation of Inputs and Outputs of the City[22].

mean of all subject areas defines the quality of the environment. (The following Tab. 5 shows information related to subject areas and indicators in details.)

Data and Coverage

The data for this study is mostly provided by the Turkish Statistical Institute and the scale of this study is local. Special Provincial Administrations and municipalities

Tab. 5: Environmental Indicators of the Study

Subject Area	Indicator	Unit
AIR	Annual Average Concentrations of PM_{10}	µg/m³
	Annual Average Concentrations of SO_2	µg/m³
	Winter Average Concentrations of PM_{10}	µg/m³
	Winter Average Concentrations of SO_2	µg/m³
	Number of days exceeding the average daily PM_{10} limit value	N
	Number of days exceeding the average daily SO_2 limit value	N
WATER	Daily water withdrawal per capita	lt/person-day
	Ratio of the population to the total population that has access to drinking water service	%
	Daily amount of discharged wastewater	lt/person-day
	Ratio of the population to the total population that has access to wastewater treatment services	%
	Ratio of the population to the total population that has access to sewerage services	%
WASTE	Ratio of the population to the total population that has access to waste services	%
	Solid waste collected per person per day	kg/person-day
LAND	Amount of the Green Area	m²/person
	Population density	person/ha

constitute local government units in Turkey. Municipalities, whose foundation, administrative structure, responsibility, and authority are determined by Law Number 5393, have been selected as the unit of this research since the availability of the reliable data. (Municipalities may be founded in the settlements with the population over 5000.)

Municipalities are the institutions that are directly relate to the management of the Environmental Law No. 5393, which designates authorities and gives them responsibilities. Municipalities are the main units in ensuring the quality of the urban environment.

Today, there are a total of 1397 municipalities, including 30 metropolitan municipalities in Turkey. Total municipal population is 72 million 505 thousand 107 people. Accordingly, (77,695,904 people) 93.3% of the total country population resides in the municipalities [23] (TUIK 2014). This means, the results of the study show in what kind of settlements 97% of the population in terms of environmental quality lives.

Tab. 6: Environmental Quality Value of Urban Areas in Turkey

Tokat	76.5	Kirşehir	68.9	Trabzon	64.6	Zonguldak	59.5
Eskişehir	76.2	Kirklareli	68.9	Isparta	64.5	Mardin	59.2
Çorum	75.7	Kütahya	68.8	Osmaniye	64.5	Hatay	58.7
Çanakkale	74.2	Ankara	68.7	Ağri	64.5	Giresun	57.4
Kilis	73.0	Bilecik	68.5	Niğde	64.4	Düzce	56.4
Nevşehir	72.2	Kastamonu	68.0	Rize	62.8	Aydin	56.3
Sinop	72.0	Burdur	67.2	Şirnak	62.6	Iğdir	56.1
Malatya	71.5	İzmir	67.0	Manisa	62.6	Sakarya	55.1
Elaziğ	71.3	Tunceli	66.7	K.Maraş	62.2	Hakkari	54.8
Kayseri	71.2	Çankiri	66.6	Yozgat	61.9	Karabük	54.8
Bursa	71.0	Bingöl	66.5	Tekirdağ	61.9	Bitlis	53.5
Kirikkale	71.0	Artvin	66.4	Antalya	61.7	Kars	53.5
Uşak	70.6	Erzincan	66.2	Siirt	61.6	Samsun	53.3
İstanbul	70.5	Aksaray	65.8	Van	61.5	Yalova	52.9
Karaman	70.4	Balikesir	65.4	Afyon	61.2	Ordu	52.6
Denizli	70.1	Sivas	65.3	Bolu	61.1	Ardahan	51.9
Konya	69.8	İçel	65.2	Şanliurfa	61.0	Muş	48.9
Amasya	69.5	Gaziantep	65.2	Gümüşhane	60.9	Muğla	40.9
Adana	69.3	Erzurum	65.0	Kocaeli	60.0		
Bayburt	69.3	Adiyaman	64.9	Batman	59.8		
Diyarbakir	68.9	Edirne	64.8	Bartin	59.7		

Findings

The results of the study show that while the highest value of the environmental quality of urban areas in Turkey is 76.5 out of 100, the lowest value of environmental quality is 40.9. While the provinces of Tokat, Eskişehir, Çorum, Çanakkale, and Kilis have the highest environmental quality values, the provinces of Muğla, Muş, Ardahan, Ordu, and Yalova have the lowest environmental quality value (Tab. 6).

The results of this research demonstrate the quality of the environment in urban areas in Turkey in accordance with each other (in some indicators) and in accordance with international standards.

When examining the environmental quality values, it is observed that the environmental quality in the 18 provinces municipalities is at a very high level, in 32 provinces in the high level, in 20 provinces in the middle level, in 10 provinces in the low level, and it is seen that the quality of the environment in 1 province is in the very low level.

When the geographical distribution of environmental quality is examined, it is observed that areas with low environmental quality concentrated in the east, southeast and southwest of the country and areas with high environmental quality are concentrated in inland areas (Fig. 2).

When the city of Tokat which has the highest quality of the environment is examined, it is seen that its most strong side is water quality, its weakest side is land quality. Muğla's weakest side which has the lowest environment quality in Turkey is water quality; its strongest side is the waste services.

In terms of environmental quality in Turkey, the lowest total average belongs to land quality. The land subject area's values demonstrate the conditions of urban areas in terms of ecosystem services as well as the pressure of population on the land due to urbanization.

Water quality is an indicator that shows the pressure on water resources and is low compared to other subject areas in Turkey. The solution developed to meet the increasing demand of water in Turkey so far has been the supply of new resources. The results of the study show that new strategies should be developed to ensure efficient, rational, and sustainable use of water resources with demand management approach in Turkey. Arlosoroff [24] indicates that ensured reduction of 30%-50 % in water consumption with demand management does not cause any change/reduction in the quality of life. The technical infrastructure directly affects the public health in terms of pollution of natural resources, and this issue contrary to the air quality is directly among the duties and responsibilities of local authorities. When the waste subject area is examined, it is observed that the technical infrastructure for waste is not adequate, especially in the north and northeastern Anatolia and southeastern Anatolia regions.

The air quality factors decrease the quality of the environment in most cities. For the improvement of air quality, Turkey has created and continues to create a number of legal mechanisms in the framework of the EU integration process and is in the process of combating climate change. However, it has been demonstrated by analysis that these are not sufficient; there is a need to increase the mechanism of inspection and control.

5 Conclusion

The environment can no longer be viewed simply as a remote or theoretical concept when a host of issues continue to plague the nation's urban centers such as poor air quality; unsafe drinking water; lack of access to public green, open, or park spaces; a variety of human health problems; crime; and numerous others [18].

There is considerable work done by international organizations and institutions related to the quality of the environment until today. While some of these works are intended for measuring directly the quality of the environment, some other parts handle the quality of the environment as a component of life quality or

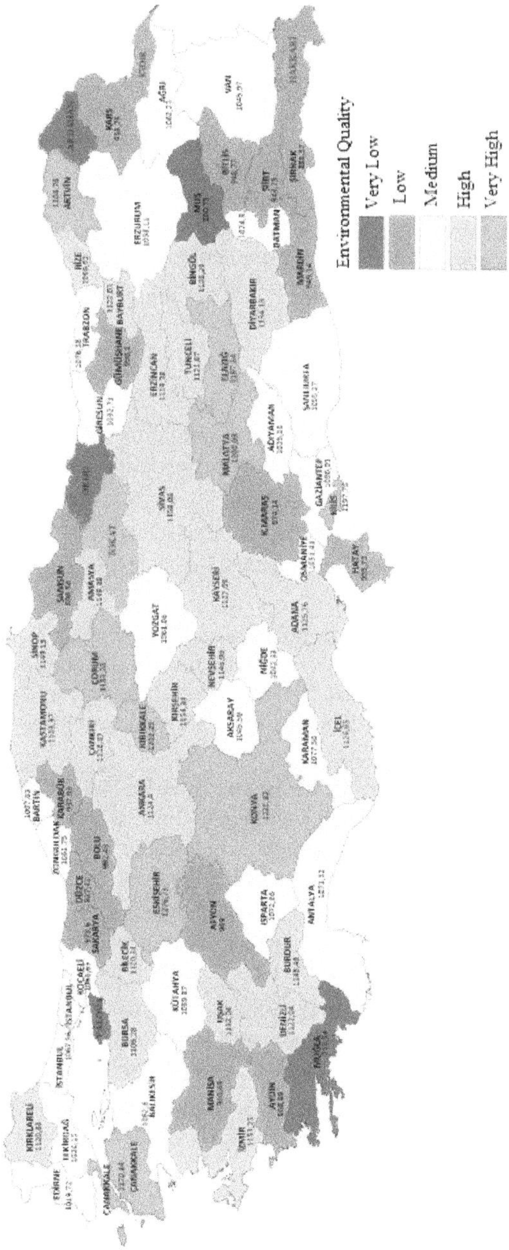

Fig. 2: Environmental quality of urban areas in Turkey.

sustainability and livability. The common point of these studies is involving different indicators and different subject areas depending on the scope of the study.

Monitoring the quality of the environment in cities is considerably important in terms of present and future generations that can live in a healthy and balanced environment. With this study (1) current environmental pressures and problems in Turkey are analyzed, (2) problematic areas and regional differences in terms of environmental quality are revealed, (3) a tool for local governments, central government and the public to monitor the quality of the environment is provided. In this study also, areas with priorities and issues are demonstrated in terms of environmental protection and sustainable development by indicating the quality of the environment of urban settlements in Turkey. However, some basic data to perform such studies are found to be missing in Turkey.

References

1 Meadows, Donella H., et al. The limits to growth: A report to the club of Rome (1972). Google Scholar, 1972. Universe Books, New York, ISBN: 0-87663-165-0

2 UNDESA., 2018 Revision of World Urbanization Prospects (2018) [cited 2018, 12 October]; Available from: https://population.un.org/wup/

3 Geray, Cevat. Kentsel Yaşam Kalitesi ve Belediyeler. Türk İdare Dergisi, 1998, 70.421: 323–346.

4 Rabinowitz, Phil, Protecting Environmental Quality, Chapter 26, Section 9, The Community Tool Box, The Center for Community Health and Development at the University of Kansas, [cited 2018, 2 October]; Available from: https://ctb.ku.edu/en/table-of-contents/implement/physical-social-environment/environmental-quality/main

5 Van Kamp, Irene, Leidelmeijer, K., Marsman, G., & De Hollander, A. "Urban environmental quality and human well-being: Towards a conceptual framework and demarcation of concepts; a literature study." Landscape and Urban Planning 65.1–2 (2003): 5–18.

6 Verhoef, Erik, and Nijkamp P. "Urban environmental externalities, agglomeration forces, and the technological 'deus ex machina'." Environment and Planning A 40.4 (2008): 928–947.

7 NRTEE (National Round Table on the Environment and the Economy), The State of the Debate on the Environment and the Economy: Environmental Quality in Canadian Cities, 2003: Ontario, Canada, ISBN 1-894737-08-3

8 WHO. The world health report 2002: Reducing risks, promoting health life. 2002: Geneva, World Health Organisation: 248. ISBN 92 4 156207. ISSN 1020-3311.

9 Narayan, KM Venkat; Ali, Mohammed K.; Koplan, Jeffrey P. Global noncommunicable diseases—where worlds meet. New England Journal of Medicine, 2010, 363.13: 1196–1198.

10 US Environmental Protection Agency EPA National Ambient Air Quality Standards, 2016 [cited 2016, 15 August]; Available from: https://www.epa.gov/criteria-air-pollutants/naaqs-table

11 US Environmental Protection Agency, Water in: Report on the Environment [cited 2018, 1 November]; Available from: https://www.epa.gov/report-environment/water

12 The message of Secretary-General Kofi Annan on the World Day for Water, observed 22 March 2002: SG/SM/8139-OBV/262 [cited 2018, 20 November]; Available from: https://www.un.org/press/en/2002/sgsm8139.doc.htm

13 Schwarzenbach, René P. et al. Global water pollution and human health. Annual Review of Environment and Resources, 2010, 35: 109–136.

14 Foley, Jonathan A. et al. Global consequences of land use. Science 309.5734 (2005): 570–574.

15 Bilsky, Edgardo; Ochoa, Karina Blanco; Osorio, Efrén. Indicators for measuring urban sustainability and resilience. The Urban Planet: Knowledge Towards Sustainable Cities, 2018, 163.

16 United Nations Human Settlements Programme, Urban Indicators Guidelines Monitoring the Habitat Agenda and the Millennium Development Goals, 2004:Nairobi, Kenya, Available from: https://unhabitat.org/urban-indicators-guidelines-monitoring-the-habitat-agenda-and-the-millennium-development-goals/ , 2018, 15 October

17 The Asian Development Bank 2001, The Cities Data Book: Urban Indicators for Managing Cities- Matthew S. Westfall and Victoria A. de Villa (Editors), Publication Stock No. 090100,505p, 13-14 ISBN 971-561-312-8, Manila, Philippines

18 Earth Day Network, Urban Environment Report Methodology 2008 [cited 2008, 20 October].

19 Science for Environment Policy, Indicators for sustainable cities. In-depth Report 12. Produced for the European Commission DG Environment by the Science Communication Unit (2018): UWE, Bristol. Available at: http://ec.europa.eu/science-environment-policy

20 Li et al., The China Urban Sustainability Index, The Urban China Initiative, 2014, Available from:http://www.mckinseychina.com/wp-content/uploads/2014/04/china-urban-sustainability-index-2013.pdf, 2018, 14 September

21 TC. Çevre ve Şehircilik Bakanlığı, Türkiye Çevre Durum Raporu/2011, ÇED İzin Denetim Genel Müdürlüğü Yayın no: 11, Altan Matbaacılık, 2011,Ankara. ISBN: 978-605-5294-01-4

22 Minx, J.C., Creutzig, F., Medinger, V., Ziegler, T., Owen, A. and Baiocchi, G., Developing a prgamatic approach to assess urban metabolism in Europe—A report to the environment agency prepared by Technische Universität Berlin

and Stockholm Environment Institute, Climatecon Working Paper 01/2011, 2011, Technische Universität Berlin.

23 TUIK, Turkey in Statistics, 2015, Available from: https://ec.europa.eu/eurostat/documents/7330775/7339623/Turkey+_in_statistics_2015.pdf/317c6386-e51c-45de-85b0-ff671e3760f8

24 Arlosoroff, Saul. Water demand management. In: International symposium on efficient water use in urban areas, IECT-WHO, 1999, Kobe, Japan.

Prof. Dr. Fatih Aydın and Assist. Prof. Dr. Yasin Dönmez

42 Investigating the Seismicity of Urban Areas Using Definite Parametres: City Center of Karabuk

Abstract: Many factors are considered in the planning of urban areas. One of these factors is earthquake zones. In this study, geology, distance to fault lines, information about topography and groundwater, with the help of geographic information systems, is aimed to determine suitable settlement areas. When the earthquake damage risk map of the study area is examined, the middle and western sections of the region are High Risk, the middle sections of the region, along the western boundary, the south section and, to a lesser extent, the eastern section, High Risk, the south-eastern part of the western part is less risky and the eastern part of the zone is the least risky area. In this study, the analysis was conducted by using ArcGIS 10.1 Geographic Information System software, an earthquake risk map was constructed and measures to be taken against the damage caused by a possible earthquake were assessed.

Keywords: Urban areas, Earthquake, Damage risk analysis, Karabük, GIS

1 Introduction

Due to the increase in human population, rapid urbanization gained a global dimension. Today, nearly half of the world's population lives at urban areas. The urban population is increasing at a faster pace than the overall world population and it is estimated that until 2030, more than 60 % of the global population would live in the cities. These dramatic changes in the global urbanization process have found its reflection also in the case of Turkey. Urbanization, which had become visible in the1950s, has now gained its pace in Turkey. Karabük is one of the places that rapid urbanization took place, dependent upon its appropriate geographical conditions, dense population and the industrialization attempts after the formation of the Turkish Republic [1,2,3].

Unplanned and rapid urbanization, which are consequences of the global changes, led the Turkish authorities to overlook the fact that earthquakes may hit many of the Turkish cities. Various national and international studies have proven that the earthquakes may have physical, chemical, biological, hydrological, and ecological negative effects on urbanization [4–10].

In addition to the unplanned urbanization, recent studies have been conducted on solutions in order to minimize the effects of earthquakes that may lead to the destruction of cities. Earthquake is an event in which seismic waves caused by the energy that occurs in the earth crust shake the terrain. Another theory on the earthquakes proposes that earthquakes are caused by elastic rebounds. Elastic Rebound theory, which was first introduced following the great 1906 San Francisco earthquake, proposes that an earthquake is a result of the elastic rebound of previously stored strain energy [11–13].

The necessity of more efficient urban planning has become more important. All factors should be taken into consideration while preparing urban plans. Within this context, prior analysis of the damage that may be caused by natural disasters such as the earthquake, is highly important to make appropriate decisions. With this aim, crucial studies have been conducted by using geographical information systems (GIS) techniques. For example, GIS techniques were used to prepare the ecological sustainability map of Lanzhou city in China and to determine the landslides that may harm the urban ecology [14–17].

According to the seismic zones map prepared by the Ministry of Public Works and Settlement in 1996, 81 % of the total area of Karabük province is located in the 1st degree earthquake zone and 19 % is found in the 2nd degree earthquake zone. On the other hand, 93 % of the total Karabük population lives in the 1st degree earthquake zone whereas 7 % lives in the 2nd degree earthquake zone.

Karabük central district and the districts of Safranbolu, Eskipazar, and Yenice are located in the 1st degree earthquake zone whereas the districts in northern Karabük, namely Eflani and Ovacuma, are found in the 2nd degree earthquake zone. The active North Anatolian fault and the Karabük tectonic fault located at the north of the province are the tectonic faults that may lead to an earthquake in the region. Gerede segment of the North Anatolian Fault System starts in Gerede, which is 55 km away from the province center, and passes from Eskipazar, İsmetpaşa, and Kastamonu-Tosya at the provincial borders. The second tectonic fault passes from Amasra, Abdipaşa, Safranbolu, Karabük provincial center and Eskipazar districts of Karabük and Bartın provinces [18,19].

2 Materials and Methods

The area of study comprises the region within the borders of the provincial center of Karabük province. The coordinates of the study area are 40^0 57'–41^0 34'N and 32^0 04'–33^0 06' E (Fig. 1). The neighboring provinces are Kastamonu at the east, Çankırı at the southeast, Bolu at the southwest, and Zonguldak at the west. Tab. 1 shows the distances between the provincial center, neighboring provinces, and the major Turkish cities.

The neighboring districts of the provincial center, the study area, are Safranbolu at the east and northeast, Ovacık at the southeast, Eskipazar at the south, and Eflani at the west. The provincial center is located at the Ankara-Zonguldak railroad, where the Araç and the Soğanlı rivers meet each other. It is 230 km from Ankara, the capital city of Turkey, and 400 km from Istanbul (Tab. 1).

The major factor that causes earthquakes in the study area is the North Anatolia Fault (NAF). The NAF is 55 km away from the provincial center. Gerede segment of the NAF Zone that has been leading earthquakes in the study area starts at Gerede district of Bolu province and passes from the İsmetpaşa site of Eskipazar district and Tosya district of Kastamonu provinces at the borders of Karabük province.

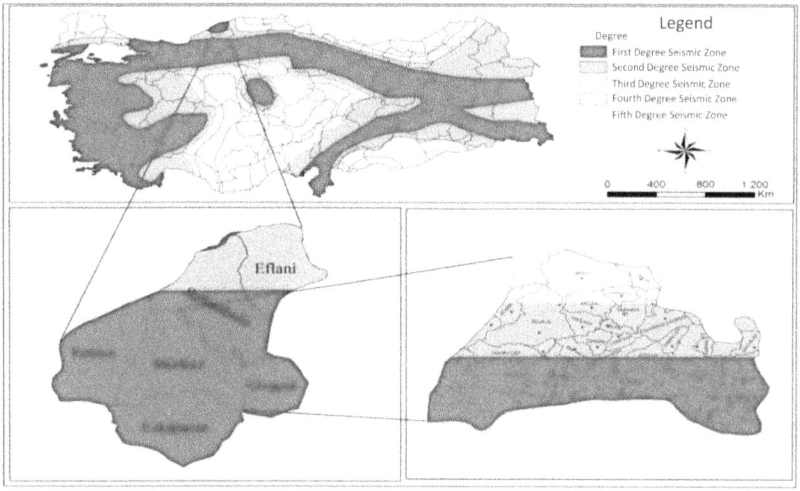

Fig. 1: Location map of the study area.

Tab. 1: Distances of the Study Area to the Neighboring Provinces and Major Cities

Provinces	Distances	Districts	Distances
Ankara	217 km	Eskipazar	35 km
Istanbul	385 km	Eflani	48 km
Çankırı	195 km	Ovacık	37 km
Kastamonu	125 km	Yenice	30 km
Zonguldak	167 km	Safranbolu	10 km
Bolu	132 km		
Bartın	87 km		

The NAF has caused disastrous earthquakes throughout time. There are various records of earthquakes that dates back to the period before 1900, the year that earthquakes started to be monitored and measured. Tab. 2 shows the characteristics of the earthquakes that hit the study area in history.

According to Topay and Parladir, during the planning process, the GIS software products may be used in the complex process of determination, storage, analysis, collection, management, and presentation of the data on natural and cultural characteristics of the study area [20,21].

GIS tools effectively support the decision-making process and help the researchers to conduct conformance analysis. Furthermore, the GIS tools enable

Tab. 2: Earthquakes That Occurred in the Gerede Segment of the NAF (Karabük Provincial Disaster and Emergency Directorate)

Location	Coordination		Moment Magnitude
	Latitude	Longitude	
Gerede	31.05	40.5	-
Bolu-Gerede	-	-	-
Gerede-Çerkeş	-	-	-
Tosya-Çankırı	-	-	-
Bolu-Kastamonu	40.70	31.60	VIII
Bolu-Kastamonu	41.30	33.80	VII
Kastamonu-Çankırı	41.20	33.80	VII
Çankırı	40.60	33.60	V
Çankırı	40.60	33.60	VIII
Tosya-İskilip -Kastamonu	41.00	34.00	VI
Kastamonu-Çankırı	41.00	33.70	VI
Çankırı			VI
Kastamonu	41.30	33.80	VI

the researchers to apply models developed in accordance with the parameters that are effective in the planning process and to make projections for the future [22–25]. In this study, we also used Arc GIS 10.1 GIS software.

This study used factors such as geological structure, groundwater levels, slope, and distance to faults in order to reveal the earthquake risks for provincial center of Karabük. These factors have been analyzed by using GIS software and earthquake damage risk map has been produced by taking the weight value of the factors on impact value for earthquake. The study was concluded in two main stages. In the first stage, the effective factors were classified and were assigned values. We analyzed the following factors:

- Damage Risk and Fault Relationship,
- Damage Risk and Geology Relationship,
- Damage Risk and Ground Water Relationship,
- Damage Risk and Slope Relationship.

The factor of geology is the factor with the highest effect, and consequently, it has been assigned the GIS weighted value of 10. We assigned the GIS weighted value of 8 to the fault factor. Given that nearly half of the study area has a slope over 50 %, slope was considered as an important risk factor and was assigned the weighted value of 6. Since the number of drilling in the study area is few, we do not have adequate information about the depth of the underground water. Therefore, we assigned 4, the lowest weighted value, to the factor of groundwater.

During the second stage, all effective factors were analyzed in relation to each other and were graded. Distribution of these effective factors in the study was shown on maps and all these factors were analyzed by using GIS for their reciprocal relationship (Tab. 3).

3 Findings and Discussion

3.1 Damage Risk and Fault Zone Relationship

When other factors are kept constant, the risk of damage that may be caused by the earthquake decreases from the hypocenter of the earthquake towards the periphery. Therefore, possible damage risk of an area to be caused by an earthquake is primarily related with the distance of the area to the faults [26].

The primary structural risk factor that may cause seismicity in the study area is the NAF. The NAF is 55 km away from the provincial center of Karabük. Gerede segment of the NAF Zone starts at Gerede district of Bolu province and passes from the southern borders of Karabük province, from İsmetpaşa site of Eskipazar district and Tosya district of Kastamonu province (Fig. 2).

The distance between Mehterler at the southern edge and Başköy at the northern edge of the study area is nearly 36 km as the crow flies. Nearly 100 % of the area is found in the 1st degree earthquake zone and is under the influence of the NAF. Consequently, any possible earthquake that may occur in the NAF is able to seriously damage the study area.

The distance between the nearest and the furthest points of the study area to the fault zone is about 36 km. The distance between the fault and the nearest point of the study area to the fault is nearly 19 km. On the other hand, the distance between the fault and the furthest point of the area to the fault is about 73 km. Consequently, the damage to be caused by a possible earthquake will not be similar at all points of the study area. Due to this situation, while forming the relationship between risk and the fault, the provincial center has been divided into three subareas with a length of 12 km per area, which extended from north to south (Fig. 3).

In case of a possible earthquake at a point that is close to the NAF, the point will be close to the hypocenter of the earthquake. Consequently, the energy released from the hypocenter will be higher and the rate of damage will increase. Therefore, while preparing the earthquake risk map for the study area, closeness to the fault zones is considered as the second most important factor other than the lithologic characteristics and is assigned the weight value of 8.

When the other factors that may influence damage risk are kept constant, the southern region of the area is considered as the highest risk zone, the middle region is the middle risk zone, and the northern region of the study area is considered as the low-risk zone.

Tab. 3: Effective Factors, Classes, GIS Value Classes, and GIS Weighted Values

Name of Effective Factor	Factor Classes	GISValue Class	GISWeightedValue
Lithology	Talus and other alluviums	10	10
	Terrace Sediments 2	8	
	Terrace Sediments 1 (Açma Formation)	6	
	Limestone-marl-siltstone(Armeşe Formation)	4	
	Pebble-Nummulitic limestone (Küpler Formation)	3	
	Sandstone-siltstone-limestone (Gazez Formation)	2	
	Nummulitic limestone (Hastarlı Formation)	1	
Distance to fault (km)	36–48	5	8
	48–60	3	
	60–72	1	
Slope (%)	0–10	1	6
	10–20	2	
	20–40	3	
	40–60	4	
	60 >	5	
Depth of groundwater (m)	0–5	1	4
	5–10	2	
	10–15	3	
	15–20	4	
	20–25	5	

3.2 Damage Risk and Geology Relationship

In 1999, a disastrous earthquake caused great damage to Avcılar district of Istanbul even though the district was 110 km away from the hypocenter of the earthquake, known as the Gölcük earthquake. The event was explained with reference to liquefaction on the ground. Due to this reason, higher levels of groundwater and the wide-spread lithology that is suitable for liquefaction will be influential over the impact of a possible earthquake in the study area [26].

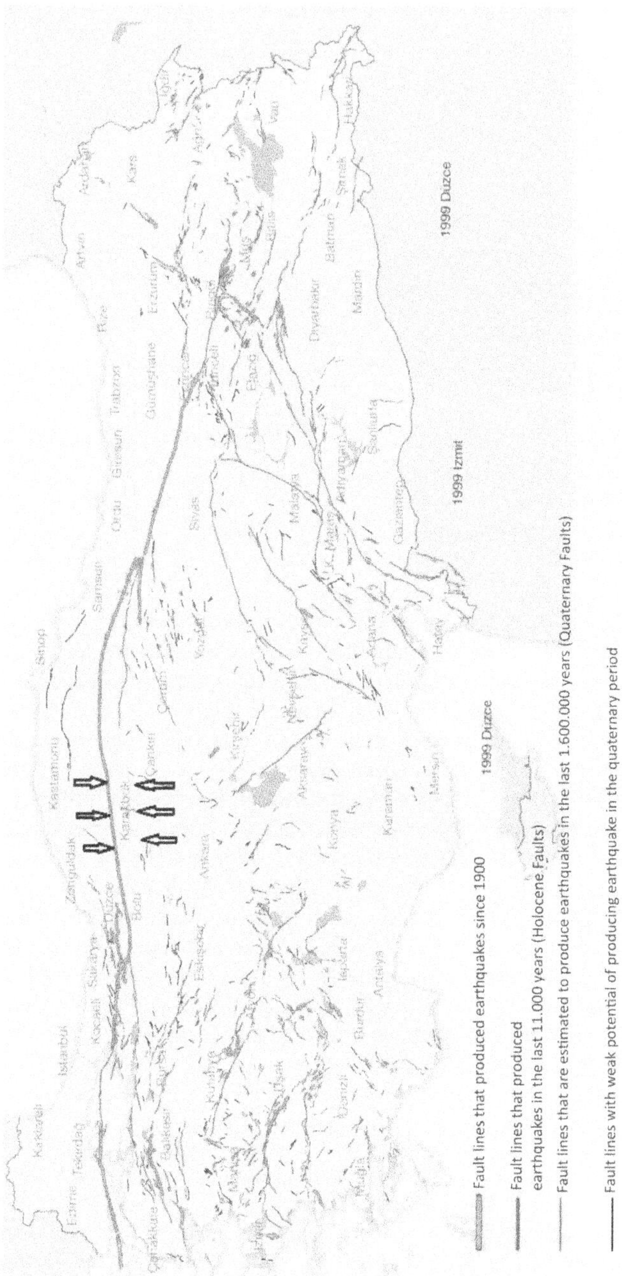

Fig. 2: Active fault zones in Turkey (MTA).

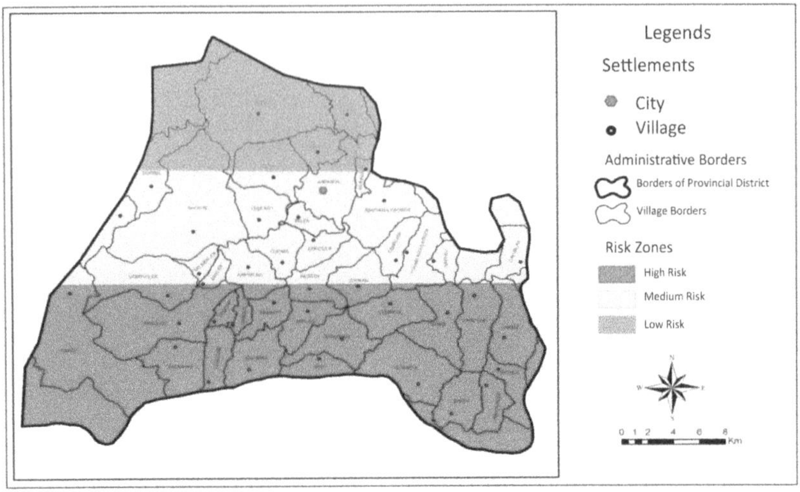

Fig. 3: The position of the study area according to the NAF.

Soil liquefaction describes a physical process whereby saturated sediments of sand, clay, or silt lose strength temporarily and move like a liquid. As a consequence of the collapses of the gaps between soil grains, the pressure of the groundwater increases, soil resistance decreases, and the soil behaves as if it is a liquid. Liquefaction is mostly observed in areas with sandy soils or where the groundwater is close to the surface. Due to this reason, fill areas, river and lake banks, and the beds of ancient rivers and lakes are prone to soil liquefaction [26–28].

In our study area, talus is found at the middle of the study area. Formation and Küpler Formation meet talus. On the other hand, the alluviums are located nearby Araç, Filyos, and Soğanlı rivers, which divide the study area into two parts. Terrace sediments are found at the south of Araç River, nearby Yenişehir and Yeşil neighborhoods, at the south of the area where Filyos and Soğanlı Rivers meet, and at the south of Yenice neighborhood. Since the areas listed above are suitable for solid liquefaction, they are under higher risk in case of an earthquake. Due to this reason, while preparing the earthquake risk map, lithology is assigned as 10, the highest weight value of the GIS.

3.3 Damage Risk and Groundwater Relationship

One of the most important factors that facilitate liquefaction and landslides after earthquakes is the depth and volume of groundwater [26]. Due to the effects of the surface water and groundwater over the surface and the rocks, while selecting an appropriate settlement area that will be less affected from the earthquakes,

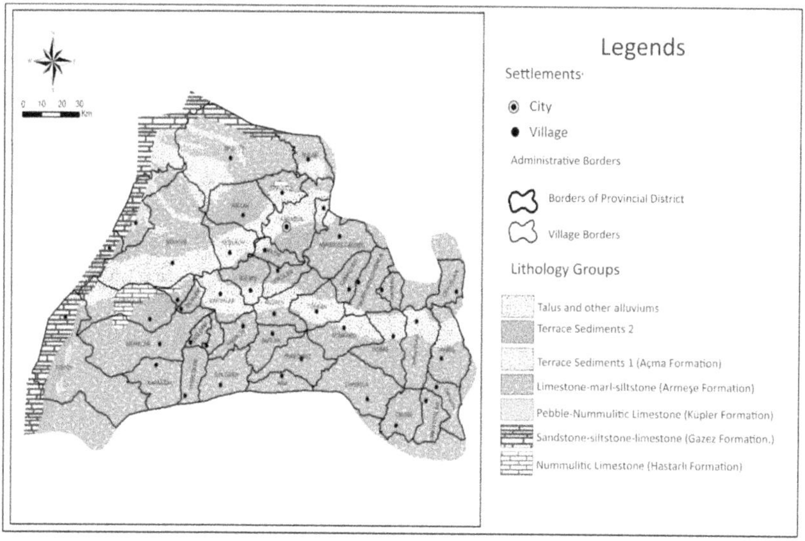

Fig. 4: Lithology map of the study area.

the analysis of the hydro geological characteristics of the area is essential. The amount of the rainfall leads to a change in the groundwater levels, creates rain-wash, contributes to the partial or full soil saturation, and massive movements of the ground by increasing the pressure of the groundwater. Surface water increases the density of the surface. Furthermore, surface water and groundwater lead to chemical changes in the components and the rocks and decrease their resistance.

The analysis of the hydro geological characteristics of the terrain and the rocks of the study area shows that the terrace debris and talus near to the surface include water depending on the seasonal rain falls. However, this percolating water leaks in short term. On the other hand, Yörük Formation, which occupies a great area, is covered with clay sediment so that the basin has an impermeable ground. Map of the study area's ground water depth (Fig. 4) has been prepared by using data obtained from the drilling activities previously conducted in the region. In the study area, nearly 20 drilling activities have been conducted. The deepest drilling descended as deep as 25 m whereas the shallowest drilling reached down to 6 m (Fig. 5).

The analysis of the map showing the groundwater levels reveals that the depth of the groundwater changes parallel to the height and the distribution of river basins. The depth of the groundwater is the lowest, between 0-to-5 and 5-to-10 meters in the southeast, southwest, and partially the east of the study area. On the other hand, the depth of groundwater increases in the middle and northern regions of the study area. In these regions, depth of underground water starts at

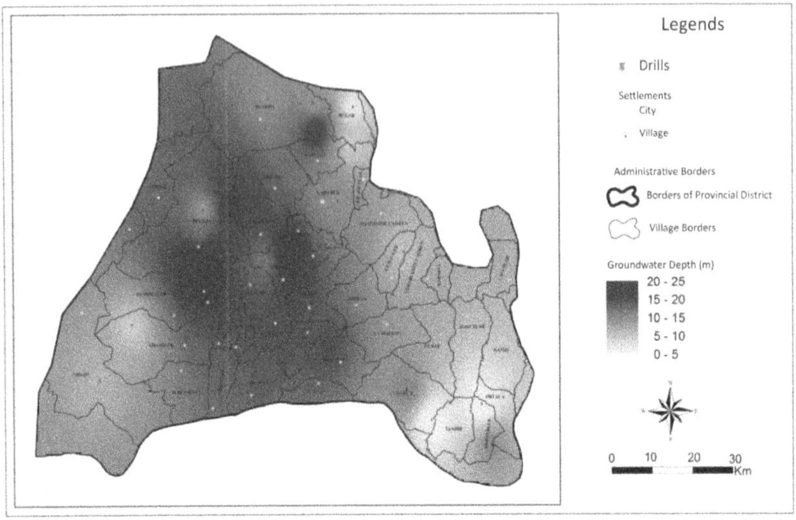

Fig. 5: Groundwater depth map of the study area.

10 meters and may increase to 25 meters. Since sufficient drilling activities were not conducted in the research site, we do not have accurate information about the depth of the groundwater. Due to this reason, we assigned the GIS weighted value of 4 as the factor of groundwater.

3.4 Damage Risk and Slope Relationship

Overall slope in the research area of Karabük provincial center is very high. The slope in the provincial center, the ridges between valleys, and the flood plains ranges between 0%–10 %. Easy hills have a slope of 10–20 %, whereas the slope of the valley slopes is higher than 20 % (Fig. 6).

The analysis of the slope map of the study area shows that the slope is very high at the west, north, northwest regions, and relatively high at the southwest and southeast regions. On the other hand, the slope is low in the middle of the study area, where the neighborhoods are concentrated. Additionally, the slope is low in the regions where alluvium carried by the Filyos, Araç, and Soğanlı rivers settle down.

Nearly 30 % of the study area has a slope of 0–10 degrees, 20 % has a slope ranging between 10–20 degrees, 15 % has a slope ranging between 20–40 degrees, 20 % has a slope of 40–60 degrees, and 15 % has a degree more than 60 %. In areas with higher slope levels, a possible earthquake may hit the areas with clayed land and cause damage by triggering landslide. In our case, a possible earthquake in

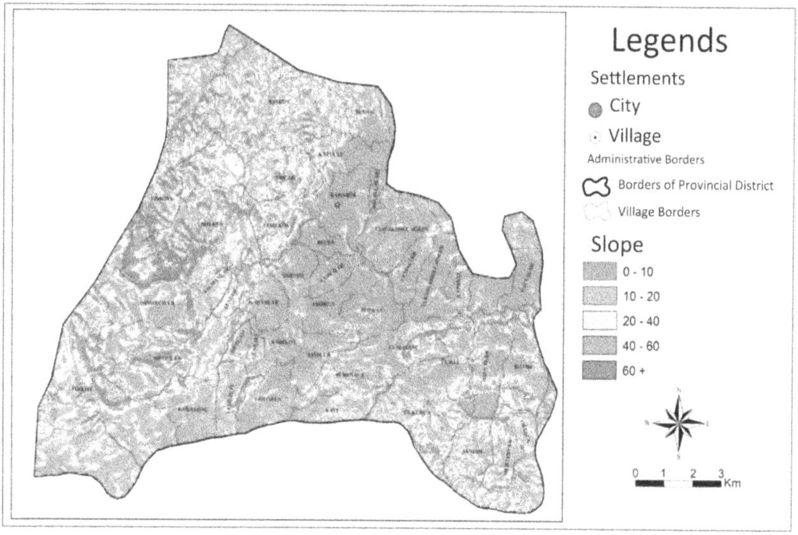

Fig. 6: Slope map of the study area.

the study area, which has high slope, may trigger damage risk. Due to this reason, while conducting the damage risk analysis for the study area, we assigned the GIS weighted value of 6 as the factor of slope.

4 Conclusions and Suggestions

Fig. 7 provides the damage risk map in case of a possible earthquake that may hit the provincial center and was prepared by using the data evaluated.

While rating and analyzing the factors that are effective, we used the GIS that placed special emphasis to the use of objective standards in order to analyze the sensitivity of the study area to a possible earthquake.

The analysis of the constructed earthquake damage risk map shows that the middle regions of the study are under very high risk, western borders of the area and the southern regions are under high risk, and, northeast region is the lowest risk zone. Fig. 7 demonstrates the earthquake risk map. Based on the findings of this research, we listed the following suggestions about the study area:

- Earthquakes are inescapable. Today, conscious and effective methods, which are successfully used by the countries where earthquakes frequently occur, may help the authorities to decrease the damage caused by earthquakes. One possible method is to monitor the changes in the areas close to the active fault systems and to use these changes in nature to evaluate and predict the possibility

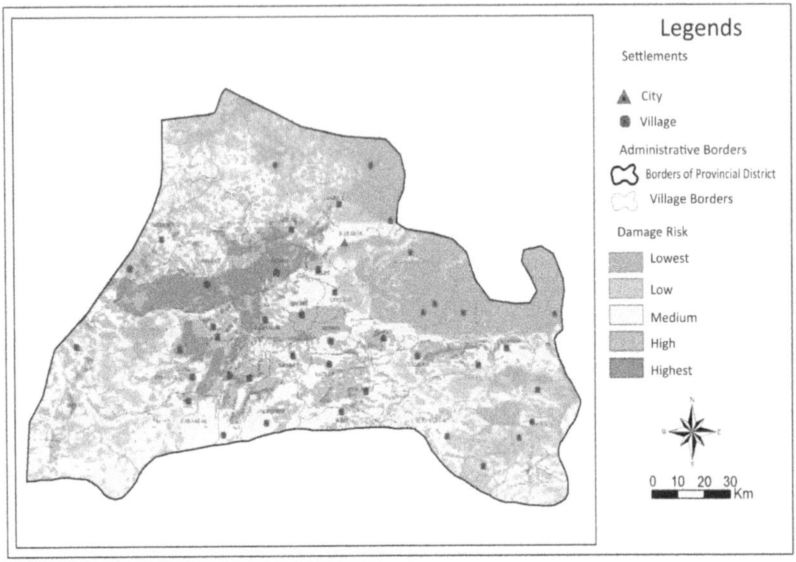

Fig. 7: Earthquake damage risk map of the study area.

of an impending earthquake. Since earthquakes are uncontrollable, the authorities should take measures in order to mitigate possible human life losses and other kinds of damage. The most important and visible natural changes that cause an alarm about an earthquake are the vertical and horizontal movements inside the earth crust. Geodetic studies are among the methods to determine these movements. Turkey, which can be considered as a natural laboratory in terms of earth movements, should support geodetic studies in order to predict possible earthquakes.

- In order to determine the areas under high risk in case of a possible earthquake, the risk potential of the buildings and infrastructure should be analyzed at neighborhood level, new projects should be developed in order to decrease the risks that an earthquake may create, and necessary measures should be taken in the short, medium, and the long-terms.
- During the preparation of urban plans, the earthquake risks should be considered while opening an area for settlement. In case of the settlement areas, the regions under high risk should be showed on a map and secure area usage and settlement decisions should be taken based on these maps. Finally, urban renewal projects should be supported.
- The construction quality of most of the buildings in the study area is low or the life cycle of the buildings has expired. An important share of the structural risks in the study area stems from the fact that the standards and the building

regulations are not taken into consideration during the construction process. In order to prevent loss of life and property in case of a possible earthquake, the buildings under risk should be reinforced or reconstructed.
- Buildings should not be constructed on areas nearby rivers and on former river beds, where soil liquefaction may take place. Before the constructions, soil investigation should be carried out and necessary measures should be taken.
- Public buildings under intense use, such as hospitals and schools, should be reinforced. Analysis of the structure of such buildings shows a serious threat of collapse.
- Conferences and seminars should be conducted in order to inform the public about disaster prevention. In order to increase individual and public consciousness about earthquakes, the primary role should be assigned to the formal education institutions. On the other hand, institutions that provide non-formal education should educate the masses that are excluded from the scope of formal education system.
- A financial assistance plan should be developed in order to help the citizens, who intend to construct earthquake-resistant buildings. Rates of credits provided with this purpose should be lower and real estate taxes should be re-evaluated in order to help the building contractors who construct earthquake-resistant buildings.
- In order to enable sufficient communication in case of an earthquake in the area, an institutional disaster management system should be developed.
- Given the fact that nearly all of the study area is at the 1st degree earthquake zone, earthquake recording networks should be increased and early warning and control systems for natural disaster reduction should be established.
- Knowledge levels of the personnel who are responsible in case of an earthquake should be improved; centers of equipment should be established in the region and the critical equipment should be stored.
- The characteristics of the public and private buildings and parks should be evaluated in order to determine the conditions of the meeting grounds after a possible earthquake. Appropriate places should be selected and planned for emergency situation after an earthquake.

References

1 Paul, M. J., Meyer, J. L., *Streams in the Urban Landscape*, Annual Review of Ecological Systems, Volume: 32, pp.: 333–365, 2001.

2 Erdem, M., Serdar Kaya, H., *İstanbul'un Dere ve Nehirlerinin Ekolojik Tasarım Yaklaşımları Doğrultusunda Irdelenmesi*, Coğrafyacılar Derneği Yıllık Kongresi Bildiriler Kitabı, S.: 390–398, 2013, Istanbul.

3 Özşahin, E. ve Kaymaz, K., *Cbs ve Ahs Kullanılarak Doğal Çevre Bileşenleri Açısından Kentsel Mekânın Yerleşime Uygunluk Analizine Bir Örnek: Antakya (Hatay)*, Doğu Coğrafya Dergisi - 33, s:111–134, 2015.

4 Hynes, H. B. N., *The Biology of Polluted Waters*, Liverpool University Press, 1960, Liverpool, UK.

5 Klein, R. D., *Urbanization and Stream Quality Impairment*, Water Resources Bulletin, Volume: 15, pp.: 948– 963, 1979.

6 Porcella, D. B., Sorensen, D. L., *Characteristics of Non-Point Source Urban Runoff and Its Effects on Stream Ecosystems*, Epa-600/3–80–032, Epa, Washington, Usa, 1980.

7 Duda, A. M., Lenat, D. R., Penrose, D. L., *Water Quality in Urban Streams-What We Can Expect*, Journal Water Pollution Control Federation, Volume: 54, pp.: 1139–1147, 1982.

8 Ellis, J. B., Marsalek, J., *Overview of Urban Drainage: Environmental Impacts and Concerns, Means of Mitigation and Implementation Policies*. Journal of Hydraulic Research, Volume: 34, pp.: 723–731, 1996.

9 Baer, K. & E., Pringle, C. M., *Special Problems of Urban River Conservation: the Encroaching Megalopolis. In Global Perspectives on River Conservation: Science, Policy, And Practice* (edit.: boon, p. J., davies, b. R., petts, g. E.), pp.: 385–402, 2000: Wiley, New York.

10 Suren, A. M., *Effects of Urbanisation, In New Zealand Stream Invertebrates: Ecology and Implications for Management* (Edit.: Collier, K. J., Winterbourn, W. J.) Pp.: 260–288. N.z. Limnol. Soc., Hamilton, 2000.

11 Heaney, J. P., Huber, W. C., *Nationwide Assessment of Urban Runoff on Receiving Water Quality*, Water Resources Bulletin, volume: 20, pp.: 35–42, 1984.

12 House, M. A., Ellise, J. B., Herricks, E. E., Huitved-Jacobsen, T., Seager, J., Lijklema, l., Aalderink, H., Clifforde, I. T., *Urban Drainage-Impacts on Receiving Water Quality*. Water Science & Technology, Volume: 27, pp. 117–58, 1993.

13 Bayar, R., *Cbs Yardımıyla Modern Alışveriş Merkezleri Için Uygun Yer Seçimi: Ankara Örneği*, Coğrafi Bilimler Dergisi, Sayı: 3 (2), s. 19–38, 2005.

14 Dai, F. C., Lee, C. F., Zhang, X. H., *Gis-Based Geo-Environmental Evalution For Urban Land-Use Planning: a Case Study*, Engineering Geology, Volume: 61, pp.: 257–271, 2001.

15 Lee, S., Ryu, J. H., Lee, M. J., Won, J. S., *Use of An Artificial Neural Network For Analysis of The Susceptibility To Landslides at Boun*, Korea, Environmental Geology, Volume: 44, pp.: 820–833, 2003.

16 Lee, S., Sambath, T., *Landslide Susceptibility Mapping in The damreiromel Area, Cambodia Using Frequency Ratio and Logistic Regression Models*, Environmental Geology, Volume: 50, pp.: 847–855, 2006.

17 Angillieri, M. Y., *Application of Frequency Ration and Logistic Regression to Active Rock Glacier Occurrence in the Andes of San juan*, Argentina, *Geomorphology*, volume: 114, pp.: 396–405, 2010.

18 Gençoğlu, S., Inan, E. ve Güler, H., *Türkiye'de Deprem Tehlikesi*, Jeofizik Mühendisleri Odası Yayını, 701s, 1996.

19 Kütükçüoğlu, M., *Türkiye'nin Ilk Ağır Sanayi Kenti* Karabük, Karabük Valiliği Yayınları, 2012.

20 Collins M. G., Steiner F., R. & Rushman M. J., *Land Use Suitability Analysis in the United States: Historical Development and Promising Technological Achievements.* Environmental Management 28(5): 611- 621, 2001.

21 Malczewski, J., *Gis-Based Multi criteria Decision Analysis: A Survey of The Literature.* International Journal of Geographical Information Science 20(7): 703–726, 2006.

22 Topay, M. *The Importance of Climate For Recreational Planning of Rural Areas; Case Study of Muğla Province, Turkey*, In: a Matzarakis, de Freitas C. R. & Scott D (Eds), Developments in Tourism Climatology, Alexandroupolis, Greece, PP. 19–22, 2007.

23 Cengiz T., Akbudak C., Özcan H. ve Baytekin h., *Gökçeada'da Optimal Arazi Kullanımının Belirlenmesi*, Tarım Bilimleri Dergisi-journal of Agricultural Sciences 19: 148–162, 2013.

24 Özyavuz, M., and Dönmez, Y., *"Monitoring the changing position of coastlines using information technologies, an example of Tekirdag."* Journal of Environmental Protection and Ecology 15.3 2014: 1051–1058.

25 Gökyer, E., Öztürk, M., Dönmez, Y., & Çabuk, S. *"Bartın Ili Dağlık Alanlarında Coğrafi Bilgi Sistemleri Kullanılarak Ekoturizm Faaliyetlerinin Değerlendirilmesi."* İnönü Üniversitesi Sanat ve Tasarım Dergisi 5.12, 2015, pp: 25–35.

26 Sönmez, M.E., *Coğrafi Bilgi Sistemleri (CBS) Tabanlı Deprem Hasar Riski Analizi: Zeytinburnu (Istanbul) Örneği*, Türk Coğrafya Dergisi, Sayı 56: 11–22, 2011, Istanbul.

27 Atabey, E., *Depremler, (Maden Tetkik ve Arama Genel Müdürlüğü Yayınlarından, Eğitim Serisi*, No. 34, 2010 Ankara.

28 Topay, M. ve Parladir, Ö. M., *Isparta Ili Örneğinde CBS Yardımıyla Alternatif Turizm Etkinlikleri Için Uygunluk Analizi*, Tarım Bilimleri Dergisi, 21(2), S:309–309, E-ISSN:2148-9297, Ankara, 2015.

Assoc. Prof. Dr. Zühal Dilaver

43 Use of Medicinal and Aromatic Plants

Abstract: Medicinal plants have been used to relieve pain and treat various diseases by indigenous communities in different parts of the world, since the origin of humanity. According to records of WHO, in 1970s there were 21,000 identified medicinal plants. Moreover, it is estimated that this number is above 70,000 today. "Aromatic plants" is a special class of plants used for their aroma and flavor. In various medicines and aromatherapy, many of the aromatic plants are used for their medicinal and aromatic features. Furthermore, they are also used as nutrition, cosmetic, spice and herbal tea. The parts of plants, such as leaf, flower, fruit, fruit peel, seed, root, tuber, bulb, branch or stem bark, branch tips, stalk and wood are used to provide medicinal and aromatic benefits. In this chapter, the concept of medicinal and aromatic plants and the development of these plants throughout history are examined. In addition, along with the commercial use of medicinal and aromatic plants, botanical gardens, medicinal and aromatic plant gardens, and therapeutic gardens are being created to conserve, exhibit, recognize, register, and also to create good environments for people are mentioned. Also, some examples of these kinds of gardens from different parts of the world are provided.

Keywords: Medicinal plants, Aromatic plants, Traditional medicine, Medicinal and aromatic plant gardens, Therapeutic gardens

1 Introduction

Medicinal plants have been used to relieve pain and treat various diseases since the origin of humanity by indigenous communities in different parts of the world [1, 2, 3]. In every period of history, new plant species identified as medicinal and their use in treatments of diseases are recorded and conveyed to following generations [2]. Although, the pharmaceutical industry has grown significantly, the prevalence of traditional medicines is still incontrovertible [1]. According to records of WHO, in 1970s there were 21,000 identified medicinal plants. Moreover, it is estimated that this number is above 70,000 today [4].

Popularity of traditional medicines in both developing and developed countries can be mentioned. According to a study conducted by WHO in 2002, diversity, flexibility, easy accessibility, broad continuing acceptance are some of the prominent features of traditional medicines in developing countries. On the other hand, in developed countries, the main reasons of usage of these plants are increasing popularity, relatively low cost, low levels of technological input, relatively low side effects and growing economic importance [5]. It can also be noted that medicinal plants are widely used in pharmaceutical industry and modern medicine [3].

Medicinal plants are an indispensable part of human life and are present in innumerable forms. In Indian traditions, all plants in Earth are considered as medicinal [6]. "Medicinal plants are plants that provide people with medicines–to prevent disease, maintain health or cure ailments" [6,7,8,9,10]. One can say that everyone on Earth benefits from medicinal plants in various ways. The following can be given as examples of areas that people use these plants for; nutrition, toiletry, body care, incense, natural pesticide, ritual healing, and ornamentals [6,10,11]. "Aromatic plants" is a special class of plants used for their aroma and flavor. In various medicines and aromatherapy, many of the aromatic plants are used for their medicinal and aromatic features. Furthermore, they are also used for nutrition and as cosmetics, spices and herbal tea. Likewise, some of the medicinal plants are used for their essential oils to use in perfumery. In literature, all of these medicinal, aromatic, essential oil plants are merged under the name of medicinal and aromatic plants (MAP) [6]. At this point, it may be beneficial to use a specific definition for traditional medicines: "Traditional medicine refers to the knowledge, skills and practices based on the theories, beliefs and experiences indigenous to different cultures, used in the maintenance of health and in the prevention, diagnosis, improvement or treatment of physical and mental illnesses." This description is often used as alternative or complementary medicine in many countries [12].

The parts of plants, such as leaf, flower, fruit, fruit peel, seed, root, tuber, bulb, branch or stem bark, branch tips, stalk and wood are used to provide medicinal and aromatic benefits [13].

2 History of Medicinal and Aromatic Plants

Throughout the human history medicinal plants are used in treatments of diseases. Numerous archeological evidences indicate that medicinal plants were widely used by people in prehistoric ages [14]. One of the oldest prescriptions in the history was found on the Hittite tablets [15,16]. In those prescriptions, there is mention about several medicinal plants that were grown in Anatolia and some other drugs that came from abroad [3]. As well as the Hittite tablets, in Sumerian and Egyptian tablets, some information on medicinal plants and their use can be found [15,16]. The ancient Egyptians are known for developing a large detailed and effective pharmacological collection of curing materials from natural sources about 3000 to 6000 years ago [14]. Among all the medical papyruses of the ancient Egypt, the most important one, Ebers papyrus, which is estimated to be written in 1550 BC, contains 77 vegetables, animals and mineral drugs in more than 800 recipes [16]. In ancient Greece, health-care practitioners continued to spread plant-based therapeutic treatments [14]. Mesopotamian civilizations were used around 250 herbal drugs during the Greek period, and about 600 medicinal plants were known [16].

In the first century, Dioscorides, who is an authority on herbs, conducted a study and wrote "De Materia Medica" which consists of five volumes. In this study, he classified more than 500 different plant species according to their shapes and effects on diseases. This work is the most significant source of information on drugs of ancient ages [17]. After those developments, exploration of medicinal plants proceeded in China, India, and Tibet as results of innumerable experimentations. Around 1000 years ago healers in the Aztec and Maya Indian cultures of Mexico and Central America were conducting experimental studies with natural curing substances. In these societies herbal specialists were recognized as reputable and influential professionals [14].

When we look at more recent years, 18th century witnessed many significant studies. An important development was achieved in the 18th century with the revolutionary taxonomic work of Swedish naturalist Carolus Linnaeus. He classified thousands of botanical species which led to the foundation of the standardized documentation of relationships and evolutionary histories of medicinal plants. *Systema Naturae* (1735) written by Linnaeus established the fundamentals of modern biological taxonomy. Also, *Genera Botanica* and *Critica Botanica* (both published in 1737) and *Philosophica Botanica* (1751) by Linnaeus deal with accurate identification of plants and their characteristics, including catalogs with Latin terminology of all species known back then. In another work, *Species Plantarum* (1757), Linnaeus recorded 5900 plant species descriptions [14].

The amount of medicinal plants used during the Arab/Persian civilizations increased up to about 4000. According to Baytop (1999), in the early nineteenth century, the number of known medicinal plants reached 13,000 [16].

According to records of WHO, in 1970s there were 21,000 identified medicinal plants. Moreover, it is estimated that this number is above 70,000 today [4].

3 Use of Medicinal and Aromatic Plants

As mentioned before, throughout the human history, medicinal and aromatic plants have been used in various areas. Today, these plants are important assets with high commercial values. In addition, botanical gardens, medicinal and aromatic plants gardens, and therapeutic gardens are being created to conserve, exhibit, recognize, register, and also to create good environments for people.

3.1 Commercial Use of Medicinal and Aromatic Plants

It is a known fact that the importance of medicinal and aromatic plants and their industrial uses are increasing day by day. Out of 3000 medicinal plants that are being traded internationally, only 900 of them are being cultivated and the rest is harvested from the wild. Besides, global warming, habitat loss, excessive harvesting and climate change have a great negative impact on population of medicinal plants and threaten the future existence of 15,000 species. One of the focuses of the IUCN Medicinal Plant Specialist Group is to identify the species that are endangered as a result of non-sustainable harvest and other factors. Recently, Bramwell (2003) estimated that 21% of the world's flora is under threat [4].

Due to violation of process and product standards, resource and risk management tools, wild nature is a risk to nature as well as to humans [18].

Cultivation is rising as an economically reasonable option to conserve the global wealth of medicinal plants and their biodiversity [19]. However, in general, the materials obtained from natural plants are cheaper than the ones extracted from cultivated plants. This causes a decrease in production. Cultivation should be done primarily in the ecological circumstances that the plant naturally grows in accordance with the secondary material contents, genetics, and heredity. Hereby, yield obtained from unit area will increase and profitable, pure, clean drugs that conform to standards will be acquired [9].

Özhatay ve Koyuncu (1998) reported that there are 347 plants in Turkey being traded internally and externally after being collected form nature [4]. Among the medicinal and aromatic plants, cumin, anise, thyme, fenugreek, poppy, fennel, mint, and coriander are cultivated widely. Plants such as laurel, linden, sage, rosemary, rose hips are collected from nature, mainly from Aegean, Marmara, Mediterranean, Eastern Black Sea, and Southeastern Anatolia Regions [20].

3.2 Use of Medicinal and Aromatic Plants in Medicinal and Aromatic Plant Gardens

Studies have been conducted and cultivations of medicinal plants have been made on botanic gardens for more than 500 years. Currently, there are more than 2500 botanic gardens in 150 different countries that are home to six million accessions of living plants representing around 80,000 species. These botanic gardens altogether are an important source to medicinal plants and the information about them. They are now centers for the research and study of many disciplines–which inform medicinal plant conservation–such as taxonomy, ecology, agronomy, horticulture, ethnobotany, and habitat restoration [21]. Most of the botanical gardens have medicinal and aromatic plant parts, but some of them focus specifically on medicinal plants.

There is a botanical garden in Padua, which was founded in 1545 and is one of the oldest botanic gardens in the world. It was created by the Vatican Republic particularly for medicinal herb cultivation. It is still a center for scientific researches and preserves its original layout [22]. The garden allowed students to learn how to use and distinguish between medicinal plants, improving scientific understanding as well as local healthcare [21].

Likewise, in 1673 the Chelsea Physic Garden was established in England by the Worshipful Society of Apothecaries to train apprentices with the aim of identifying plants and helping in the cultivation of exotic plants for medicinal purposes. These gardens satisfied a research, economic, and health goal [21]. It is home to approximately 5000 types of edible, medicinal, herbal, and useful plants [23].

The garden at Regional Science Centre Guwahati in Assam, India, spans an area of 800 m^2 and houses around 80 medicinal and aromatic plant species. Exclusive feature of this garden is that it allows visually handicapped people to experience these plants. Visitors can see these plants and learn their names, parts that are used for medicinal and aromatic purposes. About 33% of the visually handicapped people live in India. It is thought that ten million blind people live in this country and more than 33% of them are children. In the medicinal and aromatic plant garden, special provisions have been made in the form of "Touch and Smell" plants in order to inform these children about their heritage. Plants with textured leaves and smell are kept which are very important in terms of medicine and aroma [24].

In Žalec, Slovenia, an organization named the Institute of Hop Research and Brewing (IHPS) was established in 1976 to set up a pan-Slovenian center for the production, processing, and application of medicinal and aromatic plants. It serves as a public research, developmental, advisory and educational organization. Its primary functions are research and advisory services in growing industrial plants (hops, hemp, biofuel plants), brewing, and the production of medicinal and aromatic plants [25].

At the beginning, it was established as a garden of medicinal and aromatic plants with a collection of 40 different traditional, autochthonic medicinal and aromatic plants in Slovenia. As the years went by, the collection has increased via the exchange of seeds and plants from the botanical gardens or similar organizations across the world. Currently, the garden spans an area of 0.30 ha and houses about 250 different species of medicinal and aromatic plants [25].

In the garden of medicinal and aromatic herbs, various workshops are organized on the production and production processing of herbs. The garden was always an important educational and research facility from the beginning. Guided tours are being organized around the garden for large groups. In the garden, seedlings of organic seeds and medicinal and aromatic plants are on sale and consultancy services are provided to grow medicinal and aromatic plants [25].BAU Herbal Garden is home to Lebanese medicinal and aromatic plants, and it considers biodiversity as one of the pillars of sustainable development. "RCED Herbal Garden" was jointly established in Lebanon (BAU Campus, 11,000 m^2) by RCED in partnership with the Faculty of Pharmacy at BAU and Green Hand Organization to conserve some native medicinal and aromatic plants of high ecological, economic, and social importance. In addition to conservation, the garden also serves to provide educational information and plant material to viable research and healthy herbal products [26].

The first botanical garden in Turkey was established in 1839 in Galata Palace. In 1935, the Istanbul University Botanical Garden was established and in 1959 Sarıyer Atatürk Arboretum was opened. Zeytinburnu Medicinal Plant Garden was opened in 2005, and it is the first botanical garden for medicinal plants established in Turkey [27].

Zeytinburnu Medicinal Plants Garden, which has an area of 14 acres, has more than 700 medical plants and each of these plants has a brief information tag next to them. Training activities such as seminars, workshops, and kids' programs are being organized under the name of Health Environment School in the garden. Home Medical seminars encourage the production of natural and simple health care remedies at home, where appropriate, instead of relying solely on the professional health system. One of the activities for public education was the Symposium on Treatment with Plants in 2010 [27].

Another garden that houses medicinal and aromatic plants is "Medical Aromatic Plants Collection Garden" and was established after "Turkey Medicinal and

Aromatic Plants Research Center." It was opened in Antalya, Turkey. It is home to 450 species of medicinal and aromatic plants. Each of them was observed in various phonological periods and then photo albums were prepared [7].

3.3 Use of Medicinal and Aromatic Plants in Therapeutic Gardens

Most of the research on relationship between nature and humanity carried out to date has focused on stress reduction and attention restoration. In other words, they focused on the ability of nature to restore the body to a healthy condition or to improve an unhealthy condition. However, scientific advances, such as the ability to see and measure neurological functions of brain, allow scientists to see that nature can actively support health and pleasure and how it does so [28]. As the gardens, the representative of nature in cities, are considered to be a healing factor, the therapeutic gardens have become a serious issue for academic research in many countries and have been included in health systems [29].

"Therapeutic gardens" is a garden designed to encourage people, visitors, and hospital staff to be away from the stressful environment. According to Elings (2006), healing gardens can also be considered as open space therapy units, where people feel themselves psychologically and physiologically good. In addition, it is known that the therapeutic gardens have a positive effect on self-confidence, personality, and personality development of children and create an environment of socialization, renewal of memories, motivation increase, and participation in physical activities in elderly people [30].

In studies conducted in Western countries, medical dictionary defines "therapeutic" as "the healing powers of nature." On the other hand, in China, Jiang (2009) defines "healing landscape" to "green spaces in healthcare facilities" while Wang and Li (2012) define it as "landscape which has therapeutic effect on physical and mental health." However, the most popular definition of "healing garden" among Chinese scholars is presented by Eckerling (1996): "healing garden is a garden in a healing setting designed to make people feel better." Lei et al. (2011) have grouped "healing gardens" into two categories: gardens in healthcare facilities which are created to ameliorate the recovery process of patients and public parks for people suffering from "life-style depression." Wang and Li (2012) stated that healing gardens are such places where horticultural therapy activities happen. Therapeutic landscapes include various natural settings with therapeutic effects, including healing gardens, rehabilitation gardens, meditation gardens, and memorial gardens [31].

Therapeutic gardens have been used as such places where people have been seeking cures since the Middle Ages, and today they have become an important part of hospitals, rehabilitation centers, and nursing homes where the elderly and disabled are cared. In modern sense, therapeutic gardens began to be designed in 1798 in the USA by the Institute of Medicine and Clinical Practice, in order to emphasize the healing the effects of individuals with mental disorders [32,33].

The therapeutic effects of nature to enhance patients' recovery have been examined comprehensively in "Notes on Nursing" by Florence Nightingale in 1860. She believed that visual remainders of nature, such as nature view through window and bedside flowers assist the treatment of patients [31]. Then, for the first time, a greenhouse was used as a therapy garden in 1879 in Philadelphia Associates Hospital. In 1973, the first professional establishment in this area, the Plant Development and the National Council for Treatment and Rehabilitation, today's American Horticultural Therapy Association (AHTA) was constituted [32]. Hence, several schools with different knowledge have emerged and have establishing a relationship between landscape and health to explore the healing mechanisms of nature [31].

Therapeutic gardens should not be perceived as medicinal gardens in which plants with therapeutic virtues are cultivated. Therapeutic gardens are complete gardens which have special designs of environments and layouts to help sick people to get well or simply to stimulate their feelings, their focus of interests, their memory, or their bodies to improve their mental states [34].

At the designing step of therapeutic gardens, the needs of the intended type of patient or user should be taken into account. The aim for the elderly and Alzheimer's patients is to stimulate their senses. For younger patients, for example, those who suffer from autism, it should draw their attention but not to overstimulate them as it may make them distressed [34].

Therapeutic gardens can be custom-designed areas as well as complementary places in nursing homes, health institutions' gardens, urban parks, and green areas etc. [29,34].

Therapeutic gardens improve the virtues of classic gardens through the integration of the following:

- Sense-stimulating plants: the scent of aromatic plants, the taste of fruit, touch with plants having a texture, view of vivid colors and various shapes and hearing with plants that attract birds
- Horticultural therapy areas to make use of the therapeutic benefits

It should be noted that the use of medicinal and aromatic plants in the design of therapeutic gardens is not mandatory. However, especially in order to provide these effects, medicinal and aromatic plants are important in these gardens with their aesthetic and functional properties [35]. Medicinal and aromatic plants used within the therapeutic gardens contribute with their fragrances, different textures, aesthetically valuable leaves, flowers, and fruits. In addition, the bodily movements during the care of the plants and the collection of their usable parts and the pleasure in consuming the products are also notable [29].

4 Result

In this study, the concept of medicinal and aromatic plants and the development of these plants throughout history are examined. In addition, along the commercial use of medicinal and aromatic plants; botanical gardens, medicinal and aromatic plant gardens, and therapeutic gardens are being created to conserve, exhibit, recognize, register and also to create good environments for people are mentioned.

- The major problem that medicinal and aromatic plants face is the increasing demand on them, because their habitats are threatened since many species are being harvested from the wild. Firstly, unconscious gathering from nature should be prevented and the production of the demanded species should be made.
- To properly benefit from the sustainable production and market potential of medicinal and aromatic plants, these products must be of the desired quantity and quality. To obtain the desired quality and standards, it is necessary to develop improved varieties, to determine the appropriate ecological conditions, to collect the natural plants in time without harming the nature, to determine the post-harvest processes and the processing technology.
- With the use of developing technologies, the effective substances in the plants should be identified and converted into drugs.
- Medical and aromatic plant analysis criteria should be determined in accordance with international standards and active substance analysis should be done in accredited laboratories.
- MAP should be correctly identified with scientific names, produced, marketed and used.
- Extra researches in the field of medicinal plants will be needed on issues of safety, toxicity, proper dosages, contamination, and potential interactions with synthetic and other natural drugs.
- Herbal products are known to have a positive effect on health. However, misuse of these products can lead to serious health problems. Training should be given for the recognition and correct use of medicinal and aromatic plants. There is a great need of medicinal and aromatic herb gardens in Turkey and therefore should be expanded and used for these purposes.
- There aren't any therapeutic garden examples in Turkey yet, but applications can be seen in certain areas in the world and should be in Turkey as well, especially in big cities.

References

1 Güler, B., G. Kümüştekin and E. Uğurlu, *Contribution to the Traditional Uses of Medicinal Plants of Turgutlu (Manisa – Turkey)*, Journal of Ethnopharmacology, 2015. 176: pp. 102–108.

2 Petrovska, B.B., *Historical Review of Medicinal Plants' Usage*, Pharmacogn Rev. 2012. Jan-Jun; 6(11): pp. 1–5. doi: [10.4103/0973-7847.95849]

3 Kırıcı, S., *Türkiye'de Tıbbi ve Aromatik Bitkilerin Genel Durumu.* TÜRKTOB, Türkiye Tohumcular Birliği Dergisi, 2015. 4(15), Ankara.

4 Schippmann, U., D. Leaman and A.B. Cunningham, *A Comparison of Cultivation and Wild Collection of Medicinal and Aromatic Plants under Sustainability Aspects* Chapter 6 In: R.J. Bogers, L.E. Craker, and D. Lange (eds.), Medicinal and Aromatic Plants. 2006: pp. 75–95.

5 Payyappallimana, U., *Role of Traditional Medicine in Primary Health Care: An Overview of Perspectives and Challenges*, Yokohama J Soc Sci 2006. 14: pp. 57–78.

6 Maiti, S. and K.A. Geetha., *Medicinal and Aromatic Plants in India.* In e-book Horticulture Floriculture (Ornamental, Medicinal & Aromatic Crops) 2007.

7 Mert, A. and E. Dağıstan, *Tıbbi ve Aromatik Bitkilerin Ekonomik Önemi*, in *XII. Ulusal Tarım Ekonomisi Kongresi*, Bildiriler Kitabı, Cilt 3, Isparta, 2016: pp. 1525–1532,

8 Marshall, M., *Health and Wealth from Medicinal Aromatic Plants*, Diversification Booklet Number 17, Rural Infrastructure and Agro-Industries Division Food and Agriculture Organization of the United Nations Rome 2011, Isbn 978-92-5-107070-3, Fao 2012.

9 Bayram, E., et al., *Tıbbi ve Aromatik Bitkiler Üretiminin Arttırılması Olanakları. In Türkiye Ziraat Mühendisliği VII. Teknik Kongresi* Bildiriler Kitabı-I, 11–15 Ocak, Ankara, 2010: pp. 437–456.

10 Dağdelen, A.F., et al., *Mikrobiyolojik ve Aflatoksin Yönünden Bazı Tıbbi ve Aromatik Bitkiler ve Çaylarının İncelenmesi, II. Tıbbi ve Aromatik Bitkiler Sempozyumu,* 23–25 Eylül, Yalova, Bildiriler Kitabı, Ed. Dr. Filiz Pezikoğlu, Ahmet Bircan Tinmaz 2014 Isbn: 978–605–4672–85–1

11 Çınar, N., et al., *Batem Tıbbi Aromatik Bitkiler Koleksiyon Bahçesi: Türlerin Adaptasyonu ve Fenolojik Gözlemleri, II. Tıbbi ve Aromatik Bitkiler Sempozyumu* 23–25 Eylül, Yalova, Bildiriler Kitabı, Ed. Dr. Filiz Pezikoğlu, Ahmet Bircan Tinmaz 2014 Isbn: 978–605–4672–85–1

12 WHO, *Traditional Medicine* [cited 2018, 22 September]; Available from: https://Afro.Who.Int/Health-Topics/Traditional-Medicine

13 Özkan, G., *Fitoterapi ve Homeopati Uygulamaları, II. Tıbbi ve Aromatik Bitkiler Sempozyumu* 23–25 Eylül, Yalova, Bildiriler Kitabı, Ed. Dr. Filiz Pezikoğlu, Ahmet Bircan Tinmaz 2014 Isbn: 978–605–4672–85–1

14 Halberstein, R. A., *Medicinal Plants: Historical and Cross-Cultural Usage Patterns*, Aep Vol. 15, No. 9 October 2005: pp.686–699.

15 Dogan, H., S. Alnıak Sezer, and S. Akan, *Historical Importance of Medicinal Aromatic Plants*, in International Congress of Medicinal and Aromatic Plants, May 10–12, 2017.

16 Faydaoğlu, E. and M.S. Sürücüoğlu, *Geçmişten Günümüze Tıbbi ve Aromatik Bitkilerin Kullanılması ve Ekonomik Önemi*, Kastamonu Üni., Orman Fakültesi Dergisi, 2011, 11 (1): pp. 52 – 67.

17 Büyüknisan, E., *Dioskurides Pedanios'un "De Materia Medica" Adlı Eseri*, Lokman Hekim Journal 2011; 1 (1): pp. 55–60.

18 Dürbeck K. and T. Hüttenhofer *International Trade of Medicinal and Aromatic Plants*. In: Máthé Á. (Eds) Medicinal and Aromatic Plants of the World. Vol 1. Springer, Dordrecht, 2015.

19 Rajeswara Rao, B.R. and D.K. Rajput, *Global Scenario of Medicinal Plants*. – Cmphpu Conference Paper: 20 pp. 2010.

20 Gül, A., and A.D. Çelik, *Tıbbi ve Aromatik Bitki Yetiştiriciliği ve Dış Ticareti: Hatay İli Örneği*. Mustafa Kemal Üniversitesi Ziraat Fakültesi Dergisi, 21(2) 2016: pp. 227–235.

21 Hawkins, B., *Plants for Life: Medicinal Plant Conservation and Botanic Gardens*, 2008 Isbn: 1-905164-21-1 Published by Botanic Gardens Conservation International.

22 Unesco, Botanical Garden (Orto Botanico), *Padua* [cited 2018, 25 September]; Available from: https://whc.unesco.org/en/list/824

23 Chelsea Physic Garden, *Chelsea Physic Garden* [cited 2018, 3 October]; Available from: https://www.chelseaphysicgarden.co.uk

24 Regional Science Centre, *Medicinal and Aromatic Plant Garden* [cited 2018, 3 October]; Available from: http://www.rscguwahati.gov.in/mag.htm

25 Slovenian Institute of Hop Research and Brewing, *Medicinal and Aromatic Plants*, [cited 2018, 03 October]; Available from: http://www.ihps.si/en/medicinal-aromatic-plants/introduction/

26 BAU Herbal Garden for Lebanese Medicinal and Aromatic Plants, *BAU Herbal Garden for Lebanese Medicinal and Aromatic Plants* [cited 2018, 22 September]; Available from: http://www.bau.edu.lb/Research-Center-for-Environment-and-Development/Projects-BAU-Herbal-Garden-for-Lebanese-Medicinal-and-Aromatic-Plants

27 Alpınar, M.K., *Zeytinburnu Tıbbi Bitkiler Bahçesi*, Zeytinburnu Belediyesi Kültür Yayınları Kitap No: 35, Ed. Murat D. Çekin Isbn: 978 605 859 889 8, İstanbul. 2015.

28 Marcus, C.C and N.A. Sachs, *Therapeutic Landscapes An Evidence-Based Approach to Designing Healing Gardens and Restorative Outdoor Spaces*, E-Book. 2014.

29 Arslan, M. and E. Ekren, *Yaşlı Kişilerin Sağlığı ve Etkinlikleri İçin Terapi Bahçeleri*, Ahi Evran Üniversitesi Sosyal Bilimler Enstitüsü Dergisi (Aeüsbed) 2017, Cilt 3, Sayı 2, Sayfa 361–373

30 Pouya, S., E. Bayramoğlu, and Ö. Demirel, *Şifa Bahçesi Tasarım Yöntemlerinin Araştırılması*, Kastamonu Üni., Orman Fakültesi Dergisi, 2015, 15 (1).

31 Jiang, S., *Therapeutic Landscapes and Healing Gardens: A Review of Chinese Literature in Relation to the Studies in Western Countries*, Frontiers of Architectural Research 2014. 3,: pp. 141–153.

32 Arslan M., Z. Kalaylioglu, and E. Ekren *Use of Medicinal and Aromatic Plants in Therapeutic Gardens*. Indian J of Pharmaceutical Education and Research. 2018;51(4s): pp. 151–154.

33 Arslan, M., and E. Katipoğlu, *Kentsel Çevrede Yaşlı Kişilerin Sağlığı ve Etkinlikler İçin Bitki Yetiştiriciliğinin Önemi*, Yaşlı Sorunları Araştırma Dergisi, Cilt: 4, Sayı: 1–2, 2011: s: 44–52.

34 Paysalia, *Therapeutic Gardens: Making the Most of a Garden's Virtues*, [cited 2018, 25 September]; Available from: http://blog.paysalia.com/en/sport-health/therapeutic-gardens-making-the-most-of-a-gardens-virtues-2/

35 Arslan, M., and M. Peng., *Taiwan ve Türkiye'de Tıbbi ve Aromatik Bitki Türlerinin Kullanımı*, V. Süs Bitkileri Kongresi, 06–09 Mayıs 2013, Yalova, Türkiye. 2013.

List of Figures

Assoc. Prof. Dr. Funda K. Açıcı
Contemporary Approaches in Museum Space Design

Assist. Prof. Dr. Hülya Öztürk Tel
An Evaluation of Urban Texture Components in the Example of Şanliurfa Hasimiye Square with Regards to Urban Identity and Urban Aesthetics

Assoc. Prof. Dr. Kemal Demir
An Analysis of the Spatial Needs and Flexibility in Postgraduate Centers of Theology in the Framework of Kayseri Case

List of Tables

Assoc. Prof. Dr. Barış Ergen and Assoc. Prof. Dr. Kemal Demir

Evaluation of Erciyes University Eastern Campus Social Center Project from Lefebvrian Perspective

Prof. Dr. Bülent Cengiz and Oğuz Erdi Yakan

Visual Landscape Assessment in Güzelcehisar Village-Bartin (Western Black Sea Region of Turkey)

www.ingramcontent.com/pod-product-compliance
Ingram Content Group UK Ltd.
Pitfield, Milton Keynes, MK11 3LW, UK
UKHW020122290926
15471UKWH00003B/223

9 783631 782743